Nuclear Matter and Heavy Ion Collisions

NATO ASI Series

Advanced Science Institutes Series

A series presenting the results of activities sponsored by the NATO Science Committee, which aims at the dissemination of advanced scientific and technological knowledge, with a view to strengthening links between scientific communities.

The series is published by an international board of publishers in conjunction with the NATO Scientific Affairs Division

A	**Life Sciences**	Plenum Publishing Corporation
B	**Physics**	New York and London
C	**Mathematical and Physical Sciences**	Kluwer Academic Publishers Dordrecht, Boston, and London
D	**Behavioral and Social Sciences**	
E	**Applied Sciences**	
F	**Computer and Systems Sciences**	Springer-Verlag
G	**Ecological Sciences**	Berlin, Heidelberg, New York, London,
H	**Cell Biology**	Paris, and Tokyo

Recent Volumes in this Series

Volume 198—Mechanisms of Reactions of Organometallic Compounds with Surfaces
edited by D. J. Cole-Hamilton and J. O. Williams

Volume 199—Science and Technology of Fast Ion Conductors
edited by H. L. Tuller and M. Balkanski

Volume 200—Growth and Optical Properties of Wide-Gap II–VI Low Dimensional Semiconductors
edited by T. C. McGill, C. M. Sotomayor Torres, and W. Gebhardt

Volume 201—Physics of Highly-Ionized Atoms
edited by Richard Marrus

Volume 202—Point and Extended Defects in Semiconductors
edited by G. Benedek, A. Cavallini, and W. Schröter

Volume 203—Evaluation of Advanced Semiconductor Materials by Electron Microscopy
edited by David Cherns

Volume 204—Techniques and Concepts of High-Energy Physics V
edited by Thomas Ferbel

Volume 205—Nuclear Matter and Heavy Ion Collisions
edited by Madeleine Soyeur, Hubert Flocard, Bernard Tamain, and Madeleine Porneuf

Series B: Physics

Nuclear Matter and Heavy Ion Collisions

Edited by

Madeleine Soyeur

CEN–Saclay
Gif-sur-Yvette, France

Hubert Flocard

Institute of Nuclear Physics
Orsay, France

Bernard Tamain

Institut des Sciences de la Matière et du Rayonnement
Caen, France

and

Madeleine Porneuf

CEN–Saclay
Gif-sur-Yvette, France

Plenum Press
New York and London
Published in cooperation with NATO Scientific Affairs Division

Proceedings of a NATO Advanced Research Workshop on
Nuclear Matter and Heavy Ion Collisions,
held February 7–16, 1989,
in Les Houches, France

Library of Congress Cataloging in Publication Data

NATO Advanced Research Workshop on Nuclear Matter and Heavy Ion Collisions
(1989: Les Houches, Haute-Savoie, France)
   Nuclear matter and heavy ion collisions / edited by Madeleine Soyeur...
[et al.].
      p.   cm.—(NATO ASI series. Series B, Physics; vol. 205)
   "Proceedings of a NATO Advanced Research Workshop on Nuclear Matter and
Heavy Ion Collisions, held February 7–16, 1989, in Les Houches, France"—T.p.
verso.
   Includes bibliographical references.
   ISBN-13: 978-1-4684-5717-9   e-ISBN-13: 978-1-4684-5715-5
   DOI: 10.1007/978-1-4684-5715-5
   1. Matter, Nuclear—Congresses. 2. Heavy ion collisions—Congresses. I.
Soyeur, Madeleine. II. Title. III. Series: NATO ASI series. Series B, Physics; v. 205.
QC793.3.S8N374   1989   89-22901
539.7—dc20   CIP

© 1989 Plenum Press, New York
Softcover reprint of the hardcover 1st edition 1989

A Division of Plenum Publishing Corporation
233 Spring Street, New York, N.Y. 10013

CENTRE FOR PHYSICS

LES HOUCHES
(FRANCE)

Session : Nuclear Matter and Heavy Ion Collisions
February 7-16, 1989
NATO Advanced Research Workshop

Scientific Director : Madeleine Soyeur, CEN-Saclay

Organizing Committee :
Hubert Flocard, Institute for Nuclear Physics, Orsay
Madeleine Soyeur, Institute for Fundamental Research, Saclay
Bernard Tamain, University of Caen

International Advisory Committee :
Gerald E. Brown (Stony-Brook), Joseph Cugnon (Liege),
Adriano Gobbi (GSI), Rudi Malfliet (Groningen)

PREFACE

The Winter School "Nuclear Matter and Heavy Ion Collisions", a NATO Research Workshop held at Les Houches in February 89, has been devoted to recent developments in nuclear matter theory and to the study of central heavy ion collisions in which quasi-macroscopic nuclear systems can be formed at various temperatures and densities. At incident energies below 100 MeV per nucleon, the kinematic conditions are favourable for producing transient hot nuclei with temperatures of the order of a few MeV. At higher energies (100 MeV < E/A < 1-2 GeV) heavy ion collisions offer the possibility of investigating the properties of hot and dense nuclear systems.

The Workshop has been motivated by important theoretical developments in transport equations which make it possible to relate microscopic descriptions of heavy ion collisions to nuclear matter theory and by the need to review the large body of data available on heavy ion collisions and discuss future experimental programs. This discussion was especially timely a few months before the new SIS/ESR Heavy Ion Facility starts operating in Darmstadt.

The School consisted mostly of series of lectures on nuclear matter, transport equations and the dynamics of heavy ion collisions. The data and their interpretation were extensively discussed; the information carried by the various types of particles emitted during the collisions (photons, lepton pairs, pions, kaons, nucleons, fragments) has been particularly emphasized. Specialized topics were presented as shorter contributions by participants. Three discussion sessions proved very helpful to summarize the results of the meeting and to suggest future directions of research.

For this successful Winter School, we thank all the lecturers, the members of the International Advisory Committee and our sponsors: NATO, the Commission of the European Communities, the Ministère des Affaires Etrangères and the Université Scientifique et Médicale de Grenoble. It is a pleasure to thank also Nicole Leblanc and Anny Glomot who took care of all practical and administrative problems. The help of Dany Bunel for the preparation of these Proceedings is very gratefully acknowledged.

Hubert Flocard
Madeleine Soyeur
Bernard Tamain

CONTENTS

Invited Papers

NUCLEAR MATTER WITH NON-RELATIVISTIC POTENTIALS:
PRESENT STATUS 1
    C. Mahaux and R. Sartor

A RELATIVISTIC THEORY OF NUCLEAR MATTER 37
    B.D. Serot

ELEMENTS OF A RELATIVISTIC MICROSCOPIC THEORY OF HADRONIC
MATTER IN EQUILIBRIUM AND NON-EQUILIBRIUM 73
    R. Malfliet

HOT NUCLEAR MATTER 103
    V.R. Pandharipande and D.G. Ravenhall

INTRODUCTION TO NON-RELATIVISTIC TRANSPORT THEORIES 133
    C. Grégoire

RELATIVISTIC TRANSPORT THEORY OF FLUCTUATING
FIELDS FOR HADRONS 165
    P.J. Siemens

PION ABSORPTION IN NUCLEI 167
    M. Thies

FORMATION AND DECAY OF HOT NUCLEI: THE EXPERIMENTAL
SITUATION 187
    D. Guerreau

MULTIFRAGMENTATION OF NUCLEI 231
    C. Ngô

NUCLEAR MATTER AND FRAGMENTING NUCLEI 267
    J.P. Bondorf and K. Sneppen

COLLECTIVE PHENOMENA IN RELATIVISTIC HEAVY ION
COLLISIONS: THE EXPERIMENTAL SITUATION 287
    K.H. Kampert

INCLUSIVE EXPERIMENTS, CORRELATIONS AND PION
PRODUCTION DATA 323
    D. L'Hôte

CONFRONTATION OF THEORETICAL APPROACHES AND
EXPERIMENTAL DATA ON HIGH ENERGY HEAVY ION COLLISIONS . . . 343
    M. Berenguer, C. Hartnack, G. Peilert, A. Rosenhauer,
    W. Schmidt, J. Aichelin, J.A. Maruhn, W. Greiner
    and H. Stöcker

MOMENTUM DEPENDENT MEAN FIELDS IN THE BUU MODEL OF HEAVY
ION COLLISIONS 391
    G.M. Welke

HARD PHOTONS AND SUBTHRESHOLD MESONS FROM NUCLEUS-NUCLEUS
COLLISIONS 405
    E. Grosse

MEASUREMENTS OF e^+e^- PAIR PRODUCTION AT THE BEVALAC 417
    J. Carroll

STATUS OF THE SIS/ESR-PROJECT AT GSI 429
    P. Kienle

Contributed Papers

HEAVY IONS AT SATURNE 443
    J. Arvieux

PION POLARISATIONS IN THE RELATIVISTIC DIRAC-BRUECKNER
MODEL 447
    F. de Jong, B. ter Haar and R. Malfliet

CUT-OFF AND EFFECTIVE MESON FIELD THEORY 451
    M. Jaminon, G. Ripka and P. Stassart

A NON-PERTURBATIVE TRANSPORT THEORY FOR NUCLEAR
COLLECTIVE MOTION 453
    H. Hofmann

TIMESCALE OF PARTICLE EMISSION USING NUCLEAR
INTERFEROMETRY 457
    D. Ardouin, P. Lautridou, D. Durand, D. Goujdami,
    F. Guilbault, C. Lebrun, A. Peghaire, J. Quebert and
    F. Saint-Laurent

DETERMINATION OF TIME SCALES IN INTERMEDIATE
ENERGY REACTIONS 461
    U. Lynen

FRAGMENT PRODUCTION AND NUCLEAR FREEZEOUT 465
    W.G. Lynch

Δ EXCITATION IN NUCLEI BY HEAVY ION CHARGE-EXCHANGE
REACTIONS 469
    M. Roy-Stéphan

MOMENTUM DEPENDENT POTENTIALS IN RELATIVISTIC HEAVY
ION COLLISIONS 471
    V. Koch

CLASSICAL MODELS OF HEAVY ION COLLISIONS 475
    T.J. Schlagel

RELATIVISTIC MOLECULAR DYNAMICS AND THE RELATIVISTIC
VLASOV EQUATION 477
    M. Schönhofen, H. Feldmeier and M. Cubero

STABILITY CONDITIONS AND VIBRATION MODES OF
SELF-CONSISTENT VLASOV SOLUTIONS WITH DIFFUSE
SURFACE 481
    S.J. Lee, E.D. Cooper, H.H. Gan and S. Das Gupta

SIDEWARD FLOW OF CHARGED PARTICLES AND NEUTRONS
IN HEAVY ION COLLISIONS 483
    D. Keane, B.D. Anderson, A.R. Baldwin, D. Beavis,
    S.Y. Chu, S.Y. Fung, M. Elaasar, G. Krebs, Y.M. Liu,
    R. Madey, J. Schambach, G. VanDalen, M. Vient,
    S. Wang, J.W. Watson, G.D. Westfall, H. Wieman
    and W.M. Zhang

DILEPTON RADIATION FROM HOT NUCLEAR MATTER AND
NUCLEON-NUCLEON COLLISIONS 487
    C. Gale

KAON AND PION PRODUCTION IN THE REACTION SILICON
(14.5 AGeV) ON GOLD 491
    H. Sorge

Participants 495

Index 499

NUCLEAR MATTER WITH NONRELATIVISTIC POTENTIALS :

PRESENT STATUS

C. Mahaux and R. Sartor
Institut de Physique B5, Université de Liège
4000 Liège 1, Belgium

1. INTRODUCTION

The main purpose of these lectures is twofold. Firstly, we describe theoretical methods that have been devised to study the properties of nuclear matter, with particular emphasis on the ground state energy and on rearrangements of the perturbation expansion in powers of the strength of the nucleon-nucleon interaction. Secondly, we critically survey numerical results which have been published during the last few years.

Our presentation is as follows. In Sec. 2, we define nuclear matter, enumerate some of its empirical properties and specify the basic assumptions of the theoretical model. Section 3 is devoted to the various methods than can be used to calculate the quantities of interest, in particular the average binding energy per nucleon. In the subsequent sections, we focus on the perturbation expansion in powers of the strength of the nucleon-nucleon interaction. The first order term of this expansion is called the Hartree-Fock approximation; it is discussed in Sec. 4. Section 5 is devoted to the second and third order terms, with emphasis on their physical interpretation. The ladder summation that leads to the Brueckner reaction matrix is described in Sec. 6, where the Brueckner-Hartree-Fock approximation is introduced. This approximation involves an "auxiliary" potential energy $U(k)$, attached to a nucleon with momentum k. One usually tries to choose $U(k)$ in such a way as to optimize the accuracy of the Brueckner-Hartree-Fock approximation. Many debates concern the "best" choice of $U(k)$. In the "standard choice", $U(k)$ is large and attractive for k smaller than the Fermi momentum k_F but is set equal to zero when k is larger than k_F; $U(k)$ then has a large discontinuity at k_F. Numerical results derived from this "standard" Brueckner-Hartree-Fock approximation are presented in Sec. 7. Higher order terms of the standard Brueckner expansion are discussed in Sec. 8. Section 9 describes results obtained from the so-called "continuous choice" of $U(k)$, which is characterized by the fact that $U(k)$ is calculated self-consistently for all values of k. Another choice, recently advocated by Kuo and collaborators, is discussed in Sec. 10. Finally, Sec. 11 contains our conclusions.

2. DEFINITION AND EMPIRICAL PROPERTIES

2.1. Definition

Nuclear matter is a uniform medium in which pointlike nucleons interact via a realistic two-body interaction; the electromagnetic effects are omitted. The word "realistic" means that this interaction is required to yield good fits to the nucleon-nucleon scattering cross sections and to the properties of the deuteron. In these lectures we shall consider symmetric nuclear matter, in which the neutrons and protons have equal density.

Nuclear matter has received and is still receiving much attention, for the following main reasons. Firstly, nuclear matter provides *the* bridge between the two-nucleon system and many-body nuclei : methods are first developed and tested in the case of nuclear matter; they were adapted to nuclei afterwards. Secondly, some empirical properties of nuclear matter can be estimated by extrapolating the properties of nuclei towards large mass numbers. Thirdly, it is of interest to predict the properties of nuclear matter at large densities in order to analyze collisions between heavy ions with high energy as well as for astrophysical purposes; an extrapolation towards large densities may be reliable only if the theory accurately reproduces the empirical properties of nuclear matter at densities close to those encountered near the centre of heavy nuclei. Fourthly, the requirement that the empirical properties of nuclear matter be reproduced might enable one to distinguish among various realistic nucleon-nucleon potentials, or to exhibit the necessity of taking into account three-body forces or the inner structure of nucleons : this type of conclusion is justified only if the nuclear matter problem can be solved accurately.

2.2. Empirical properties

(1) *Saturation density.* Medium-weight and heavy nuclei approximately have the same density near the nuclear centre, namely

$$\rho_0 \approx (0.17 \pm 0.02) \text{ nucleon / fm}^3 \; ; \tag{2.1}$$

the error bar mainly derives from uncertainties in the neutron density distribution and from the fact that the measured density has slight wiggles in the nuclear interior. The density ρ is often expressed in terms of the *Fermi momentum* k_F, which is the maximum value of the nucleon momentum in a free Fermi gas with density ρ. In symmetric nuclear matter, one has

$$\rho = \frac{2}{3\pi^2} k_F^3 \; . \tag{2.2}$$

At saturation, the Fermi momentum is equal to

$$k_F^{(0)} = (1.36 \pm 0.05) \text{ fm}^{-1} \; . \tag{2.3}$$

(2) *Average binding energy per nucleon.* The measured binding energy of nuclei can be fitted with the following expression (Z , N and A denote the number of protons, neutrons and nucleons, respectively) :

$$\mathcal{B}_0 \approx a_1 A + a_2 A^{2/3} + a_3 Z^2 A^{-1/3} + a_4 (N - Z)^2 / A + \ldots \ ;$$

the first term on the right-hand side is associated with the nuclear interior, the second with the nuclear surface, the third with the Coulomb energy and the fourth with the symmetry energy. In the limit of symmetric nuclear matter, the average binding energy per nucleon is given by a_1 , whose empirical value is

$$B_0 = \mathcal{B}_0 / A = (-16 \pm 1)\,\text{MeV} \ . \tag{2.4}$$

The empirical values (2.1) and (2.4) imply that, if one calculates the average binding energy per nucleon (in short "the binding energy", B) as a function of the *Fermi momentum* k_F , the function $B(k_F)$ should have a minimum equal to ≈ -16 MeV at $k_F \approx 1.36$ fm^{-1} *if the nuclear matter model is a realistic description of the physical system.* We return to this proviso in Sec. 2.3.

(3) *Compression modulus.* It is defined by

$$K = 9 \left[\frac{d}{d\rho} (\rho^2 \frac{dB}{d\rho}) \right]_{\rho = \rho_0} . \tag{2.5}$$

Empirical information on K is derived from the excitation energy of the monopole resonance and from the isotope shift. The generally accepted value is[1]

$$K \approx (210 \pm 30)\,\text{MeV} \ . \tag{2.6}$$

The error bar is probably an underestimate because of the difficulty of disentangling bulk from surface effects; in the literature, the empirical values which are quoted range from 350 MeV (Ref. 2) to 100 MeV (Ref. 3). The value of K is of great practical interest because it significantly influences the extrapolation from $\rho \approx \rho_0$ to larger ρ . One should note, however, that the definition of K involves derivatives of $B(\rho)$ at $\rho = \rho_0$. Thus, K can be used for an extrapolation only if one makes a simple assumption on the dependence of $B(\rho)$, for instance that it is a parabola. This type of assumption becomes unreliable when ρ sizeably differs from ρ_0 . One should thus view K as a constraint of a calculated functional $B(\rho)$, which can then be used for extrapolation purposes.

The first derivative of $B(\rho)$ yields the *pressure* P :

$$P = \rho^2 \frac{d}{d\rho} B \ . \tag{2.7}$$

Of course, the pressure vanishes at the equilibrium density ρ_0, where $B(\rho)$ has a minimum.

(4) *Momentum distribution*. It gives the probability that a plane wave state with momentum k is occupied. It is defined by

$$n(k) = < \Psi \mid a^+(k)\, a(k) \mid \Psi > , \qquad (2.8)$$

where $a(k)$ and $a^+(k)$ denote annihilation and creation operators and Ψ the *correlated* (exact) normalized ground state wavefunction. In the independent particle model (free Fermi gas), the momentum distribution is given by

$$n_0(k) = < \Phi_0 \mid a^+(k)\, a(k) \mid \Phi_0 > , \qquad (2.9)$$

where Φ_0 is a Slater determinant. In the uncorrelated system, $n_0(k)$ is equal to one for $k < k_F$, and equal to zero for $k > k_F$; this is represented by the dashed line in Fig. 1. In the correlated ground state, $n(k)$ has the shape schematically depicted by the solid curve in Fig. 1 : its slope is vertical at $k = k_F$, where it has a discontinuity : [4, 5]

$$n(k_F - 0) - n(k_F + 0) = Z(k_F) . \qquad (2.10)$$

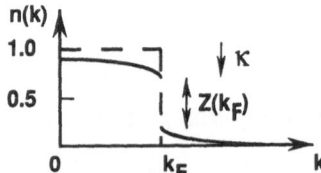

Fig. 1. Schematic representation of the momentum distribution in the uncorrelated (dashed line) and correlated (solid curve) system.

The average *depletion of the Fermi sea* will be denoted by κ . This depletion is of great importance, since it provides a measure of the validity of the independent particle model. We shall moreover see later on that κ *is believed to determine the rate of convergence of the theoretical expansions*.

Empirical information on $n(k_F - 0)$ and on $Z(k_F)$ are becoming available from the analysis of nucleon knockout and pickup reactions,[6, 7] as well as of nucleon-nucleus scattering.[8, 9] However, the results of these analyses are not yet quantitatively reliable. Moreover, the extracted value is affected by finiteness effects; reactions (e,e'p) in which a proton from an inner (deeply bound) shell is ejected will be more directly related to the properties of nuclear matter. The published information yields[6, 8, 9]

$$\kappa \approx 0.15 \pm 0.05 \quad , \tag{2.11}$$

but larger values are suggested by preliminary analyses of (e,e'p) reactions in which the ejected proton originates from a deeply bound shell.[7]

(5) *The single-particle potential* felt by a nucleon in the correlated ground state can be identified with the real part of the so-called *mass operator* (or *self-energy*). At positive nucleon energy ε, this potential $V(\varepsilon)$ is the central value of the real part of the optical-model potential which can be determined from analyses of nucleon-nucleus scattering cross sections. Empirically, one has (all energies are in MeV)

$$V(\varepsilon) \approx -55 + 0.32\,\varepsilon \quad , \text{ for } 20 < \varepsilon < 80 \text{ MeV} \quad . \tag{2.12}$$

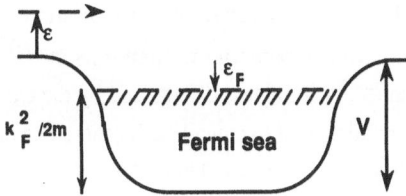

Fig. 2. *Schematic representation of the single-particle potential V, of the Fermi energy ε_F ($\varepsilon_F \approx -16$ MeV in nuclear matter) and of the kinetic energy at the Fermi surface $k_F{}^2 / 2m$. The short dashed lines illustrate a scattered nucleon with energy $\varepsilon > 0$. Bound nucleons have a negative energy, $\varepsilon < 0$.*

The *Fermi energy* ε_F is the energy of a nucleon located at the Fermi surface, i.e. with momentum $k = k_F$. It is given by (we set $\hbar = 1$)

$$\varepsilon_F = k_F{}^2 / 2m + V(\varepsilon_F) \quad , \tag{2.13}$$

where $k_F{}^2 / 2m = 38.4$ MeV is the kinetic energy of a nucleon located at the Fermi surface, see Fig. 2. The value of $V(\varepsilon_F)$ for $\rho = \rho_0$ is known because of the Hugenholtz-Van Hove theorem[10] which states that, at equilibrium, the value of ε_F is equal to the average binding energy per nucleon. In nuclear matter, ε_F is thus equal to -16 MeV at $\rho = \rho_0$. From Eqs. (2.3) and (2.13) one obtains

$$V(\varepsilon_F) = -54.4 \text{ MeV} \quad . \tag{2.14}$$

In heavy nuclei, ε_F is close to - 8 MeV ; the extrapolation of the linear law (2.12) would yield $\approx$ - 57.6 MeV , which is somewhat deeper than the value determined by Eq. (2.14). The deviation reflects the fact that at small (positive or negative) energy, the empirical linear law (2.12) is not valid.[8, 9] In nuclei, the energy dependence of the depth $V(\varepsilon)$ for ε close to ε_F is moreover influenced by finiteness effects.

(6) The *effective mass* $m^*(\varepsilon)$ is defined by

$$m^*(\varepsilon) / m = 1 - dV(\varepsilon) / d\varepsilon \ . \tag{2.15}$$

It thus characterizes the energy dependence of $V(\varepsilon)$. The empirical law (2.12) indicates that

$$m^*(\varepsilon) / m \approx 0.68 \ \text{for} \ 20 < \varepsilon < 80 \ \text{MeV} \ . \tag{2.16}$$

This value cannot be used at small (positive or negative) energy since, there, the energy dependence of $V(\varepsilon)$ is not linear. In particular, the value of $m^*(\varepsilon_F) / m$ is larger than 0.68 . Recent analyses[8, 9] of the average potential for neutrons in ^{208}Pb yields $m^*(\varepsilon_F) / m \approx 0.80$. This value is influenced by finiteness effects and by the existence of a neutron excess in ^{208}Pb .These two corrections tend to cancel each other. The effective mass at the Fermi energy is intimately related to one of the Landau parameters. If one uses the constraint that a sum rule connects these parameters, $m^*(\varepsilon_F)$ strongly influences the estimated value of the compression modulus.[3]

(7) *Mean free path.* The optical model potential is complex. Its imaginary part $W(\varepsilon)$ is due to nucleon-nucleon collisions and determines the mean free path of a nucleus.[11] This quantity thus has a meaning at negative as well as at positive energy. The imaginary part vanishes at the Fermi energy

$$W(\varepsilon) \sim c \, (\varepsilon - \varepsilon_F)^2 \ \text{for} \ \varepsilon \approx \varepsilon_F \ . \tag{2.17}$$

The empirical knowledge of c is poor because, in a nucleus, the empirical mean free path of a low energy nucleon is mainly determined by surface effects. A rough estimate yields[9]

$$c \approx - (0.003 \pm 0.001) \ \text{MeV}^{-1} \ . \tag{2.18}$$

2.3. Discussion

We surveyed a few empirical properties of nuclear matter. For historical as well as for technical reasons, most theoretical investigations focus on the binding energy $B(k_F)$,[12-14] on which we shall also devote most of our attention in the following. We emphasize, however, that this limitation is detrimental to the internal consistence of the field. It appears more satisfactory to investigate on the same footing all the quantities enumerated above.[15,16]

We mentioned that the calculated $B(k_F)$ should have a minimum ($\approx$ - 16 MeV) at $k_F \approx 1.36$ fm^{-1} *if* the nuclear matter model is a realistic description of the physical system. The latter proviso must be kept in mind. Indeed, the nuclear matter model contains many simplifications. For instance, nucleons are not structureless, three-body forces exist, relativistic effects may be sizeable, etc. Accordingly, it is important to keep in mind the following warning by Day :[17] *"The problem of devising a reliable method of calculation is distinct from the question of agreement between theory and experiment. (...) The accuracy of the calculation is determined not by comparing the result with experiment but rather by estimating the magnitude of the higher-order terms that have been neglected".*

3. METHODS OF CALCULATION

3.1. Input

One makes the assumption that nuclear matter is a *normal* system, i.e. a system whose properties can be calculated by *starting from* a perturbation expansion in powers of the strength of the nucleon-nucleon interaction. In practice, this perturbation expansion diverges because of the strong nature of the nuclear force. This is why the perturbation expansion is only a starting point : one has to rearrange the series by performing partial summations. One should always keep in mind that the original series is singular, with poorly known convergence properties.[18] It appears likely that one deals with asymptotic series. It might happen that the rearranged series apparently converge when only a few terms are calculated but nevertheless yield an incorrect answer.

The many-body Hamiltonian reads

$$H = \mathcal{T} + \mathcal{V} = \sum_i \frac{1}{2m} \nabla_i^2 + \sum_{i<j} v\,(i,j) \quad , \tag{3.1}$$

where $\mathcal{T}$ and $\mathcal{V}$ are the kinetic and potential energy operators. One can always write

$$H = H_0 + H_1 = (\mathcal{T} + U) + (\mathcal{V} - U) \quad , \tag{3.2}$$

where U is an *auxiliary single-particle potential*. Here, the word "auxiliary" is meant to emphasize that U is an external potential whose choice is in principle arbitrary and which need not have a physical meaning. The auxiliary potential U should be chosen in such a way as to optimize the rate of convergence of the rearranged perturbation expansion while being relatively easy to calculate and convenient to use. One should keep in mind that some choices of U might lead to an apparent convergence towards an incorrect answer, because of the poorly known nature of the convergence properties of the original perturbation series. This has been emphasized, in particular, by Baker and collaborators.[18,19] This is one of the reasons why it is of interest to investigate different theoretical approaches to the calculation of the properties of nuclear matter. These approaches are briefly surveyed in the present section.

3.2. Perturbation expansion

Let Φ_0 and Ψ denote the uncorrelated and correlated ground state wave functions :

$$H_0 \, \Phi_0 \; = \; E_0 \, \Phi_0 \;\; , \;\; H \, \Psi \; = \; E \, \Psi \;\; . \tag{3.3}$$

Equation (3.2), (3.3) yield

$$E \; = \; E_0 + D \;\; , \;\; D \; = \; <\Phi_0 | \, H_1 \, | \, \Psi > / <\Phi_0 | \, \Psi > \;\; . \tag{3.4}$$

A theorem due to Gell-Mann and Low[20] enables one to expand the second term in powers of the strength of H_1 :

$$| \, \Psi > / <\Phi_0 | \, \Psi > \; = \; \sum_{q=0}^{\infty} \, [\, (E_0 - H_0)^{-1} \, H_1 \,]^q \, | \, \Phi_0 >_{\ell} \;\; , \tag{3.5}$$

where the lower index ℓ refers to a *"linked" cluster expansion*.

Equation (3.4) gives the binding energy per nucleon as the sum of the free Fermi gas plus a "correction" D , which is the quantity to be evaluated. Unless otherwise specified, we set $U = 0$. Then, the average energy per nucleon of the free Fermi gas is the average kinetic energy per nucleon, namely $T_0 = 0.6 \, k_F^2 / 2m$. For $k_F = 1.36 \, fm^{-1}$, one has $T_0 = 23 \, MeV$. Since the empirical binding energy of the correlated system is equal to $- 16$ MeV , the magnitude of the "correction" D that is to be calculated is about $- 39 \, MeV$. This is often taken as reference for evaluating the relative accuracy of a theoretical calculation : 1 MeV approximately corresponds to a 2.5 per cent accuracy.

3.3. Kinetic energy of the correlated system

Instead of using Eq. (3.4), one can write

$$E \; = \; <\mathcal{T}> + <\mathcal{V}> \;\; , \tag{3.6}$$

where $<\mathcal{T}>$ and $<\mathcal{V}>$ are the kinetic and potential energies in the *correlated* ground state Ψ , e.g. $<\mathcal{T}> = <\Psi | \, \mathcal{T} | \, \Psi >$. Then, the binding energy per nucleon is given as the sum of the kinetic energy per nucleon plus the potential energy per nucleon of the *correlated* system. Equation (3.5) yields perturbation expansions for $<\mathcal{T}>$ and $<\mathcal{V}>$. The relationship between the resulting expansion of E and that considered in Sec. 3.2 has been discussed in Ref. 21.

3.4. Hierarchy of Green functions

The one-body Green function is defined by

$$G_1(k;\tau) = -i < \Psi | T[a(k;\tau)a^\dagger(k;0)] | \Psi > , \qquad (3.7)$$

where T is the time ordering operator and where we explicitly wrote the time dependence of the creation and annihilation operators upon time (Heisenberg representation). For $\tau > 0$, $G_1(k;\tau)$ is seen to be related to the probability amplitude of finding, at time τ, a nucleon with the same momentum k as the one that it had when it was created at time zero. Thus, G_1 describes single-particle motion, and it is not surprising that it is intimately related to the mean field, i.e. the mass operator.

One can also define a two-body Green function G_2 which involves two creation and two annihilation operators, a three-body Green function G_3, etc. These Green functions are related by an infinite set of coupled equations. In practice, this set has to be truncated. This is performed by replacing G_3 by an approximation which only involves G_1 and G_2. There exist various ways of doing this.[22] They yield the energy in the form (3.6) rather than (3.4). The accuracy of the results is hard to evaluate in the light of the warning quoted at the end of Sec. 2.3. One of the truncations, called Λ_{11}, yields a "conserving" approximation in which, for instance, the Hugenholtz-Van Hove theorem[10] is automatically fulfilled. This property may be of interest for some purposes. On the other hand *nonconserving schemes are also of interest because the amount of violation of some theorems may provide an estimate of the accuracy of the approximation.* This remark applies to all the theoretical approaches based on rearrangements of the perturbation series.

3.5. Exp S method

The exact ground state wavefunction can be written in the form $\Psi = (\exp S) \Phi_0$. An expansion $S = S_2 + S_3 + ...$ yields

$$\Psi = \Phi_0 + \Phi_2 + \Phi_3 + ... \qquad (3.8)$$

where the indices refer to the number of holes in the corresponding term : Eq. (3.8) expresses Ψ as a sum of a Slater determinant (Φ_0), a two particle-two hole component (Φ_2), etc. (in nuclear matter there exists no one particle-one hole component because of momentum conservation). One may thus call (3.8) a "hole-line expansion" of the wavefunction. The components Φ_j are coupled by an infinite set of equations. When this set is truncated, the resulting approximation to Ψ can be inserted in Eq. (3.4), or in Eq. (3.6), to calculate the binding energy.[14] This approach is closely related to the hole-line expansion which will be discussed in Sec. 8.

3.6. Correlated basis function

Because the nuclear interaction is strongly repulsive at short distance, two nucleons cannot be close to one another. Jastrow[23] introduced a procedure for approximately taking this feature into account. Let us represent the corresponding *model* normalized wavefunction by $\Psi_J = J\Phi_0$, where J is a correlation operator. Sophisticated techniques[24] have recently been developed for calculating the expectation value

$$E_J = <\psi_J| H |\psi_J> \, , \tag{3.9}$$

even for quite involved forms of the correlation operator J. The accuracy of these techniques has been checked by comparison with a direct Monte-Carlo calculation of the expectation value (3.9).[25]

The variational principle states that E_J is an upper bound for the true energy E. In practice the correlation operator J contains parameters which are varied in order to make E_J as small as possible. Hopefully, the corresponding upper bound is close to (but larger than) the true energy E :

$$E = E_J + \Delta E_J \, , \quad \Delta E_J < 0 \, . \tag{3.10}$$

The *correlated basis function approach* consists in evaluating the correction ΔE_J.

4. HARTREE-FOCK APPROXIMATION

In the rest of these notes we shall focus on the perturbation expansion of Sec. 3.2, and on rearrangements of it. We shall occasionally refer to results obtained from other methods only occasionally, for comparison. The present section is devoted to the first order approximation, in which in Eq. (3.4) one approximates Ψ by Φ_0. This corresponds to the *Hartree-Fock* (HF) approximation. The binding energy is given by

$$B_{HF} = T_0 + \frac{1}{2} \sum_{hh'} <hh'| v | hh'>_{\mathcal{A}} \, . \tag{3.11}$$

The labels h,h',... correspond to *"hole states"*, i.e. to momenta smaller than k_F. The index $\mathcal{A}$ indicates that the ket is an antisymmetrized product of plane wave states; *we shall henceforth omit it*, for simplicity.

In the HF approximation, the single-particle potential reads

$$V_{HF}(k) = \sum_{h} <kh| v | kh> \, ; \tag{3.12}$$

Note that, in Eq. (3.12), k can take any positive value, larger as well as smaller than k_F. Equations (3.11) and (3.12) yield

$$B_{HF} = T_0 + \frac{1}{2} \sum_{hh'} <hh'|\, v\, | \, hh'> \; = \; T_0 + \frac{1}{2} \sum_{h} V_{HF}\,(h) \; . \qquad (3.13)$$

In the HF approximation, the ground state wavefunction is the Slater determinant Φ_0 : the Fermi sea is unperturbed. Equation (3.12) expresses that the potential energy of a nucleon with momentum k results from its interaction with the nucleons of the Fermi sea, and Eq. (3.13) that the binding energy is the sum of the interactions between the nucleons of the sea.

It is very convenient to represent expressions by diagrams. The Hartree-Fock approximation to D and to the mean field are represented by the diagrams shown in Fig. 3. Lines with arrows are momenta over which sums are performed. Downgoing arrows are associated with hole states : these have momentum h,h',... smaller than k_F . In the follo-

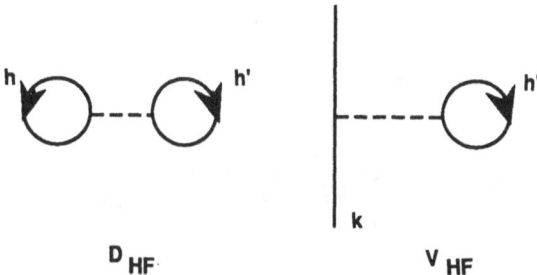

Fig. 3. Diagrams which represent the Hartree-Fock approximation to the binding energy (D_{HF}) and to the single-particle potential (V_{HF}) . The dashed lines represent antisymmetrized interactions.

wing we shall also encounter diagrams with upgoing arrows : these are associated with particle states; they have momenta p,p',... larger than k_F .

The Hartree-Fock approximation is not meaningful if one uses a realistic nucleon-nucleon interaction v . In practical applications it is used with an *effective* interaction whose parameters are adjusted to reproduce the empirical values of $B(k_F)$ in the vicinity of the saturation. The corresponding Hartree-Fock potential typically has a Gaussian-type shape $V_{HF}\,(k) \approx V_0 \exp\,(-\frac{1}{4}\,a^2\,k^2)$; for Skyrme-type interactions, however, it has the form $V_0 + \alpha\,k^2$, i.e. increases linearly with k^2 .

5. SECOND AND THIRD ORDER CONTRIBUTIONS

5.1. Second order contribution

The contribution that is of second order in the strength of H_1 is obtained by retaining the contribution associated with $q = 1$ in Eq. (3.5). This contribution is represented by the diagrams drawn in Fig. 4. In contrast to D_{HF} , the value of D_2 depends on the choice of the auxiliary potential U(k) . If U(k) is set equal to zero, the expression of D_2 reads

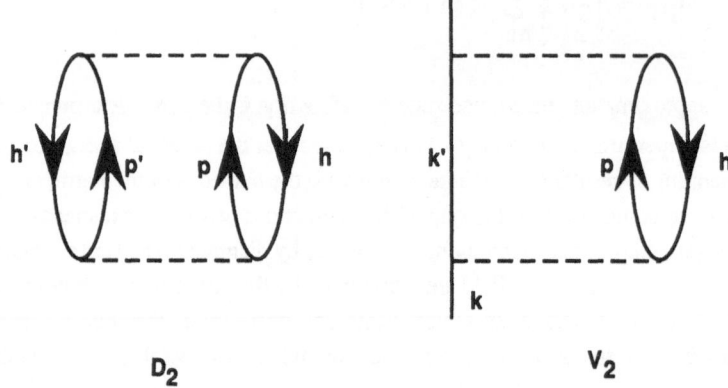

D_2 $\qquad\qquad\qquad\qquad$ V_2

Fig. 4. Second order contributions to the binding energy (D_2) and to the single-particle potential (V_2).

$$D_2^{(0)} = \frac{1}{4} \sum_{\substack{p\,p' \\ h\,h'}} \frac{|<hh'|\ v|\ pp'>|^2}{t\,(h) + t\,(h') - t\,(p) - t\,(p')} \ , \qquad (3.14)$$

where $t\,(k) = k^2 / 2m$ is the kinetic energy of a nucleon with momentum k. If $U(k)$ differs from zero, the expression B_2 becomes

$$D_2^{(U)} = \frac{1}{4} \sum_{\substack{p\,p' \\ h\,h'}} \frac{|<hh'|\ v|\ pp'>|^2}{e\,(h) + e\,(h') - e\,(p) - e\,(p')} \ , \qquad (3.15)$$

where

$$e\,(k) = k^2 / 2m + U\,(k) \qquad (3.16)$$

is the energy associated with a nucleon with momentum k (any k). This energy $e\,(k)$ need *not* have a physical meaning, since $U\,(k)$ can be chosen at one's will. However, problems would arise if $e\,(k_F + 0) < e\,(k_F - 0)$, since the energy that appears in Eq. (3.15) then could vanish. We exclude such choices of $U\,(k)$. Then, D_2 is negative (attractive). Since Eq. (3.15) is a particular case of Eq. (3.14), in the following we give expressions that can correspond to $U\,(k) \neq 0$ as well as to $U\,(k) = 0$.

The second order contribution to the physical single-particle potential energy is represented by the diagram labelled V_2 in Fig. 4. It consists of a sum of two terms, respectively associated with $k' = p' > k_F$ and with $k' = h' < k_F$: $V_2\,(k) = V_{2a}\,(k) + V_{2b}\,(k)$, where V_{2a} and V_{2b} are the real parts of the following quantities : [26]

$$M_{2a}(k) = \frac{1}{2} \sum_{pp'h} \frac{|<kh| v| pp'>|^2}{e(k) + e(h) - e(p) - e(p') + i\delta} \quad , \quad (3.17a)$$

$$M_{2b}(k) = \frac{1}{2} \sum_{hh'p} \frac{|<hh'| v| kp>|^2}{e(k) + e(p) - e(h) - e(h') + i\delta} \quad . \quad (3.17b)$$

For k close to k_F, $V_{2a}(k)$ is negative (attractive) while $V_{2b}(k)$ is positive (repulsive).

In second order, the ground state wavefunction deviates from Φ_0 : it contains two particle-two hole admixtures. The corresponding momentum distribution is given by (we recall that $h, h' < k_F$ and $p, p' > k_F$)

$$n_2(h) = 1 - \frac{1}{2} \sum_{pp'h'} \frac{|<hh'| v| pp'>|^2}{[e(h) + e(h') - e(p) - e(p')]^2} = 1 - \kappa_2(h) \quad , \quad (3.18a)$$

$$n_2(p) = \frac{1}{2} \sum_{hh'p'} \frac{|<hh'| v| pp'>|^2}{[e(p) + e(p') - e(h) - e(h')]^2} \quad . \quad (3.18b)$$

The expressions of $\kappa_2(h)$ and of $n_2(p)$ are represented by the diagrams shown in Fig. 5.

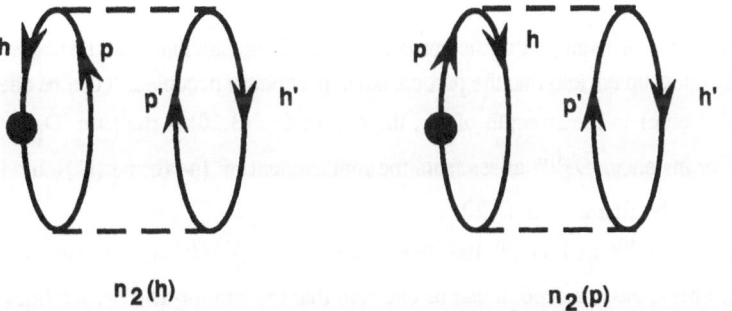

$$n_2(h) \qquad\qquad\qquad\qquad n_2(p)$$

Fig. 5. Diagrammatic representation of the second order contribution to the momentum distribution.

Note the intimate relationship between $\kappa_2(h)$ and the partial derivative of $V_{2a}(h)$ with respect to $e(h)$, and between $n_2(p)$ and the partial derivative of $V_{2b}(p)$ with respect to $e(p)$. Note also that

$$D_2 = \frac{1}{2} \sum_h V_{2a}(h) \quad . \quad (3.19)$$

Most papers only deal with the binding energy and pay little attention to the momentum distribution and to the single-particle potential. The interconnections between the expressions of D_2, V_2 and n_2 confirm the usefulness of simultaneously considering these various quantities.

5.2. Third order contributions

The number of contributions rapidly increases with the order of the perturbation series. For instance, Fig. 6 represents the diagrams associated with the third order term. This brings to one's mind the following remark made by Peierls : [27] *"Often, one can write down an infinite series for a quantity, which can be done very elegantly in terms of diagrams, and then sum a partial series of these diagrams. Sometimes, the physical significance of such subsets of diagrams is clear (and then the answer usually could have been written down without using the series in the first place), but often it is not".* In the light of this remark, we exhibit the physical meaning of some of the diagrams shown in Fig. 6, and show that subsets of diagrams can and should be summed.

We first consider the case when no auxiliary potential is used. Then, the second order approximation to the momentum distribution is given by Eqs. (3.18a,b) with $e(k) = k^2/2m$ for all k. Since the Fermi sea is partly depleted, it is physically clear that it would be preferable to replace D_{HF} by the following approximation

$$\frac{1}{2} \sum_{hh''} \{n_2(h) \, n_2(h'')\} < hh'' \mid v \mid hh'' > + \sum_{ph''} n_2(p) \, n_2(h'') < h''p \mid v \mid h''p > \quad (3.20)$$

which takes into account part of the corrections due to the fact that the hole states h and h'' are partly unoccupied, and that the particle state p is partly occupied. It can be checked that, up to third order in the strength of v, the expression (3.20) is the sum $D_{HF} + D_3{}^{hb} + D_3{}^{pb}$. For instance, $D_3{}^{hb}$ arises from the replacement of $[n_2(h) \, n_2(h'')]$ by $[1 - \kappa_2(h) - \kappa_2(h'')]$ in the first term of (3.20).

Since $D_3{}^{hb}$ and $D_3{}^{pb}$ have a physical meaning, it is advisable to take them into account. This is easy. Indeed, it can be checked that the sum of $D_2{}^{(0)}$, see Eq. (3.14), and of all the hole-bubble diagrams and particle-bubble diagrams can be written in the form (3.15) where

$$U(k) = V_{HF}(k) = \sum_{h'} < kh' \mid v \mid kh' > \quad (3.21)$$

for all k. As noted, this $U(k)$ is identical to the HF single-particle potential, see Eq. (3.12). *It is continuous at $k = k_F$*. It has the property that the corresponding second order term $D_2{}^{(HF)}$ (see Eq. (3.14)) includes the sum of $D_2{}^{(0)}$ and of the diagrams (see Fig. 7) $D_3{}^{(0)hb}$, $D_3{}^{(0)pb}$, $D_3{}^{(0)hbb}$, $D_3{}^{(0)pbb}$, etc. ; as in Eq. (3.14), the upper index (0) indicates that the diagrams are calculated with $U(k) = 0$, i.e. with kinetic energies in the

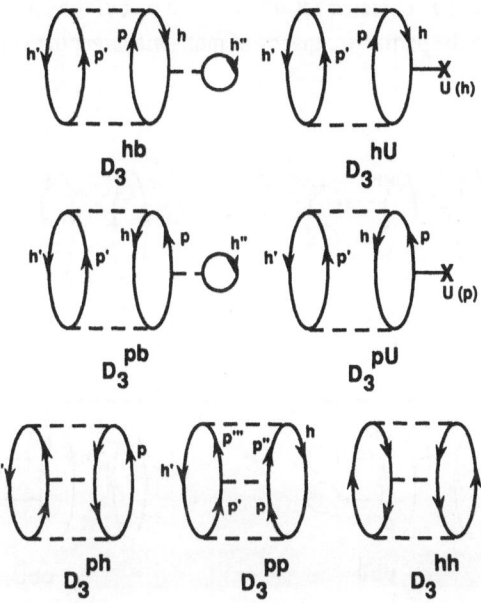

Fig. 6. Diagrammatic representation of the third order contributions to the binding energy. A cross is associated with a "U-insertion", where U (k) is the auxiliary potential.

denominator. Once one introduces an auxiliary potential $U(k)$, one generates diagrams with U-insertions, which arise from the appearance of U in H_1 in Eq. (3.5). Such diagrams are shown in Figs. 6 and 7. When one takes $U(k) = V_{HF}(k)$, the diagrams of $D_3{}^{hb}$ and $D_3{}^{hU}$, $D_3{}^{pb}$ and $D_3{}^{pU}$, $D_3{}^{hbb}$ and $D_3{}^{hbU}$, $D_3{}^{pbb}$ and $D_3{}^{pbU}$ cancel each other. More generally, *all the diagrams with bubble- and U-insertions cancel each other when the auxiliary potential U is taken equal to the Hartree-Fock potential.* Then only three diagrams of third order remain, namely $D_3{}^{ph}$, $D_3{}^{pp}$ and $D_3{}^{hh}$; $D_3{}^{pp}$ is discussed in Sec. 6 below.

In conclusion, we have shown that an infinite series of diagrams, namely those with bubble insertions on particle or on hole lines, disappear if one introduces the HF single-particle potential as auxiliary potential. This illustrates the possibility of performing partial summations, and the importance of adopting an educated choice for the auxiliary potential. In the present case, the HF choice takes into account part of the effects due to the depletion of the Fermi sea and to the partial occupancy of momenta larger than the Fermi momentum.

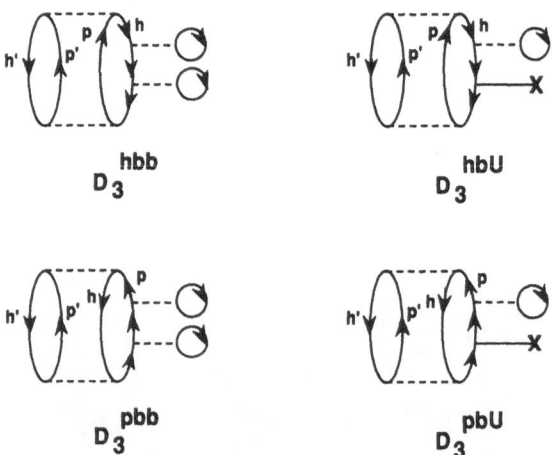

Fig. 7. Diagrams which cancel each other if the auxiliary potential $U(k)$ is taken equal to the HF potential $V_{HF}(k)$.

6. BRUECKNER-HARTREE-FOCK APPROXIMATION

Among the third order diagrams that remain, $D_3{}^{pp}$ plays an important role. Its physical meaning is the following (see Fig. 8). Two nucleons of the Fermi sea, with momenta h and h', interact and jump outside the Fermi sea, with momenta p and p'. Once they are in these states, they interact again and can either fall back in the holes (this yields D_2) or jump into new momentum states p'' and p'''. The latter process is the one that is described by the diagram $D_3{}^{pp}$. It can easily occur because the Pauli principle does not hinder the interaction between two nucleons that lie above the Fermi sea. Hence, many interactions can occur between nucleons that lie above the Fermi surface. These processes are represented by

p"
p'''
p
p'
h
h'

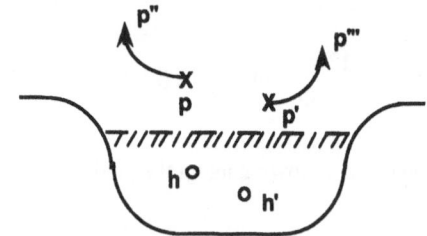

Fig. 8. Physical interpretation of the diagram B_3^{pp} .

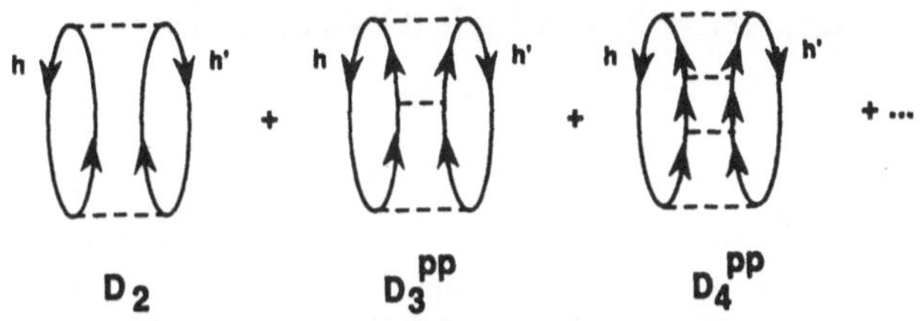

h
h'
+
h
h'
+
h
h'
+ ...
D2
D3pp
D4pp

Fig. 9. Summation of particle-particle ladder which yield the Brueckner-Hartree-Fock approximation D_{BHF} shown in Fig. 10. Note that all the diagrams have two hole lines (h and h').

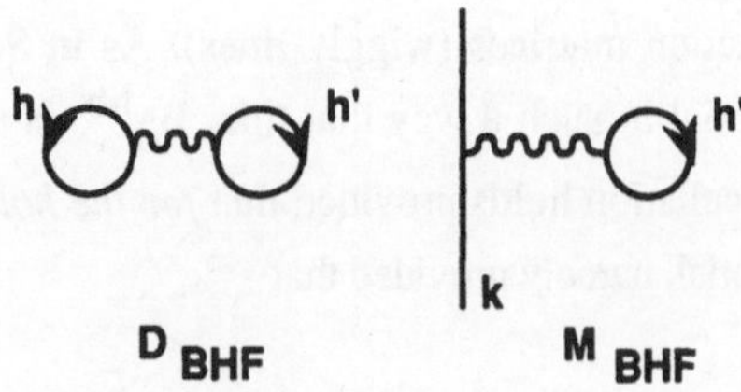

h
h'
h'
k
D BHF
M BHF

Fig. 10. Diagrammatic representation of the Brueckner-Hartree-Fock approximation to the binding energy and to the mean field.

the diagrams in Fig. 9. These diagrams are characterized by the fact that they all have two hole lines and an arbitrary number of interactions between the particle lines.

Such diagrams with repeated interactions between the two particles, namely D_2, $D_3{}^{pp}$, $D_4{}^{pp}$, ... can be summed in closed form and yield the following Brueckner-Hartree-Fock (BHF) approximation to the binding energy :

$$B_{BHF} = T_0 + \frac{1}{2} \sum_{hh'} < hh' \mid g\,[\,e(h) + e(h')\,]\mid hh' > \quad , \qquad (3.22)$$

where $g\,[w]$ is the solution of the following integral equation

$$g\,[w] = v + v \sum_{pp'} \frac{\mid pp' > < pp' \mid}{w - e\,(p) - e\,(p') + i\delta}\, g\,[w] \quad . \qquad (3.23)$$

Likewise, the BHF approximation to the mean field is given by

$$M_{BHF}\,(k) = \sum_{h'} < kh' \mid g\,[e\,(k) + e\,(h')\,]\mid kh' > \quad . \qquad (3.24)$$

The comparison between Eqs. (3.13), (3.12) and Eqs. (3.22), (3.24) show that the BHF approximation is formally identical to the HF approximation, with the difference that the original interaction v is replaced by the operator g. The latter is called the *reaction matrix*. This analogy between the HF and the BHF approximation is illustrated by Fig. 3 and Fig. 10. The reaction matrix plays the role of an effective interaction. It has the merit of being well-behaved and rather weak even if v has a hard repulsive core.

It can be shown that the original perturbation expansion (Sec. 3.2) can be expressed in terms of the reaction matrix. This rearranged series still contains an infinite series of terms. These are represented by diagrams similar to those used for the original perturbation expansion, with the main (but not sole) difference that the nucleon-nucleon interaction is replaced by a reaction matrix, see Fig. 11. For instance, a diagram $B_3{}^{hb}$ appears which is the same as the diagram $D_3{}^{hb}$ in Fig. 6, except that the interactions v (horizontal dashed lines) are replaced by reaction matrices (wiggly lines). As in Sec. 5.2, it is possible to choose the auxiliary potential in such a way that this $B_3{}^{hb}$ is exactly cancelled by the diagram $B_3{}^{hU}$. This cancellation holds provided that *for the hole states* $U(h)$ is chosen identical to the BHF potential, namely provided that

$$U\,(h) = \sum_{h'} < hh' \mid g\,[\,e\,(h) + e\,(h')\,]\mid hh' > \quad \text{for } h < k_F \quad . \qquad (3.25)$$

This prescription is used in practically all papers, see however Ref. 28.

For particle states, it is unfortunately not possible to choose U (p)   in such a way that the contributions represented by the diagrams B_3^{pb} and B_3^{pU} cancel each other. The "best" choice of U (p) for particle states has been the subject of many discussions. Two choices will be considered below, in Sec. 7 and in Sec. 9, respectively. Here, we note that even though the BHF approximation apparently involves a sum only over hole states (h,h') , it actually depends upon the choice of U (p) for particle states because the latter enter in the integral equation (3.23) that determines the reaction matrix.

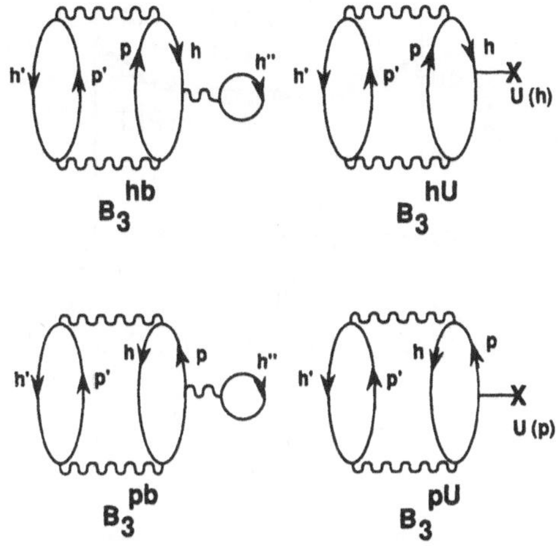

Fig. 11. Diagrams analogous to those shown in the upper part of Fig. 6, but in which reaction matrices g (wiggly lines) replace the nucleon-nucleon interaction v .

7. STANDARD CHOICE

Several choices have been used for U (p > k_F) . The simplest prescription is to set it to zero; some arguments give semi-quantitative support to this prescription.[17] *In the so-called "standard" BHF approximation, U (h) for hole states is defined by Eq. (3.25) and U (p) for particle states is set equal to zero.* This introduces a large discontinuity in the value of U (k) at k =k_F , see Fig. 20 below. It has been argued that this discontinuity is inconvenient, and possibly dangerous.[18, 19] Here, we only note that this *standard choice* of U (k) does not enable one to treat on the same footing the empirical quantities enumerated in Sec. 2.[15, 29, 30] We return to this point in Sec. 9.

In order to illustrate typical results obtained from the standard BHF approximation, we consider two realistic interactions, namely the "Paris potential" [31] and "Argonne V14 potential".[32] Both of these are realistic (Sec. 2.1) and incorporate constraints related to our present microscopic understanding of the nuclear interaction. Figure 12 shows the binding energy as calculated from the standard BHF approximation. The two interactions practically yield the same results. The minimum of the calculated curves lie at a too high energy ($\approx$ - 11.3 MeV) and at a too large Fermi momentum ($k_F \approx 1.53$ fm^{-1}, which corresponds to $\rho \approx 0.24$ fm^{-3}) as compared to the empirical saturation area. One may wonder whether this disagreement is due to the inaccuracy of the standard BHF approximation, or to the inadequacy of the Paris and Argonne V14 interactions.

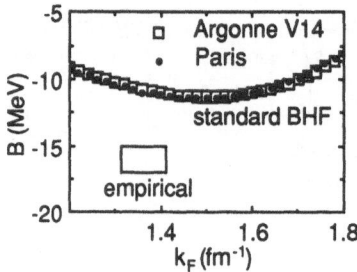

Fig. 12. *Dependence upon k_F of the binding energy as calculated from the standard BHF approximation. The solid dots (Ref. 33) are associated with the Paris potential and the open squares (Ref. 34) with the Argonne V14 interaction. The rectangle represents the empirical saturation area (Eqs. (2.1), (2.3)).*

We first discuss the second possibility : are the saturation properties calculated from the standard BHF approximation in disagreement with empirical evidence also for other realistic nucleon-nucleon interactions as well ? The squares in Fig. 13 illustrate the saturation points calculated from the standard BHF approximation, for a variety of realistic interactions. They are seen to lie in the vicinity of a smooth curve.[35] The location of the square on this so-called "Coester line" is mainly determined by the strength of the tensor part of the interaction. The Coester line does not come close to the empirical saturation area. Two explanations are possible, namely (i) the sBHF approximation is inaccurate, and/or (ii) the nuclear matter model with two-body interactions between structureless nucleons is not a faithful description of the physical reality.

In the last ten years, it has become clear that the standard BHF approximation is inaccurate. In order to exhibit this, we consider the example of the Argonne V14 interaction. In that case, it is possible to calculate the expectation value (3.9), and thus a reliable *upper bound* to the binding energy. This upper bound is represented by the full squares in Fig. 14. There, the open squares are the same as in Fig. 2 : they show the binding energy B_{sBHF} calculated from the standard BHF approximation. Clearly, B_{sBHF} lies *above* the upper bound. This demonstrates that the standard BHF approximation is inaccurate. Hence, one must calculate corrections to it; some of these are discussed in the next section.

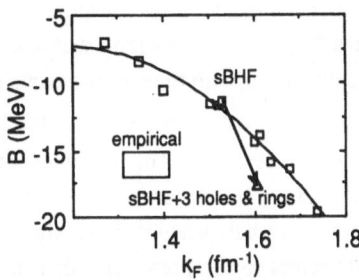

Fig. 13. The squares (taken from Refs. 33, 34, 36) represent saturation points calculated from the standard BHF approximation for various nucleon-nucleon interactions (from low to large k_F : the Hamada and Johnston, the soft core Reid as modified by Bethe and Johnson, the soft core Reid as modified by Day, the Bonn, the Paris, the Argonne V14, the Ueda and Green 1, the Bryan and Scott, the de Tourreil and Sprung, the Lagaris and Pandharipande, and the Ueda and Green 3 interactions). The arrow points to the saturation point obtained, from the Paris or Argonne V14 interaction, when three-hole line and generalized ring corrections are included (Refs. 36, 37). The rectangle is the empirical saturation area.

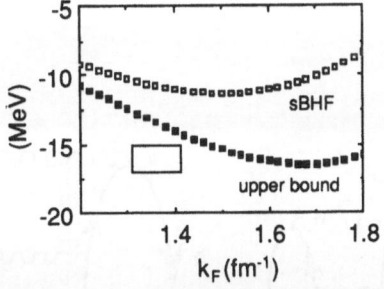

Fig. 14. The open squares represent the binding energy as calculated from the standard BHF approximation;[34] the full squares shown an upper bound, calculated by means of a two-body cluster plus Fermi hypernetted chain-single operator approximation.[37] The Argonne V14 interaction has been used as input for both calculations. The fact that the open squares lie above the full squares exhibit that the variational principle is violated by the standard BHF approximation.

8. HOLE LINE EXPANSION

The analogy between the HF and the BHF approximation suggests that the reaction matrix g plays the role of an effective interaction. One might therefore believe that one could use a rearranged perturbation series in which the terms would be grouped according to the number of g-operators that they contain. The term with one g-operator yields the BHF approximation; then would come terms with three g-operators, those with four g-operators etc. It turns out that this rearranged series diverges.[38] For instance the diagrams shown in Fig. 15 all have comparable magnitudes.[17, 39] The magnitude of a contribution is not determined by the number of g-operators, but rather by the *number of independent hole lines* that it contains, i.e. by the number of independent summations over momenta smaller than k_F. The word "independent" refers to the possibility that, because of the constraint of momentum conservation, the number of downgoing arrows may be larger than the number of independent hole lines; for instance, the diagram D_3^{hh} in Fig. 6 involves three, not four, independent hole lines.

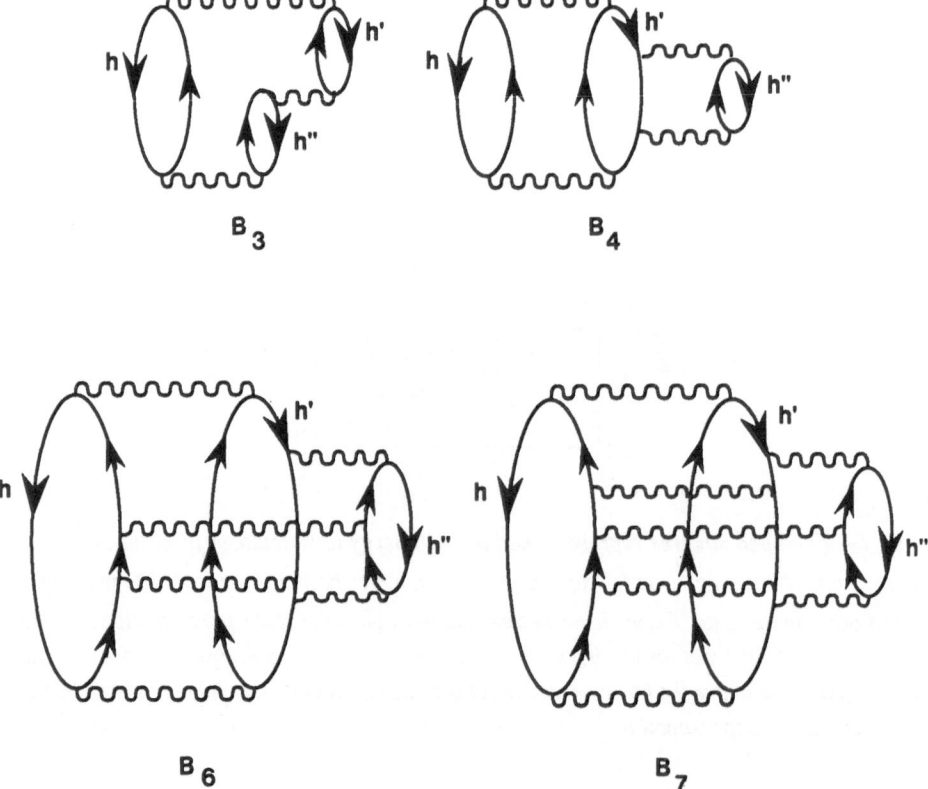

B_3 B_4

B_6 B_7

Fig. 15. *Examples of diagrams with 3, 4, 6 and 7 g-operators. They all have comparable magnitudes, because they all involve three hole lines* (h, h', h'').

This leads to the concept of a *hole-line expansion*, in which all one groups all the diagrams with two hole lines (this yields the BHF approximation), all those with three independent hole lines (among these, the diagrams which appear in Figs. 11 and 15), all those with four independent hole lines, etc. Physically, the three-hole line contribution corresponds to the sum of the processes depicted in Fig. 16. Nor surprisingly, the hole-line expansion is akin to the exp S (Sec. 3.5) since in the latter the wavefunction is expanded in terms of contributions with 2-, 3-, ... hole lines.

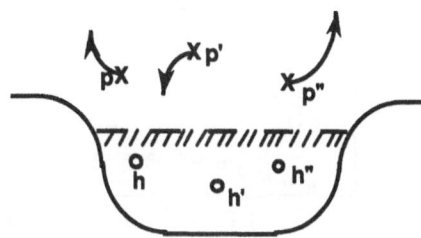

Fig. 16. Physical interpretation of the three hole line contributions : three nucleons which initially belonged to the Fermi sea (momenta h , h' , h'') have been excited above the Fermi surface (momenta p , p' , p'') . Then, they can interact an arbitrary number of times.

Semi-quantitative estimates [17, 40] suggest that the magnitude of the diagrams with H hole lines is smaller than that of a diagram with (H - 1) hole lines by a factor κ , where κ is the average probability that one nucleon be excited above the Fermi surface (Fig. 1). Note that κ should be identified with the average of the exact depletion, and *not* the average (κ_{BHF}) of the BHF approximation to the depletion. Figure 17 shows that the quantities κ and κ_{BHF} are sizeably different. The average depletion κ is model independent : it is an intrinsic property of nuclear matter. In contrast, κ_{BHF} depends upon the choice of the auxiliary potential. In the BHF approximation, κ_{BHF} is given by an expression analogous to Eq. (3.18a), except that the interaction v is replaced by the operator g[e(h) + e(h')] ; the expression (3.18a) can be made very small by increasing the denominator. This is precisely what is being done when one chooses the standard choice for U (k) , since e (k_F - 0) $\approx$ - 16 MeV , while e (k_F + 0) $\approx$ + 38 MeV . With the standard choice of U (k), the minimum value of the denominator in Eq. (3.18a) is thus equal to about 100 MeV , while the actual two particle-two hole virtual states may lie at zero excitation energy. Thus, the standard choice of U (k) yields a κ_{BHF} which underestimates the true depletion. Since the three-hole line contribution to the binding energy is smaller than the two-hole line contribution by (approximately) a factor κ_{BHF} , see Refs. 17, 41, the sole consideration of the two- and three-hole line contributions may incorrectly give the impression that the rate of convergence of the hole-line expansion is fast. This is an artefact of the standard choice of U (k) .

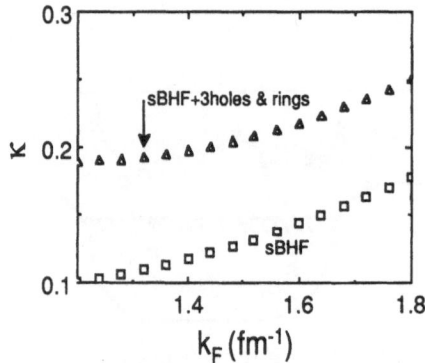

Fig. 17. Modification of the average depletion of the Fermi sea calculated from the standard BHF approximation (lower curve) when three hole line and generalized ring contributions are taken into account (upper curve), in the case of the Argonne V14 interaction (derived from Ref. 37).

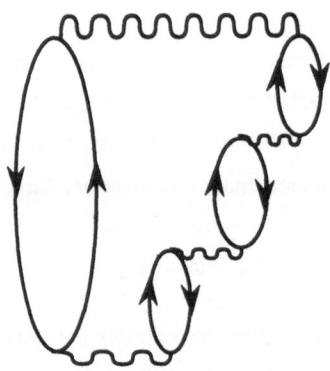

Fig. 18. A particle-hole ring diagram that is included in the generalized (forward-going) ring series summed in Ref. 36.

One concludes that it is necessary to evaluate contributions with three, four and more hole lines. This program has, in part, been carried out by Day[36, 37, 41] who summed the three hole line contributions and also the so-called generalized ring series, which contains diagrams of which Fig. 18 gives an example. In the case of the Argonne V14 interaction, the resulting values of the average depletion are represented in Fig. 16, and those of the binding energy in Fig. 19. The calculated binding energy (triangles) is no longer in contradiction with the calculated upper bound (full squares). The calculated saturation properties are still quite different from the empirical ones. In particular, the calculated saturation density is approximately equal to 0.28 fm^{-3}, while the empirical value is 0.17 fm^{-3}. We also note that an "empirical" curve $B\,(k_F)$ based on the empirical values (2.1), (2.3) and (2.6) would violate the upper bound already at $k_F \approx 1.45$ fm^{-1}. These two observations support the conclusion that *the empirical saturation properties of nuclear matter cannot be reproduced by a model in which structureless nucleons interact via realistic two-body interactions.*

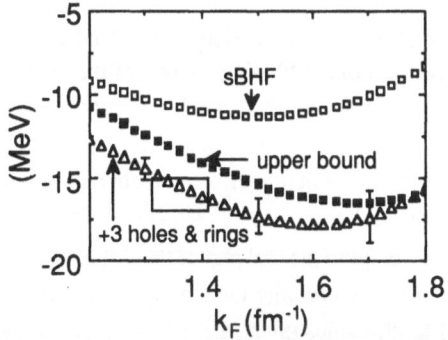

Fig. 19. *The triangles (with typical error bars) represent the binding energy obtained when three- and four-hole line diagrams and the generalized ring contributions are added to the standard BHF approximation (open squares), in the case of the Argonne V14 interaction. The full squares show an upper bound (see Fig. 14).*

9. CONTINUOUS CHOICE

We have seen that the standard BHF approximation is unreliable since it violates the variational principle (Fig. 14). It would be of great practical interest to devise another version of the BHF approximation that would be more accurate. Indeed, this would enable one to construct a two-body effective interaction, namely the reaction operator g, that would account for the two-body component of the nuclear interaction. Three-body components could then be added a posteriori, possibly in the form of *effective* two-body density-dependent interactions.[42]

The available freedom in the BHF approximation (3.22) lies in the choice of the auxiliary potential $U(k)$. We mentioned that *for hole states* $U(h)$ is practically always taken as in Eq. (3.25), so that the bubble- and U-insertions on hole lines cancel each other. No such simple prescription can be made *for particle states*. The reason is that the middle reaction matrix that connects the lines labeled p and h" in graph of Fig. 11 is given by $g[e(h') + e(h) + e(h'') - e(p')]$: it thus depends on momenta (h, h', p') of lines that are different from those that this g matrix connects. This is referred to as an "off the energy shell" effect. It hinders one from finding a simple prescription for $U(p)$ that would have the property that the graphs B_3^{ph} and B_3^{pU} cancel each other. Since no obvious prescription exists for $U(p)$, one has most often set $U(p) = 0$ for $p > k_F$: this is the "standard choice" discussed in Secs. 7 and 8. It has been pointed out long ago[43] that this standard choice is inappropriate for performing a hole line expansion of the mean field and of the momentum distribution. For instance, we already mentioned in connection with Fig. 17 that the average depletion of the Fermi sea, κ_{BHF}, is underestimated when this standard choice is adopted. Relatedly, the corresponding momentum distribution $n(k)$, calculated from the standard choice of $U(k)$, does not have a vertical slope at $k = k_F$, whereas the exact one has a vertical slope (Fig. 1); the imaginary part of the mean field does not vanish at the Fermi energy as it should (Eq. (2.17)), etc. Hence, the standard choice does not lend itself to a simultaneous investigation of the binding energy, of the mean field, of the momentum distribution and of related quantities, for instance the spectral function.

This was the original motivation for proposing[43] the use of an auxiliary potential $U(k)$ that would be a continuous function of k, in particular that would have no large discontinuity at $k = k_F$. Analytical arguments[15, 16, 30, 43, 44-46] exist in favour of a continuous choice that should be close to the real part of the full (physical) mean field.

In the present section, we consider the simplest *continuous choice*, which consists in extending Eq. (3.25) to all values of k, i.e. in using the following prescription *for k larger as well as smaller than k_F* :

$$U(k) = Re \sum_{h'} < hh' \mid g[e(h) + e(h')] \mid hh' > \text{ for all } k, \qquad (9.1)$$

Note that this $U(k)$ is the real part (Re) of the BHF approximation $M_{BHF}(k)$ to the mean field (Eq. (3.24)) : this $M_{BHF}(k)$ is complex for $k > k_F$). Figure 20 illustrates the difference between the standard and this continuous choice of $U(k)$. The continuous choice is similar to the HF field that would be obtained from an effective interaction (Sec. 4). Since it is continuous at k_F, it leads to plausible results for the momentum distribution.[44, 47] It has moreover been argued in Refs. 29, 48-51 that the binding energy calculated from the corresponding BHF approximation is more reliable than that derived from the standard BHF approximation (note, however, that the reliability of the numerical results presented in Ref. 50 has been questioned[52]). In other words, it has been argued that the "continuous BHF approximation" includes part of the contributions that appear only in third order in the hole line expansion based on the standard choice.

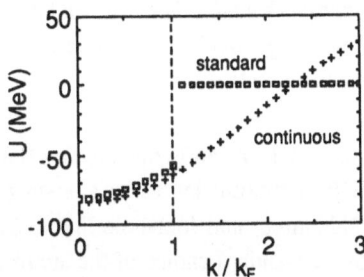

Fig. 20. *Dependence of the auxiliary potential upon k / k_F, for the standard (squares) and continuous (plusses) choices, in the case of the Reid hard core interaction, for $k_F = 1.4$ fm^{-1} (adapted from Ref. 48).*

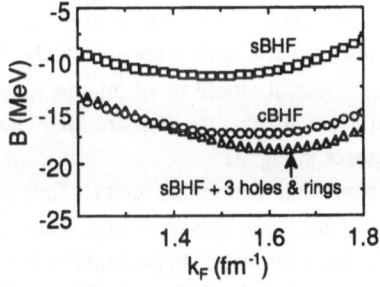

Fig. 21. *The open circles show the binding energy calculated from the continuous BHF approximation (Eq. (9.1)), using as input a finite rank form of the Paris interaction.[28] The squares and the triangles have the same meaning as in Fig. 19 : they represent, respectively, the result of the standard BHF approximation and its modification when three- and four-hole line diagrams and generalized ring contributions are included, with the standard choice for the auxiliary potential and for the Paris interaction.*

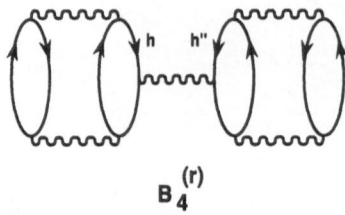

$$B_4^{(r)}$$

Fig. 22. One four-hole line diagram that takes into account part of the effect of the depletion of the Fermi sea on the binding energy. In Fig. 21, this correction has been included in the results shown by the triangles but not in those represented by the circles.

This is confirmed by the results recently obtained by Baldo et al.[28] Indeed, Fig. 21 shows that the continuous BHF approximation yields a binding energy that lies quite close to that obtained when one adds three- and four-hole line diagrams and generalized ring contributions (evaluated with the standard choice of the auxiliary potential U (k)) to the standard BHF approximation. This result is quite encouraging. The agreement would become even more striking if one would include the correction associated with the contribution κ_{BHF} (h) κ_{BHF} (h") to the product { n_2 (h) n_2 (h") } in the first term on the right-hand side of Eq. (3.20). This correction is represented by the diagram shown in Fig. 22. It is approximately equal to $\kappa^2_{BHF} D_{BHF}$ .It is thus attractive; its magnitude increases with density since both κ_{BHF} and the modulus of D_{BHF} increase with density.

One should keep in mind that the interaction adopted by Baldo et al.[28] is a finite rank representation of the original Paris interaction used to obtain the squares and triangles in Fig. 21. Moreover, the latter include the effect of high partial waves, that have been omitted by Baldo et al. (the contribution of these high partial waves differs in the continuous and in the standard BHF approximations, see Ref. 51). It appears likely that these differences yield only small effects.[53] Nevertheless, it would be of interest to check this, for instance by using the code of Baldo et al. for evaluating the standard BHF approximation, and by comparing the result with the squares in Fig. 21.

In the case of a finite rank interaction, the integral equation for the g-matrix can be solved with ease and with great accuracy. This advantage is of practical interest because the evaluation of the g-matrix is time consuming, in particular if the continuous choice (9.1) is adopted for the auxiliary potential. Indeed, Eqs. (3.23) and (9.1) must be solved self-consistently for all the values of k , while the standard choice (3.25) only involves self-consistency for momenta smaller than k_F .

Various methods exist for solving the integral equation (3.23) for the g-matrix. The accuracy of these various codes should always be checked and exhibited , for instance by evaluating the standard BHF approximation. *Reproducing known benchmark results is the only satisfactory means of testing a code.* This reminder may be of relevance for elucidating the difference between the results shown in Fig. 23, which is quite puzzling. Indeed, the two curves correspond to the continuous BHF approximation. The difference between the two calculations is twofold. Firstly, the circles use as input a finite rank representation of the original Paris interaction. Secondly, the numerical methods used for solving Eq. (3.23) are not the same. It appears rather unlikely that the difference between the results shown in Fig. 23 can be ascribed to the slight difference between the input interactions.[53] One may then have to question the accuracy of the code used in Ref. 54 (that code was originally developed and checked for a local interaction and may be less reliable for the nonlocal Paris

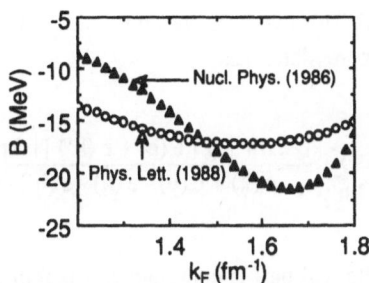

Fig. 23. Binding energy calculated from the continuous BHF approximation. The circles[28] are the same as in Fig. 21; they use as input a finite rank representation of the Paris potential. The triangles[54] use as input the original Paris potential but a possibly less accurate code for computing the reaction matrix.

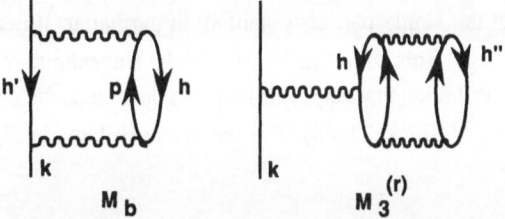

Fig. 24. The two-hole line contribution M_b and one three-hole line contribution $M_3^{(r)}$ to the mean field felt by a nucleon with momentum k.

interaction) or the accuracy of implementing the self-consistent condition (9.1). It would be instructive to calculate the standard BHF with the two codes, and to compare the results with the squares in Fig. 21. Until these checks are performed, the success of the continuous BHF approximation shown in Fig. 21 remains somewhat questionable.

We mentioned that the continuous choice (9.1) of $U(k)$ is identical to the real part $V_{BHF}(k)$ of the BHF approximation to the mean field. Analytic arguments indicate that it would be preferable to use for $U(k)$ the real part of the *full* mean field, and not only the real part of the BHF approximation to it. As a first step in that direction, one may use as auxiliary potential the sum of $V_{BHF}(k)$ and of the real part of the two-hole line contribution to the mean field. This two-hole line contribution is represented by the diagram at the left of of Fig. 24. Its expression reads

$$M_b(k) = \frac{1}{2} \sum_{hh'p} \frac{|<hh'| \, g[\,e(h)+e(h')\,]\,|\,kp>|^2}{e(k)+e(p)-e(h)-e(h')-i\delta} \, , \qquad (9.2)$$

compare with Eq. (3.17b). Its real part is continuous; it is repulsive for all k; it rapidly decreases with increasing k for $k > k_F$. If the real part of $M_b(k)$ is included in the prescription for the auxiliary potential, the corresponding continuous BHF approximation to the binding energy yields more binding that the one used in Fig. 21 (see Ref. 28). Then, however, the graphs B_3^{hb} and B_3^{hU} of Fig. 11 do not cancel, in contrast to all the cases considered heretofore. It would thus be more consistent to add the sum $B_3^{hb} + B_3^{hU}$ to this new BHF approximation; it has been argued in Ref. 55 that this would yield back approximately the same result as those represented by the open circles in Fig. 21, i.e. as those obtained from the continuous BHF in which $U(k)$ does not include the real part of (9.2). Although the inclusion of the real part of the graph $M_b(k)$ in the auxiliary potential would most probably have no sizeable effect on the sum $B_{BHF} + B_3^{hb} + B_3^{hU}$ for the binding energy, it does affect the single-particle potential; in particular, it decreases the absolute value of the Fermi energy. This helps fulfilling[28, 56] the Hugenholtz-Van Hove theorem.[10] Another important contribution that helps fulfilling this theorem is the one represented by the graphs $M_3^{(r)}$ and $B_4^{(r)}$ (Figs. 22, 24) : this has been exhibited in Fig. 40 of Ref. 15.

10. MODEL SPACE APPROACH

We mentioned above that the reaction matrix approximately plays the role of an effective two-body interaction. However, caution must be exercised if one wishes to use this g operator in calculations of nuclear spectra. Indeed, these calculations explicitly evaluate the coupling between configurations that have, qualitatively, already been included as intermediate states in the definition (3.23) of the g-matrix. This double counting problem has led Kuo and collaborators[57, 58] to divide the set of particle momenta into two categories, namely those with $k_F < p < k_M$, and those with momenta $k_M < q < \infty$. The *model space* consists in the configurations in which all the lines have momenta in the range $(0, k_M)$.

The *eliminated space* contains the configurations in which *at least one* momentum lies in the range (k_M, ∞). Thus, the cut-off momentum k_M determines the size of the model space. Recent numerical calculations[34] use $k_M = 3.2 \text{ fm}^{-1}$.

In the model space approach of Kuo and collaborators,[33, 34, 57, 58] a new reaction matrix is defined by the following integral equation

$$K[w] = v + v \sum_{Q,Q'} \frac{|QQ'><QQ'|}{w - \varepsilon(Q) - \varepsilon(Q') + i\delta} K[w] \quad . \tag{10.1}$$

The difference between the equation (3.23) for the g-matrix and the equation (10.1) for the K-matrix is that *in Eq. (10.1) the summation extends over momenta Q,Q' among which at least one must be larger than the cut-off momentum* k_M. As in Eq. (3.16), the single-particle energies are written in the from

$$\varepsilon(k) = k^2/2m + \mathcal{U}(k) \quad , \tag{10.2}$$

where $\mathcal{U}(k)$ is an auxiliary potential that remains to be defined. The authors of Ref. 34 (see also Refs. 33, 57) take

$$\mathcal{U}(k) = \mathcal{V}_{mBHF}(k) \text{ for } 0 < k < k_M$$

$$\mathcal{U}(k) = 0 \qquad \text{for } k_M < k < \infty \quad . \tag{10.3}$$

Here, $\mathcal{V}_{mBHF}(k)$ is the real part of the following "modified" BHF approximation to the mean field

$$M_{mBHF}(k) = \sum_{h'} < kh' | K[\varepsilon(k) + \varepsilon(h')] | kh' > \quad . \tag{10.4}$$

The modified BHF approximation to the binding energy reads[34]

$$B_{mBHF} = T_0 + \frac{1}{2} \sum_{hh'} < hh' | K[\varepsilon(h) + \varepsilon(h')] | hh' > \quad . \tag{10.5}$$

Note that the modified and the standard BHF approximation (Sec. 7) become identical if one sets $k_M = k_F$. The diamonds in Fig. 25 show the binding energy as calculated from the modified BHF approximation (10.5), in the case of the Argonne V14 interaction and for $k_M = 3.2 \text{ fm}^{-1}$. They are seen to lie below the results obtained from the standard BHF approximation. However, they still lie above the upper bound. This exhibits the need of considering corrections to the modified BHF approximation.

In Eq. (10.1) at least one of the momenta Q, Q' must be larger than the cut-off momentum k_F. Therefore, the modified BHF approximation does not include those parts of the diagrams shown in Fig. 9 in which the momenta associated with the two upgoing arrows lie in the range (k_F, k_M). It would thus appear natural to add these missing parts to the modified BHF approximation. This would yield a typical BHF approximation, whose sole difference with the standard or with the continuous BHF approximation lies in the prescription (10.3) used here for the auxiliary potential. Here, $\mathcal{U}(k)$ is continuous at $k = k_F$ as in the continuous choice, but it is discontinuous at $k = k_M$. If one sets $k_M \approx 3.2$ fm^{-1}, as done in practice, the value of $\mathcal{U}(k_M)$ is small and the *shape* of $\mathcal{U}(k)$ is similar to that adopted in the continuous choice. However, the modified and the continuous prescriptions are quantitatively different since $M_{mBHF}(k)$ (Eq. (10.4)) differs from $M_{BHF}(k)$ (Eq. (3.24)). *This remains true even if the particle-particle ladder is summed. If one would let k_M go to infinity, the K-matrix would approach the bare nucleon-nucleon interaction and the modified Brueckner-Hartree-Fock approximation would become the Hartree-Fock approximation. This illustrates the sensitivity of the modified BHF approximation upon the choice of the cut-off momentum k_M.*

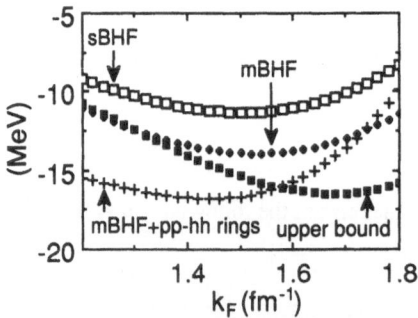

Fig. 25. *Energies calculated from the Argonne V14 interaction. The open squares show the binding energy as calculated from the standard BHF approximation and the full squares give the upper bound (see Fig. 14). The diamonds have been calculated[34] from the modified BHF approximation (Eq. (10.5)). The plusses have been obtained[34] by adding particle-particle and hole-hole ring diagrams to this modified BHF approximation.*

We argued that the modified BHF approximation would become a typical BHF approximation if one would add to it the missing parts of the particle-particle (pp) ladder of Fig. 9. It is possible[59] to sum not only these missing parts, but also hole-hole ladders, of which D_3^{hh} in Fig. 6 is an example. This "summation of the particle-particle and hole-hole ring diagrams to all orders" includes terms with an arbitrary number of hole lines. It yields the binding energy represented by the plusses in Fig. 25. It is seen that these results violate the upper bound for Fermi momenta larger than $1.6\,\mathrm{fm}^{-1}$. Even below $1.6\,\mathrm{fm}$ the calculated saturation properties appear to be incompatible with the upper bound. It would be of interest to see whether this discrepancy persists if one sets $k_M = k_F$. This would indeed be a blow to the convergence of the hole line expansion. If $k_M \gg k_F$, it appears dangerous to perform the summation of diagrams with many hole lines as done when summing the particle-particle and hole-hole ring diagrams as in Refs. 33, 34. Indeed, the graphs with bubble- and U-insertions on hole lines (see e.g. B_3^{hb} and B_3^{hU} in Fig. 11) do not cancel each other if one adopts the prescription (10.3) for the auxiliary potential.

11. CONCLUSIONS

The calculations performed during the last ten years strongly suggest that the empirical saturation properties (Eqs. (2.3), (2.4), (2.6)) cannot be reproduced with realistic *two-body* interactions between *structureless* nucleons. This conclusion is mainly based on the following two types of results. (i) The hole line expansion (Sec. 8) has been carried out up to four-hole line diagrams, for the standard choice (Sec. 7) of the auxiliary potential. It is believed that convergence has then reached, mainly because the four-hole line contribution is small and because the result is not much changed when one adds a series of diagrams with many hole lines, namely the generalized particle-hole ring diagrams. The calculated binding energy (triangles in Fig. 19) saturates at a density (≈ 0.28 nucleon/fm^3) that is much larger than the empirical value (≈ 0.17 nucleon/fm^3). (ii) Upper bounds to the binding energy have been evaluated (squares in Fig. 19). They are incompatible with the empirical saturation properties, in the sense that a parabola $B(\rho)$ that would reproduce the empirical saturation properties would violate the upper bound somewhat above ρ_0.

One should nevertheless remain cautious. In the past, many results that had been believed to be reliable finally turned out to be inaccurate. At present, it is for instance not yet absolutely certain that the hole line expansion converges towards the correct answer when one adopts the standard choice for the auxiliary potential.[18] Also, one would like to have an estimate of the error bar that is attached to the evaluated "upper bound".

If two-body interactions between structureless nucleons fail to reproduce the empirical properties, one must modify the starting assumptions of the nuclear matter model. The most natural modification consists in introducing three-body interactions. It could also happen that the virtual excitation of internal degrees of freedom of the nucleon may be different in nuclear matter than in a two-body system. Besides, much effort is being devoted to the study of relativistic models of nuclear matter, as described by other lecturers.

From the results presented in Sec. 7, it can be concluded that the standard Brueckner-Hartree-Fock approximation is inaccurate. The continuous Brueckner-Hartree-Fock approximation (Sec. 9) appears to be more reliable. Indeed, it yields a binding energy that lies close to the one found in fourth order in the hole line expansion with the standard choice of the auxiliary potential (Fig. 21). However, it faces two problems. Firstly, a sizeable difference exists between two existing numerical calculations (Fig. 23). Secondly, the rate of convergence of the hole line expansion for the continuous choice has not yet been investigated in a reliable way; this investigation would be quite difficult to carry out.

One of the main merits of the continuous choice is that it enables one to investigate several physical quantities on the same footing, in particular the binding energy, the momentum distribution, the spectral functions, the mean field, etc. In the past, one has probably focused too much attention on the calculation of the binding energy. Thereby, one has somewhat lost sight of the fact that, originally, the main aim of the study of nuclear matter was to explain why the shell model is successful despite the strong and complicated nature of the nuclear interaction. Theoretical interest in this problem is strongly revived by recent experimental information on the validity of the shell model for inner orbits,[60] and on the occupancy of shell-model orbits in the nuclear ground state.[6,7]

REFERENCES

1. J.P. Blaizot, D. Gogny and B. Grammaticos, Nucl. Phys. A265:315 (1976).
2. G. Co' and J. Speth, Phys. Rev. Lett. 57:547 (1986).
3. G.E. Brown and E. Osnes, Phys. Lett. 159B:233 (1985).
4. A.B. Migdal, Sov. Phys. JETP 5:333 (1957).
5. J.M. Luttinger, Phys. Rev. 119:1153 (1960).
6. E.N.M. Quint, B.M. Barnett, A.M. van den Berg, J.F.J. van der Brand, H. Clement, R. Ent, B. Frois, D. Goutte, P. Grabmayr, J.W.A. den Herder, E. Jans, G.J. Kramer, J.B.J.M. Lanen, L. Lapikas, H. Nann, G. van der Steenhoven, G.J. Wagner and P.K.A. de Witt Huberts, Phys.Rev.Lett. 58:1088 (1987).
7. P.K.A. de Witt Huberts, in: "Perspectives in Nuclear Physics at Intermediate Energies", S. Boffi, C. Ciofi degli Atti and M.M. Giannini, Eds., World Scientific Publ. Comp., Singapore, (1988), p. 381.
8. C.H. Johnson, D.J. Horen and C. Mahaux, Phys. Rev. C36:2252 (1987).
9. C. Mahaux and R. Sartor, Nucl. Phys. (in press).
10. N.M. Hugenholtz and L. Van Hove, Physica 24:363 (1958).
11. J.W. Negele and K. Yazaki, Phys. Rev. Lett. 47:71 (1981).
12. H.A. Bethe, Ann. Rev. Nucl. Sci. 21:93 (1971).
13. H.S. Köhler, Phys. Reports 18:217 (1975).
14. H. Kümmel, K.H. Lührmann and J.G. Zabolitzky, Phys. Reports C36:1 (1978).
15. J.P. Jeukenne, A. Lejeune and C. Mahaux, Phys. Reports C25:83 (1976).
16. C. Mahaux in "Jastrow versus Brueckner Theory", R. Guardiola and J. Ros, Eds., Lecture Notes in Physics, Springer Verlag (1981), p. 50.
17. B.D. Day, Revs. Mod. Phys. 39:719 (1967).
18. G.A. Baker, Jr. and J.L. Gammel, Phys. Rev. C6:403 (1972).
19. G.A. Baker, Jr., Phys. Rev. C17:1253 (1978).
20. M. Gell-Mann and F.E. Low, Phys. Rev. 84:350 (1951).
21. C. Mahaux and R. Sartor, Phys. Rev. C19:229 (1979).
22. M.K. Weigel and G. Wegmann, Fortschr. Phys. 19:451 (1971).
23. R. Jastrow, Phys. Rev. 98:1479 (1955).
24. S. Rosati and S. Fantoni, in "Jastrow versus Brueckner Theory", R. Guardiola and J. Ros, Eds., Lecture Notes in Physics, Springer Verlag (1981), p. 1.
25. D. Ceperley, G.V. Chester and M.H. Kalos, Phys. Rev B16:3081 (1977).
26. C. Mahaux, P.F. Bortignon, R.A. Broglia and C.H. Dasso, Phys. Reports 120:1 (1985).

27. R.E. Peierls, in "New Directions in Physics", N. Metropolis, D.M. Kerr and G.-C. Rota, Eds., Academic Press, Inc. (1987), p. 95.

28. M. Baldo, I. Bombaci, L.S. Ferreira, G. Giansiracusa and U. Lombardo, Phys. Lett. 209B:135 (1988).

29. A. Lejeune and C. Mahaux, Nucl. Phys. A295:189 (1978).

30. R. Sartor, Phys. Rev. C30:2036 (1984).

31. M. Lacombe, B. Loiseau, J.M. Richard, R. Vinh Mau, J. Côté, P. Pires and R. de Tourreil, Phys. Rev. C21:861 (1980).

32. R.B. Wiringa, R.A. Smith and T.L. Ainsworth, Phys. Rev. C29:1207 (1984).

33. T.T.S. Kuo, Z.Y. Ma and R. Vinh Mau, Phys. Rev. C33:717 (1986).

34. M.F. Jiang, T.T.S. Kuo and H. Müther, Phys. Rev. C38:2408 (1988).

35. F. Coester, S. Cohen, B. Day and C.M. Vincent, Phys. Rev. C1:769 (1970).

36. B.D. Day, Phys. Rev. Lett. 47:226 (1981).

37. B.D. Day and R.B. Wiringa, Phys. Rev. C32:1057 (1985).

38. H.A. Bethe, Phys. Rev. 138:B804 (1965).

39. R. Rajaraman and H.A. Bethe, Revs. Mod. Phys. 39:745 (1967).

40. B.D. Day, Revs. Mod. Phys. 50:495 (1978).

41. B.D. Day, Phys. Rev. 24:1203 (1981).

42. J. Carlson, V.R. Pandharipande and R.B. Wiringa, Nucl. Phys.A401:59 (1983).

43. J. Hüfner and C. Mahaux, Ann. Phys. (N.Y.) 73:525 (1972).

44. R. Sartor, Nucl. Phys. A289:329 (1977).

45. C. Mahaux, Nucl. Phys. A328:24 (1979).

46. R. Sartor, Phys. Rev. C27:899 (1983).

47. P. Grangé, J. Cugnon and A. Lejeune, Nucl. Phys. A473:365 (1987).

48. J.P. Jeukenne, A. Lejeune and C. Mahaux, Nucl. Phys. A245:411 (1975).

49. A. Lejeune and C. Mahaux, Nucl. Phys. A317:37 (1979).

50. P. Grangé and A. Lejeune, Nucl. Phys. A327:335 (1979).

51. P. Grangé, A. Lejeune and C. Mahaux, Nucl. Phys. A319:50 (1979).

52. B.D. Day, private communication (December 1980).

53. H. Lampl and M.K. Weigel, Phys. Rev. C33:1834 (1986).

54. A. Lejeune, P. Grangé, M. Martzolff and J. Cugnon, Nucl. Phys. A453:189 (1986).

55. C. Mahaux, Nucl. Phys. A163:299 (1971).

56. K.A. Brueckner and D.J. Goldman, Phys. Rev. 116:424 (1959).

57. Z.Y. Ma and T.T.S. Kuo, Phys. Lett. 127B:137 (1983).

58. T.T.S. Kuo and Z.Y.Ma, in "Nucleon-Nucleon Interaction and Nuclear Many-Body Problems", S.S. Wu and T.T.S. Kuo, Eds., World Scientific Publ., Singapore, (1984) 178-202.

59. H.Q. Song, S.D.Yang and T.T.S. Kuo, Nucl. Phys. A462:491 (1987).

60. B. Frois, J.M. Cavedon, D. Goutte, M. Huet, Ph. Leconte, C.N. Papanicolas, X.H. Phan, S.K. Platchkov, S.E. Williamson and W. Boeglin, Nucl. Phys. A396:409c (1983).

A RELATIVISTIC THEORY OF NUCLEAR MATTER

Brian D. Serot

Physics Department and Nuclear Theory Center
Indiana University
Bloomington, IN 47405

ABSTRACT

Nuclear matter is studied in the framework of quantum hadrodynamics (QHD). As a simple first example, the nuclear equation of state is computed in the mean-field approximation to the Walecka model (QHD–I). General principles of covariant thermodynamics and thermodynamic consistency are introduced, and these principles are illustrated by recomputing the mean-field nuclear matter properties in an arbitrary reference frame. The loop expansion is proposed as a candidate for performing reliable calculations beyond the mean-field approximation, and the one-loop vacuum corrections to the mean-field results are discussed. The two-loop corrections for nuclear matter, including vacuum polarization, are then calculated. The size and nature of the two-loop corrections indicate that the loop expansion is apparently not a useful procedure in this model. Prospects for alternative expansion schemes are discussed.

INTRODUCTION

The accurate description of hot, dense matter is an important problem in theoretical physics. Shortly after the Big Bang, the universe was composed entirely of hot, dense matter. Calculations based on quantum field theory at finite temperature and density have been used to compute the properties of the pervasive quark–gluon plasma[1-4] and to investigate symmetry restoration in the early universe.[5-7] In the present universe, the nuclear equation of state as a function of temperature, density, and the ratios of protons to neutrons and nucleons to hyperons is needed to study stellar collapse through a possible supernova phase into a neutron star. On a smaller scale, we can explore the same nuclear dynamics through energetic collisions of heavy ions, where at least part of the hot matter can be described in terms of its hadronic constituents.

In this series of lectures, I discuss systematic techniques for computing the properties of hot, dense nuclear matter. The traditional approach to this problem relies on the Schrödinger equation for nonrelativistic nucleons interacting through static, two-body potentials. Some of the existing techniques are discussed by Professors Pandharipande and Mahaux elsewhere in this volume. Although the Schrödinger equation has been useful in this program for more than fifty years, newly completed experimental facilities and existing initiatives will force us to go beyond the Schrödinger equation to compare calculations with the data of the future. A more complete treatment of hadronic systems should include relativistic motion of the nucleons, dynamical mesons and baryon resonances, modifications of the nucleon structure in the nucleus, and the dynamics of the quantum vacuum, while maintaining general properties of quantum mechanics, covariance, gauge invariance, and causality. These physical effects will be relevant regardless of the degrees of freedom used to describe the system, and they must be studied simultaneously and consistently to draw definite conclusions about nuclear dynamics at high temperatures and high densities.

Our basic goal is therefore to formulate a consistent microscopic treatment of strongly interacting, relativistic, quantum statistical mechanical systems. This will allow for investigation of a wide variety of phenomena. For example, we can compute both static thermodynamic properties (like energy, pressure, and entropy) and dynamical characteristics (such as viscosity, transport coefficients, and collective modes and their damping). In addition, we can study the production and absorption of particles under extreme conditions and map out the nuclear matter phase diagram. Ultimately, we would like to extend the techniques to deal with nonequilibrium systems, so that we can describe the development of two isolated nuclei into a single system in equilibrium, which may be particularly relevant for heavy-ion collisions.

Since the physical phenomena of interest are relativistic and involve particle production and absorption, the only existing consistent framework for their description is relativistic quantum field theory. I will use a relativistic quantum field theory based on mesons and baryons, which is known as quantum hadrodynamics (QHD).[8] QHD is consistent in the sense that the dynamical assumptions (such as the relevant degrees of freedom, the form of the lagrangian, and the normalization conditions) are made at the outset, and one then attempts to extract concrete results from the implied formalism. In principle, the assumptions permit the formulation of systematic, "conserving" approximations[9,10] that maintain the important general properties mentioned above. Calculations can then be compared to data to see if the framework is related to the real world and to decide where QHD succeeds and where it fails.

There are several reasons for considering only hadronic degrees of freedom. First, these variables are the most efficient at low densities and temperatures and for describing particle absorption and production, as hadrons are the particles observed experimentally. Second, hadronic calculations can be calibrated by comparing to observed hadron-hadron scattering and empirical nuclear properties; we can then extrapolate to extreme conditions and test model predictions. Most importantly, we must understand the limitations of QHD to deduce true signals of QCD behavior in nuclear matter.

Moreover, if we build QHD with renormalizable lagrangians, the number of parameters in any model is finite.[11,12] Calculations can be carried out beyond the tree level without introducing additional parameters determined solely by short-distance phenomena; thus, the sensitivity of calculated results to short-distance input is minimized. The dynamical assumption underlying renormalizability is that the short-distance behavior and the quantum vacuum can be described in terms of hadronic degrees of freedom only. This assumption must ultimately break down, since hadrons are actually composed of quarks and gluons. For QHD to be useful, nuclear observables of interest must not be dominated by contributions from short distances (where QHD is inappropriate). This conjecture must be tested (and its limitations uncovered) by performing detailed calculations in a consistent relativistic framework.

In addition to concentrating on hadronic variables, I will emphasize a covariant formulation. This allows for a direct computation of the quantities that appear in the equations of General Relativity (for example, the energy-momentum tensor). More importantly, a covariant formalism can be used in any convenient reference frame, which can simplify calculations in some cases and is essential in others.[13]

I will begin these lectures by focusing on some general aspects of relativistic many-body systems at finite temperature and density. In particular, I discuss covariant formulations of thermodynamics and the preservation of thermodynamic consistency in microscopic calculations. To provide a concrete example of these ideas, I will illustrate a covariant finite-temperature calculation in a simple context: the mean-field theory of the Walecka model (QHD–I).[14,8] This model contains some basic elements of hadronic theories of nuclei, namely, baryons coupled strongly to neutral scalar and vector fields. There are several advantages to this calculation. First, it is a calculation that can be done by all of you. Second, although the calculation is simple, it produces some informative results, such as the Lorentz transformation properties of the temperature and chemical potential. Finally, it verifies explicitly that a preferred frame (namely, that of the "thermal bath" in contact with our system) does not destroy the covariance of the results.

We must remember, however, to keep this calculation in context. It is just a simple approximation to a simple model. In fact, much of the previous work in QHD has been performed in either the mean-field theory (MFT) or at the one-loop level ("relativistic Hartree approximation" or RHA), which includes the shift in the baryon vacuum energy.[14-16] The original motivation for these studies was that the MFT should become increasingly valid as the density increases, and that the MFT and RHA could be good *nonperturbative* starting points for calculations at normal nuclear density. These MFT and RHA calculations are successful in describing a variety of nuclear systems and phenomena.[8,17,18] For example, the bulk properties of spherical nuclei can be described accurately with parameters similar to those obtained by fitting mesonic potentials to nucleon-nucleon scattering observables. These results have stimulated many successful phenomenological calculations that use Dirac nucleons and that are modeled after QHD.

In spite of these phenomenological successes, however, the QHD framework is hollow if it cannot make predictions that are subject to definitive experimental tests. Moreover, it is still unclear that the simple physical picture obtained the MFT and RHA is a useful starting point for systematic treatments of nuclear phenomena at ordinary densities. Detailed calculations are needed to justify the successes of the MFT and RHA, or to relegate them to pure phenomenology. One advantage of the Schrödinger formalism is that systematic approximation schemes have been developed and used to produce reliable conclusions. Such schemes have yet to be developed for QHD.

The major obstacle to this development is that QHD is a strong-coupling field theory. Unlike QED, there is no obvious asymptotic expansion to use to obtain results and refine them systematically. In fact, it is presently unknown whether QHD permits *any* expansion for systematic computation and refinement of theoretical results. Such an expansion (or expansions) must be found if we are to compute reliably and make definitive comparisons with precision experimental data. My purpose in the second part of these lectures is to investigate a systematic procedure for going beyond the MFT with the intent of performing reliable calculations in QHD.

I will again focus on the Walecka model and, consistent with the philosophy of QHD, treat both the mesons and baryons as true quantum degrees of freedom.

For simplicity, I will restrict discussion to zero-temperature nuclear matter in its rest frame. The basic theoretical tools will be path integrals,[19-22] the quantum generating functional for the Green's functions,[19,23] Feynman diagrams,[8,19] and the loop expansion.[24-28] My emphasis here is not on quantitative accuracy in describing experiment, but rather on the formulation and investigation of reliable approximations in strong-coupling relativistic quantum field theory. I will show how the one-loop approximation (which contains the MFT) arises naturally and then discuss the rules for the loop expansion of the energy density. (At finite temperature, this expansion produces the thermodynamic potential.) After illustrating the form of the two-loop corrections, I evaluate these terms in nuclear matter with several sets of parameters to determine the size of the corrections. We will see that the two-loop contributions are large, and the loop expansion is not a useful procedure in this model. Finally, I explain why this result is not surprising and present prospects for alternative expansion schemes.

THE WALECKA MODEL

Quantum hadrodynamics is a general framework for the relativistic nuclear many-body problem.[8] The detailed dynamics must be specified by choosing a particular renormalizable lagrangian density. To illustrate the nuclear matter formalism as simply as possible, I will consider the Walecka model[14] (sometimes called QHD–I), which contains baryons (ψ) and neutral scalar (ϕ) and vector (V^μ) mesons.

The lagrangian density in the Walecka model is given by[8]

$$\mathcal{L} = \overline{\psi}\left[\gamma_\mu(i\partial^\mu - g_\mathrm{v}V^\mu) - (M - g_\mathrm{s}\phi)\right]\psi + \tfrac{1}{2}(\partial_\mu\phi\partial^\mu\phi - m_\mathrm{s}^2\phi^2)$$
$$- \tfrac{1}{4}F_{\mu\nu}F^{\mu\nu} + \tfrac{1}{2}m_\mathrm{v}^2 V_\mu V^\mu + \delta\mathcal{L}\,, \tag{1}$$

where $F^{\mu\nu} = \partial^\mu V^\nu - \partial^\nu V^\mu$ and $\delta\mathcal{L}$ contains counterterms. The parameters M, g_s, g_v, m_s, and m_v are phenomenological constants that may be determined (in principle) from experimental measurements. The counterterms are for renormalization purposes.

The motivation for this model has evolved considerably since it was introduced. At present, there is significant empirical evidence that when the nucleon-nucleon (NN) interaction is described in a Lorentz covariant fashion, it contains strong Lorentz scalar and four-vector pieces.[29-31] These must be reproduced in any relativistic theory of nuclear structure, and the simplest way to do this is through the exchange of scalar and vector mesons. These two components are the most important for describing bulk nuclear properties, which is my main concern here. The other Lorentz components of the NN interaction, in particular the terms arising from pion exchange, average essentially to zero in spin-saturated nuclear matter and may be incorporated as refinements to the present model.[8] (Some models containing additional degrees of freedom are described in this volume by Professors Malfliet and Siemens.) The important point is that even in more refined models, the dynamical features generated by scalar and vector mesons remain; thus it is important to first understand the consequences of these degrees of freedom for relativistic descriptions of nuclear systems. Note that eq. (1) resembles massive QED with an additional scalar interaction, so the relativistic quantum field theory generated by this Lorentz invariant lagrangian is *renormalizable*.[32]

The field equations for the model follow from the Euler–Lagrange equations and can be written as

$$(\partial_\mu \partial^\mu + m_s^2)\phi = g_s \overline{\psi}\psi \,, \tag{2}$$

$$\partial_\nu F^{\nu\mu} + m_v^2 V^\mu = g_v \overline{\psi}\gamma^\mu \psi \,, \tag{3}$$

$$[\gamma^\mu(i\partial_\mu - g_v V_\mu) - (M - g_s\phi)]\psi = 0 \,. \tag{4}$$

Equation (2) is simply the Klein–Gordon equation with a scalar source. Equation (3) looks like massive QED with the conserved baryon current

$$B^\mu \equiv (\rho_B, \underset{\sim}{B}) = \overline{\psi}\gamma^\mu\psi \,, \qquad \partial_\mu B^\mu = 0 \tag{5}$$

rather than the (conserved) electromagnetic current as source. Finally, eq. (4) is the Dirac equation with scalar and vector fields introduced in a minimal fashion. These field equations also imply that the canonical energy-momentum tensor

$$T^{\mu\nu} = i\overline{\psi}\gamma^\mu\partial^\nu\psi - \tfrac{1}{2}[\partial_\sigma\phi\partial^\sigma\phi - m_s^2\phi^2]g^{\mu\nu} + \partial^\mu\phi\partial^\nu\phi$$

$$+ \tfrac{1}{2}[\partial_\sigma V_\lambda\partial^\sigma V^\lambda - m_v^2 V_\sigma V^\sigma]g^{\mu\nu} - \partial^\mu V_\lambda\partial^\nu V^\lambda \tag{6}$$

is conserved ($\partial_\mu T^{\mu\nu} = \partial_\nu T^{\mu\nu} = 0$). Note that to write eq. (6), I discarded several total divergences (which are irrelevant for evaluating observables in infinite matter) and used $\partial_\mu V^\mu = 0$, which follows from the vector meson field equation (3) and the conservation of the baryon current (5).

I emphasize that eqs. (2)–(4) are *nonlinear quantum field equations*, and their exact solutions are very complicated. In particular, they describe mesons and baryons *that are not point particles*, but rather objects with intrinsic structure due to the implied (virtual) mesonic and baryon-antibaryon loops. It is here that the dynamical input of renormalizability is apparent, since we are assuming that (at least the long-range part of) this intrinsic structure can be described using hadronic degrees of freedom. The validity of this input and its limitations have yet to be tested consistently within the framework of QHD, and we will return to this question (much) later in these lectures.

We also expect the coupling constants in eqs. (2)–(4) to be large, and thus perturbative solutions are not useful. We have therefore made little progress by writing down these equations without a suitable method for solving them. Fortunately, there is an approximate solution that should become increasingly valid as the nuclear density increases. Consider a system of B baryons in a box of volume V at zero temperature. For simplicity, take the total momentum to be zero, so that we are in the system's rest frame; we will generalize later by allowing for uniform motion. As the baryon density B/V increases, so do the source terms on the right-hand sides of eqs. (2) and (3). When the sources are large, the meson field operators can be replaced by their expectation values, which are classical fields:

$$\phi \equiv \langle\phi\rangle \,, \qquad \delta^{\mu 0} V_0 \equiv \langle V^\mu\rangle \qquad \text{(classical)} \,. \tag{7}$$

For our static, uniform system, ϕ and V_0 are *constants* that are independent of space and time. Rotational invariance implies that the expectation value of the three-vector piece of V^μ vanishes.

It is important to emphasize the strategy involved in the preceding "mean-field" approximation. First, the resulting mean-field theory (MFT) should give the correct

41

solution to the field equations in the high-density limit. More importantly, however, the MFT serves as a starting point for consistently calculating corrections within the framework of QHD, using Feynman diagrams, path-integral methods, and so forth, as I will discuss later in these lectures.

When the meson fields in eq. (4) are approximated by the classical fields of eq. (7), the Dirac equation is linear,

$$[i\gamma_\mu \partial^\mu - g_\mathrm{v}\gamma^0 V_0 - (M - g_\mathrm{s}\phi)]\psi = 0 \,, \tag{8}$$

and can be solved exactly. [It is this *linearization* of the full field equation (4) that allows the baryons to be interpreted now as point particles.] The resulting baryon solutions have a mass that is shifted by the scalar field:

$$M^\star \equiv M - g_\mathrm{s}\phi \,, \tag{9}$$

and an energy spectrum that is shifted by the vector field:

$$\epsilon^{(\pm)}(k) = g_\mathrm{v}V_0 \pm (\underset{\sim}{k}^2 + M^{\star 2})^{1/2} \equiv g_\mathrm{v}V_0 \pm E^\star(k) \,, \tag{10}$$

but otherwise look precisely like free-particle Dirac solutions.[33] As expected, there are solutions with both positive and negative square roots characteristic of the Dirac equation. These solutions can be used to define quantum field operators, and the hamiltonian density for the system can be constructed in the canonical fashion. (For the details of these procedures, the reader is directed to ref. 8.) The result is

$$\hat{H} = \hat{H}_\mathrm{MFT} + \delta H \,, \tag{11}$$

$$\hat{H}_\mathrm{MFT} = g_\mathrm{v}V_0\hat{B} + \sum_{\underset{\sim}{k}\lambda} E^\star(k)(A^\dagger_{\underset{\sim}{k}\lambda}A_{\underset{\sim}{k}\lambda} + B^\dagger_{\underset{\sim}{k}\lambda}B_{\underset{\sim}{k}\lambda}) + V(\tfrac{1}{2}m_\mathrm{s}^2\phi^2 - \tfrac{1}{2}m_\mathrm{v}^2V_0^2) \,, \tag{12}$$

$$\hat{B} = \sum_{\underset{\sim}{k}\lambda} A^\dagger_{\underset{\sim}{k}\lambda}A_{\underset{\sim}{k}\lambda} - B^\dagger_{\underset{\sim}{k}\lambda}B_{\underset{\sim}{k}\lambda} \,, \tag{13}$$

$$\delta H = -\sum_{\underset{\sim}{k}\lambda} \left[(\underset{\sim}{k}^2 + M^{\star 2})^{1/2} - (\underset{\sim}{k}^2 + M^2)^{1/2}\right] \,. \tag{14}$$

Here $A^\dagger_{\underset{\sim}{k}\lambda}$, $B^\dagger_{\underset{\sim}{k}\lambda}$, $A_{\underset{\sim}{k}\lambda}$, and $B_{\underset{\sim}{k}\lambda}$ are creation and destruction operators for (quasi)baryons and (quasi)antibaryons with shifted mass and energy, and $\hat{B}$ is the baryon number operator, which clearly counts the number of baryons minus the number of antibaryons. (The index λ denotes both spin and isospin projections.) The correction term δH arises from placing the operators in $\hat{H}_\mathrm{MFT}$ in "normal order."[8] This correction is easily interpreted in the context of Dirac hole theory. The spectrum of the infinite Dirac sea of occupied states shifts in the presence of the surrounding nucleons at finite density. Since all energies are measured relative to the vacuum, the energy shift must be computed by subtracting the total energy of the Dirac sea in the vacuum, where the nucleons have their free mass M. This leads to the result in eq. (14). I will return later to discuss this "zero-point energy" correction; for now I will concentrate on the MFT hamiltonian defined by eq. (12).

Since $\hat{H}_\mathrm{MFT}$ is diagonal, this model mean-field problem has been solved *exactly*. The solution retains the essential features of QHD: relativistic covariance, explicit meson degrees of freedom, and the incorporation of antiparticles. Furthermore, it yields a simple solution to the field equations that should become increasingly valid as the baryon density increases. This solution and model problem thus provide a useful starting point for describing the nuclear many-body system as well as a consistent basis for computing corrections using relativistic quantum field theory and standard many-body techniques.

THE NUCLEAR MATTER EQUATION OF STATE

For uniform nuclear matter in its rest frame, the ground state of the hamiltonian (12) is obtained by filling states with wavenumber $\underset{\sim}{k}$ and spin-isospin degeneracy γ up to Fermi level k_F. For nuclear matter, $\gamma = 4$ (neutrons and protons with spin up and down). The baryon density ρ_B and energy density $\mathcal{E}$ may be readily evaluated by taking ground-state expectation values of the operators in eqs. (13) and (12):

$$
\rho_B = \frac{\gamma}{(2\pi)^3} \int_0^{k_F} \mathrm{d}^3 k = \frac{\gamma}{6\pi^2} k_F^3 , \tag{15}
$$

$$
\mathcal{E} = \frac{g_v^2}{2m_v^2} \rho_B^2 + \frac{m_s^2}{2g_s^2} (M - M^\star)^2 + \frac{\gamma}{(2\pi)^3} \int_0^{k_F} \mathrm{d}^3 k \, E^\star(k) , \tag{16}
$$

and the pressure p follows from similar manipulations on the trace of the stress tensor $(\frac{1}{3} T_{ii})$:

$$
p = \frac{g_v^2}{2m_v^2} \rho_B^2 - \frac{m_s^2}{2g_s^2} (M - M^\star)^2 + \frac{1}{3} \frac{\gamma}{(2\pi)^3} \int_0^{k_F} \mathrm{d}^3 k \, \frac{\underset{\sim}{k}^2}{E^\star(k)} . \tag{17}
$$

Here V_0 has been eliminated in terms of the conserved baryon density using the mean-field equation [see eqs. (3) and (7)]

$$
V_0 = \frac{g_v}{m_v^2} \langle \psi^\dagger \psi \rangle = \frac{g_v}{m_v^2} \rho_B , \tag{18}
$$

and eq. (9) serves to eliminate ϕ. The first two terms in eqs. (16) and (17) arise from the classical meson fields. The final terms in these equations are those of a relativistic gas of baryons of mass $M^\star$. These expressions give the nuclear matter equation of state in parametric form: $\mathcal{E}(\rho_B)$ and $p(\rho_B)$.

The effective mass $M^\star$ can be determined by minimizing $\mathcal{E}(M^\star)$ with respect to $M^\star$, leading to the *self-consistency condition*

$$
M^\star = M - \frac{g_s^2}{m_s^2} \frac{\gamma}{(2\pi)^3} \int_0^{k_F} \mathrm{d}^3 k \, \frac{M^\star}{E^\star(k)} = M - \frac{g_s^2}{m_s^2} \langle \overline{\psi} \psi \rangle \equiv M - \frac{g_s^2}{m_s^2} \rho_s , \tag{19}
$$

which also defines the scalar density ρ_s. This equation is equivalent to the MFT scalar field equation for ϕ. Note that the scalar density is smaller than the baryon density [eq. (15)] due to the factor $M^\star/E^\star(k)$, which is an effect of Lorentz contraction. Thus the contribution of rapidly moving baryons to the scalar source is significantly reduced. Most importantly, eq. (19) is a *transcendental self-consistency equation* for $M^\star$ that must be solved at each value of k_F. This illustrates the *nonperturbative* nature of the mean-field solution.

An examination of the analytic expression (16) for the energy density shows that the system is unbound ($\mathcal{E}/\rho_{\text{B}} > M$) at either very low or very high densities.[14] At intermediate densities, the attractive scalar interaction will dominate if the coupling constants are chosen properly. The system then *saturates*. The empirical equilibrium properties of nuclear matter are reproduced if the couplings are chosen as

$$ C_{\text{s}}^2 \equiv g_{\text{s}}^2 \left(\frac{M^2}{m_{\text{s}}^2}\right) = 267.1 , \qquad C_{\text{v}}^2 \equiv g_{\text{v}}^2 \left(\frac{M^2}{m_{\text{v}}^2}\right) = 195.9 , \qquad (20) $$

which leads to an equilibrium Fermi wavenumber $k_{\text{F}}^0 = 1.42 \, \text{fm}^{-1}$ and an energy/nucleon $(\mathcal{E}/\rho_{\text{B}} - M) = -15.75$ MeV. (I use this somewhat large saturation density merely for illustration, since it yields results consistent with those in refs. 8 and 14.) Note that only the *ratios* of coupling constants to masses enter in eqs. (16) and (17). The resulting saturation curve is shown in fig. 1. In this approximation, the relativistic properties of the scalar and vector fields are responsible for saturation; a Hartree–Fock variational estimate built on the nonrelativistic (Yukawa) potential limit of the interaction shows that such a system is unstable against collapse.

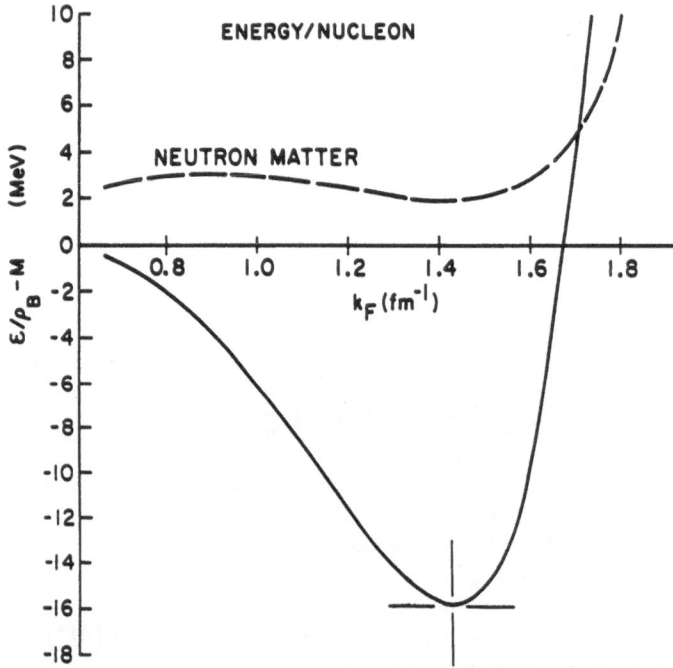

Fig. 1. Saturation curve for nuclear matter. These results are calculated in the relativistic mean-field theory with baryons and neutral scalar and vector mesons (QHD–I). The coupling constants are chosen to fit the value and position of the minimum. The prediction for neutron matter ($\gamma = 2$) is also shown.

The solution of the self-consistency condition (19) for $M^\star$ yields an effective mass that is a decreasing function of the density, as illustrated in fig. 2. Note that $M^\star/M$ becomes small at high density and is significantly less than unity at ordinary nuclear densities. This is a consequence of the large scalar field $g_{\text{s}}\phi$, which is approximately 400 MeV and produces a large attractive contribution to the energy/baryon. There

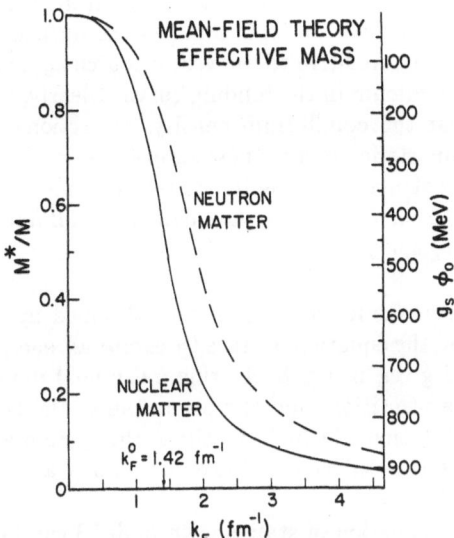

Fig. 2. Effective mass as a function of density for nuclear ($\gamma = 4$) and neutron ($\gamma = 2$) matter based on fig. 1.

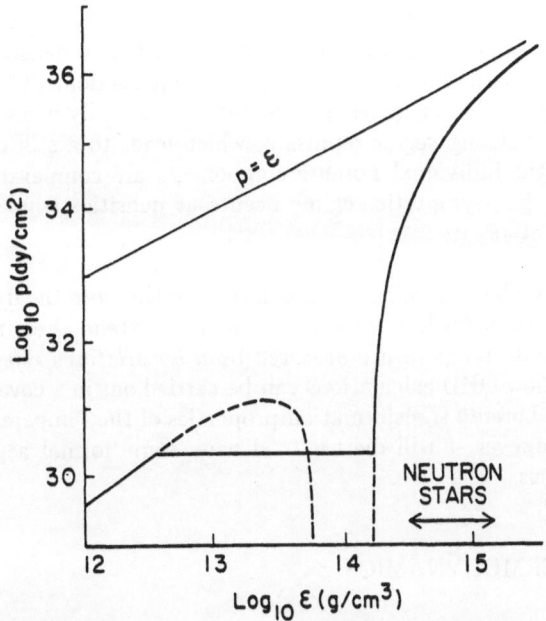

Fig. 3. Predicted equation of state for neutron matter at all densities. The solid and dashed curve shows the result for QHD–I based on fig. 1. A Maxwell construction is used to determine the equilibrium (horizontal) curve in the region of the phase transition. The density regime relevant for neutron stars is also shown.

is also a large repulsive energy/baryon from the vector field $g_v V_0 \approx 350$ MeV. Thus *the Lorentz structure of the interaction leads to a new energy scale in the problem*, and the small nuclear binding energy (≈ 16 MeV) arises from the cancellation between the large scalar attraction and vector repulsion. As the nuclear density increases, the scalar source ρ_s becomes small relative to the vector source ρ_B, and the attractive forces saturate, producing the minimum in the binding curve. Clearly, because of the sensitive cancellation involved near the equilibrium density, corrections to the MFT must be calculated before the importance of this new saturation mechanism can be assessed. Nevertheless, the Lorentz structure of the interaction provides an *additional saturation mechanism that is not present in the nonrelativistic potential limit*, as this limit ignores the distinction between ρ_s and ρ_B.

The corresponding curves for neutron matter obtained by setting $\gamma = 2$ are also shown in figs. 1 and 2, and the equation of state (pressure *vs.* energy density) for neutron matter at all densities is given in fig. 3. In this mean-field model, there is a van der Waals (liquid–gas) phase transition, and the properties of the two phases are deduced through a Maxwell construction. At high densities, the system approaches the "causal limit" $p = \mathcal{E}$ representing the stiffest possible equation of state.

The neutron matter equation of state shown in fig. 3 can be used in the Tolman–Oppenheimer–Volkoff equation for the general-relativistic metric to give the masses of neutron stars as a function of the central density.[34] This MFT gives a maximum neutron star mass of 2.59 solar masses at a central density approximately six times larger than the central density in ^{208}Pb. Note that the asymptotic approach of the equation of state to the causal limit is already relevant in this regime. I emphasize that although the low-density behavior of nuclear matter is sensitive to the nearly exact cancellation between attractive scalar and repulsive vector components, the stiff high-density equation of state is determined simply by the Lorentz structure of the interaction. (The scalar attraction saturates completely at high densities, producing an essentially massless gas of baryons interacting through a strong vector repulsion, which leads to a stiff equation of state.) Moreover, because the individual Lorentz components are comparable to the nucleon mass, the onset of the asymptotic regime occurs at densities similar to those in the interiors of neutron stars.

At this point, we have developed some basic intuition into the dynamics contained in a relativistic theory of nuclear matter. Let us now extend these results to uniform nuclear matter at finite temperature *observed from an arbitrary reference frame*. This will allow us to see how QHD calculations can be carried out in a covariant fashion, and we will discover the Lorentz transformation properties of the temperature and chemical potential. First, however, I will digress to discuss some formal aspects that will be needed in the analysis.

COVARIANT THERMODYNAMICS

Here I summarize some important formulas for the covariant description of a uniform, isolated system in equilibrium. These results are not new; my purpose is to define notation, to establish working definitions of Lorentz covariance and thermodynamic consistency, and to clarify which calculated quantities provide nontrivial tests of the consistency of a particular approximation. For a more complete discussion, see refs. 35, 36, and 2.

Let me begin by defining *primary thermodynamic functions* for the equilibrium system:

$$\text{energy-momentum tensor:} \quad T^{\mu\nu}\,, \qquad\qquad (21)$$
$$\text{entropy flux vector:} \quad S^{\mu}\,, \qquad\qquad (22)$$
$$\text{baryon current density vector:} \quad B^{\mu}\,. \qquad\qquad (23)$$

These quantities are covariant and involve no specification of a particular reference frame. I assume the conservation laws

$$\partial_\mu T^{\mu\nu} = 0\,, \quad \partial_\mu B^\mu = 0\,, \qquad\qquad (24)$$

and the symmetry of $T^{\mu\nu}$. The conventions are those of ref. 8, with a metric $g^{\mu\nu} = \text{diag}(+,-,-,-)$, and natural units with $\hbar = c = k_{\mathrm{B}} = 1$ will be used.

The main objective is to compute these thermodynamic functions by taking ensemble averages of the corresponding quantum-mechanical operators, which I will denote with carets, for example, $\hat{T}^{\mu\nu}$. In QHD–I, the conservation laws (24) are satisfied automatically as a consequence of the field equations (2)–(4).[8]

The primary thermodynamic quantities (21)–(23) are functions of six variables:

$$\text{baryon thermal potential:} \quad \alpha\,, \qquad\qquad (25)$$
$$\text{inverse temperature:} \quad \beta\,, \qquad\qquad (26)$$
$$\text{fluid four-velocity:} \quad u^\mu\,, \qquad\qquad (27)$$
$$\text{volume:} \quad V\,. \qquad\qquad (28)$$

The volume V will be taken large and *fixed* throughout the calculation; I let $V \to \infty$ at the end to define the "thermodynamic limit" and restore invariance under translations. The quantities α and β are Lorentz scalars defined by

$$\beta \equiv \frac{1}{T'}\,, \qquad \alpha \equiv \frac{\mu'}{T'}\,, \qquad\qquad (29)$$

where T' and μ' are the temperature and baryon chemical potential in the *comoving frame*, where the fluid three-velocity $\underset{\sim}{v} = 0$. When I refer to a quantity that may be defined by an observer in any frame, the value taken in the comoving frame will be denoted with a prime. The primed value is, by definition, a Lorentz scalar, and these scalars are commonly used as the thermodynamic parameters for the system.[2,34,36] Here it will be more convenient to allow all observers to define their own temperature, chemical potential, *etc.*, and a prime simply distinguishes the value observed in the comoving frame.

The fluid four-velocity vector u^μ can be written in terms of the three-velocity $\underset{\sim}{v}$ as

$$u^\mu = \eta(1,\underset{\sim}{v})\,, \qquad \eta \equiv (1 - \underset{\sim}{v}^2)^{-1/2}\,. \qquad\qquad (30)$$

There are only three independent components in u^μ, since $u_\mu u^\mu = 1$. In the comoving frame, $u^{\mu'} = (1,\underset{\sim}{0})$. It is also convenient to introduce a timelike thermal four-vector

$$\beta^\mu \equiv \beta u^\mu \equiv \frac{1}{T'} u^\mu\,, \qquad\qquad (31)$$

which depends on the four independent variables β and $\underset{\sim}{v}$.

One now introduces *secondary thermodynamic functions* that are defined in the comoving frame and are thus Lorentz scalars:

$$\text{pressure:} \quad p' = p, \tag{32}$$

$$\text{proper energy density:} \quad \mathcal{E}', \tag{33}$$

$$\text{proper entropy density:} \quad \sigma', \tag{34}$$

$$\text{proper baryon density:} \quad \rho'_B, \tag{35}$$

$$\text{scalar density:} \quad \rho'_s = \rho_s. \tag{36}$$

The pressure p and the scalar density of baryons ρ_s are the same in all frames,[35] so the primes are superfluous. In the thermodynamic limit, these secondary quantities are functions of α and β (or μ' and T') only.

The secondary thermodynamic functions can be used to construct the primary functions in any frame:

$$T^{\mu\nu} = (\mathcal{E}' + p)u^\mu u^\nu - pg^{\mu\nu}, \tag{37}$$

$$S^\mu = \sigma' u^\mu, \tag{38}$$

$$B^\mu = \rho'_B u^\mu. \tag{39}$$

In the thermodynamic limit, the primary quantities are functions of α, β, and u^μ, or equivalently, α and β^μ.

I will now make the following working definition:

> *If the theory (or approximation) is Lorentz covariant, cal-*
> *culation of the secondary quantities and insertion into eqs. (37)–*
> *(39) should give the same result as the direct evaluation of the*
> *primary quantities (21)–(23) in a frame where $\underset{\sim}{v} \neq 0$.*

With the preceding definitions, we can discuss the thermodynamics of the system. All equilibrium results follow from the First Law of thermodynamics, which is written covariantly as[36]

$$\beta_\nu \, dT^{\nu\mu} = dS^\mu + \alpha \, dB^\mu. \tag{40}$$

This important expression can be recast in more familiar form by inserting eqs. (37)–(39). After taking the indicated differentials and realizing that u^μ and du^μ are orthogonal vectors, we find the two scalar equations

$$d\mathcal{E}' = T' \, d\sigma' + \mu' \, d\rho'_B, \tag{41}$$

$$\mathcal{E}' = -p + T'\sigma' + \mu'\rho'_B. \tag{42}$$

These will be recognized as the usual First Law and Gibbs' relation,[37] written in the comoving frame at fixed volume V'. Equation (41) produces the familiar thermodynamic results at fixed volume:

$$\left(\frac{\partial \mathcal{E}'}{\partial \sigma'}\right)_{V', \rho'_B} = T', \quad \left(\frac{\partial \mathcal{E}'}{\partial \rho'_B}\right)_{V', \sigma'} = \mu'. \tag{43}$$

These results allow me to make another working definition:

48

If the theory (or approximation) is thermodynamically consistent, secondary quantities calculated from operator ensemble averages should satisfy Gibbs' relation and the corresponding differential laws.

This definition of consistency is most important when applied to the pressure, since p can be computed from either the thermodynamic potential[37] or from the stress tensor.[8] I will call an approximation thermodynamically consistent only if the two results for the pressure agree.

If I rewrite eq. (42) as

$$pu^\mu = -\mathcal{E}'u^\mu + T'\sigma'u^\mu + \mu'\rho_B'u^\mu = -u_\nu T^{\nu\mu} + T'S^\mu + \mu'B^\mu , \qquad (44)$$

division by T' gives the covariant form of Gibbs' relation:

$$p\beta^\mu = -\beta_\nu T^{\nu\mu} + S^\mu + \alpha B^\mu . \qquad (45)$$

This expression can be recast as a differential relation using eq. (40), with the result

$$d(p\beta^\mu) = -T^{\mu\nu}\,d\beta_\nu + B^\mu\,d\alpha . \qquad (46)$$

If a calculation is both Lorentz covariant and thermodynamically consistent (as defined above), then both (45) and (46) should hold. Since the (constant volume) thermodynamic functions depend only on α and β^μ, eq. (46) implies

$$\left(\frac{\partial(p\beta^\mu)}{\partial\beta_\nu}\right)_\alpha = -T^{\nu\mu} , \qquad \left(\frac{\partial(p\beta^\mu)}{\partial\alpha}\right)_{\beta_\nu} = B^\mu . \qquad (47)$$

To compute the thermodynamic functions in a particular theory, one must relate them to ensemble averages of quantum-mechanical operators. This is achieved by defining a *grand partition function* Z and a four-vector thermodynamic potential $\Phi^\mu = \Phi^\mu(\alpha, \beta^\nu)$ through

$$Z \equiv \exp\left(-\int d\Lambda_\mu \Phi^\mu(\alpha, \beta^\nu)\right) \equiv \text{Tr}\left\{ \exp\left[-\int d\Lambda_\mu (\beta_\nu \hat{T}^{\nu\mu} - \alpha \hat{B}^\mu)\right]\right\} . \qquad (48)$$

Here Λ_μ is a spacelike hypersurface, and these expressions are manifestly Lorentz invariant. In the comoving frame, eq. (48) reduces to the familiar result[8]

$$Z = \text{Tr}\left\{ \exp\left[-\beta(\hat{H} - \mu'\hat{B})\right]\right\} , \qquad (49)$$

where $\hat{H}$ is the hamiltonian and $\hat{B}$ is the baryon number operator. The motivation and justification for eq. (48) will be discussed below, where I show how this definition arises naturally from a canonical calculation of the grand partition function in an arbitrary reference frame.

If one makes small variations in β^μ and α, it follows immediately from eq. (48) that

$$d\Phi^\mu = T^{\mu\nu}\,d\beta_\nu - B^\mu\,d\alpha . \qquad (50)$$

Here the classical quantities on the right-hand side are defined as usual in terms of ensemble averages:

$$A \equiv \langle\!\langle \hat{A} \rangle\!\rangle = Z^{-1} \, \mathrm{Tr} \Big\{ \hat{A} \, \exp\Big[-\int d\Lambda_\mu (\beta_\nu \hat{T}^{\nu\mu} - \alpha \hat{B}^\mu) \Big] \Big\} . \tag{51}$$

Thus we have

$$\left(\frac{\partial \Phi^\mu}{\partial \beta_\nu} \right)_\alpha = T^{\mu\nu} , \qquad \left(\frac{\partial \Phi^\mu}{\partial \alpha} \right)_{\beta_\nu} = -B^\mu . \tag{52}$$

To relate the thermodynamic four-potential Φ^μ to more familiar quantities, recall that the conventional thermodynamic potential[37] is defined by

$$\Omega \equiv \Omega(T', V', \mu') = (\mathcal{E}' - T'\sigma' - \mu'\rho_{\mathrm{B}}')V' = -pV' . \tag{53}$$

A comparison of eqs. (47) and (52) implies that

$$\Phi^\mu(\alpha, \beta^\nu) = -p\beta^\mu = \frac{\Omega(T', V', \mu')}{V'}\beta^\mu = \frac{\Omega(T', V', \mu')}{V'T'}u^\mu , \tag{54}$$

so that

$$\Phi^\mu = \beta_\nu T^{\nu\mu} - S^\mu - \alpha B^\mu . \tag{55}$$

If a calculation is Lorentz covariant, the computation of Φ^μ from eq. (54) should agree with the result determined directly from (48) in an arbitrary frame. Moreover, if the calculation is also thermodynamically consistent, the thermodynamic pressure determined from eq. (53) or (54) should agree with the "hydrostatic" pressure defined by the ensemble average of the stress-tensor operator in the comoving frame:

$$p = \frac{1}{3} \langle\!\langle \hat{T}'_{i'i'} \rangle\!\rangle . \tag{56}$$

HOT, FLOWING NUCLEAR MATTER

To illustrate the preceding ideas and to provide a concrete example of a covariant calculation, consider a uniform system of hot, flowing nuclear matter. I will work in the mean-field approximation to the Walecka model,[14,8] in which Dirac nucleons interact with classical scalar and vector fields. Vacuum corrections will be ignored for the remainder of this section.

The starting point is the mean-field theory (MFT) lagrangian for the Walecka model. Since the nuclear medium is homogeneous, the classical meson fields ϕ and V^μ are constant, but since the matter is flowing with velocity $\underset{\sim}{v}$, there is now a uniform baryon flux $\underset{\sim}{\mathcal{B}}$. Thus the classical vector field has both temporal and spatial components: $V^\mu = (V_0, \underset{\sim}{V})$, and the mean-field lagrangian density is

$$\mathcal{L}_{\mathrm{MFT}} = \overline{\psi}[\gamma_\mu(i\partial^\mu - g_{\mathrm{v}}V^\mu) - (M - g_{\mathrm{s}}\phi)]\psi - \tfrac{1}{2}m_{\mathrm{s}}^2\phi^2 + \tfrac{1}{2}m_{\mathrm{v}}^2 V^\mu V_\mu . \tag{57}$$

The conserved baryon four-current and energy-momentum tensor can be derived in the usual fashion,[38] resulting in

$$B^\mu = \overline{\psi}\gamma^\mu\psi , \tag{58}$$

$$T^{\mu\nu} = i\overline{\psi}\gamma^\mu\partial^\nu\psi - \tfrac{1}{2}(m_{\mathrm{v}}^2 V^\lambda V_\lambda - m_{\mathrm{s}}^2\phi^2)g^{\mu\nu} . \tag{59}$$

As discussed by Freedman,[39] there is no need to symmetrize $T^{\mu\nu}$ if we consider only homogeneous nuclear matter.

Since the meson fields are classical, only the fermion field must be quantized. The Dirac field equation follows from $\mathcal{L}_{\mathrm{MFT}}$:

$$[i\gamma_\mu\partial^\mu - g_\mathrm{v}\gamma_\mu V^\mu - (M - g_\mathrm{s}\phi)]\psi(t,\underset{\sim}{x}) = 0\,, \tag{60}$$

and since this equation is linear, it can be solved exactly. We look for normal-mode solutions of the form

$$\psi^{(+)}_{\underset{\sim}{k}\lambda}(t,\underset{\sim}{x}) = U(\underset{\sim}{k},\lambda)\,e^{i\underset{\sim}{k}\cdot\underset{\sim}{x} - i\epsilon^{(+)}(\underset{\sim}{k})t}\,,$$

$$\psi^{(-)}_{\underset{\sim}{k}\lambda}(t,\underset{\sim}{x}) = V(\underset{\sim}{k},\lambda)\,e^{-i\underset{\sim}{k}\cdot\underset{\sim}{x} - i\epsilon^{(-)}(-\underset{\sim}{k})t}\,, \tag{61}$$

where [compare eq. (10)]

$$\begin{aligned}
\epsilon^{(\pm)}(\underset{\sim}{k}) &= g_\mathrm{v}V_0 \pm [(\underset{\sim}{k} - g_\mathrm{v}\underset{\sim}{V})^2 + M^{*2}]^{1/2} \\
&\equiv g_\mathrm{v}V_0 \pm [\underset{\sim}{\kappa}^2 + M^{*2}]^{1/2} \\
&\equiv g_\mathrm{v}V_0 \pm E^*(\kappa)\,.
\end{aligned} \tag{62}$$

The second line defines the *kinetic momentum* $\underset{\sim}{\kappa} \equiv \underset{\sim}{k} - g_\mathrm{v}\underset{\sim}{V}$, and the final line defines $E^*(\kappa)$. Here $M^* \equiv M - g_\mathrm{s}\phi$.

The baryon field operator can now be written as

$$\begin{aligned}
\psi(t,\underset{\sim}{x}) = V^{-1/2}\sum_{\underset{\sim}{k}\lambda}\Big[&A_{\underset{\sim}{k}\lambda}U(\underset{\sim}{k},\lambda)\,e^{i\underset{\sim}{k}\cdot\underset{\sim}{x} - i\epsilon^{(+)}(\underset{\sim}{k})t} \\
&+ B^\dagger_{\underset{\sim}{k}\lambda}V(\underset{\sim}{k},\lambda)\,e^{-i\underset{\sim}{k}\cdot\underset{\sim}{x} - i\epsilon^{(-)}(-\underset{\sim}{k})t}\Big]\,,
\end{aligned} \tag{63}$$

where V is the volume of the system and the spinors have the (noncovariant) normalization $U^\dagger(\underset{\sim}{k},\lambda)U(\underset{\sim}{k},\lambda') = V^\dagger(\underset{\sim}{k},\lambda)V(\underset{\sim}{k},\lambda') = \delta_{\lambda\lambda'}$. I now impose the familiar equal-time anticommutation relations[8] and construct the baryon number operator $\hat{B} \equiv \int \mathrm{d}^3x\,\overline{\psi}\gamma^0\psi$ and the four-momentum operators $\hat{P}^\mu = (\hat{H},\underset{\sim}{\hat{P}}) \equiv \int \mathrm{d}^3x\,\hat{T}^{0\mu}$, with the results [compare eqs. (12) and (13)]

$$\hat{B} = \sum_{\underset{\sim}{k}\lambda}(A^\dagger_{\underset{\sim}{k}\lambda}A_{\underset{\sim}{k}\lambda} - B^\dagger_{\underset{\sim}{k}\lambda}B_{\underset{\sim}{k}\lambda})\,, \tag{64}$$

$$\begin{aligned}
\hat{H} = \sum_{\underset{\sim}{k}\lambda}E^*(\kappa)(A^\dagger_{\underset{\sim}{k}\lambda}A_{\underset{\sim}{k}\lambda} + B^\dagger_{-\underset{\sim}{k}\lambda}B_{-\underset{\sim}{k}\lambda}) + g_\mathrm{v}V_0\hat{B} \\
+ \tfrac{1}{2}(m_\mathrm{s}^2\phi^2 + m_\mathrm{v}^2\underset{\sim}{V}^2 - m_\mathrm{v}^2V_0^2)V\,,
\end{aligned} \tag{65}$$

$$\underset{\sim}{\hat{P}} = \sum_{\underset{\sim}{k}\lambda}\underset{\sim}{k}\,(A^\dagger_{\underset{\sim}{k}\lambda}A_{\underset{\sim}{k}\lambda} + B^\dagger_{\underset{\sim}{k}\lambda}B_{\underset{\sim}{k}\lambda})\,. \tag{66}$$

Note that eq. (65) corrects a misprint in eq. (3.82) of ref. 8 and that $\underset{\sim}{\hat{P}}$ involves a sum over canonical momenta $\underset{\sim}{k}$. To obtain these expressions, the operator products have been *normal ordered*, so that all destruction operators are to the right, and c-number pieces have been omitted. As discussed in ref. 8, the c-number contribution to the hamiltonian leads to vacuum corrections, which I will neglect for now. Note that, in principle, there are also contributions to the energy from thermal excitation of real mesons. Since these do not become appreciable in this model until very high temperatures ($T \gtrsim m_\mathrm{s},\ m_\mathrm{v}$), I will omit these terms.

The MFT hamiltonian (65) is diagonal, so we have solved this model problem exactly (once we have specified the meson fields). Since $\hat{B}$ and $\hat{P}$ are also diagonal, the baryon number and total momentum are constants of the motion, as are their corresponding densities ρ_B and $\mathcal{P}$, since the volume is fixed. At zero temperature, the ground state of the moving medium is obtained[8,40] by filling energy levels up to a nonspherical Fermi surface k_F. The shape of the Fermi surface is determined *thermodynamically* by minimizing the mean-field energy density $\mathcal{E}$ at fixed baryon density ρ_B and momentum density $\mathcal{P}$. This is achieved by introducing Lagrange multipliers for the chemical potential and flow velocity, so that the quantity to be minimized is

$$\mathcal{E}(k_F; \phi, V_0, V) - \mu\rho_B(k_F) - v \cdot \mathcal{P}(k_F). \tag{67}$$

Note that the thermodynamic parameters appearing in this expression are those observed in the "laboratory" frame, where the fluid has velocity v.

To describe the system at finite temperature, we need a thermodynamic potential and partition function that will select the correct ground state in the $T \to 0$ limit. Thus we are naturally led to define

$$Z = \mathrm{Tr} \exp\left\{-(\hat{H} - \mu\hat{B} - v \cdot \hat{P})/T\right\} \equiv \exp\left\{-\Omega(T, V, \mu, v)/T\right\}. \tag{68}$$

As before, all thermodynamic parameters in this expression are defined by an observer in the laboratory frame.

Since the operators appearing in eq. (68) are all diagonal, the thermodynamic potential can be evaluated exactly in this MFT. The results are analogous to those for noninteracting fermions, and with proper care regarding signs, we find

$$\Omega(T, V, \mu, v) = -T \ln Z$$

$$= \tfrac{1}{2}(m_s^2\phi^2 + m_v^2 V^2 - m_v^2 V_0^2)V - T\sum_{k\lambda}\left\{ \ln\left(1 + e^{-[E^*(\kappa) - v\cdot\kappa - \nu]/T}\right)\right.$$

$$\left. + \ln\left(1 + e^{-[E^*(\kappa) + v\cdot\kappa + \nu]/T}\right)\right\}. \tag{69}$$

The sum runs over all single-particle states labeled by momentum k and intrinsic quantum numbers λ. The *effective chemical potential* ν is defined by

$$\nu \equiv \mu - g_v(V_0 - v \cdot V). \tag{70}$$

The ensemble average of an operator $\hat{A}$ is given by

$$A = \langle\!\langle \hat{A} \rangle\!\rangle = Z^{-1}\,\mathrm{Tr}\left[\hat{A}\,e^{-(\hat{H} - \mu\hat{B} - v\cdot\hat{P})/T}\right]. \tag{71}$$

For example, the baryon density is

$$\rho_B \equiv \frac{\langle\!\langle \hat{B} \rangle\!\rangle}{V} = -\frac{1}{V}\left(\frac{\partial\Omega}{\partial\mu}\right) = \frac{1}{V}\sum_{k\lambda}(n_\kappa - \bar{n}_\kappa). \tag{72}$$

Here the partial derivative is taken with all other thermodynamic variables and field parameters held fixed, and I have identified the particle and antiparticle occupation numbers

$$n_{\underset{\sim}{k}} \equiv n_{\underset{\sim}{k}}(T, \nu, \underset{\sim}{v}) \equiv \langle\!\langle A^\dagger_{\underset{\sim}{k}\lambda} A_{\underset{\sim}{k}\lambda} \rangle\!\rangle = \left\{ 1 + e^{[E^\star(\kappa) - \underset{\sim}{v}\cdot\underset{\sim}{\kappa} - \nu]/T} \right\}^{-1}, \tag{73}$$

$$\bar{n}_{\underset{\sim}{k}} \equiv \bar{n}_{\underset{\sim}{k}}(T, \nu, \underset{\sim}{v}) \equiv \langle\!\langle B^\dagger_{-\underset{\sim}{k}\lambda} B_{-\underset{\sim}{k}\lambda} \rangle\!\rangle = \left\{ 1 + e^{[E^\star(\kappa) + \underset{\sim}{v}\cdot\underset{\sim}{\kappa} + \nu]/T} \right\}^{-1}. \tag{74}$$

With these results, I can derive the equations that determine the meson fields. For a system in equilibrium, these should be chosen to make the thermodynamic potential Ω stationary. For example, $\partial\Omega/\partial\phi = 0$ leads to

$$\phi = \frac{g_s}{m_s^2} \frac{1}{V} \sum_{\underset{\sim}{k}\lambda} \frac{M^\star}{E^\star(\kappa)} (n_{\underset{\sim}{k}} + \bar{n}_{\underset{\sim}{k}}) = \frac{g_s}{m_s^2} \langle\!\langle \hat{\rho}_s \rangle\!\rangle. \tag{75}$$

Similarly, the vector field equations become

$$V_0 = \frac{g_v}{m_v^2} \frac{1}{V} \sum_{\underset{\sim}{k}\lambda} (n_{\underset{\sim}{k}} - \bar{n}_{\underset{\sim}{k}}) = \frac{g_v}{m_v^2} \langle\!\langle \hat{\rho}_B \rangle\!\rangle, \tag{76}$$

$$\underset{\sim}{V} = \frac{g_v}{m_v^2} \frac{1}{V} \sum_{\underset{\sim}{k}\lambda} \frac{\underset{\sim}{\kappa}}{E^\star(\kappa)} (n_{\underset{\sim}{k}} + \bar{n}_{\underset{\sim}{k}}) = \frac{g_v}{m_v^2} \langle\!\langle \hat{\underset{\sim}{B}} \rangle\!\rangle. \tag{77}$$

Thus, by making the thermodynamic potential stationary with respect to the fields, they automatically satisfy the ensemble averages of the normal-ordered field equations resulting from the lagrangian (57).

These relations are extremely useful, for they imply that ϕ and V^μ can be held *fixed* in computing thermodynamic functions as derivatives of the thermodynamic potential through the relations

$$B = -\left(\frac{\partial}{\partial\mu} \Omega(T, V, \mu, \underset{\sim}{v}) \right)_{T, V, \underset{\sim}{v}}, \quad \underset{\sim}{P} = -\left(\frac{\partial}{\partial\underset{\sim}{v}} \Omega(T, V, \mu, \underset{\sim}{v}) \right)_{T, V, \mu},$$

$$S = -\left(\frac{\partial}{\partial T} \Omega(T, V, \mu, \underset{\sim}{v}) \right)_{V, \mu, \underset{\sim}{v}}, \quad p = -\left(\frac{\partial}{\partial V} \Omega(T, V, \mu, \underset{\sim}{v}) \right)_{T, \mu, \underset{\sim}{v}}. \tag{78}$$

Notice that the first two of these relations are satisfied trivially from the definition of the thermodynamic potential in eq. (68). That is, they will be valid for *any* approximations to the exact hamiltonian, number operator, and momentum operator. Moreover, since

$$S = -\left(\frac{\partial\Omega}{\partial T} \right) = \ln Z + \frac{T}{Z} \left(\frac{\partial Z}{\partial T} \right) = \frac{-\Omega + E - \mu B - \underset{\sim}{v} \cdot \underset{\sim}{P}}{T}, \tag{79}$$

Gibbs' relation is satisfied automatically. Finally, since Ω of eq. (69) is linear in V (once the sum has been converted into an integral), eq. (78) merely *defines* the "thermodynamic" pressure. Thus, the only real check of the thermodynamic consistency of this approximation is to verify that p defined by eq. (78) agrees with the result computed from the stress tensor *in the comoving frame*. Since this requires a discussion of the covariance of the preceding results, I postpone this verification until later.

The calculation of the thermodynamic functions in the laboratory frame can now be carried out straightforwardly in this MFT, leading to

$$\rho_{\rm s} = \frac{\gamma}{(2\pi)^3} \int {\rm d}^3\kappa\, \frac{M^\star}{E^\star(\kappa)}(n_{\underset{\sim}{\kappa}} + \bar{n}_{\underset{\sim}{\kappa}})\,, \tag{80}$$

$$\rho_{\rm B} = \frac{\gamma}{(2\pi)^3} \int {\rm d}^3\kappa\, (n_{\underset{\sim}{\kappa}} - \bar{n}_{\underset{\sim}{\kappa}})\,, \tag{81}$$

$$\underset{\sim}{\mathcal{B}} = \frac{\gamma}{(2\pi)^3} \int {\rm d}^3\kappa\, \frac{\underset{\sim}{\kappa}}{E^\star(\kappa)}(n_{\underset{\sim}{\kappa}} + \bar{n}_{\underset{\sim}{\kappa}})\,, \tag{82}$$

$$\underset{\sim}{\mathcal{P}} = \frac{g_{\rm v}^2}{m_{\rm v}^2}\rho_{\rm B}\underset{\sim}{\mathcal{B}} + \frac{\gamma}{(2\pi)^3} \int {\rm d}^3\kappa\, \underset{\sim}{\kappa}(n_{\underset{\sim}{\kappa}} - \bar{n}_{\underset{\sim}{\kappa}})\,, \tag{83}$$

$$\mathcal{E} = \frac{g_{\rm v}^2}{2m_{\rm v}^2}\rho_{\rm B}^2 + \frac{m_{\rm s}^2}{2g_{\rm s}^2}(M - M^\star)^2 + \frac{g_{\rm v}^2}{2m_{\rm v}^2}\underset{\sim}{\mathcal{B}}^2$$
$$+ \frac{\gamma}{(2\pi)^3} \int {\rm d}^3\kappa\, E^\star(\kappa)(n_{\underset{\sim}{\kappa}} + \bar{n}_{\underset{\sim}{\kappa}})\,, \tag{84}$$

$$p = \frac{g_{\rm v}^2}{2m_{\rm v}^2}\rho_{\rm B}^2 - \frac{m_{\rm s}^2}{2g_{\rm s}^2}(M - M^\star)^2 - \frac{g_{\rm v}^2}{2m_{\rm v}^2}\underset{\sim}{\mathcal{B}}^2$$
$$- T\frac{\gamma}{(2\pi)^3} \int {\rm d}^3\kappa\, \left\{ \ln(1 - n_{\underset{\sim}{\kappa}}) + \ln(1 - \bar{n}_{\underset{\sim}{\kappa}}) \right\}\,, \tag{85}$$

$$\sigma = -\frac{\gamma}{(2\pi)^3} \int {\rm d}^3\kappa\, \left\{ n_{\underset{\sim}{\kappa}} \ln n_{\underset{\sim}{\kappa}} + (1 - n_{\underset{\sim}{\kappa}}) \ln(1 - n_{\underset{\sim}{\kappa}}) \right.$$
$$\left. + \bar{n}_{\underset{\sim}{\kappa}} \ln \bar{n}_{\underset{\sim}{\kappa}} + (1 - \bar{n}_{\underset{\sim}{\kappa}}) \ln(1 - \bar{n}_{\underset{\sim}{\kappa}}) \right\}\,. \tag{86}$$

Here γ is the spin-isospin degeneracy, the occupation number distributions are given by eqs. (73) and (74), and the meson field equations can be used to write

$$\underset{\sim}{\kappa} = \underset{\sim}{k} - \frac{g_{\rm v}^2}{m_{\rm v}^2}\underset{\sim}{\mathcal{B}}\,, \qquad \nu = \mu - \frac{g_{\rm v}^2}{m_{\rm v}^2}(\rho_{\rm B} - \underset{\sim}{v} \cdot \underset{\sim}{\mathcal{B}})\,. \tag{87}$$

Note that eq. (75) can be recast as

$$M^\star \equiv M - g_{\rm s}\phi = M - \frac{g_{\rm s}^2}{m_{\rm s}^2}\rho_{\rm s} = M - \frac{g_{\rm s}^2}{m_{\rm s}^2}\frac{\gamma}{(2\pi)^3} \int {\rm d}^3\kappa\, \frac{M^\star}{E^\star(\kappa)}(n_{\underset{\sim}{\kappa}} + \bar{n}_{\underset{\sim}{\kappa}})\,. \tag{88}$$

This is a transcendental *self-consistency condition* that determines ϕ (or $M^\star$) and is a generalization of eq. (19).

To compute the thermodynamic functions, one first specifies T, ν, and $\underset{\sim}{v}$. The self-consistency condition (88) is then solved to determine $M^\star$. (There may be several solutions for fixed T, ν, and $\underset{\sim}{v}$.) These solutions specify the distribution functions $n_{\underset{\sim}{\kappa}}$ and $\bar{n}_{\underset{\sim}{\kappa}}$, and the remaining integrals in eqs. (81)–(86) can be evaluated directly. At the end of the calculation, one can (in principle) invert these relations to find μ and $\underset{\sim}{v}$ in terms of $\rho_{\rm B}$ and $\underset{\sim}{\mathcal{P}}$, but in practice, desired values of $\rho_{\rm B}$ and $\underset{\sim}{\mathcal{P}}$ are found by searching on values of ν and $\underset{\sim}{v}$.

Let us discuss the Lorentz covariance of the expressions (80)–(86). In principle, one can perform the integrals and verify that the results are the properly transformed counterparts of the corresponding quantities calculated in the comoving frame. This calculation must be done numerically, since the integrals over the nonspherical Fermi distribution functions cannot, to my knowledge, be expressed in closed form.

Fortunately, this tedious procedure is unnecessary, since the covariance can be proven directly by making a suitable change of integration variables. Not surprisingly, this change of variables looks like a Lorentz transformation to the comoving frame. Let us define two new momenta $\underset{\sim}{t}$ and $\underset{\sim}{q}$ related to $\underset{\sim}{\kappa}$ by

$$\underset{\sim}{\kappa} \equiv \underset{\sim}{t} + \frac{\eta^2 \underset{\sim}{v}}{1+\eta} \underset{\sim}{v} \cdot \underset{\sim}{t} + \eta \underset{\sim}{v} E^*(t), \quad \underset{\sim}{\kappa} \equiv \underset{\sim}{q} + \frac{\eta^2 \underset{\sim}{v}}{1+\eta} \underset{\sim}{v} \cdot \underset{\sim}{q} - \eta \underset{\sim}{v} E^*(q). \tag{89}$$

The variable $\underset{\sim}{t}$ will be used to rewrite the integrals over the particle distributions, while $\underset{\sim}{q}$ will be used for the antiparticle distributions. The parameter M^* appearing in E^* can be given its value in the laboratory frame, but we will find that this is a scalar, as expected. It is a straightforward matter of algebra to show that

$$E^*(\kappa) = \eta(E^*(t) + \underset{\sim}{v} \cdot \underset{\sim}{t}), \quad E^*(\kappa) = \eta(E^*(q) - \underset{\sim}{v} \cdot \underset{\sim}{q}), \tag{90}$$

and to compute the Jacobians

$$d^3\kappa = \eta\left(1 + \frac{\underset{\sim}{v} \cdot \underset{\sim}{t}}{E^*(t)}\right) d^3 t, \quad d^3\kappa = \eta\left(1 - \frac{\underset{\sim}{v} \cdot \underset{\sim}{q}}{E^*(q)}\right) d^3 q. \tag{91}$$

Consider now the Fermi distribution for particles under the change of variables $\underset{\sim}{\kappa} \rightarrow \underset{\sim}{t}$. Since eqs. (89) and (90) imply that $E^*(\kappa) - \underset{\sim}{v} \cdot \underset{\sim}{\kappa} = E^*(t)/\eta$, we can rewrite the distribution as

$$n_{\underset{\sim}{\kappa}}(T, \nu, \underset{\sim}{v}) = \left\{1 + e^{[E^*(t) - \eta\nu]/\eta T}\right\}^{-1} \equiv n'_t(\eta T, \eta\nu). \tag{92}$$

With this change of variables, the particle distribution function looks just like a comoving-frame distribution function (i.e., no angular dependence) with the thermodynamic parameters

$$T' \equiv \eta T, \quad \nu' \equiv \eta\nu. \tag{93}$$

In other words, if we *define* the transformation properties of the temperature and chemical potential as in eq. (93), the distribution function $n_{\underset{\sim}{\kappa}}(T, \nu, \underset{\sim}{v})$ is a Lorentz scalar. This is as it should be, since all observers must agree on the occupation probability of a given single-particle state. Similarly, the antiparticle distribution function becomes

$$\bar{n}_{\underset{\sim}{\kappa}}(T, \nu, \underset{\sim}{v}) = \left\{1 + e^{[E^*(q) + \eta\nu]/\eta T}\right\}^{-1} \equiv \bar{n}'_q(\eta T, \eta\nu). \tag{94}$$

The proof of Lorentz covariance now proceeds easily. To illustrate the procedure with a nontrivial example, consider the baryon flux $\underset{\sim}{B}$. To simplify the expressions, take $\underset{\sim}{v} \parallel \hat{z}$, so that

$$B_z = \frac{\gamma}{(2\pi)^3} \int d^3\kappa \frac{\kappa_z}{E^*(\kappa)} (n_{\underset{\sim}{\kappa}} + \bar{n}_{\underset{\sim}{\kappa}})$$

55

$$= \frac{\gamma}{(2\pi)^3} \left\{ \int d^3t \, n'_t \, \eta \left(1 + \frac{vt_z}{E^\star(t)}\right) \eta(t_z + vE^\star(t)) \left[\eta E^\star(t)\left(1 + \frac{vt_z}{E^\star(t)}\right)\right]^{-1} \right.$$

$$\left. + \int d^3q \, \bar{n}'_q \, \eta \left(1 - \frac{vq_z}{E^\star(q)}\right) \eta(q_z - vE^\star(q)) \left[\eta E^\star(q)\left(1 - \frac{vq_z}{E^\star(q)}\right)\right]^{-1} \right\}$$

$$= \frac{\gamma}{(2\pi)^3} \left\{ \eta v \int d^3t \, n'_t - \eta v \int d^3q \, \bar{n}'_q \right\}$$

$$= \eta v \rho'_{\rm B} \,. \tag{95}$$

Since the transformed distribution functions n'_t and $\bar{n}'_q$ are spherically symmetric, all integrals linear in $\underset{\sim}{t}$ or $\underset{\sim}{q}$ vanish, and I have identified the proper baryon density $\rho'_{\rm B}$ in the final expression. With similar techniques it is easy to show that

$$\rho_{\rm s} = \rho'_{\rm s} \,, \quad \rho_{\rm B} = \eta \rho'_{\rm B} \,, \quad \sigma = \eta \sigma' \,,$$
$$\mathcal{E} = \eta^2(\mathcal{E}' + v^2 p) \,, \quad \underset{\sim}{\mathcal{P}} = \eta^2 \underset{\sim}{v}(\mathcal{E}' + p) \,. \tag{96}$$

Equations (95) and (96) verify that $\rho_{\rm s}$ is a scalar, $\rho_{\rm B}$ and $\underset{\sim}{B}$ are the components of a four-vector B^μ, σ is the timelike component of a four-vector S^μ, and $\mathcal{E} = T^{00}$ and $\mathcal{P}^i = T^{0i}$ are components of the energy-momentum tensor $T^{\mu\nu}$, all defined correctly in terms of comoving-frame ("secondary") quantities by eqs. (30) and (37)–(39). (The transformation of the spacelike components T^{ij} can be verified analogously.) Moreover, the pressure is indeed a Lorentz scalar, which follows by inspection of eq. (85).

The proof of Lorentz covariance is now complete. The primary thermodynamic functions computed for the moving fluid are correctly described in terms of the transformed secondary thermodynamic functions. This also verifies the thermodynamic consistency of the MFT in all frames, since we know that the pressure evaluated from the stress tensor in the comoving frame[8] agrees with Gibbs' relation (42). Most importantly, since the MFT is both thermodynamically consistent and Lorentz covariant, the covariant thermodynamic relations (40), (45), and (46) are all guaranteed to hold. The verification of these relations in terms of the explicit expressions (81)–(86) is left as an exercise for the reader.

I turn now to the partition function defined in eq. (68) and show that this is consistent with the covariant definition given in eq. (48). Since the system is uniform, the partition function (68) can be written in terms of density operators as

$$Z = \operatorname{Tr} \exp \left\{ -T^{-1} \int d^3x \left[\hat{\mathcal{H}} - \mu \hat{\rho}_{\rm B} - \underset{\sim}{v} \cdot \hat{\underset{\sim}{\mathcal{P}}}\right] \right\} \,. \tag{97}$$

All quantities in this expression are defined in the laboratory frame. In particular, the density operators are constructed from baryon fields that satisfy equal-time anticommutators in this frame. By expressing the inverse temperature as $1/T = \eta/T' \equiv \eta\beta$ and defining the thermal potential as $\alpha \equiv \mu/T = \mu'/T'$, eq. (97) becomes

$$Z = \operatorname{Tr} \exp \left\{ -\int d^3x \left[\beta\eta(\hat{T}^{00} - v^i \hat{T}^{i0}) - \alpha \hat{\rho}_{\rm B}\right] \right\}$$

$$= \operatorname{Tr} \exp \left\{ -\int d^3x \, \delta_{\mu 0}(\beta_\nu \hat{T}^{\nu\mu} - \alpha \hat{B}^\mu) \right\} \,. \tag{98}$$

Here I have identified the hamiltonian density and momentum density operators as components of the laboratory-frame energy-momentum tensor and have defined a thermal four-vector β^μ as in eq. (31). If I now define the purely spacelike hypersurface element in the laboratory frame as $d\Lambda_\mu \equiv d^3x\,\delta_{\mu 0}$, eq. (98) reproduces eq. (48).

There are several important points to note. First, the preceding derivation shows that the canonical evaluation of the partition function yields a Lorentz scalar, which is the reason for the covariance of the MFT results presented above. Moreover, the partition function defined by eq. (48) is a scalar for *any* choice of spacelike hypersurface Λ_μ. Indeed, since quantization can be performed on any Λ_μ with equivalent results,[38] one can even choose *different hypersurfaces in different reference frames*, as long as the operators $\hat{B}^\mu$ and $\hat{T}^{\mu\nu}$ are quantized on the appropriate hypersurface in each frame. Thus Λ_μ can be chosen to make the computation as simple as possible. This last result is usually overlooked in the literature.[2,41]

FINITE TEMPERATURE EQUATION OF STATE

Since the mean-field calculation can be carried out in any convenient reference frame with covariant results, I can study the properties of finite temperature nuclear matter in this model by working in the comoving frame and boosting when necessary. Thus, for the remainder of these lectures, *I will drop the primes on the thermodynamic quantities*, all of which are to be interpreted as comoving-frame variables unless otherwise stated.

In the comoving frame, the thermodynamic observables can still be calculated from eqs. (80)–(86) and (88), except that all the integrations are now spherically symmetric, and the distribution functions have the simple forms

$$n_t(T,\nu) = \left\{1 + e^{[E^\star(t)-\nu]/T}\right\}^{-1}, \qquad \bar{n}_t(T,\nu) = \left\{1 + e^{[E^\star(t)+\nu]/T}\right\}^{-1}. \tag{99}$$

For example, the self-consistency condition (88) becomes

$$M^\star = M - \frac{g_s^2}{m_s^2}\rho_s = M - \frac{g_s^2}{m_s^2}\frac{\gamma}{(2\pi)^3}\int d^3t\,\frac{M^\star}{E^\star(t)}\left[n_t(T,\nu) + \bar{n}_t(T,\nu)\right], \tag{100}$$

and the three-vector quantities $\underset{\sim}{B}$ and $\underset{\sim}{P}$ vanish.

To determine the equation of state, one chooses a value for ν and solves eq. (100) for $M^\star$ at fixed ν and T. These values of $M^\star$, ν, and T specify n_t and $\bar{n}_t$ through eq. (99) and can then be used to compute ρ_B, $\mathcal{E}$, and p through eqs. (81), (84), and (85). The result is a thermodynamic surface $p(\mathcal{E}, T)$ that is the desired equation of state. [Note that the equation of state is *defined* through the pressure and *proper* energy density, which can be used to construct the energy-momentum tensor in any frame with eq. (37).] In fig. 4, the isotherms $p(\mathcal{E}, T = \text{constant})$ are shown for neutron matter with $\gamma = 2$. As we have already seen, there is a phase transition in this model, and the region of the phase coexistence is indicated. (The isothermal spinodal curves can also be calculated easily.) The theoretical value of the critical temperature is $T_c = 9.1 \pm 0.2$ MeV, and the effective mass at the critical point is $M_c^* = 0.88M$.

Several limiting cases of the equation of state are of interest:

- As $T \to 0$ at any $\rho_B > 0$, the baryon distribution becomes a step function $n_t \to \theta(k_F - t)$. The results of eqs. (15)–(17) are then reproduced. This conclusion also obtains as $\rho_B \to \infty$ at any T.

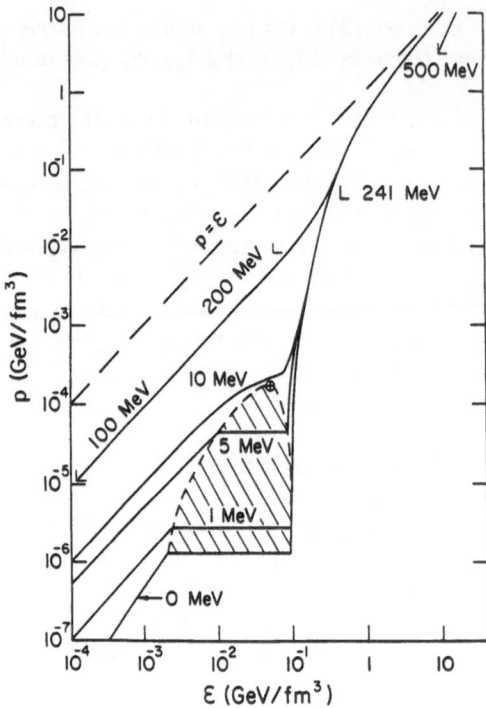

Fig. 4. Isotherms of the neutron matter equation of state, as calculated in the mean-field theory of QHD–I. The curves are labeled by the temperature, and the left hand "notches" correspond to zero baryon density. The shaded area shows the region of the phase separation, and the critical point is indicated by $\oplus$.

- For $T \ll M$ and $\rho_B \to 0$, the equation of state is that of a classical nonrelativistic gas: $\mathcal{E} = \rho_B(M + 3T/2)$, $p = \rho_B T = 2(\mathcal{E} - M\rho_B)/3$.

- As $T \to \infty$ at any ρ_B, baryon-antibaryon pairs will be produced, and the equation of state resembles that of a black body: $\mathcal{E} = 7\pi^2\gamma T^4/120$, $p = \mathcal{E}/3$.

The solution to the self-consistency equation (100) shows that the effective mass of the nucleon decreases as the temperature is raised due to $N\overline{N}$ pair formation. The values of $M^\star$ at vanishing baryon density are shown in fig. 5. The most striking feature is the sudden decrease in the mass of the nucleon well below $T = M$. Thus, at high temperatures (as at high density), the baryons are essentially massless. As the temperature is lowered, the baryons acquire mass suddenly from the self-consistent "freezing out" of the vacuum pairs and the corresponding decrease in ϕ.

As a consequence of pair formation, the high-temperature isotherms in fig. 4 will terminate as the density is decreased. Thus, at a given temperature, there is a finite, limiting value of the energy density as $\rho_B \to 0$. One simply has a vanishingly dilute solution of baryons in a sea of pairs. These limiting points on the isotherms are indicated for a representative set of temperatures in fig. 4. The isotherms start at these limiting values (at $\rho_B = 0$) and approach the causal limit $p = \mathcal{E}$ as the density is increased.

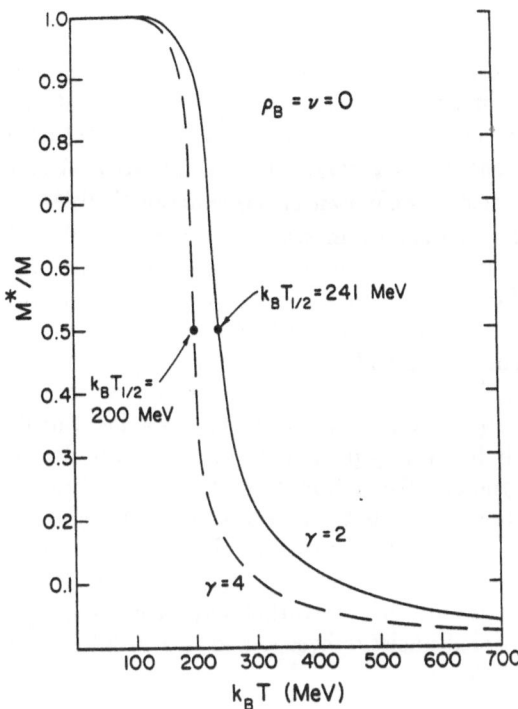

Fig. 5. Self-consistent nucleon mass as a function of temperature at vanishing baryon density. Results are indicated for both neutron matter ($\gamma = 2$) and nuclear matter ($\gamma = 4$) in QHD–I.

We have now studied the properties of nuclear matter in the simplest approximation to the Walecka model. As I emphasized in the Introduction, however, it is crucial to extend these calculations to more sophisticated approximations, in order to produce reliable predictions and to verify that the simple physical picture obtained in the MFT is a true consequence of the model QHD lagrangian.

There are many techniques available for studying relativistic, interacting many-body systems with quantum field theory. For example, Feynman diagrams allow corrections to the MFT to be included systematically, as described for zero temperature in chapter 5 of ref. 8. Clearly, since QHD is a strong-coupling theory, perturbative approaches are of no use, and various classes of diagrams must be summed to infinite order in the couplings. Unfortunately, diagrammatic techniques are of limited use in carrying out this program: while the diagrams tell us *how* to calculate, they give no indication of *what* diagrams are important. We must therefore turn to more powerful approaches that allow us to include various infinite classes of diagrams in a well-defined fashion.

Path integral methods are efficient for this purpose, and I will use path integrals for the remainder of these lectures to discuss the loop expansion of the Walecka model energy density. The path integral techniques required for this analysis are discussed in a number of texts[19–22] and applications to QHD–I are given in chapter 6 of ref. 8; the reader is directed to these references for background material.

THE LOOP EXPANSION

In this section, I examine the loop expansion as a systematic procedure for calculating in QHD. For simplicity, I consider zero temperature, but the procedure at finite temperature is analogous.[42] One starts with the exact ground-state to ground-state generating functional of the theory, which can be written as a path integral.[43,19−22] This path integral is used to define an effective action,[24−28,44−46] which is the field-theoretic analog of the free energy in statistical mechanics.[24,25] The effective action can be systematically expanded around its classical value by considering terms with increasing numbers of quantum loops. The "loop expansion" is formally a power series in $\hbar$, but as pointed out by several authors,[24,28,47] $\hbar$ is merely a bookkeeping parameter and need not be considered "small."

The effective action is a functional of the meson fields; when it is extremized with respect to the fields, it becomes proportional to the energy density of the system.[25,47] The extremization of the effective action generates contributions to the energy density that contain the coupling constants to all orders. Thus the loop expansion is inherently *nonperturbative*. Nevertheless, this expansion places primary importance on the mean fields, and only mean fields are included nonperturbatively; all correlations, both long-range and short-range, are included perturbatively. It is therefore unlike the expansion schemes used in traditional nuclear physics, and it is of interest to see if it provides a useful procedure in QHD.

I will again specialize to the Walecka model, whose lagrangian is given in eq. (1). The generating functional $Z[j]$ is defined as

$$Z[j] \equiv \exp\left(iW[j]/\hbar\right) \equiv \mathcal{N}^{-1} \int D(\overline{\psi})D(\psi)D(\phi) \exp\left\{\frac{i}{\hbar} \int \mathrm{d}^4x \left[\mathcal{L}(x) + j(x)\phi(x)\right]\right\}. \tag{101}$$

For simplicity, I will write only the scalar and baryon fields explicitly, although the vector mesons are included in the calculations, as described in ref. 48. (Note that Planck's constant $\hbar$ has been restored momentarily.) The path integral is evaluated at finite density in the presence of an external classical source j. The normalization factor $\mathcal{N}$ is given by a similar path integral with no source that is evaluated at zero density, so that it provides the vacuum subtraction.

$W[j]$ generates all totally connected n-point amplitudes for the ϕ field and is a functional of the external source j. For any j, there is a corresponding expectation value of the scalar field

$$\phi_e(x) \equiv \frac{\langle 0^+|\hat{\phi}(x)|0^-\rangle_j}{\langle 0^+|0^-\rangle_j} = -i\hbar\frac{\delta \ln Z[j]}{\delta j(x)} = \frac{\delta W[j]}{\delta j(x)} \tag{102}$$

that reduces to the ground-state expectation value $\overline{\phi}$ for a vanishing source: $\lim_{j\to 0} \phi_e(x) = \overline{\phi} = \mathrm{const.}$

It is more useful, however, to treat the field ϕ_e as the independent variable, which is accomplished through a functional Legendre transformation.[45] Define the effective action $\Gamma[\phi_e]$ by

$$\Gamma[\phi_e] \equiv W[j] - \int \mathrm{d}^4x\, j(x)\phi_e(x)\,, \tag{103}$$

where $j(x)$ is to be eliminated in favor of $\phi_e(x)$ by solving eq. (102). (The vector field

V_e^μ has been suppressed, but enters in a similar fashion.) This transformation ensures that $\Gamma[\phi_e]$ is independent of j so that

$$\frac{\delta\Gamma[\phi_e]}{\delta j(x)} = 0\,, \qquad \frac{\delta\Gamma[\phi_e]}{\delta\phi_e(x)} = -j(x)\,. \tag{104}$$

If I set $j = 0$, ϕ_e must become a constant, equal to the ground-state expectation value $\overline{\phi}$. This implies that $\overline{\phi}$ is obtained by extremizing the effective action with respect to ϕ_e.

Since the baryon number B and baryon density ρ_B are conserved (I work in the comoving frame at fixed volume V and let $V \to \infty$ at the end of the calculation), the mean value $\overline{V}^0$ of the vector field, obtained by extremizing $\Gamma[\phi_e, V_e^\mu]$ with respect to $V_e^{\,0}$, is given by the tree-level result *at all orders in loops*:

$$\overline{V}^0 = \hbar\frac{g_v}{m_v^2}\rho_B \equiv \hbar\frac{g_v}{m_v^2}\frac{\gamma}{(2\pi)^3}\int \mathrm{d}^3k\,\theta(k_F - |\underset{\sim}{k}|) = \hbar\frac{g_v}{m_v^2}\frac{\gamma k_F^3}{6\pi^2}\,. \tag{105}$$

Here k_F is the Fermi momentum and γ is the spin-isospin degeneracy. [This result actually depends on the renormalization scheme; I follow ref. 48, where eq. (105) is obtained.]

For application to infinite matter, it is most convenient to expand the effective action Γ in powers of derivatives of the fields, as we only need the term involving uniform fields. Since $\overline{V}^0$ is always given by (105), the extremized effective action can be expressed in terms of the mean scalar field $\overline{\phi}$ and baryon density ρ_B, and I write the result for uniform fields in terms of the energy density $\mathcal{E}$:

$$\Gamma[\overline{\phi}, \rho_B] = -\int \mathrm{d}^4x\,\mathcal{E}(\overline{\phi}, \rho_B)\,. \tag{106}$$

At zero baryon density, there are no valence nucleons and the vector field vanishes; the resulting $\mathcal{E}$ is then called the scalar *effective potential* $U_{\text{eff}}(\overline{\phi})$.

The loop expansion in QHD–I now proceeds by defining classical meson fields that make the action in eq. (101) stationary. We can expand in the quantum fluctuations about these fields by defining

$$\phi(x) \equiv \phi_0(x) + \hbar^{1/2}\sigma(x) \tag{107}$$

and similarly for the vector meson. At the one-loop level, the resulting path integrals are gaussian and can be performed exactly[8,21] producing an explicit expression for the one-loop effective action. Since the classical fields make the action stationary, the Legendre transformation (103) is trivial at this order. After adding suitable counterterms (see below), the one-loop energy density defined by eq. (106) becomes [compare eq. (16)]

$$\mathcal{E}^{(1)}(M^\star, \rho_B) = \frac{g_v^2}{2m_v^2}\rho_B^2 + \frac{m_s^2}{2g_s^2}(M - M^\star)^2 + \frac{\gamma}{(2\pi)^3}\int_0^{k_F} \mathrm{d}^3p\,E^\star(p) + \Delta\mathcal{E}(M^\star)\,, \tag{108}$$

where

$$\Delta\mathcal{E}(M^\star) \equiv -\frac{1}{4\pi^2}\Big\{ M^{\star 4}\ln(M^\star/M) + M^3(M-M^\star) - \frac{7}{2}M^2(M-M^\star)^2$$
$$+ \frac{13}{3}M(M-M^\star)^3 - \frac{25}{12}(M-M^\star)^4 \Big\}, \tag{109}$$

which has no explicit dependence on $k_{\rm F}$, and $E^\star(p) \equiv (\underline{p}^2 + M^{\star 2})^{1/2}$. Here I have suppressed factors of $\hbar$, and $M^\star \equiv M - g_{\rm s}\overline{\phi}$ is determined at each $\rho_{\rm B}$ by minimization, which produces the well-known one-loop (RHA) self-consistency condition[16]

$$M^\star = M - \frac{g_{\rm s}^2}{m_{\rm s}^2}\frac{\gamma}{(2\pi)^3}\int_0^{k_{\rm F}}{\rm d}^3 p\,\frac{M^\star}{E^\star(p)} + \frac{g_{\rm s}^2}{m_{\rm s}^2}\frac{1}{\pi^2}\Big\{ M^{\star 3}\ln(M^\star/M)$$
$$- M^2(M^\star - M) - \frac{5}{2}M(M^\star - M)^2 - \frac{11}{6}(M^\star - M)^3 \Big\}. \tag{110}$$

Note that the solution to this equation contains all orders in the coupling $g_{\rm s}$.

Although $\Delta\mathcal{E}$ has historically been called the "vacuum fluctuation correction," this appellation is unfortunate, since it does not involve any fluctuations. More precisely,

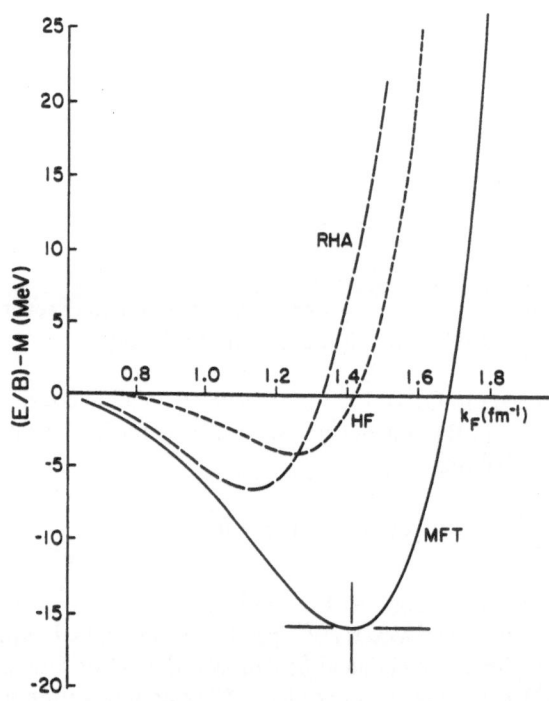

Fig. 6. Energy/nucleon in nuclear matter computed with the coupling strengths in eq. (20). The mean-field theory (MFT) results are shown as a solid line. The relativistic Hartree approximation (RHA), which includes the one-loop vacuum correction, produces the long-dashed line. Relativistic Hartree–Fock results, which include exchange effects (but which are not discussed here), are indicated by the short dashed line.

$\Delta\mathcal{E}$ is the finite shift in the baryon zero-point energy that occurs at finite density, and thus it represents the correction to the MFT that I labeled δH in eq. (14). Indeed, $\Delta\mathcal{E}$ can also be calculated by adding counterterms to that divergent expression, resulting in

$$\Delta\mathcal{E}(M^\star) = -\frac{1}{V}\sum_{\underset{\sim}{k}\lambda}\left[(\underset{\sim}{k}^2 + M^{\star 2})^{1/2} - (\underset{\sim}{k}^2 + M^2)^{1/2}\right] + \sum_{n=1}^{4}\frac{c_n}{n!}\overline{\phi}^n. \qquad (111)$$

Here the counterterms appear as a quartic polynomial in $\overline{\phi}$, and the (infinite) coefficients c_n are determined by specifying appropriate renormalization conditions on the energy. As in refs. 8 and 16, I will choose the counterterms to cancel the first four powers of $\overline{\phi}$ appearing in the expansion of the infinite sum over energies. This has the virtue of minimizing the many-body forces arising from this vacuum correction, and it is easy to verify that only the first four terms in this expansion produce divergent results. After removing these divergences with the counterterms, the remaining terms are finite, and reconstructing the sum leads to the result in eq. (109).

I emphasize that the zero-point (one-loop) vacuum correction is insensitive to the short-distance structure of the baryons, as it arises solely from the change in the baryon mass in the presence of the *uniform* scalar field $\overline{\phi}$. In addition, this correction cannot be calculated in nonrenormalizable meson–baryon models that are regularized by inserting *ad hoc* form factors at the meson–baryon vertices, since the uniform scalar field involves only zero-momentum-transfer components of the interaction between baryons, and the usual form factors have no effect on these contributions.

To discuss the size of the one-loop vacuum correction, we can compare predicted quantities using a fixed set of parameters determined from the empirical saturation properties of nuclear matter in the MFT [see eq. (20)]. (An alternative comparison can be made by determining new couplings that reproduce nuclear matter saturation in the RHA calculation. See table I below.) Figures 6 and 7 show the energy/nucleon and the equation of state for the MFT and RHA approximations. Observe that the equilibrium Fermi wavenumber k_F^0 shifts by roughly 0.25 fm^{-1}, and the binding energy decreases by about 10 MeV when the one-loop vacuum correction is included. Although the latter is small on the scale of the large scalar and vector fields ($\approx$ 250 MeV in the RHA), the modification to the binding energy is significant, reflecting the sensitive cancellation between the attractive and repulsive components in the potential energy.

The one-loop vacuum corrections are a direct consequence of the relativistic treatment of the nuclear many-body problem and are absent in a nonrelativistic approach. The nuclear matter equation of state at low densities (fig. 7) also changes because the saturation point is different in the two approximations, but for $\mathcal{E} \gtrsim 0.5$ GeV/fm^3 $\approx 10^{15}$ g/cm^3, the RHA results are in essential agreement with the MFT, signaling the dominance of the vector repulsion and the onset of a stiff equation of state. Moreover, although I have not discussed this approximation in detail here, figs. 6 and 7 show that similar results are obtained when exchange corrections are added to the MFT, as indicated by the curves labeled "HF". (See refs. 8, 49, and 50 for a discussion of the relativistic Hartree–Fock approximation.)

To compute the two-loop corrections, carry out the loop expansion of $W[j]$ to the next order in $\hbar$, isolate the effective action, and perform the Legendre transformation.[48] The result is exactly what is obtained by using second-order perturbation theory around the one-loop (RHA) ground state described by eq. (108) and keeping only the "one-particle irreducible" (1-PI) diagrams. (These diagrams cannot be broken into two pieces

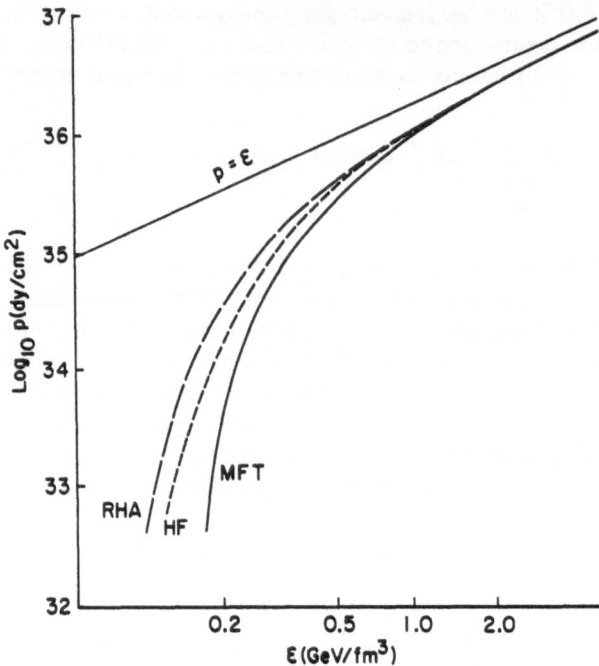

Fig. 7. Nuclear matter equation of state. The mean-field (MFT), one-loop (RHA), and Hartree–Fock (HF) curves are calculated as in fig. 6.

by severing a single meson line.) Because of the presence of the scalar field, the baryon propagator for this expansion is[48]

$$G^\star(p) \equiv (\gamma_\mu p^\mu + M^\star)\left(\frac{1}{p^2 - M^{\star 2} + i\eta} + \frac{i\pi}{E^\star(p)}\delta(p^0 - E^\star(p))\theta(k_F - |\underline{p}|)\right) \quad (112)$$

$$\equiv G_F^\star(p) + G_D^\star(p). \quad (113)$$

The first term $G_F^\star$ resembles the Feynman propagator, and the second term $G_D^\star$ contributes at finite density. The second term allows for the propagation of holes inside the Fermi sea and corrects $G_F^\star$ for the Pauli exclusion principle.[8] (The meson propagators are noninteracting Feynman propagators.) Since each power of the coupling constant is accompanied by $\hbar^{1/2}$, the loop expansion is perturbative in the couplings, but nonperturbative in the mean fields, as it involves the propagator $G^\star$.

Only 1-PI diagrams arise in the expansion since reducible diagrams will be generated by minimizing the energy density with respect to $M^\star$. The two-loop contributions to the energy density are shown in fig. 8, and the full two-loop corrections, including vacuum polarization, will be calculated exactly. Each baryon propagator contains both $G_F^\star$ and $G_D^\star$, as in eq. (113), so that the energy density through two-loop order can be written as

$$\mathcal{E}^{(2)}(M^\star, \rho_B) = \mathcal{E}^{(1)}(M^\star, \rho_B) + \mathcal{E}_{ex}^{(2)}(M^\star, \rho_B) + \mathcal{E}_{LS}^{(2)}(M^\star, \rho_B) + \mathcal{E}_{VF}^{(2)}(M^\star), \quad (114)$$

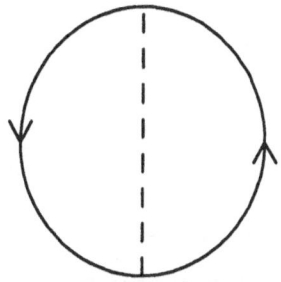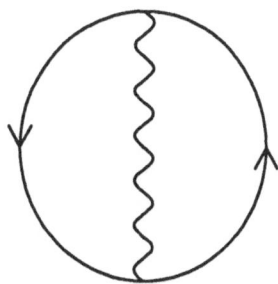

Fig. 8. Two-loop contributions to the nuclear matter energy density. The solid line represents the baryon propagator G^*, and the dashed and wiggly lines denote scalar and vector mesons, respectively. Counterterm contributions and vacuum subtractions, which are needed to produce finite results, are not shown.

and I can discuss each of the two-loop pieces separately. The first piece $\mathcal{E}_{\text{ex}}^{(2)}$ is finite, since it contains two G_D^* propagators that restrict the integration to the filled Fermi sea. This term generates the exchange of identical fermions in occupied states, just as in nonrelativistic Hartree–Fock calculations. These exchange contributions are precisely those studied by Chin.[16]

The second two-loop contribution $\mathcal{E}_{\text{LS}}^{(2)}$ contains the cross-terms between the G_F^* and G_D^* propagators. The integrals over G_F^* generate nested divergences, which must be removed by suitable renormalization conditions on the baryon self-energy and baryon–scalar vertex. $\mathcal{E}_{\text{LS}}^{(2)}$ is analogous to the Lamb shift in atomic physics, since it involves particles in occupied states whose spectrum is shifted by interaction with the fluctuating meson fields at finite density. The counterterm subtractions simply remove the fluctuations that would occur in free space, where $M^* = M$.

The final two-loop contribution $\mathcal{E}_{\text{VF}}^{(2)}$ is a true vacuum fluctuation correction; it involves virtual excitation of both mesons and baryons. Since both baryon propagators are G_F^*, the integrals contain nested, overlapping, and overall divergences, which can be removed by adding suitable counterterms and imposing renormalization conditions. This intricate procedure is described in detail in ref. 48, where all the two-loop results are given in a form suitable for numerical computation. At each baryon density ρ_B, the energy density $\mathcal{E}^{(2)}(M^*, \rho_B)$ must be minimized with respect to M^* to determine the correct value of the mean scalar field.

TWO-LOOP RESULTS

First, I evaluate the two-loop contributions with parameters that reproduce nuclear matter saturation at the one-loop level (RHA), shown in table I, and the values of M^* that minimize the *one-loop* energy at each density. For the purposes of this discussion, the parameters are chosen to produce nuclear matter saturation at $k_F^0 = 1.30 \text{ fm}^{-1}$ with a binding energy/nucleon of 15.75 MeV, and the scalar mass was fitted using the experimental charge radius of ^{40}Ca as an additional input.[18] The nuclear matter binding curve for this "perturbative" two-loop calculation is compared to the corresponding curve for the one-loop energy in fig. 9. To set the scale, I also show the MFT part of the one-loop energy defined by $\mathcal{E}^{(1)}(M^*, \rho_B) \equiv \mathcal{E}_{\text{MFT}}(M^*, \rho_B) + \Delta \mathcal{E}(M^*)$. Evidently, the

Table I. Walecka Model Parameters

	g_s^2	m_s (MeV)	g_v^2	m_v (MeV)
MFT	109.6	520.	190.4	783.
RHA	54.3	458.	102.8	783.
Set A	183.0	893.	55.0	783.
Set B	290.0	1300.	0.0	783.

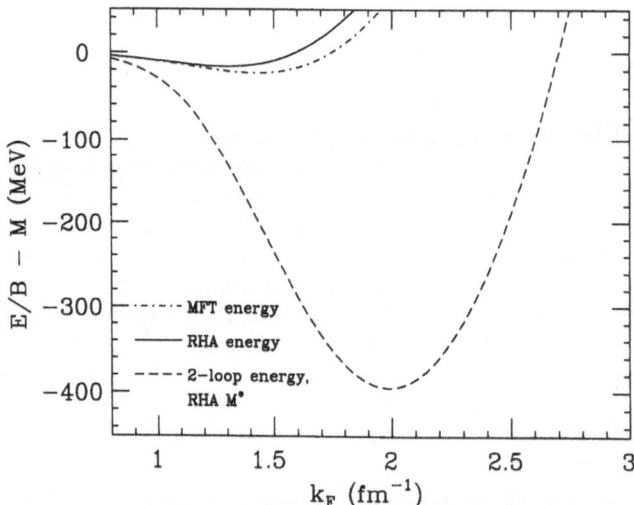

Fig. 9. Nuclear matter saturation curve for the mean-field (dot-dashed), one-loop (solid), and two-loop (dashed) energies, using RHA parameters and one-loop M^*.

two-loop corrections are huge when evaluated perturbatively; nuclear matter saturates at $k_F \simeq 2$ fm^{-1} with a binding energy of 400 MeV/nucleon.

To explain this result, I show in fig. 10 the separate contributions to the energy/nucleon at the RHA equilibrium density ($k_F = 1.3$ fm^{-1}). The bars on the left correspond to the full two-loop energy (the dashed line in fig. 9) while those on the right are for the one-loop energy (the solid line in fig. 9). The two-loop contributions are defined in eq. (114) and are further separated into pieces proportional to the scalar

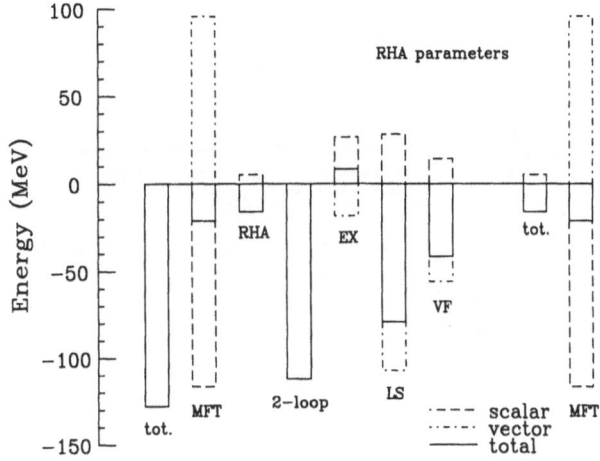

Fig. 10. Contributions to the two-loop (left) and one-loop (right) energies for the perturbative calculation (RHA parameters and one-loop $M^\star$) at empirical nuclear matter saturation density ($k_F = 1.3$ fm^{-1}).

and vector couplings. The two-loop contributions are clearly dominated by the strongly attractive vector Lamb shift and vacuum fluctuation terms, and their sum is comparable to the separate scalar and vector contributions to the mean-field energy. The exchange terms are relatively unimportant. The additional attraction overwhelms the sensitive cancellation of the MFT pieces at the empirical equilibrium point, driving saturation to higher density and much greater binding energy.

If I now try to minimize the two-loop energy with respect to $M^\star$ near nuclear matter density, *I find no minima for positive $M^\star$*. To understand this situation, first consider the effective potential $U_{\rm eff}$, which has a ϕ^2 piece from the classical (mean) field, a one-loop contribution $\Delta\mathcal{E}(M^\star)$, and a two-loop vacuum fluctuation contribution $\mathcal{E}_{\rm VF}^{(2)}(M^\star)$. Because of the renormalization conditions on the energy density, there is always a local minimum at $M^\star = M$. (See fig. 11.) At the one-loop level, the effective potential turns over at some $M^\star > M$ and goes to $-\infty$ as $M^\star \to \infty$, which is typical of one-loop effective potentials with fermions.[25] At the two-loop level, the new effective potential goes to large negative values as $M^\star \to 0$. Thus, with either one or two loops, the "vacuum" we work with is only a local minimum.

At finite density, the overall slope of the energy density as a function of $M^\star$ becomes more positive (*i.e.*, the curves in fig. 11 "rotate" counterclockwise). In the RHA (one-loop), the local minimum moves toward lower $M^\star$, further away from the maximum at $M^\star > 1$, which slowly moves toward larger $M^\star$. Thus there is always a (locally) well-defined minimum. With RHA parameters in the *two-loop* calculation, however, the local minimum becomes shallower as the density increases and then *disappears completely* for

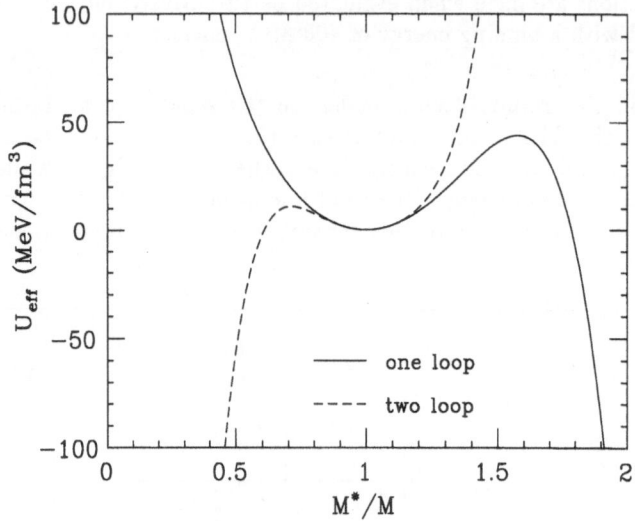

Fig. 11. One-loop and two-loop effective potentials as a function of M^*, using RHA parameters from table I.

$k_F \gtrsim 0.9$ fm^{-1}. At higher densities, a new minimum appears at small positive M^*, but this corresponds to an unphysical solution with an acausal equation of state. Thus the two-loop corrections change the nature of the system *qualitatively*.

It is possible, however, to find new parameters that reproduce nuclear saturation in the full two-loop calculation (sets A and B in table I). The resulting nuclear binding curves (and effective potential) are similar to the one-loop results. Nevertheless, the qualitative features of the description are very different from the RHA because the parameters must be tuned to minimize corrections from the quantum vacuum at the two-loop level. Note that two-loop terms involving vector mesons are much too attractive for values of the vector coupling used in all previous QHD studies and values implied by modern boson-exchange potentials.[51] After fine tuning, nuclear matter saturation is achieved almost entirely from the dynamics of scalar-meson exchange—in particular, from an interplay of attractive one-loop terms and repulsive scalar two-loop terms that involve occupied states in the Fermi sea.[48] In fact, it is possible to reproduce nuclear matter saturation with no vector meson at all (set B)!

Thus the appealing RHA picture of intermediate-range attraction from scalar exchange and short-range repulsion from vector exchange is apparently destroyed by the two-loop corrections. These qualitative changes lead us to conclude that the two-loop contributions produce unacceptably large corrections to the RHA results.

DISCUSSION

The preceding results demonstrate that two-loop corrections in the Walecka model produce large quantitative changes in the description of nuclear matter if the parameters are held fixed and large qualitative changes in the physics if they are adjusted to reproduce empirical saturation properties.

To understand these results in general terms, it is useful to focus on $\mathcal{E}(M^*, \rho_B)$ as a function of both M^* and ρ_B. In the Walecka model, the effective potential in the MFT is simply $\frac{1}{2} m_s^2 \phi^2$ and thus has an absolute minimum at $\phi = 0$ or $M^* = M$. When the one-loop correction is added, the renormalization conditions on the potential ensure that this minimum survives, but as is clear from fig. 11, the minimum is now local. As discussed by Coleman and Weinberg,[24] however, when the one-loop correction to the MFT is large, it is reasonable to expect that neglected higher-loop corrections are (at least) comparable. Thus *the one-loop correction can be valid only where it is small*, namely, near the local minimum. Fortunately this regime is sufficient. Since the overall slope of the effective potential is negative, increasing the density improves the status of the local minimum, and reasonable one-loop solutions exist at all densities; it is easy to verify that no other local minima exist in the Walecka model RHA. Moreover, although the location of the minimum differs from the MFT, the resulting RHA description is not qualitatively different from the MFT, and as has been emphasized repeatedly,[8] the RHA equation of state approaches that of the MFT in the high-density limit. (See fig. 7.)

When the two-loop corrections are included with the RHA parameters, several important changes occur. Although the local minimum at $M^* = M$ remains intact (as guaranteed by the renormalization conditions), it becomes less pronounced, and the regime where the loop corrections are small is reduced. This is certainly not characteristic of a useful expansion, in which higher-order corrections would presumably enlarge the regime of applicability. It could even imply that this regime gets smaller and smaller as further loop corrections are calculated, producing an increasingly unstable situation. In addition, since the attractive vector contributions dominate, the overall slope of U_{eff} is now positive. Increasing the density causes the local minimum to disappear before equilibrium nuclear matter density is reached, and this behavior persists at higher densities.

This disaster can be avoided by readjusting the parameters to reduce the two-loop vacuum corrections, which allows the local minimum to be maintained through the regime of normal density. With the new parameters, one finds an effective potential and saturation curve that qualitatively resemble the one-loop results. Unfortunately, the situation is still unsatisfactory, since the new parameters imply a nucleon-nucleon interaction that is qualitatively different from that at the one-loop level. We are therefore forced to conclude that unless miraculous cancellations occur when higher-loop diagrams are included, the loop expansion is not useful in this model.

The failure of the loop expansion is, perhaps, not surprising. The two-loop contributions are essentially *perturbative* corrections to the one-loop results; the nonperturbative aspects reside solely in the determination of the new mean field. Unfortunately, in models relevant for nuclear physics, the couplings are large. This leaves open the strong possibility that there are important corrections to the RHA that are not included efficiently in the loop expansion. These corrections could reduce the strength of the interaction, particularly in the vacuum loops. These considerations are especially important for calculations at low density (*e.g.*, the effective potential), since even a reasonable starting point is not known in this regime. There is no reason at all to believe that the loop expansion is meaningful at low density for large couplings.

Thus the failure of the loop expansion encountered here does not necessarily imply that the MFT and RHA results are inaccurate representations of the underlying quantum field theory. These simple approximations may still be useful starting points for computing corrections in QHD, especially at high density. We have simply found that the loop expansion is apparently not a useful way to obtain these corrections.

If we assume that the loop expansion cannot be rescued, then we must search for alternatives; these are needed to justify the phenomenological successes of the MFT and RHA, or to relegate them to pure phenomenology. The calculations presented here make two things reasonably clear: First, a useful expansion scheme (if it exists) must reduce the importance of the vacuum corrections that enter beyond one-loop order. Second, it is likely that the RHA will serve as a starting point in any successful scheme, since it allows us to incorporate the mean fields in the propagators. Moreover, in any scheme for calculating corrections to the RHA, the corresponding mean fields can be determined *at the end of the calculation* by extremizing the energy density. The difficult part is deciding how to group the corrections, which must undoubtedly involve an infinite summation of classes of Feynman diagrams computed with the G^* propagator of eq. (112). This will not be an easy task, since the simplest (perturbative) grouping is not useful. The burden is on the QHD practitioners to discover viable alternatives.

SUMMARY

Quantum hadrodynamics is a consistent framework for studying the relativistic nuclear many-body problem. By specifying the interactions with a local, renormalizable lagrangian density, we may include the effects of meson exchange, relativistic propagation, retardation, causality, and the dynamical quantum vacuum. In addition, the correct Lorentz structure of the NN interaction can be maintained. Renormalizable theories keep the number of parameters to a minimum, and since they require no additional *ad hoc* cutoffs, they are minimally sensitive to input from short-distance physics.

In these lectures, I examined some simple results in the Walecka model, which incorporates the dominant features of the observed NN force relevant for bulk nuclear systems through the exchange of neutral scalar and vector mesons. This Lorentz structure leads naturally to nuclear saturation in the mean-field approximation, and the model parameters were chosen to reproduce empirical saturation properties. The small binding energy of nuclear matter arises from a sensitive cancellation between large attractive and repulsive components in the potential energy. These large components introduce a new energy scale into the nuclear matter problem and lead to new physical effects from the shifted mass of the nucleon in nuclear matter. They also imply a stiff equation of state for nuclear matter at energy densities greater than approximately 0.5 GeV/fm^3.

I then generalized these results by computing the finite temperature equation of state in an arbitrary reference frame. I verified that the mean-field calculation satisfies the important thermodynamic identities and that the calculated results are consistent with Lorentz covariance. Although I illustrated these ideas in a simple mean-field calculation, the general features will remain in more sophisticated treatments. Thus QHD provides a covariant framework that allows calculations to be carried out in any convenient reference frame.

The mean-field theory is a simple approximation to the full quantum field equations of QHD. Because QHD is a consistent many-body theory, corrections to the MFT may be examined systematically, and I introduced the loop expansion as a possible method for computing nonperturbative corrections to the MFT. I evaluated the one-loop vacuum correction and found that it has a small effect on the MFT equation of state at energy densities above about 0.5 GeV/fm^3. The effects on the nuclear saturation curve are larger, due to the sensitive cancellations that occur near equilibrium density.

I then turned to the two-loop corrections, which include contributions from nucleon exchange, the nuclear Lamb shift, and true vacuum fluctuations. These corrections produce large quantitative changes in nuclear matter if the parameters are held fixed and large qualitative changes in the underlying physics if the parameters are adjusted to reproduce empirical saturation properties. I argued that this failure of the loop expansion is not surprising, since the expansion is essentially perturbative, except for the determination of the mean fields. Our experience in nuclear physics suggests, however, that there are important nucleon-nucleon correlations that must also be included nonperturbatively, and the loop expansion is simply not an efficient way to do this in models with strong scalar and vector fields. The search for useful expansion schemes in the strongly coupled, relativistic quantum field theory of QHD is an important problem that is currently under active investigation.

Acknowledgement

I am pleased to acknowledge productive collaborations and valuable discussions with Dick Furnstahl, Chuck Horowitz, Tetsuo Matsui, Robert Perry, and Dirk Walecka during the course of this work and the preparation of these lectures. This work was supported in part by DOE contract DE–FG02–87ER40365.

REFERENCES

1. E. V. Shuryak, Phys. Rep. **61**, 71 (1980).
2. H. A. Weldon, Phys. Rev. **D26**, 1394 (1982); **D28**, 2007 (1983).
3. E. V. Shuryak, Phys. Rep. **115**, 151 (1984).
4. H. Satz, Ann. Rev. Nucl. Part. Sci. **35**, 245 (1985).
5. L. Dolan and R. Jackiw, Phys. Rev. **D9**, 3320 (1974).
6. S. Weinberg, Phys. Rev. **D9**, 3357 (1974).
7. M. B. Kislinger and P. D. Morley, Phys. Rev. **D13**, 2765, 2771 (1976).
8. B. D. Serot and J. D. Walecka, Adv. Nucl. Phys. **16**, 1 (1986).
9. G. Baym and L. P. Kadanoff, Phys. Rev. **124**, 287 (1961).
10. G. Baym, Phys. Rev. **127**, 1391 (1962).
11. W. E. Caswell and A. D. Kennedy, Phys. Rev. **D25**, 392 (1982).
12. J. C. Collins, *Renormalization* (Cambridge University Press, New York, 1984).
13. C. J. Horowitz and B. D. Serot, Nucl. Phys. **A464**, 613 (1987).
14. J. D. Walecka, Ann. Phys. (N.Y.) **83**, 491 (1974).
15. S. A. Chin and J. D. Walecka, Phys. Lett. **52B**, 24 (1974).
16. S. A. Chin, Ann. Phys. (N.Y.) **108**, 301 (1977).
17. C. J. Horowitz and B. D. Serot, Nucl. Phys. **A368**, 503 (1981).
18. C. J. Horowitz and B. D. Serot, Phys. Lett. **140B**, 181 (1984).
19. C. Itzykson and J. Zuber, *Quantum Field Theory* (McGraw–Hill, New York, 1980).
20. L. D. Faddeev and A. A. Slavnov, *Gauge Fields: Introduction to Quantum Theory* (Benjamin/Cummings, Reading, MA, 1980).
21. P. Ramond, *Field Theory: A Modern Primer* (Benjamin, Reading, MA, 1981).

22. K. Huang, *Quarks, Leptons, and Gauge Fields* (World Scientific, Singapore, 1982).

23. F. A. Berezin, *The Method of Second Quantization* (Academic Press, New York, 1966).

24. S. Coleman and E. Weinberg, Phys. Rev. **D7**, 1888 (1973).

25. T. D. Lee and G. C. Wick, Phys. Rev. **D9**, 2291 (1974).

26. R. Jackiw, Phys. Rev. **D9**, 1686 (1974).

27. J. M. Cornwall, R. Jackiw, and E. Tomboulis, Phys. Rev. **D10**, 2428 (1974).

28. J. Iliopoulos, C. Itzykson, and A. Martin, Rev. Mod. Phys. **47**, 165 (1975).

29. J. A. McNeil, J. R. Shepard, and S. J. Wallace, Phys. Rev. Lett. **50**, 1439 (1983).

30. J. R. Shepard, J. A. McNeil, and S. J. Wallace, Phys. Rev. Lett. **50**, 1443 (1983).

31. B. C. Clark, S. Hama, R. L. Mercer, L. Ray, and B. D. Serot, Phys. Rev. Lett. **50**, 1644 (1983).

32. D. G. Boulware, Ann. Phys. (N.Y.) **56**, 140 (1970).

33. J. D. Bjorken and S. D. Drell, *Relativistic Quantum Mechanics* (McGraw-Hill, New York, 1964).

34. C. W. Misner, K. S. Thorne, and J. A. Wheeler, *Gravitation* (Freeman, San Francisco, 1973).

35. R. C. Tolman, *Relativity, Thermodynamics, and Cosmology* (Oxford U. Press, Oxford, 1946).

36. W. Israel, Ann. Phys. (N.Y.) **100**, 310 (1976); Physica **106A**, 204 (1981).

37. A. L. Fetter and J. D. Walecka, *Quantum Theory of Many-Particle Systems* (McGraw–Hill, New York, 1971).

38. J. D. Bjorken and S. D. Drell, *Relativistic Quantum Fields* (McGraw-Hill, New York, 1965).

39. R. A. Freedman, *Relativistic Many-Body Studies of Nuclear Mattter at Finite Temperature* (Ph.D. Thesis, Stanford University, 1978).

40. R. J. Furnstahl and B. D. Serot, Nucl. Phys. **A468**, 539 (1987).

41. N. P. Landsman and Ch. G. van Weert, Phys. Rep. **145**, 141 (1987).

42. R. J. Furnstahl and B. D. Serot, in preparation.

43. R. P. Feynman, Rev. Mod. Phys. **20**, 367 (1948).

44. J. Schwinger, Proc. Nat'l. Acad. Sci. (US) **37**, 452; **37**, 455 (1951).

45. G. Jona-Lasinio, Nuovo Cim. **34**, 1790 (1964).

46. Y. Nambu, Phys. Lett. **26B**, 626 (1968).

47. S. Coleman, "Secret Symmetry," in *Laws of Hadronic Matter*, Proceedings of the 11th Course of the International School in Physics "Ettore Majorana," A. Zichichi, ed. (Academic, New York, 1975).

48. R. J. Furnstahl, R. J. Perry, and B. D. Serot, Phys. Rev. **C40**, 321 (1989).

49. C. J. Horowitz and B. D. Serot, Phys. Lett. **109B**, 341 (1982).

50. C. J. Horowitz and B. D. Serot, Nucl. Phys. **A399**, 529 (1983).

51. R. Machleidt, K. Holinde, and Ch. Elster, Phys. Rep. **149**, 1 (1987).

ELEMENTS OF A RELATIVISTIC MICROSCOPIC THEORY OF HADRONIC MATTER IN EQUILIBRIUM AND NON EQUILIBRIUM

Rudi Malfliet

Kernfysisch Versneller Instituut
Zernikelaan 25, 9747 AA Groningen, The Netherlands

1. INTRODUCTION

Nuclear matter is a prime example of a strongly interacting many-hadron system. The understanding of its properties like saturation and compressibility is a first necessary step towards the description of finite nuclei and their properties. A unified treatment of equilibrium and non-equilibrium phenomena in nuclear matter might lead to a consistent approach to study properties of static nuclei and the behaviour of colliding nuclei.

In these lectures we will treat the relativistic many-body problem microscopically. This means that we begin with a nucleon-nucleon interaction which reproduces nucleon-nucleon observables very well. This fixes once and for all the parameters of the theory. Having constrained our theory in a well-known sector we combine it with a prescription to handle the many-nucleon situation. In this way one can expect new insights on the procedures used as well as predictions for interesting observables. The microscopic approach is quite different from effective models like the QHD-approach of Walecka, Serot and Chin[1,5]. Here one does not start at the N-N interaction level but, within a definite many-body prescription (mean field or Hartree-Fock), one adjusts the parameters such as to reproduce certain nuclear matter properties.

We will focus on the microscopic theory and overview its present status. For reference and for further details see also a recent review[2].

2. THE RELATIVISTIC MANY-BODY PROBLEM

The starting point of a microscopic treatment of the nuclear many-body problem is the nucleon-nucleon or more generally the hadron-hadron interaction. The nucleon-nucleon interaction is constructed based on a field-theoretical approach. The One-Boson-Exchange (OBE) interaction, of which the Bonn potential presents a typical example, is based on mesons (π, ω, ρ, σ, η, δ) as the exchanged quanta. The long-range part is governed by pion exchange while in the intermediate range (1-2 fm) the heavier mesons (ρ, ω) are important, as well as two-pion exchange. This 2π-exchange can be either a twice-iterated one-pion exchange or a more complicated diagram. Another important second-order diagram is the two-pion exchange either crossed or with a delta (instead of a nucleon) as an intermediate particle. These processes are represented by a fictitious scalar meson, the σ-meson. For a further discussion see Brown and Jackson[4]. The very short-range part of the N-N interaction, where multi-meson (multi-quark) exchanges are possible, where also the nucleon cannot be treated as a point particle and vertex corrections will play a role, is handled by a "phenonemological" formfactor to be applied to different vertex couplings. It is usually parametrised as

$$F(k) = \left[\frac{\Lambda^2}{\Lambda^2 + k^2} \right]^n \tag{1}$$

where k is the momentum carried away by the exchanged meson. Λ is the so-called cut-off mass and n=1 (monopole) or n=2 (dipole). In table 1 we collect some meson properties and the type of NN-meson coupling. What

Table 1. Meson properties and type of NN-meson coupling

meson α	mass (MeV/c^2)	I, J^P	$\mathscr{L}_I$
π	139	$1,0^-$	$\lambda_\pi \, \bar{\psi} \, \gamma_5 \, \gamma^\mu \, \psi \, \partial_\mu \, \vec{\tau}.\vec{\phi}_\pi$
σ	571	$0,0^+$	$\lambda_\sigma \, \bar{\psi} \, \psi \, \phi_\sigma$
ω	784	$0,1^-$	$\lambda_\omega \, \bar{\psi} \, \gamma_\mu \, \psi \, V^\mu_\omega + \mu_\omega \, \bar{\psi} \, \sigma_{\mu\nu} \, \psi \, (\partial^\mu V^\nu_\omega - \partial^\nu V^\mu_\omega)$
ρ	764	$1,1^-$	$\lambda_\rho \, \bar{\psi} \, \gamma_\mu \, \vec{\tau}.\vec{V}^\mu_\rho + \mu_\rho \, \bar{\psi} \, \sigma_{\mu\nu} \, \psi \, \vec{\tau} \, (\partial^\mu \vec{V}^\nu_\rho - \partial^\nu \vec{V}^\mu_\rho)$
η	550	$0,0^-$	$\lambda_\eta \, \bar{\psi} \, \gamma_5 \, \gamma^\mu \, \psi \, \partial_\mu \, \phi_\eta$
δ	962	$1,0^+$	$\lambda_\delta \, \bar{\psi} \, \psi \, \vec{\tau}.\vec{\phi}_\delta$

remains to be determined now are the coupling constants and the cut-off masses in the formfactor. This will be discussed further on.

A powerful tool to study properties of many interacting particles is the Green's function formalism. It is especially suited to field-theoretical approaches, physical observables can be expressed directly in terms of Green's functions and it offers a unified framework for both equilibrium and non-equilibrium situations. The equilibrium formulation as applied to a simple σ–ω exchange only, can be found for instance in the work of Chin[5]. Also the studies of Brown, Puff and Wilets present important initiations in the formalism[6]. The non-relativistic treatment is explained in ref. 8.

2.1. Green's Function

We denote the fields which describes the fermions (nucleons) as $\psi(x)$ and the mesons as $\phi(y)$, where x,y ... are 4-vectors $(t,\vec{r})$ in time- and coordinate space. For a non-interacting system the fermions obey the Dirac equation and the mesons either a Klein-Gordon equation or a Proca equation (vector mesons). For the basic ingredients and notations we refer to the textbook of Itzykson and Zuber[7]. The fields obey the equal-time, anti-commutation (fermions) or commutation (bosons) relations:

$$\left\{\psi_\alpha(x),\ \bar{\psi}_\beta(y)\right\}_+\ \delta(x_0-y_0) = \gamma^0_{\alpha\beta}\ \delta^4(x-y)$$

(2)

$$\left\{\phi(x),\ \phi^+(g)\right\}_-\ \delta(x_0-y_0) = \delta^4(x-y)$$

where α,β refer to the spinor components, $\bar{\psi} \equiv \psi^+\gamma^0$ and ψ^+ the complex conjugate. The symbol γ refers to the Dirac γ-matrices.

Consider the following interaction Lagrangian density $\mathcal{L}_I$:

$$\mathcal{L}_I = \bar{\psi}(x)\ \Gamma^j\psi(x)\ \phi_j(x)$$

(3)

where j is the type of meson and Γ^j the corresponding vertex coupling (see table 1). The equations for the fields $\psi(x)$ and $\phi(x)$ can be obtained from the Euler-Lagrange equations:

$$(i\gamma^\mu\partial_\mu - M)\ \psi(x) = -\left[\frac{\partial\mathcal{L}_I}{\partial\bar{\psi}(x)}\right] = -\Gamma^j(x,y,\zeta)\psi(y)\phi_j(\zeta)$$

$$(\Box_\zeta + m_j^2)\phi_j(\zeta) = \left[\frac{\partial\mathcal{L}_I}{\partial\phi_j^+(\zeta)}\right] = \Gamma^j(x,y,\zeta)\bar{\psi}(x)\psi(y)$$

(4)

$$\Gamma^j(x,y,\zeta) = \Gamma^j\delta(x-y)\ \delta(x-\zeta)$$

where the latter relation implies local interactions.

Green's functions are expectation values of (bilinear) products of field operators i.e. $\langle\psi(x)\bar\psi(y)\rangle$. The field operators are in the Heisenberg representation and therefore the full time dependence is contained in the field operators. The expectation value $\langle\ \rangle$ is then taken over the initial interacting state of the system. In equilibrium it is usually taken as the thermodynamical (grand canonical) expectation $\langle * \rangle = \mathrm{Tr}\{\rho_0 *\}$, $\rho_0 = \exp-\beta(H-\mu N)$ and $\mathrm{Tr}\rho_0 = 1$.

The solutions of the Dirac equation allow for both positive and negative energy states. The latter are interpreted as follows: the absence of a negative-energy particle corresponds to an anti-particle with positive energy. Initially all negative-energy states are filled. Physical observables will usually be expressed in terms of causal Green's functions. They propagate particles from the time they appear t_1 to the time where they disappear t_2 with $t_2 > t_1$. The anti-particles with $t_1 > t_2$. It is therefore crucial to pinpoint the direction of the time-evolution wether forward or backward. In non-equilibrium situations this is even more important since there the time-evolution forward or backward is not the same. Thus $\langle\psi(x)\bar\psi(x')\rangle$ will correspond to 4 different expressions dependent on whether $t\rightarrow,t'\rightarrow$ forward or $t,t'\rightarrow$ backward. They are most conveniently defined by introducing a directed time-contour as in fig. 1.

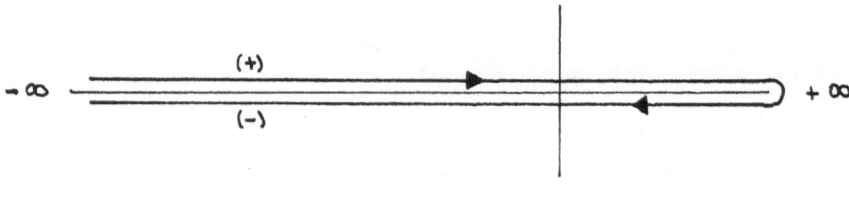

Fig. 1

The corresponding Green's function (see ref.2 for further references) we call the Keldysh or path-ordered Green's function $\underline{G}$:

$$\underline{G} \equiv \underline{G}(x,x') \equiv (-i)\langle P(\psi(\underline{x})\bar\psi(\underline{x}'))\rangle \tag{5}$$

Here P is the temporal path-ordering operator: field operators further on the contour are put to the left of the ones which lie behind on the contour. P also includes a sign change (for fermions only) for every permutation to achieve the proper ordering. The Keldysh contour (fig. 1) contains an upper (+) part and a lower (-) part. Specifically:

$$t,t' \to (+,+) \quad G(x_+,x'_+) \equiv G^c(x,x') = (-i)\langle T(\psi(x)\bar{\psi}(x'))\rangle$$

$$t,t' \to (-,-) \quad G(x_-,x'_-) \equiv G^a(x,x') = (-i)\langle \tilde{T}(\psi(x)\bar{\psi}(x'))\rangle$$

$$(6)$$

$$t,t' \to (-,+) \quad G(x_-,x'_+) \equiv G^>(x,x') = (-i)\langle \psi(x)\bar{\psi}(x')\rangle$$

$$t,t' \to (+,-) \quad G(x_+,x'_-) \equiv G^<(x,x') = (+i)\langle \bar{\psi}(x')\psi(x)\rangle$$

where we recognise the usual causal (Wick time ordering operator T) and anti-causal (anti-chronological) operator $\tilde{T}$ corresponding to the Green's functions G^c and G^a respectively together with two others which are commonly denoted as $G^<$ and $G^>$. Clearly $\underline{G}$ should be considered as a 2×2 matrix with components as in (6). All these Green's functions are not independent of each other and one has for instance:

$$G^c(x,x') = \theta(t-t')G^>(x,y) + \theta(t'-t)G^<(x,y)$$

$$(7)$$

$$G^a(x,x') = \theta(t'-t)G^>(x,y) + \theta(t-t')G^<(x,y)$$

Also combinations of them are frequently used like the retarded and advanced Green's functions $G^+(x,x')$ and $G^-(x,x')$ which are rigorously zero if $t'>t$ or $t>t'$ respectively.

$$G^+ = G^c - G^< = -G^a + G^>$$

$$(8)$$

$$G^- = G^c - G^> = -G^a + G^<$$

While this zoo of different Green's functions is at first sight confusing they were only introduced in order to do the proper time-evolution bookkeeping. Some of them are familiar (causal, advanced and retarded) from conventional approaches to the many-body problem (see ref. 8). For non-interacting fermions one can deduce explicit expressions for the different Green's functions which makes their physical interpretation more transparent[2].

We will now derive equations for the temporal path-ordered Green's functions. The first step is to transcribe the field equations (4) on a Keldysh contour C (fig. 1). This is achieved as follows. All fields are defined as having 2 components with either their time argument on the upper (+) or lower branch (-) of the contour C. The vertex-function $\Gamma(x,y,\zeta)$ which couples fields is now generalised to

$$\underline{\Gamma}^j = \underline{\Gamma}^j(x,y,\zeta) = \Gamma^j \ \underline{\delta}(x-y) \ \underline{\delta}(x-\zeta) \tag{9}$$

where $\underline{\delta}$-functions are ordinary δ-functions in coordinate space, but in time-space they behave as:

$$\underline{\delta}(t-t') = \delta(t-t') \text{ for } t,t' \text{ both on } (+) \text{ contour}$$

$$= -\delta(t-t') \text{ for } t,t' \text{ both on } (-) \text{ contour} \tag{10}$$

$$= 0 \text{ otherwise.}$$

The Euler-Lagrange equation for the fermion field $\psi(x)$, eq. (4), then takes the form

$$(i\gamma^\mu \partial_\mu - M)\psi(\underline{x}) = - \int_c d\underline{y} \int_c d\underline{\zeta} \ \underline{\Gamma}^j(x,y,\zeta) \ \psi(\underline{y}) \ \phi_j(\underline{\zeta}) \tag{11}$$

$$\int_c d\underline{z} \equiv \left\{ \underbrace{\int_{-\infty}^{+\infty} dt_z}_{(+) \text{branch}} - \underbrace{\int_{-\infty}^{+\infty} dt_z}_{(-) \text{branch}} \right\} \int d^3 z$$

Because of the $(-)$ sign in $\int_c d\underline{z}$ we had to take the convention for $\underline{\delta}$ as above, eq. (10).

The Green's functions are defined in terms of fields and thus from the field equations (4) one can obtain the equations for the Green's functions. For instance multiplying the first of eqs. (4) from the right with $\bar\psi(\underline{x}')$ and taking the temporal path ordered product P one obtains (verify this):

$$(i\gamma^\mu \partial_\mu - M)\underline{G}(x,x') = \underline{\delta}(x-x') - \underline{\Gamma}^j(x,y,\zeta)\underline{G}_{1\frac{1}{2}}(y,x',\zeta) \tag{12}$$

$$\underline{G}_{1\frac{1}{2}}(x,x',\zeta) \equiv (-i)\langle P(\psi(\underline{x})\bar\psi(\underline{x}')\phi_j(\underline{\zeta}))\rangle$$

with $\underline{G}_{1\frac{1}{2}}$ a "mixed" Green's function.

Similarly multiplying the second of eqs.(4) by $\psi(\underline{y})\bar\psi(\underline{x}')$ from the left and taking the P-ordering gives:

$$(\square_\zeta + m_j^2)\underline{G}_{1\frac{1}{2}}(y,x',\zeta) = i\underline{\Gamma}^j(u,v,\zeta)\underline{G}_2(y,v|u,x') \tag{13}$$

$$\underline{G}_2(x,x'|y,y') \equiv (-i)^2 \langle P(\psi(\underline{x})\psi(\underline{x}')\bar\psi(\underline{y}')\bar\psi(\underline{y}))\rangle$$

with G_2 the two-nucleon Green's function.

Finally multiply the third of eqs. (4) by $\phi^j(\underline{\zeta}')$ from the right and taking the P-product results in:

$$(\square_\zeta + m_j^2)\underline{g}(\zeta,\zeta') = \underline{\delta}(\zeta-\zeta') + \underline{\Gamma}^j(u,v,\zeta)\underline{G}_{1\frac{1}{2}}(v,u,\zeta')$$

(14)

$$\underline{g}(\zeta,\zeta') = (-i)\langle P(\phi_j(\underline{\zeta})\phi_j(\underline{\zeta}'))\rangle$$

with $\underline{g}$ the meson Green's function.

In all foregoing equations repeated variables are integrated and summed over. The time integration goes over the contour C as indicated in (11). The equations express for instance the one-fermion Green's function $\underline{G}(x,y)$ in terms of higher-order Green's functions. The latter obey also equations which can be obtained from the basic field equations and are also expressed in still more complicated Green's functions. One obtains an hierarchy of equations which is equivalent to the well-known Martin-Schwinger hierarchy[9]. In order to solve for $\underline{G}(x,y)$ one has either to solve them all or find some suitable truncation scheme. This forms the essence of the theory of many-particle interactions.

First of all the "mixed" Green's function $\underline{G}_{1\frac{1}{2}}$ can be eliminated in favour of the two-particle one $\underline{G}_2$. This is achieved as follows. We denote the unperturbed solution of eq. (14) as $g^0(\zeta,\zeta')$:

$$(\square_\zeta + m_j^2)\underline{g}^0(\zeta,\zeta') = \underline{\delta}(\zeta-\zeta')$$

(15)

Then, $\underline{G}_{1\frac{1}{2}}$ can be expressed as:

$$\underline{G}_{1\frac{1}{2}}(1,2,\zeta) = (+i)\underline{g}^0(\zeta,\zeta')\underline{\Gamma}^j(3,4,\zeta')\underline{G}_2(1,4|3,2)$$

(16)

and is a solution of eq. (13) as can be verified explicitly. Substitute the expression for $\underline{G}_{1\frac{1}{2}}$, eq. (12) into the equation for the fermion Green's function and one obtains:

$$(i\gamma^\mu\partial_\mu - M)\,\underline{G}(1,1') = \underline{\delta}(1-1') + \underline{\Sigma}(1,2)\underline{G}(2,1')$$

(17)

$$\underline{\Sigma}(1,2)\underline{G}(2,1') = (-i)\left\{\underline{\Gamma}^j(1,2,\zeta)\underline{g}^0(\zeta,\zeta')\underline{\Gamma}^j(3,4,\zeta')\right\}\,\underline{G}_2(2,4|3,1')$$

The term between brackets in (17) has the form of a one-boson-exchange interaction. Similarly, the equation for the meson Green's function reads:

79

$$(\Box_\zeta + m_j^2)\underline{g}(\zeta,\zeta') = \underline{\delta}(\zeta-\zeta') + \underline{\Pi}(\zeta,\zeta'')\underline{g}(\zeta'',\zeta')$$

$$\underline{\Pi}(\zeta,\zeta'')\underline{g}(\zeta'',\zeta') \equiv (+i)\left\{\underline{\Gamma}^j(3,4,\zeta)\underline{G}_2(4,2,|1,3)\underline{\Gamma}^j(1,2,\zeta'')\right\}\underline{g}^0(\zeta'',\zeta') \tag{18}$$

Equations (17) and (18) are the Dyson equations for $\underline{G}(1,1')$ respectively $\underline{g}(\zeta,\zeta')$ and are expressed in terms of the (proper) self-energy $\underline{\Sigma}$ and (proper) polarisation $\underline{\Pi}$. These are familiar concepts from the standard treatment of the many-body problem[8]. Both depend on the two-fermion Green's function $\underline{G}_2$. If we know how to obtain the latter then we have solved the whole problem. In the following we will denote $\underline{G}_2$ as $\underline{G}_{12}$ to indicate clearly its 2-nucleon character.

2.2. Self-Energy and Polarisation

If we denote also the non-interacting fermion Green's function as $\underline{G}^0$ (which is a solution of eq. (17) without the $\underline{\Sigma}\,\underline{G}$ term) the equations (17) and (18) for $\underline{G}$ respectively $\underline{g}$ can be expressed as Dyson equations

$$\underline{G} = \underline{G}^0 + \underline{G}^0\underline{\Sigma}\,\underline{G}$$

$$\underline{g} = \underline{g}^0 + \underline{g}^0\,\underline{\Pi}\,\underline{g} \tag{19}$$

These are written in the Keldysh representation i.e. they really are 2×2 matrix equations. The self-energy $\underline{\Sigma}$ and polarisation $\underline{\Pi}$ are both expressed in terms of the two-fermion Green's function $\underline{G}_{12}$. From the Euler-Lagrange equations for the fields one can also obtain an equation for $\underline{G}_{12}$. This equation is rather complicated and expressed in terms of even high-order Green's functions. Therefore one tries to find approximated solutions for the two-particle Green's function. However, some care must be exercised since not all approximations lead to so-called conserving approximations i.e. they respect the conservation laws. This matter is discussed briefly in ref. 2 where also other references can be found. We will consider two conserving approximations which are frequently used in the literature i.e. the Hartree-Fock approximation and the Brueckner ladder approximation.

The Hartree(-Fock) or mean field (plus exchange) approximation neglects correlations in the two-particle Green's function $\underline{G}_{12}$. In general the latter has the form:

$$\underline{G}_{12} = (\underline{G}\;\underline{G}) + (\underline{G}\;\underline{G})\underline{T}(\underline{G}\;\underline{G}) \tag{20}$$

where $\underline{T}$ (properly anti-symmetrised) includes all correlations. If we truncate $\underline{G}_{12}$ to be equal to ($\underline{G}$ $\underline{G}$) we obtain the self-consistent Hartree(-Fock) approximation. Diagrammatically:

This approximation is discussed in the lectures of Brian Serot and also in Serot and Walecka[1].

Another approximation, of a more microscopic nature, is the Brueckner ladder approximation. It is guided by two observations. First of all at not too high density there is a certain type of diagrams occuring in $\underline{T}$ which is dominant. Secondly, at very low density the correlation part $\underline{T}$ should coincide with the nucleon-nucleon interaction, that is the nucleon-nucleon t-matrix. A suitable solution for $\underline{G}_{12}$ which posesses these virtues is obtained by taking for $\underline{T}$ the solution of:

$$\underline{T} = \underline{V} + i\underline{V} \, (\underline{G} \, \underline{G}) \, \underline{T} \qquad (21)$$

The nucleon-nucleon interaction term $\underline{V}$ is identical to the term in brackets at the RHS of eq. (17) for the self-energy $\underline{\Sigma}$ and has the form of a one-boson-exchange interaction. The Green's functions $\underline{G}$ are self-consistently determined by incorporating the equation (21) for $\underline{T}$ into $\underline{G}_{12}$ (eq. 20) which then leads to an expression for $\underline{\Sigma}$ and thus $\underline{G}$ through the Dyson eq. (17). The results are:

$$\underline{G}_1 = \underline{G}_1^0 + \underline{G}_1^0 \underline{\Sigma}_{1-1} \underline{G}_1$$

$$\underline{\Sigma}_1 = -i \underset{(2)}{\mathrm{Tr}} \, \underline{T}_{12} \underline{G}_2$$

$$\underline{T}_{12} = \underline{V}_{12} + i\underline{V}_{12}\underline{G}_1\underline{G}_2\underline{T}_{12} \tag{22}$$

$$\langle 12|\underline{T}_{12}|1'2'\rangle = \underline{\delta}(t_1-t_2)\underline{\delta}(t_1'-t_2') \, \langle 12|\underline{T}_{12}|1'2'\rangle$$

where we have labeled explicitly the nucleons as 1 and 2, and $\underline{G}_1$ and $\underline{G}_2$ are one-nucleon Green's functions.

The algebraic structure of the equations (22) is simple but since they are in the Keldysh representation their matrix content leads to a set of rather complicated equations which couple different components $G^<$, $G^>$, G^c, $\Sigma^<$, $\Sigma^>$, Σ^c, ... Only for the retarded respectively advanced components one obtains transparent and familiar forms[2]:

$$T_{12}^\pm = V_{12} + iV_{12}(G_1 G_2)^\pm T_{12}^\pm$$

$$G^\pm = (G^0)^\pm + (G^0)^\pm \Sigma^\pm G^\pm \tag{23}$$

$$\Sigma_1^+ = -i \underset{(2)}{\mathrm{Tr}} \, (T_{12}^+ G_2^< + T_{12}^< G_2^-) \equiv (\Sigma_1^+)_{Br} + (\Sigma_1^+)_R$$

$$T_{12}^< = iT_{12}^+ G_1^< G_2^< T_{12}^-$$

A number of equilibrium observables like the total energy, single particle energy, and many others can all be expressed all in terms of the $T_{12}^\pm$, $G^\pm$ and $G^<$, $G^>$ quantities. Therefore the set of equations (23) is all we need.

To fully understand the physical meaning of these results we give the explicit expressions for the non-interacting Green's functions $(G^0)^\pm$, $(G^0)^<$ and $(G^0)^>$. These can be obtained from the definitions by substituting the free-field solutions. Denoting by $\Lambda^\pm$ $(\pm p)$ the usual positive- and negative- energy projection operators one finds in four-momentum $(p=p_0,\vec{p})$ representation:

$$(G^0)^<(p) = 2\pi i \, \frac{M}{E_p} \left[\delta(p_0-E_p)f(p)\Lambda^+(p) - \delta(p_0+E_p)\Lambda^-(-p) \right]$$

$$\equiv G_F + G_D$$

$$(G^0)^>(p) = -2\pi i \, \frac{M}{E_p} \left[\delta(p_0-E_p)(1-f(p))\Lambda^+(p) \right] \tag{24}$$

$$(G^0)^{\pm}(p) = \frac{(\gamma^{\mu}p_{\mu}+M)}{p^2-M^2\pm i\varepsilon} \pm 2\pi i \frac{M}{E_p} \left[\delta(p_0+E_p)\Lambda^-(-p)\right]$$

Here $E_p = (M^2+p^2)^{\frac{1}{2}}$ and $f(p) = [\exp(\beta p_0)+1]^{-1}$ is the equilibrium momentum occupation distribution with $\beta = 1/kT$ (T temperature). For $T=0$ $f(p)$ corresponds to the well-known Fermi sphere $f(p) = \theta(p_F-|\vec{p}|)$.

The equations (23) are valid both for equilibrium and non-equilibrium. But, as they stand they are not equivalent to the so-called self-consistent Brueckner approximation. To obtain the latter, additional approximations are necessary: First of all, one neglects the second term $(\Sigma_1^+)_R$ in the expression for the self-energy (23) which corresponds to a higher-order contribution (rearrangement). Secondly, the intermediate nucleon-nucleon propagator $(G_1G_2)^{\pm}$ in the $T_{12}^{\pm}$-matrix equation (23) contains a particle-particle and hole-hole contribution $(G_1^{>}G_2^{>}$ respectively $G_1^{<}G_2^{<})$. The hole-hole term is dropped since it contains integrations from $0 \rightarrow k_F$ while the particle-particle term integrates from k_F up to ∞ and is dominant. Thirdly, and this will become clear in the next section, we use the quasi-particle approximation, i.e. in the Dyson equation for $G^{\pm}$ only the real part of $\Sigma^{\pm}$ contributes. In equilibrium the imaginary part is exactly zero at the Fermi surface and elsewhere rather small. In non-equilibrium situations the imaginary part of $\Sigma^{\pm}$, which is related to collision cross sections, is not necessarily small (section 3.3.). The foregoing approximations lead to a non-conserving truncation scheme which violates the Hugenholtz-Van Hove theorem[2]. The latter relates the binding energy at saturation to the single-particle enegy evaluated at the Fermi momentum k_F.

As a final remark we note that the Hartree-Fock approximation is obtained from eqs. (23) by taking $T_{12}^+=V_{12}$ and $T_{12}^{<}=0$.

2.3. Dirac-Brueckner Approach for Nuclear Matter in Equilibrium

In the self-consistent (relativistic) Brueckner approach (see eq. (23) and the discussion at the end of section 2.2)) a nucleon in the nuclear medium may be viewed as a bare nucleon that is "dressed" in consequence of its effective two-body interaction with the other nucleons in the medium. This effective two-nucleon interaction furthermore coincides with the bare two-nucleon interaction in the zero-density limit. The Brueckner scheme then gives a prescription how to construct self-consistently the properties of the dressed nucleons. The (relativistic) Dirac-Brueckner approach is formally similar to the Brueckner-Bethe-Goldstone method in non-relativistic nuclear matter theory but differs in some essential points as was emphasized first by Shakin and collaborators[10]. This will be explained in this subsection.

The proper self-energy $\Sigma^+(k)$ of the physical nucleon is given by the Dyson equation (23) which relates the bare $(G^0)^{\pm}$ and physical ("dressed") nucleon propagators $G^{\pm}$. The Dyson equation (23) for the full baryon propagator $G^{\pm}$ can be solved by expressing the retarded and advanced self-energy $\Sigma^{\pm}$ in the restframe of nuclear matter in terms of its independent (Lorentz) invariants in the following (we leave out some of the (±) labels):

$$\Sigma^{\pm}(k) = \Sigma_s(k) - \gamma_{\mu} \Sigma^{\mu}(k) + \sigma_{\mu\nu} \Sigma^{\mu\nu}_t(k) + \gamma_5 \Sigma_{ps} - \gamma_5\gamma_{\mu}\Sigma^{\mu}_a \qquad (25)$$

The last two terms (pseudoscalar (ps) and axial vector (a)) will be zero if parity is a good quantum number (as we assume here). The remaining terms are scalar (s), vector (v) and tensor (t). The invariants Σ_s, Σ^{μ}, ... which appear in the representation (25) of the baryon self-energy constitute the necessary 16 quantities (4×4) to characterise fully this self-energy. In case we would have mixing between positive- and negative-energy spinors we need 256 invariants (more precisely for on-mass-shell kinematics we need 44 independent amplitudes) to fully characterise Σ (for a discussion see Tjon and Wallace[11]). In our case we will only deal with the simplest case as written in (25) since we will neglect negative-energy states altogether.

In uniform spin-saturated nuclear matter we have only scalar and vector components in (25) and we can solve the Dyson equation to yield:

$$G^{\pm}(k) = \left[\gamma_{\mu}(k^{\mu} + \Sigma^{\mu}(k)) - (M + \Sigma^s(k)) \pm i\varepsilon \right]^{-1} \qquad (26)$$

which has the same form as the free baryon propagator. We define the "effective" quantities (which all depend on $k=(k_0,\vec{k})$):

$$M^*(k) \equiv M + \Sigma^s(k)$$

$$\left. \begin{aligned} \vec{k}^* &\equiv \vec{k} + \vec{\Sigma} \\ (k^*)^0 &\equiv k^0 + \Sigma^0 \end{aligned} \right\} \quad (k^*)^{\mu} \equiv k^{\mu} + \Sigma^{\mu}; \; \Sigma_v\vec{k} \equiv \vec{\Sigma} \qquad (27)$$

The on-shell single-particle energy E is then given as:

$$E(k) = \left[E^*_k - \Sigma^0(k) \right]_{k_0=E(k)} = \left\{ (\vec{k}+\vec{\Sigma}\,(k))^2 + (M+\Sigma^s(k))^2 \right\}^{1/2} - \Sigma^0(k) \qquad (28)$$

where E^* is defined as

$$E^*(k) \equiv \left[(\vec{k}^*)^2 + (M^*)^2 \right]^{1/2} \qquad (29)$$

The on-shell $((k^*)^0 = E^*)$ behaviour of the effective baryon is given by a Dirac equation:

$$(\gamma k^* + M^*) \, u^* = 0 \qquad (30)$$

which yields i.e. the dressed positive-energy spinor u^*:

$$u^*(\vec{k}) = \left(\frac{E^* + M^*}{2M^*} \right)^{1/2} \begin{pmatrix} 1 \\ \dfrac{\vec{\sigma} \cdot \vec{k}^*}{(E^* + M^*)} \end{pmatrix} \chi \qquad (31)$$

which looks like a free Dirac spinor but with effective mass, energy and momentum. The asterisk * should not be confused with complex conjugation but indicates that we are dealing with an effective (dressed) spinor. The spinors $u^*(\vec{k})$ in (31) furthermore are normalised as $\bar{u}^* u^* = 1$.

As explained before, in the Brueckner approximation one neglects the second term $\Sigma_R^{\pm}$ in $\Sigma^{\pm}$ (see eq. (23)). When evaluating the self-energy $\Sigma^{\pm}(k)$ one has to perform a 4-dimensional integral over the full Green's function $G^<(k)$. This Green's function has the same form as the unperturbed one (eq. (24)) but with effective quantities M^*, k^* (eqs. (27)).

The Green's function $G^<(k)$ in the expression for the Brueckner self-energy eq. (23) contains thus two parts: G_F and G_D (see eq. (24)). The part G_F describes the real nucleons in the Fermi sea and is density dependent. The antinucleon-contributions (Dirac sea) contained in G_D usually give rise to divergent integrals and therefore a renormalisation procedure is needed to render the expressions finite. In the free sector it is usually assumed that the effect of the Dirac-sea (vacuum polarisation) is absorbed into the physical masses and coupling constants. However in a many-body system this G_D also changes (due to real interactions) and therefore we have an additional effect because of the difference between G_D (vacuum) and G_D (medium). We will assume in the following that this effect depends smoothly on the density and is small. We therefore leave it out and replace $G^<(k)$ by $G_F(k)$ in the appropriate expressions. Further discussion on this assumption can be found in ref. 1, 2 and 12.

As a second remark concerning the calculation of $\Sigma^{\pm}(k)$ we point out that the occurence of the $\delta(p_0^* - E_p^*)$ factor in the expression of $G^<$ implies that we have taken the quasi-particle approximation which is valid if $\mathrm{Im}\,\Sigma^+$ can be neglected.

We now proceed to the calculation of the (retarded , advanced) T-matrix $T_{12}^{\pm}$. First of all we discuss the vacuum (zero density) case where $\Sigma=0$ and all G are replaced by G^0. In 4-momentum space (($P=p_1+p_2$, $k=\frac{p_1-p_2}{2}$) suppressing other variables like spin, isospin and leaving out the ($\pm$) indices):

$$\langle p_1' p_2'|T_{12}|p_1 p_2\rangle = \langle p_1' p_2'|V_{12}|p_1 p_2\rangle + \frac{i}{(2\pi)^4}\int d^4k\ \langle p_1' p_2'|V_{12}|\tfrac{1}{2}P + k, \tfrac{1}{2}P-k\rangle\ \times$$

$$G^0(\tfrac{1}{2}P+k)\ G^0(\tfrac{1}{2}P-k)\ \langle \tfrac{1}{2}P+k, \tfrac{1}{2}P-k|T_{12}|p_1 p_2\rangle \tag{32}$$

The four-dimensional integral equation (32) is of a tremendous complexity and therefore one reduces it to a relativistic three-dimensional equation. Denoting by $\tilde{g}$ some approximate expression for the two-nucleon propagator $G^0 G^0$ in (32), this equation can be replaced by the equivalent set of equations:

$$T_{12} = W + W\tilde{g}\ T_{12}$$
$$W = V_{12} + V_{12}(G^0 G^0 - \tilde{g})\ W \tag{33}$$

The approximate $\tilde{g}$ should also satisfy relativistic elastic unitarity (as was the case for $G^0 G^0$) implying that the resulting T_{12} fullfils relativistic unitarity whatever further approximation one introduces for W. Now $\tilde{g}$ is chosen as a three-dimensional approximation (fixing the zeroth-components of the external momenta in (32)) for which the operator $(G^0 G^0-\tilde{g})$ is sufficiently small to permit an iterative solution for W. In the two-particle center-of-mass frame $\vec{P}=0$ we can take for instance the mass-shell conditions (for a discussion see ref. 4):

$$p_1^0 = p_2^0 = \tfrac{1}{2}\sqrt{s} \qquad s = (p_1 + p_2)^2 \tag{34}$$

The condition (34) puts both particles equally off-mass shell and therefore contains no effects of meson-retardation. The resulting three-dimensional $\tilde{g}$ takes the form (in the two-particle center-of-mass frame):

$$\tilde{g}(\vec{k}, s) = -i\int dk_0\ G^0(\tfrac{1}{2}P+k)\ G^0(\tfrac{1}{2}P-k)$$
$$P = (\sqrt{s}, \vec{0}) \qquad k = (k_0, \vec{k}) \tag{35}$$

where the k_0-interaction can be performed explicitly since the condition (34), removes the other zeroth-component dependencies in (32). We thus obtain the so-called Thompson equation:

$$\langle \vec{p}_1' \vec{p}_2' | T(s) | \vec{p}_1 \vec{p}_2 \rangle = \langle \vec{p}_1' \vec{p}_2' | V(s) | \vec{p}_1 \vec{p}_2 \rangle$$

(36)

$$+ \frac{1}{(2\pi)^3} \int d\vec{k} \; \langle \vec{p}_1' \vec{p}_2' | V(s) | \tfrac{1}{2}\vec{P}+\vec{k}, \; \tfrac{1}{2}\vec{P}-\vec{k} \rangle \; \tilde{g}(\vec{k},s) \; \langle \tfrac{1}{2}\vec{P}+\vec{k}, \; \tfrac{1}{2}\vec{P}-\vec{k} | T | \vec{p}_1 \vec{p}_2 \rangle$$

where V acts as a "quasi-potential" which is equal to a constrained form of V_{12}:

$$\langle \vec{p}_1' \vec{p}_2' | V(s) | \vec{p}_1 \vec{p}_2 \rangle = \langle p_1^{\,0} \vec{p}_1' \; p_2^{\,0} \vec{p}_2' | V_{12} | p_1^{\,0} \vec{p}_1 \; p_2^{\,0} \vec{p}_2 \rangle$$

(37)

with p_1^0, p_2^0, $p_1'^{\,0}$, $p_2'^{\,0}$ fixed as in (34). We have taken W to first order i.e. $W \approx V_{12}$ in eq. (33)

The expression (35) can be evaluated analytically. Neglecting the negative-energy part leads then to:

$$\tilde{g}(\vec{k},s) = \Lambda^+(\vec{k}) \; \Lambda^+(-\vec{k}) \; \frac{M^2}{E_k^2} \; \frac{\pi}{(\tfrac{1}{2}\sqrt{s} - E_k + i\varepsilon)}$$

(38)

This expression for $\tilde{g}(\vec{k},s)$ corresponds to the Thompson approximation[4]. The neglect of the negative-energy part in G^0 is based on the following observations: first of all the use of a pseudo-vector coupling of the pion (instead of pseudo-scalar) reduces strongly the effect of negative energy states. Including them or excluding them leads to a change of less than 10% in the coupling constants when fitted to nucleon-nucleon observables. Their effect is thus taken into account effectively by the coupling constants.

The (NN-interaction) potential V_{12} in (37) is now given as a sum of one-boson-exchange contributions for different mesons j (π, ρ, σ, ω, η, δ). In the two-nucleon center-of-mass frame we have:

$$\langle \vec{p}, \; \sigma_1, \; \sigma_2 | V | \vec{p}', \; \sigma_1', \; \sigma_2' \rangle =$$

$$\sum_j \left[\bar{u}(\vec{p}',\sigma_1') \; \Gamma^j \; u(\vec{p},\sigma_1) \right] \left[\bar{u}(-\vec{p}',\sigma_2') \; \Gamma^j \; u(-\vec{p},\sigma_2) \right] \; g_j(\vec{p}-\vec{p}')$$

(39)

where $u(\vec{p},\sigma_1)$ represents the (positive-energy) nucleon spinor and the $g_j(\vec{q})$ denote the different meson propagators. Each vertex Γ_j is multiplied with a formfactor (1).

The Thompson equation (36) is then solved with the OBE-potential (39) as input. Its parameters (coupling constants, see table 1) and cut-off mass Λ (eq. (1)) are determined from a fit to nucleon-nucleon observables (deuteron properties, phase shifts, cross sections, polarisations) in an energy range (0-300 MeV). Including explicitly Δ-degrees of freedom one can even reproduce fairly well elastic and inelastic scattering data (including pion production) in the energy range 0-1000 MeV. For more details on all of this we refer to ref. 2 where also additional references can be found.

Now we discuss the solution of the $T_{12}^{\pm}$-matrix equation in nuclear matter. In the case of a nuclear medium the nucleon states are now dressed particle states. The coupled set of equations (23) is represented in fig. 2. These are now given in the Brueckner approximation as discussed at the end of the previous section. The four-dimensional T_{12}-equation is tedious to solve, and is therefore also reduced to a covariant three-dimensional form. We shall adopt the same procedure as outlined before in the free case, and thus also neglect the contributions from the dressed (effective) negative-energy states. Since $\Lambda^{+}(p^{*})$ (the dressed positive-energy projection) contains both $\Lambda^{+}(p)$ and $\Lambda^{-}(p)$ (the undressed projections), one still has a contribution from bare negative-energy states.

In Brueckner theory, in contrast to the T-matrix for the free nucleon-nucleon scattering, in the evaluation of T (usually called the G-matrix) only those intermediate states with both particles lying outside the Fermi sea are summed over (Pauli blocking). This is ordinarily accomplished with the use of an angle averaged Pauli blocking operator $\bar{Q}$. As explained before, hole-hole scattering is neglected. In the relativistic form used here the Fermi surface in the two-particle center of mass frame assumes an ellipsoidal form. Furthermore because of the relativistic kinematics, $\bar{Q}$ depends on the momentum of the intermediate state, as well as on both the total energy s^{*} and the total momentum P^{*} of the two incoming particles in the nuclear matter rest frame. See refs. 12 and 13 for an explicit expression.

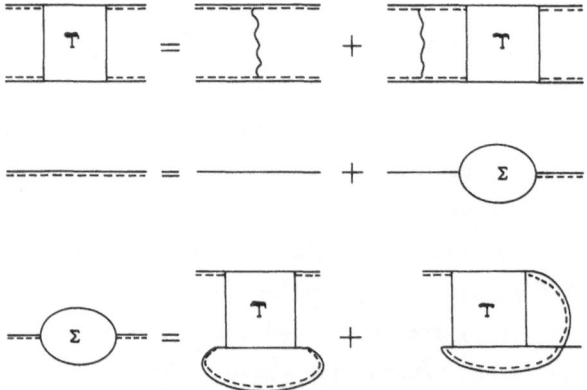

Fig. 2. Diagrammatic representation of the Dirac-Brueckner approach.

The nuclear matter Thompson equation for T in the two-particle c.m. frame can now be written as:

$$\langle \vec{p}'' \sigma''_{12} | T | \vec{p} \sigma_{12} \rangle = \langle \vec{p}'' \sigma''_{12} | V | \vec{p} \sigma_{12} \rangle + \sum_{\sigma'_{12}} \int \frac{d\vec{p}'}{(2\pi)^3} \langle \vec{p}'' \sigma''_{12} | V | \vec{p}' \sigma'_{12} \rangle \times$$

$$\times \frac{M^{*2} \bar{Q}(\vec{p}', \vec{P}, s^*)}{2E_{p'}^{*2}(E(p)-E(p')+i\varepsilon)} \langle \vec{p}' \sigma'_{12} | T | \vec{p} \sigma_{12} \rangle \tag{40}$$

where σ_{12} denotes the spin directions of nucleon 1 and 2, and $E(p)$ is the single-particle energy (28).

In order to calculate the self-energy $\Sigma^{\pm}$ (eq. 23), the interaction T has to be transformed from the two-particle c.m. frame where it is actually calculated to the nuclear matter rest frame. This can be achieved by projecting the on-shell values of Γ on five Lorentz invariant interacting matrices[12,13]

$$T = \sum_{\alpha} \Gamma^{\alpha} f^{\alpha}_{(1)} \cdot f^{\alpha}_{(2)}$$

with: $\tag{41}$

$$f^{\alpha}_{(i)} = \left\{ 1, \gamma^{\mu}_{(i)}, \sigma^{\mu\nu}_{(i)}, \gamma_5 \gamma^{\mu}_{(i)}, \gamma_5 (\gamma q_{(i)}) \right\} \qquad (i = 1,2)$$

for the scalar, vector, tensor, axial-vector and pseudo-vector coupling. We use a pseudo-vector interaction instead of pseudo-scalar, in agreement with our choice for the one-pion-exchange coupling.

The (retarded, advanced) self-energy is then given by:

$$\Sigma(k) = \int_0^{k_F} \frac{d^3 q}{(2\pi)^3} \frac{1}{2E_q^*} f^{(\alpha)}_{(1)} \left\{ Tr\left[(\gamma q^* + M^*) f^{\alpha}_{(2)} \right] \Gamma^{\alpha}_{direct} - \right.$$

$$\tag{42}$$

$$\left. (\gamma q^* + M^*) f^{\alpha}_{(2)} \Gamma^{\alpha}_{exchange} \right\}$$

It is clear that Γ must be separated into direct and exchange contributions and we shall do this from the outset. Explicit expressions for Σ_s, Σ_0 and Σ_v as obtained from eq. (42) can be found in ref. 13.

In ref. 13 we treat the inclusion of Δ-degrees of freedom. Here, one first constructs a N-N, N-Δ interaction which is then fixed by solving a set of coupled Thompson equations for the different N-N and N-Δ channels. The Δ-width is treated microscopically. This procedure is then extended for nuclear matter where both nucleon and Δ are dressed through the solution of Dyson eqs. with appropriate self-energies.

3. DIRAC-BRUECKNER RESULTS FOR NUCLEAR MATTER

3.1. The energy density and saturation properties

The total energy of an interacting system is obtained in standard fashion[7] from the energy-momentum tensor $T_{\mu\nu}$:

$$T_{\mu\nu} = -\mathscr{L}g_{\mu\nu} + \frac{\partial\mathscr{L}}{\partial(\partial_\mu\psi)}\,\partial_\nu\psi + \frac{\partial\mathscr{L}}{\partial(\partial_\mu\phi)}\,\partial_\nu\phi \tag{43}$$

where $\mathscr{L}$ is the total Lagrangian density. The energy density (total energy per particle) is defined as:

$$\varepsilon = <|T_{00}|> \tag{44}$$

where the average is over the interacting ground state. The energy-momentum tensor T_{00} has contributions from the nucleons as well as from the mesons. It can be expressed in Green's functions[5] and using the equations (17) and (18) for the Green's function the following result is obtained:

$$\varepsilon = -i\,\underset{(1)}{\text{Tr}}\left\{(i\vec{\gamma}\cdot\vec{\partial}_1 + M)\,G_1^c(t_1,t_1')\right.$$

$$\tag{45}$$

$$\left. - \frac{i}{4}\,\underset{(2)}{\text{Tr}}\,V_{12}\,[G_{12}^c(t_1 t_1|t_1' t_1) + G_{12}^c(t_1 t_1'|t_1' t_1')]\right\} \quad t_1'=t_1+\varepsilon \quad \varepsilon\to0^+$$

with V_{12} the OBE-interaction (39).

Substituting our approximation for the two-nucleon Green's function G_{12} (eq. (20,21)) and neglecting negative-energy states ($G^< = G_F$, $G_D=0$) as we have done before one gets in the Brueckner approximation:

$$\varepsilon = \underset{(1)}{(-i)\text{Tr}\,(i\vec{\gamma}\cdot\vec{\partial} + M)\,G_1^<} + \underset{(2)}{\tfrac{1}{2}(-i)^2\text{Tr Re }T_{12}^+ G_1^< G_2^<}$$

$$\tag{46}$$

$$= \underset{\alpha}{\Sigma}\int_{|\vec{k}|<k_F}d\vec{k}\,\frac{M^*}{E_k^*}\,\bar{u}^*(\vec{k},\alpha)\,(\vec{\gamma}\cdot\vec{k} + M + \tfrac{1}{2}\Sigma(k))u^*(\vec{k},\alpha)$$

with $\Sigma(k)$ defined in (42). The Dirac spinors are covariantly normalised: $\bar{u}(\vec{k},\alpha)u(\vec{k},\alpha) = \bar{u}^*(\vec{k},\alpha)u^*(\vec{k},\alpha) = 1$.

From eq. (46) one can calculate the binding energy per nucleon $\varepsilon=E/A$ for symmetric nuclear ($Z/A = \tfrac{1}{2}$) matter. In the case of asymmetric nuclear matter ($N\neq Z$) the calculation is more complex. There appear two distinct Fermi seas and the neutron and proton will have different effective masses and one has to calculate the p-p, n-n and the n-p effective interaction T separately. Since n and p have different effective masses and they induce different Pauli blocking ($\bar{Q}$ in eq. (40)) this leads in the n-p channel to a difference between direct and exchange[16]. For asymmetric nuclear matter one also calculates the symmetry energy a_4. Another interesting quantity is the compressibility K which is defined as:

$$K = k_F^2 \frac{\partial}{\partial k_F^2} \varepsilon(k_F)|_{saturation} \tag{47}$$

and reflects the curvature at saturation (the minimum of ε). We have also performed calculations[13] for temperature T≠0 which we will not discuss here.

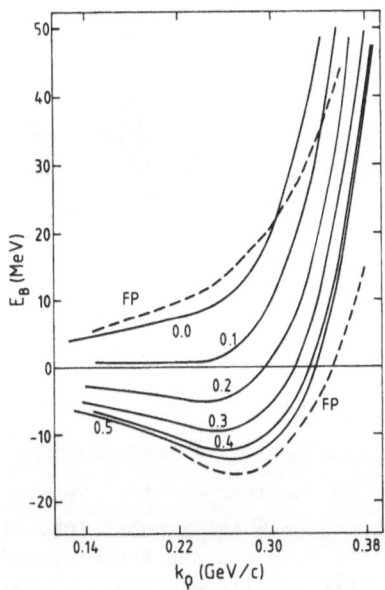

Fig. 3. Energy/nucleon in an asymmetric nuclear matter versus the "averaged" Fermi momentum k_ρ, for different Z/A ratios. The upper dashed line gives the neutron matter result (Z/A=0) of Friedman and Pandharipande[14]. The lower dashed curve the symmetric nuclear matter results (Z/A=0.5) of the same authors.

In fig. 3 we display for different Z/A ratios our results for ε as a function of density ρ through an average Fermi momentum k_ρ defined by

$$k_\rho = (\frac{3}{2} \pi^2 \rho)^{1/3} \tag{48}$$

with ρ the total baryon density. Also plotted are the symmetric nuclear matter (Z/A=0.5) and neutron matter (Z/A=0) results of Friedman and Pandharipande[14]. The latter are obtained through a non-relativistic variational calculation which includes a phenomenological three-body force adjusted to give the correct saturation. We find a remarkable equivalence between the two results. The Dirac-Brueckner calculation reproduces correctly saturation which one was never able to obtain based on a non-relativistic Brueckner calculation without an adjustable 3-body force. In fact taking the non-relativistic limit of our relativistic DB-calculation (i.e. repeating it but leaving out the small components in the nucleon spinors and modifying the two-body intermediate propagator) we obtain saturation very much at the same place where all other non-relativistic calculations lie (fig. 4), that is to say, the Coester line[13]. The reason for the difference is discussed in refs. 13 and 15.

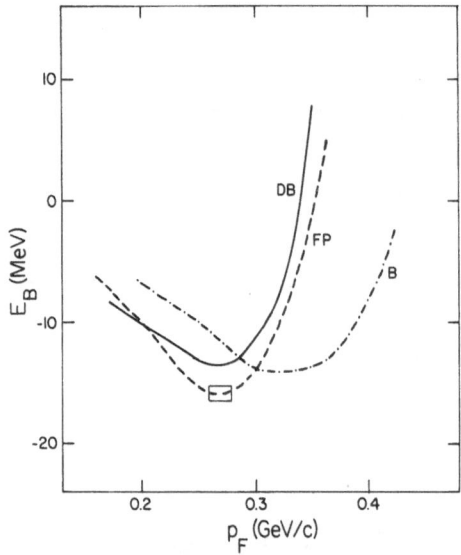

Fig. 4. Energy/nucleon in nuclear matter versus Fermi momentum p_F. Comparison of the Dirac-Brueckner result (DB, full line) with the variational calculation of Friedman and Pandharipande[14] (FP, dashed line) and a conventional non-relativistic Brueckner calculation (B, dash-dotted line).

The compressibility obtained is K=250 MeV (Z/A=0.5), K=240 MeV (Z/A=0.4), K=210 MeV (Z/A=0.3) and K=100 MeV (Z/A=0.2). The asymmetry energy obtained is a_4=32.3 MeV which is a very good value. We remark that this value is obtained based on the total energy per particle at (Z/A=0.5, 0.4, 0.3) evaluated at the different saturation points. This procedure is usually not followed and one extrapolates between Z/A=0.5 and Z/A=0. at one fixed saturation point corresponding to Z/A=0.5 (Z/A=0. has no minimum). We believe this method to be inaccurate. In ref. 16 we quoted a value of 26 MeV for a_4 based on the latter procedure, which is different from the correct value of 32.3 MeV mentioned above.

The value obtained for the compressibility K (K=250 MeV) in the case of symmetric nuclear matter is well in agreement with "experimental" values. On the other hand these are criticised by Brown[17] who also discussed modifications to be included in conventional approaches (like the one we discuss here) in order to bring the value of K down to much lower values. The "experimental" values for K are obtained[19] based on an analysis of the giant monopole resonance i.e. finite nuclei. There are valid reasons to believe that K(∞) (infinite nuclear matter) is different from K(A), the bulk compressibility of finite nuclei, where we already substracted the surface contribution. For instance[17], if we take the Landau definition for the compression modulus K (which is generally valid):

$$K = 6 \frac{k_F^2}{2M^*} (1+f_0) \qquad (49)$$

we see that it scales inversely with the effective mass M^*. Now the effective mass is energy dependent and for the giant monopole should be evaluated at $\omega=14$ MeV instead of $\omega=0$ (Fermi surface). Also M^* for finite nuclei is different as compared to infinite nuclear matter[18]. Therefore one can easily get a value for K(∞) which is different from K(A). Brown argues that K(∞)=(2/3)K(A) or if K(A)=230 MeV ("experimental" value) then K(∞)=150 MeV or even lower. This is an interesting issue since lowering K means softening the equation of state compared to what we find. This is not excluded since after all we have only included so-called ladder diagrams in the calculation of ε and at high densities other diagrams (RPA-type) become dominant. On the other hand the non-relativistic calculations of Friedman and Pandharipande[14] are based on a more general approach than the Brueckner prescription and although softer than our results, they obtain a similar compression modulus of K=250 MeV. Much more work has to be done in this area.

Inclusion of Δ-degrees of freedom shifts the saturation point to lower densities and higher saturation energy and leads to a smaller compression modules[13].

3.2. The single-particle energy in nuclear matter

The single-particle energy E(k) is given in eq. (28) and corresponds to the energy-momentum relation for a nucleon with momentum k in a nuclear medium

$$E(k) = \{\vec{k}^2(1+\Sigma^v(k))^2 + (M+\Sigma^s(k))^2\}^{1/2} - \Sigma^0(k) \qquad (50)$$

Hereby it is assumed that the imaginary parts of the self-energy Σ are small such that the quasi-particle concept is valid. If the momentum $|\vec{k}|$ is greater than the Fermi momentum we can deduce from (50) evaluated at saturation density ρ_0 the properties of the nucleon–nucleus optical potential (depth of the central part).

The equivalent nonrelativistic central potential U is obtained by putting $\Sigma_v = 0$ in (50) and writing $E(k) = E = k^2/2M + U + M$. (The actual calculations of Σ_s, Σ_0 and Σ_v show indeed that Σ_v is very small.) One obtains

$$U(E) = \left[\Sigma_s(k) - \frac{E}{M}\Sigma_0(k)\right] + \frac{\Sigma_s^2(k) - \Sigma_0^2(k)}{2M} \tag{51}$$

This result was first obtained by Jaminon, Mahaux and Rochus[20] through a different procedure which recast the Dirac equation into an apparent Schrödinger form through elimination of the small components. The first term between brackets shows that at low energies the single particle energy is simply the difference of $\Sigma_s(k)$ and $\Sigma_0(k)$. These self–energy components are both in the order of 200–300 MeV[13,28] but their difference at E=0 amounts to ~50 MeV binding which corresponds exactly to the observed central depth of the nucleon–nucleus optical potential (see fig. 5). At higher energies and in contrast to non–relativistic Brueckner

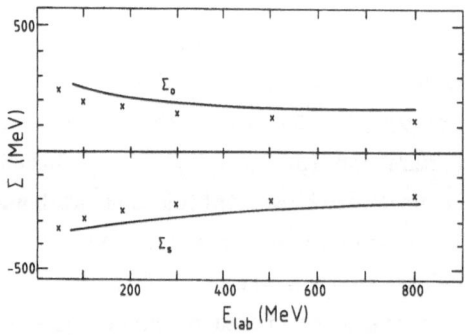

Fig. 5. Comparison of the real parts of scalar and vector fields as calculated by Tjon and Wallace[34] in full relativistic impulse approximation and by ter Haar and Malfliet[28] in the Dirac–Brueckner approach. Both calculations include NΔ–channel.

calculations U(E) has the correct E–dependence[21] i.e. U(E)=0 at E≈300 MeV (see fig. 6). I also would like to mention an interesting observation made by G. Brown[22] concerning the importance of the second term at the RHS of eq. (51). This term, typically a relativistic correction originating from so–called pair–terms, will dominate U(E) in that range of energies E where the first term (between brackets) is roughly zero. This occurs at E≈500 MeV, an energy where the so–called Dirac phenomenology is very successful. In this approach the scalar and vector self–energies Σ_s and Σ_0 are parametrised (Fermi shape) and the parameters

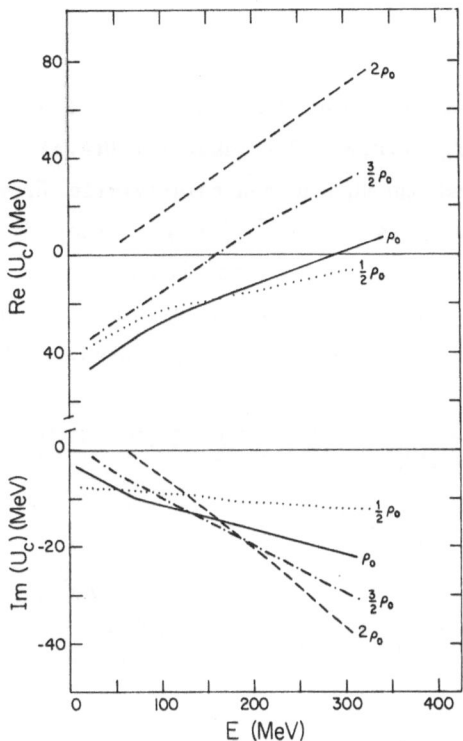

Fig. 6. The Schroedinger-equivalent optical potential[13] at different nuclear matter densities within the Dirac-Brueckner approach.

are obtained by fitting nucleon-nucleus data (elastic cross sections and polarisation data). A major component in the success of Dirac phenomenology is the fact that the experimentally determined strength of the spin-orbit force should be larger than non-relativistic calculations predict, and this is what exactly happens in relativistic approaches[2,23].

Another important result is the behaviour of the mean field or single-particle energy (50) as a function of density (see fig. 6). With increasing density and at fixed momentum E(k) becomes more and more repulsive, very much in the way as the total energy per particle ε, which corresponds more or less to a momentum-integrated single-particle energy. This has led to the identification of repulsive effects in heavy-ion observables like flow and transverse momentum distributions with a stiff equation of state[24]. The nuclear mean field is chosen to be only density dependent and in order to explain the observed trends in the data one needs a fairly repulsive behaviour which expressed in terms of the compression modulus K corresponds to K=380 MeV, an unrealistic high value. However, as we have seen, at a fixed density the momentum (or energy) dependence of the single-particle energy itself turns more and more repulsive with increasing momentum. The relativistic "definition" of a mean potential energy $U(k,\rho)$ can be given as

$$U(k,\rho) \equiv E(k) - [m_N^2 + k^2]^{\frac{1}{2}} \tag{52}$$

where we substracted the kinetic energy of a free particle from the single particle energy $E(k,\rho)$. The resulting $U(k,\rho)$ at $\rho=\rho_0$ are similar to the ones obtained through a non-relativistic Brueckner calculation except at higher energies. However for $\rho \gtrsim \rho_0$ they differ substantially since the non-relativistic one does not reproduce the correct saturation behaviour. This is one of the great virtues of the Dirac-Brueckner approach where saturation is correctly reproduced. In fig. 7 we display

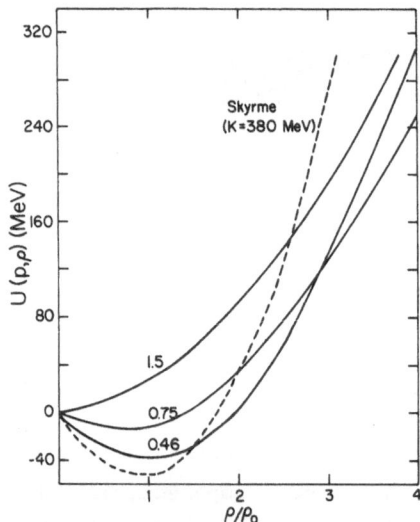

Fig. 7. The mean potential energy $U(k,\rho)$ for different densities ρ and single particle momenta p (0.46, 0.75, 1.5 GeV/c). The dashed line is the "stiff" Skyrme mean potential energy (53).

$U(k,\rho)$ for three different values of the nucleon momentum k (0.46, 0.75, 1.5 GeV/c) as a function of ρ/ρ_0 where ρ_0 is the nuclear saturation density. The dotted curve represents a static "Skyrme" parametrization:

$$U_{SK}(\rho) = A\rho + B\rho^2 \tag{53}$$

which, through fitting A and B, corresponds to a single-particle energy density $\varepsilon(\rho) = \varepsilon_F(\rho) + U_{SK}(\rho)$ where $\varepsilon_F(\rho)$ is the Fermi energy and which gives correct saturation properties ($\varepsilon(\rho) = -16$ MeV at $\rho = \rho_0$) and has a compression modulus of K = 380 MeV. This particular potential has been used extensively in VUU-calculations. It is clear from our comparison in fig. 7 that the momentum dependence which is absent in (53) (since it is averaged over a Fermi momentum distribution) is very important. In fact at k = 1.5 GeV/c and $\rho = \rho_0$ we have as much repulsion as the "Skyrme" static mean field at $\rho = 2\rho_0$. This observation was made already some time ago[25] and its implications have been substantiated in recent calculations using a modified VUU (see the lectures of H. Stöcker). Consequently a soft equation of state including a momentum dependent mean field can be mocked up by a hard equation of state without momentum dependence.

The fact that the single-particle energy depends on two variables (momentum and density) results in a possible ambiguity concerning the origin of the "replusive" behaviour of the aforementioned flow observables. As we will see in the next section, where we discuss effective (medium corrected) cross sections, there are other possible factors which come into play. All these aspects of the dynamics in a nuclear medium become apparent if one considers the microsopic origin of the kinetic equations (BUU-VUU-LVE or whatever) used to describe non-equilibrium phenomena, as in nucleus-nucleus collisions[26,31] (see also the lectures by C. Grégoire). This will be discussed briefly in the next section.

3.3. Effective nucleon-nucleon cross sections in nuclear medium: Relativistic quantum kinetic equation for nucleus-nucleus collisions

There are clear indications, that the nucleon mean free path, or related to this, the total nucleon cross section in a nuclear medium, is density dependent. We shall use the concept of effective cross section for this quantity. The effective cross section can be calculated either from the imaginary part of the nucleon-self-energy or directly by evaluating the Dirac-Brueckner G-matrix (eq. (40)) and constructing $|G|^2$ with appropriate phase-space factors. We will do this for the scattering of an "external" nucleon (momentum p) with a nucleon (momentum p_2) in a nuclear medium (averaging p_2 over the Fermi sea).

To be specific, the corresponding relativistic kinetic equation which describes these non-equilibrium processes has the form[31]:

$$(\partial_T + \nabla_p E(p) \cdot \nabla_R - \nabla_R E(p) \cdot \nabla_p) f(\vec{p}, \vec{R}, T)$$

$$= \iiint \frac{d\vec{p}_2 \, d\vec{p}_1' \, d\vec{p}_2'}{(2\pi)^9} \frac{M_p^* M_{p_2}^* M_{p_1'}^* M_{p_2'}^*}{E_p^* E_{p_2}^* E_{p_1'}^* E_{p_2'}^*} \tag{54}$$

$$\times (2\pi)^4 \delta^3(\vec{p} + \vec{p}_2 - \vec{p}_1' - \vec{p}_2') \, \delta(E(p) + E(p_2) - E(p_1') - E(p_2'))$$

$$\times \ \langle pp_2|T_{12}^{(+)}|p_1'p_2'\rangle\langle p_1'p_2'|T_{12}^{(-)}|pp_2\rangle \left\{ f_1'f_2'(1-f)(1-f_2) - (1-f_1')(1-f_2') \ ff_2 \right\}$$

Here $f(\vec{p},\vec{R},T)$ represents the one-nucleon distribution function (denoted also as f) which is governed by eq. (54). To obtain the result (54) it was necessary to make the quasi-particle approximation, i.e. restricting the 4-th component p_0^* by $\theta(p_0^*)\delta(p^{*2}-M_p^{*2})$. We will discuss the validity of this particular approximation later. In (54) E(p) denotes the single-particle energy (28) and M_p^*, E_p^* are defined in (27) and (29). In the kinetic eq. (54) we assume a spin-saturated system for which the spin-degrees of freedom can be averaged out. The $T_{12}^{(\pm)}$-matrices are the Dirac-Brueckner (G-matrix) solutions from eq. (40).

The result (54) is obtained using the formalism explained in section 2. There, we discussed the Green's function formalism in a time-dependent fashion which allows us to construct kinetic equations. The important Green's functions are $G^<$ and $G^>$ which can be expressed in terms of a spectral function and a distribution function f (see f.e. the expression (24) in the non-interacting case). The spectral function becomes a δ-function in the quasi-particle approximation. This is explained elsewhere[31] in great detail, but we like to stress here that once one has decided for a definite prescription in the equilibrium case it is straightforward to construct the non-equilibrium counterpart. We have done so for the Dirac-Brueckner prescription. If one on the other hand prefers the mean field or Hartree approximation (like in the σ-ω model) then the corresponding kinetic equation is purely Vlasov-type i.e. it has no collision term (which is always of second order or higher in the interaction).

The kinetic eq. (54) has a form similar to the VUU-BUU-LVE-type equations used in numerical treatments of nucleus-nucleus collisions. However it differs in a number of non-trivial points. In the previous section we have already discussed the mean field, which is identified as the single-particle energy E(p) and which is both momentum- and density- (and thus $\vec{R}$- and T-dependent) dependent. We consider now the collision term. Here we see the Pauli-blocking through the appearence of (1-f)(1-f)-terms. The T-matrices are not the free nucleon-nucleon T-matrices but correspond to the Dirac-Brueckner G-matrices which also determine the mean field E(p). Therefore one sees a close connection and self-consistency between mean field and collision term.

The relation between the effective differential cross section $d\sigma_{eff}$ and the collision probability can now be found easily[7,27] from inspection of eq. (54) and writing it in the form of a Boltzmann-type equation. We obtain the following total averaged effective cross section:

$$\sigma_{\bullet ff}(p) = \frac{3}{4\pi^2 k_F^3} \int_0^{k_F} d\vec{p}_2 \; \bar{Q}(P,s^*,\vec{q}) \frac{M^{*4}(\hbar c)^2}{2(2\pi)^2 s^*} \int d\Omega_{CM} \sum_{\substack{spin \\ isospin}} |T(\vec{P},s^*,\vec{q})|^2 \qquad (55)$$

$$\vec{P} = \vec{p} + \vec{p}_2 \qquad \vec{q} = \vec{p} - \vec{p}_2$$

$$s^* = (E^*(p) + E^*(p_2))^2 - \vec{P}^2$$

The symbol Σ represents summation (average) of outgoing (incoming) spin and isospin. The function $\bar{Q}$ gives the angle-averaged Pauli-blocking operator (corresponding to $(1-f)(1-f)$ in (54)). The momentum p_2 is averaged over the Fermi sea and k_F denotes the Fermi momentum. In the expression (55) we recognise three aspects of the influence of the nuclear medium. First of all the effect of Pauli-blocking, secondly a kinematical phase-space factor $R=(M^*)^4/s^*$ which, for small momenta p, scales as $(M^*)^2$, and thirdly the intrinsic medium dependence of the Dirac-Brueckner G-matrix itself $T(\vec{P},s^*,\vec{q})$. The effective mass M^* is taken as a constant here since it depends only weakly on the momentum. If we compare $\sigma_{\bullet ff}(p)$ with a similar expression used in VUU-calculations for which one takes the free nucleon-nucleon cross section, then the effects of $\bar{Q}$ are normally taken into account by a Pauli-blocking prescription but the other medium effects due to the kinematical factor and the Dirac-Brueckner t-matrix are usually neglected. The phase-space factor R introduces a global reduction of the effective cross section as compared to the free one of the order of $(M^*/M)^2$. It takes on the values 0.6 $(\rho=\frac{1}{2}\rho_0)$, 0.4 $(\rho=\rho_0)$, 0.2 $(\rho=2\rho_0)$ but at higher momenta p and increasing densities where the correct expression for R should be used it is even smaller. While at low bombarding energies this reduction factor determines more or less the difference between free and effective cross sections (disregarding the Pauli blocking) this no longer holds at high energies and at high densities. In table 2 we present a selection of results from ref. 28. (Non-relativistic Brueckner results are discussed in ref. 33). $(M^*/M)^2$ is the global kinematical reduction factor discussed before and $\sigma_{\bullet ff}/\sigma$ the ratio of the effective- versus free-total average cross section as calculated from (55) with $\bar{Q}$ omitted. (Including $\bar{Q}$ does not change the results very much[28]). Clearly at high energies where $\sigma_{\bullet ff}/\sigma$ becomes larger than $(M^*/M)^2$, the intrinsic Dirac-Brueckner G-matrix is different (in absolute magnitude larger) from the free one. This demonstrates the importance of including correctly all the medium

Table 2. Effective elastic nucleon-nucleon cross sections and single-particle energies E(p).

ρ/ρ_0	p (GeV/c)	E(p) (MeV)	$(M^*/M)^2$	σ_{eff}/σ
0.5	0.31	7.3	0.6	0.8
	0.46	81.2		0.6
	0.65	188.		0.6
	0.85	320.		0.8
	1.25	628.		0.9
	1.65	973.		0.9
1.	0.31	−6.1	0.4	0.3
	0.46	68.7		0.4
	0.65	185.		0.5
	0.85	325.		0.7
	1.25	640.		0.9
	1.65	994.		0.9
2.	0.34	50.	0.2	0.2
	0.46	108.		0.3
	0.65	225.		0.6
	0.85	365.		0.8
	1.25	696.		0.8
	1.65	1060.		0.8

effects in the kinetic equation consistently. To our knowledge, up to now, there exists no actual calculation which has done so properly, that is to say within a theoretical model (Dirac-Brueckner in our case). To emphasize this point furher, exploratory Dirac-Brueckner calculations which incorporate also pion-polarisation (section 2.2) show a tremendous increase in effective cross section (see ref. 29 and the contribution of F. de Jong et al. at this school). This seems connected with the so-called "pisobar" dynamics advocated by G. Brown[17].

We conclude this section by discussing the validity of the quasi-particle approximation which was used in all our calculations and which also underlies the derivation of the kinetic equation (54). Here the single-particle energy E(p) is taken for p_0 and E(p) is evaluated using the real parts of Σ_s, Σ_0 only. This approximation, leading to a non-decaying quasi-particle, is valid only if the imaginary parts of Σ_s and Σ_0 can be assumed to be small as compared to the real parts. The imaginary parts can be related to the cross section, i.e. from mean free path-arguments one deduces $\sigma_{eff} = -Im\Sigma/(\rho v)$, with ρ the density and v the velocity (the precise meaning of $Im\Sigma$ in terms of Σ_s and Σ_0 can be found in ref. 31, eq. (38)). Therefore if the effective cross section σ_{eff}

increases drastically (as we mentioned before in the case of Dirac-Brueckner including pion-polarisation) this implies the break down of the quasi-particle approximation. This invalidates the kinetic eq. (54) and makes it necessary to reconsider its derivation. This point was first stressed by P. Danielewicz[30].

Finally, we have also performed a fully self-consistent calculation including Δ-degrees of freedom (which are also medium-corrected) for the effective elastic and inelastic cross sections. This is discussed at length in refs. 13 and 28. We only recall from these results that the in-medium pion-production cross section $N+N \rightarrow N+N+\pi$ is quenched as compared to the free one. Here again it is demonstrated that medium effects are neither trivial nor unimportant and should be studied further.

4. CONCLUSIONS

We discussed the relativistic many-body problem with emphasis on the Dirac-Brueckner approximation. A unified treatment of equilibrium and non-equilibrium can be achieved based on the time-dependent Green's function formalism in the Keldysh representation (section 2).

The results obtained within this approach are quantitatively in agreement with a large number of experimental equilibrium observables: saturation properties, for symmetric and asymmetric nuclear matter, optical potential properties and compressibility of nuclear matter (section 3). On the other hand many open problems remain to be understood. Amongst others we mention the role of the formfactor, the neglect of negative energy states and the question of convergence of the Brueckner approximation. Also the problem of a proper treatment of non-equilibrium phenomena as they appear in nucleus-nucleus collisions is presently not satisfactory. The hope we have however is that the formalisms to treat these intriguing problems in nuclear physics are available. They can and will be used to extract microsopic information on the effective interaction of hadrons in a dense nuclear medium and lead to a fundamental understanding of the properties of strongly interacting many-hadron systems.

REFERENCES

1. B.D. Serot and J.D. Walecka, Advances in Nuclear Physics, eds. J.W. Negele and E. Vogt, Vol. 16, 1 (1986).
2. R. Malfliet, Progress in Particle and Nuclear Physics, Vol. 21 (1988) 207.

3. R. Machleidt, K. Holinde and Ch. Elster, Phys. Rep. 149 (1987) 1.
4. G.E. Brown and A.D. Jackson. The nucleon-nucleon interaction. North Holland (1976).
5. S.A. Chin, Ann. Phys. 108 (1977) 301.
6. W.D. Brown, R.D. Puff and L. Wilets, Phys. Rev. C2 (1970) 331; L. Wilets in: Mesons in Nuclei, Vol. III, eds. M. Rho and D. Wilkinson, North Holland (1979).
7. C. Itzykson and J.B. Zuber, Quantum Field Theory, Mc.Graw Hill (1985).
8. A. Fetter and J.D. Walecka, Quantum Theory of Many-Particle systems, Mc.Graw Hill (1971).
9. P.C. Martin and J. Schwinger, Phys. Rev. 115 (1959) 1342.
10. M.R. Anastasio, L.S. Celenza and C.M. Shakin, Phys. Rep. 100 (1983) 327;
    L.S. Celenza and C.M. Shakin. Relativistic Nuclear Physics. World Scientific (1986).
11. J.A. Tjon and S.J. Wallace, Phys. Rev. C35 (1987) 280.
12. C.J. Horowitz and B.D. Serot, Nucl. Phys. A464 (1987) 613.
13. B. ter Haar and R. Malfliet, Phys. Rep. 149 (1987) 207.
14. B. Friedman and V.R. Pandharipande, Nucl. Phys. A361 (1981) 502.
15. G.E. Brown, W. Weise, G. Baym and J. Speth, Comments Nucl. Part. Phys. 17 (1987) 39.
16. B. ter Haar and R. Malfliet, Phys. Rev. Lett. 59 (1987) 1652.
17. G.E. Brown, Nucl. Phys. A488 (1988) 689.
18. J.P. Jeukenne, A. Lejeune and C. Mahaux, Phys. Rep. 25C (1976) 83.
19. J.P. Blaizot, D. Gogny and B. Grammaticos, Nucl. Phys. A265 (1976) 315.
    M.M. Sharma, W.T.A. Borghols, S. Brandenburg, S. Crona, A. van der Woude and M.N. Harakeh, Phys. Rev. C38 (1988) 2562.
20. M. Jaminon, C. Mahaux and P. Rochus, Nucl. Phys. A365 (1980) 371.
21. A. Nadasen, P. Schwandt, P.P. Singh, W.W. Jacobs, A.D. Bacher, P.T. Dekvec, M.D. Kaitchuk and J.T. Meek, Phys. Rev. C23 (1981) 1023.
22. G. Brown, private communication.
23. B.C. Clark, in: "Relativistic Dynamics and Quark-Nucleon Physics", eds. M. Johnson and A. Picklesimer, Wiley (1986).
24. H. Stöcker and W. Greiner, Phys. Rep. 137 (1986) 277.
    G.F. Bertsch and S. Das Gupta, Phys. Rep. 160 (1988) 189.
25. R. Malfliet, B. ter Haar and W. Botermans, in: "Phase Space Approach to Nuclear Dynamics", ed. M. di Toro, World Scientific (Singapore, 1986) p.47.
26. W. Botermans and R. Malfliet, Phys. Lett. B171 (1986) 22.
27. W.A. van Leeuwen and S.R. de Groot, Physica 51 (1971) 1; R.G. Newton, "Scattering Theory of Waves and Particles", Springer-Verlag (1982), p. 211.
28. B. ter Haar and R. Malfliet, Phys. Rev. C36 (1987) 1611.
29. F. de Jong, B. ter Haar and R. Malfliet, Phys. Lett.
30. P. Danielewicz, Ann. Phys. 152 (1984) 239.
31. W. Botermans and R. Malfliet, Phys. Lett. B215 (1988) 617.
    W. Botermans, Ph.D. thesis (1989).
32. A. Lejeune, P. Grangé, M. Martzolff and J. Cugnon, Nucl. Phys. A453 (1986) 189.
33. J. Cugnon, A. Lejeune and P. Grangé, Phys. Rev. C35 (1987) 861.
34. J.A. Tjon and S.J. Wallace, Phys. Rev. C32 (1985) 267.

HOT NUCLEAR MATTER

V.R. Pandharipande and D.G. Ravenhall

Department of Physics,University of Illinois at
Urbana-Champaign
1110 West Green Street, Urbana, IL 61801   USA

I. INTRODUCTION

The theory of hot nuclear matter has been developed primarily to
help understand some aspects of highly excited nuclei, heavy-ion
reactions, supernova and neutron stars. In the next two sections we
review some of the basic concepts of quantum statistical mechanics and
nuclear forces. The results of rather simplistic calculations of hot
nuclear matter and its phase diagram are given in Section IV. Topics of
special interest including instabilities in hot nuclear matter and very
hot nuclear matter are briefly discussed in sections V-VIII.

II. QUANTUM STATISTICAL MECHANICS

Let us consider a system containing N nucleons in a volume V with
the "nuclear matter" assumption $N \rightarrow \infty$, $V \rightarrow \infty$ at fixed density $\rho = N/V$.
Let H be the Hamiltonian and $|I\rangle$ its eigenstates with energy E_I:

$$H|I\rangle = E_I|I\rangle \tag{2.1}$$

The basic principle of statistical mechanics is that, when the system is
in equilibrium at temperature T, i.e. the container is held at
temperature T, then the probability of finding it in a state $|I\rangle$ is
proportional to e^{-E_I/k_BT}, where k_B is Boltzmann's constant. A nice
discussion of this principle can be found in Feynman's book.[1] For
notational brevity we will set $k_B = 1$, measure T in units of energy and
define $\beta \equiv 1/T \equiv 1/k_BT$.

The partition function Q is defined as:

$$Q = \sum_I e^{-\beta E_I}, \tag{2.2}$$

and the probability of finding the system in the state $|I>$ is:

$$P_I = \frac{1}{Q} e^{-\beta E_I}, \tag{2.3}$$

The Helmholtz free energy F and the entropy S are defined as:

$$F = -T\ln Q$$

$$S = - \sum_I P_I \ln P_I = - \frac{\partial F}{\partial T} \tag{2.4}$$

The average value of an observable denoted by operator $\hat{O}$ is:

$$<\hat{O}> = \sum_I P_I <I|\hat{O}|I>, \tag{2.5}$$

thus the average energy U is given by:

$$U = <H> = \sum_I P_I E_I = \frac{T^2}{Q} \frac{\partial Q}{\partial T} = - T^2 \frac{\partial}{\partial T} \left(\frac{F}{T}\right) = F + TS. \tag{2.6}$$

The pressure P is

$$P = - \frac{\partial U}{\partial V} = - \sum_I P_I \frac{\partial E_I}{\partial V} = - \left(\frac{\partial F}{\partial V}\right)_T. \tag{2.7}$$

All thermodynamic information can be obtained from $Q(\rho,T)$ or equivalently $F(\rho,T)$. Unfortunately Q is rather difficult to calculate for most realistic Hamiltonians. An exact method of calculating Q is based on Feynman path integrals in imaginary time.[1,2] This method has been used along with Monte Carlo sampling techniques[3] to calculate the thermodynamic properties of Bose liquid ^{4}He from the interatomic potential $v(r)$. However, many problems in the application of these techniques to Fermi liquids[4] and to systems such as nuclei, where the interparticle interactions can flip spins and isospins of the particles[5], need to be resolved before they can be used to study hot nuclear matter.

Thus, most of the theoretical work on hot nuclear matter is based on approximations using either the Brueckner g-matrix[6] or variational methods.[7] We begin with a review of the calculation of $Q(\rho,T)$ of hot non-interacting Fermi gas (FG) germane to both approximations. It is most easily accomplished using a grand canonical ensemble[1,2] in which the average number of particles is controlled by choosing a proper chemical potential $\mu(\rho,T)$. Let i denote a single-particle plane-wave state with momentum $\vec{k}_i$. The eigenstates $|I]$ of FG are labeled with occupation numbers $n_i(I)$, and denoted by kets with square brackets

$$|I] = |n_1(I), n_2(I)......],$$

$$E_I = \sum_i e(k_i) n_i(I), \tag{2.8}$$

$$e(k_i) = \hbar^2 k_i^2 / 2m. \tag{2.9}$$

The $n_i(I)$ can have values 0 or 1, and the sum over I is equivalent to that over $n_i(I)$, so that:

$$Q = \sum_{n_i(I)=0,1} \exp\{-\beta[\sum_i (e(k_i)-\mu)n_i(I)]\},$$

$$= \prod_i (1 + e^{-\beta(e(k_i)-\mu)}). \tag{2.10}$$

The average occupation probability $\langle n(k_i)\rangle \equiv \langle n_i\rangle$ of the state i is given by:

$$\langle n(k_i)\rangle = \frac{1}{Q} \sum_{n_j(I)=0,1} n_i(I)\exp\{-\beta[\sum_j (e(k_j)-\mu)n_j(I)]\},$$

$$= (1 + e^{\beta(e(k_i)-\mu)})^{-1}, \tag{2.11}$$

and we have:

$$\rho = \frac{\langle N\rangle}{V} = \int \frac{d^3k}{(2\pi)^3} \langle n(k)\rangle, \tag{2.12}$$

$$\frac{S}{V} = -\int \frac{d^3k}{(2\pi)^3} \{\langle n(k)\rangle \ln\langle n(k)\rangle + (1-\langle n(k)\rangle)\ln(1-\langle n(k)\rangle)\}, \tag{2.13}$$

$$\frac{U}{V} = \int \frac{d^3k}{(2\pi)^3} e(k)\langle n(k)\rangle \tag{2.14}$$

Equations 2.12 and 14 are obvious, while 2.13 can be obtained from 2.4 by algebraic manipulations. The chemical potential μ is fixed by requiring that eq. (2.12) gives the desired density ρ at the chosen temperature T.

Fluctuations in the number of particles can be estimated from the difference between $\langle N^2\rangle$ and $\langle N\rangle^2$. We note that the operator N^2 is given by:

$$N^2 = [\sum_i n(k_i)]^2 = \sum_{i\neq j} n(k_i)n(k_j) + \sum_i n(k_i), \tag{2.15}$$

$$\therefore \langle N^2\rangle = \sum_{i\neq j} \langle n(k_i)\rangle \langle n(k_j)\rangle + \sum_i \langle n(k_i)\rangle,$$

$$= \langle N\rangle^2 + \sum_i (\langle n(k_i)\rangle - \langle n(k_i)\rangle^2). \tag{2.16}$$

Now since $\langle n(k_i)\rangle$ is between 0 and 1, $\langle n(k_i)\rangle^2$ is bounded between 0 and $\langle n(k_i)\rangle$

$$\therefore \quad \langle N^2 \rangle - \langle N \rangle^2 \leq N. \qquad (2.17)$$

Thus the fluctuations in N are of order $\sqrt{N}$, and can be neglected in the limit $N \to \infty$.

The above analysis shows that, in the infinite volume limit, all the states that contribute to Q are essentially the same. Let us consider a small volume element $(\Delta k)^3$ about $\vec{k}$ in momentum space. The number of particles $\Delta N(k)$ in it is given by $\langle n(k) \rangle V(\Delta k/2\pi)^3$ when Δk is small compared to a typical scale like the Fermi momentum k_F. However in the limit $V \to \infty$, this number is essentially large for any reasonable Δk, like $10^{-3} k_F$ for example, and hence the fluctuations in it can be ignored. Even though states that have identical values of $\Delta N(k)$ may not be truly identical because particles can occupy different single-particle states within the volume $(\Delta k)^3$, they are all the same for practical purposes, since they have same expectation values for macroscopic observables like energy.

We have so far ignored spin and isospin in our discussion of the FG. If these are included the equations 2.12-14 for ρ, S and U acquire a multiplicative factor s which gives the spin-isospin degeneracy; s=2 for neutron matter and s=4 for symmetric nuclear matter for example.

The main approximation used to simplify the theory of hot nuclear matter is in classifying its states |I) by occupation numbers $n_i(I)$. Thus one retains eq. (2.7) in the form:

$$|I) = |n_1(I) \ n_2(I), \ldots) \equiv G|I] / \sqrt{[I|G^\dagger G|I]}, \qquad (2.18)$$

where the operator G (not to be confused with the Brueckner G-matrix) takes into account the correlations induced by the strong nuclear forces. Note that kets |I) and |I] are used to denote correlated and FG states respectively. The energies of these states are approximated by:

$$E_I = \sum_i e(k_i, \rho, T) \ n_i(I) + C(\rho, T), \qquad (2.19)$$

where $e(k_i, \rho, T)$ are density and temperature dependent single-particle energies, and $C(\rho, T)$ is a constant that depends only on ρ and T. Eq. (2.19) is valid only for states I having $n_i(I)$ close to the equilibrium $\langle n(k_i, \rho, T) \rangle$. Thus the single particle energies $e(k_i, \rho, T)$ should be identified with the derivatives $\partial U(\rho, T)/\partial n(k_i)$.

The above approximations are similar to those in Landau's theory[8] of normal Fermi liquids, and the $n_i(I)$ can be said to denote the quasi-particle occupation numbers. At low enough temperatures, where the quasi-particles have long life times, such an approach becomes exact. At high temperatures the quasi-particles have short life times, and thus it is not really possible to label states with fixed occupation numbers $n_i(I)$. The collisions between quasi-particles change their momenta, or equivalently occupations, and are responsible for achieving thermal equilibrium. However, they may not have a large effect on the free energy calculated assuming thermal equilibrium. The key assumption is that if the correlated states $|I\rangle$ mix mostly with states that are degenerate with $|I\rangle$, then they can be used to calculate the partition function.

The average quasi-particle occupation probability, density and entropy are then obtained by using the $e(k_i, \rho, T)$ in equations 2.11, 12 and 13; however, the energy U is no longer given by eq. (2.14) because the quasi-particle energies are the derivatives of the total energy $U(\rho, T)$ with respect to $n(k_i)$, and eq. (2.8) is not applicable. That equation is valid when the Hamiltonian contains only one-body terms, and no two- or many-body interactions.

In the infinite volume limit, since all the states that contribute to Q are essentially the same, $U(\rho, T)$ is just the energy of a state having $\Delta N(k) = <n(k)>V (\Delta k/2\pi)^3$. It can be estimated with realistic nuclear forces, using Brueckner's two-body g-matrix. At finite temperatures the g-matrix is complex; its real part gives $U(\rho, T)$ and the imaginary part describes scatterings that change quasi-particle occupation numbers. ·The $U(\rho, T)$ depends upon the $e(k_i, \rho, T)$ which are themselves the derivatives of $U(\rho, T)$. Thus the $e(k_i, \rho, T)$ and $U(\rho, T)$ have to be calculated consistently by an iterative procedure. Calculations of hot nuclear matter, using Brueckner's theory, have been attempted recently be Lejeune et al.[9] Including the effects of many-body correlations into these calculations is a challenging task.

In strongly interacting liquids the energies $e(k_i, \rho, T)$ can be better determined, up to an additive constant, at a fixed density ρ. Hence, it is simpler to calculate $F(\rho, T)$ at a fixed ρ by absorbing the additive constant in an effective chemical potential μ'. The real chemical potential μ can then be determined from the function $F(\rho, T)$

$$\mu(\rho, T) = \frac{1}{V} \frac{\partial F(\rho, T)}{\partial \rho} \qquad (2.20)$$

The variational theory[7] is based on the principle that the free energy $F(\rho, T)$ has the minimum value.[1] In practical calculations the correlation operator G is approximated by a symmetrized product of pair correlation operators:[10]

$$G = S \prod_{i<j} F_{ij}, \qquad (2.21)$$

$$F_{ij} = \sum_{p=1,8} f^p(r_{ij}) \; O_{ij}^p , \qquad (2.22)$$

$$O_{ij}^{p=1,8} = 1, \; \tau_i \cdot \tau_j, \; \sigma_i \cdot \sigma_j, \; \sigma_i \cdot \sigma_j \; \tau_i \cdot \tau_j, \; S_{ij}, \; S_{ij}\tau_i \cdot \tau_j, \; \vec{L} \cdot \vec{S}, \; \vec{L} \cdot \vec{S}\tau_i \cdot \tau_j, \quad (2.23)$$

and the functions $f^p(r_{ij})$ and $e(k,\rho,T)$ are determined variationally by minimizing the free energy given by:

$$F_v(\rho,T) = \frac{[n(k,\rho,T)|G^\dagger HG|n(k,\rho,T)]}{[n(k,\rho,T)|G^\dagger G|n(k,\rho,T)]} - TS(\rho,T), \qquad (2.24)$$

where $|n(k,\rho,T)]$ is the FG state with occupation probabilities $\langle n(k) \rangle$. Both $|n(k,\rho,T)]$ and the entropy $S(\rho,T)$ depend upon the spectrum $e(k,\rho,T)$. Calculation of the expectation value of H in eq. 2.24 poses a major problem in these theories. It is calculated approximately using hypernetted chain and operator chain summation techniques.[10]

For such a variational calculation to succeed, the operator G must induce all the important correlations in the system, and in nuclear matter these are believed to be the central, spin-isospin, tensor and spin-orbit correlations induced by the repulsive core and the spin-isospin, tensor and spin-orbit components in nuclear forces. These considerations dictate the choice (2.23) of the pair correlation operator F_{ij}. The choice (2.21) of G tacitly assumes that correlations between many particles can be reasonably approximated by a symmetric product of pair correlation operators. In some systems, such as atomic helium liquids, products of three-body functions $f_{ijk}(\vec{r}_i, \vec{r}_j, \vec{r}_k)$ have been included in the G to improve the variational bound.[10]

III. NUCLEAR FORCES

Nuclear forces are the only input in the many-body theory of nuclear matter; however, they are not yet well understood. Nucleons are finite objects made up of quarks, and they interact in a fairly complex manner. This also implies that our description of matter in terms of interacting nucleons will break down at some density ρ_{max} dependent upon temperature T. At $\rho > \rho_{max}(T)$ matter may be more appropriately described as composed of interacting quarks and gluons. Thus our theory is limited to the low ρ, T corner of the ρ, T plane, but its limits of validity are poorly known. It is believed that ρ_{max} (T $\gtrsim$ 155 MeV) = 0 from lattice QCD studies[11], but ρ_{max} (T=0) is quite uncertain, though it is believed to be much greater than nuclear matter density.

Realistic models of two nucleon interaction potential v_{ij} are obtained in several towns like Paris[12], Urbana[13], Argonne[14], Bonn[15] and Nijmegen[16], by fitting the deuteron properties and two-nucleon scattering data at $E_{lab} \lesssim 400$ MeV. All these models have a one-pion exchange long-range part and more or less phenomenological intermediate and short range parts. The Nijmegen potential gives the best χ^2 fit to the data, those obtained with Paris and Argonne are reasonably good, while the coordinate space Bonn model is the poorest. The measured deuteron form factors A and B are sensitive the deuteron wavefunction at r ~ 1fm. The Argonne model gives the best description of $A(q^2)$ and $B(q^2)$, Paris and Nijmegen give reasonable fits and the full Bonn model fails to explain the measured $A(q^2)$.[17] It is possible that, after adding exchange current corrections[18] the Paris model may give an even better description of $B(q^2)$. The isovector exchange currents that contribute to $B(q^2)$ are not well understood.

The Urbana potential is an older version of that of Argonne. The tensor force in the Argonne model is much stronger than that in the Urbana. In other respects these two models are similar. The tensor forces in Argonne and Paris models are similar; the D-states account for 5.2, 5.8 and 6.2% of the Urbana, Paris and Argonne deuteron wave functions. A figure comparing the tensor forces in these three models is given in ref. 19. The properties of matter depend to some extent on the strength of the tensor force. For example, neutron matter, which we do not discuss here, has a transition to a phase in which π^0-condensation appears likely, when the Argonne model is used, but not if the Urbana model is used.[19]

The binding energy of the three-body nucleus ^{3}H can be very accurately calculated for each model from 34-channel Faddeev calculations[20]. The experimental energy is -8.48 MeV, while the Paris, Argonne and Nijmegen potentials give -7.47, -7.67 and -7.62 MeV respectively. The coordinate-space Bonn potential gives a better triton energy[21,20] of -8.29 MeV, but this result may not be very significant since the improvement in the energy probably comes from the unrealistic energy dependence of the coordinate space Bonn $^3S_1$-$^3D_1$ mixing phase[22] ε_1.

The energies of nuclei having four or more nucleons cannot as yet be calculated accurately, i.e. with <1% error. Quite recently Carlson[5] has made significant advances in exact calculations of ^{4}He binding energy with the Green's function Monte Carlo method. However, the effects of quadratic spin-orbit forces are still not included in these calculations. It is likely that accurate values of the binding energy of ^{4}He obtained from different realistic potential models will soon be available.

The binding energies of ^{3}H and ^{4}He nuclei have also been calculated with the variational method using the wavefunction:

$$\psi_v = \left[S \prod_{i<j} F_{ij} \right] \Phi, \qquad\qquad (3.1)$$

where Φ is an uncorrelated wavefunction and the correlation operators F_{ij} have the form given by eq. (2.18). The energy expectation value is calculated exactly by Monte Carlo methods, and the F_{ij} is varied to minimize it. The available variational calculations[23] give energies that are above the exact results for ^{3}H and ^{4}He by $\lesssim$ 7%. More recently Wiringa[24] has used better parameterization of the F_{ij}'s to reduce the error in the ^{3}H variational energy to ~3%. The energy of ^{4}He has also been studied with the Faddeev-Yakubovesky equations[25] and the coupled cluster[26] expansion. All these calculations give similar results, and underbind ^{4}He. Variational calculations with the Urbana and Argonne potentials respectively give -23 and -22 MeV for the energy of ^{4}He against the experimental value of -28.3 MeV.

There have been several studies of nuclei like ^{16}O and ^{40}Ca with realistic nuclear forces; however, much of the attention has been focussed on the nuclear matter (NM) problem. There are, as yet, no practical exact or demonstrably accurate methods to calculate the properties of cold NM from realistic forces. The major difficulties in applying the Green's function Monte Carlo methods, that have been successfully used to study atomic liquid ^{4}He[27] and ^{3}He[28], stem from the spin, isospin, tensor and spin-orbit components in the nuclear force.[5]

Brueckner theory[29] and the variational method[29,19], with the wave function (3.1), have been used to study cold nuclear matter. Day[29] obtained almost identical NM energies with the Argonne and the Paris potentials, by summing two-, three- and some four-hole line terms in Brueckner theory. His results are compared with those obtained for the Argonne and Urbana potentials by the variational method[19] in fig. 1. The two methods give rather similar results.

The main conclusions of the results obtained with the Argonne, Urbana and Paris potentials can be summarized as follows. All these potentials underbind ^{3}H and ^{4}He, they give reasonable value for the binding energy of NM, but the calculated equilibrium density (~0.3fm^{-3}) has about twice the empirical value of ~0.16fm^{-3}. Most importantly the differences between the results of different potential models are

smaller than those between experiment and theory. It thus appears
likely that the simple model of nucleons interacting with only two-body
forces is not realistic for nuclei and NM.

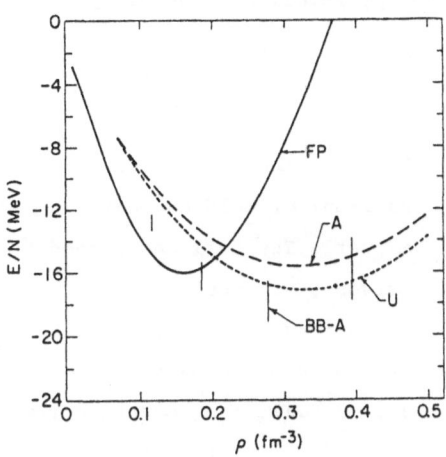

Fig. 1 The energy (per nucleon) of cold nuclear matter. The error bars show results obtained with Brueckner-Bethe theory and the Argonne V_{ij}, the long and short dashed curves show results of variational calculations with the Argonne and Urbana V_{ij}. The curve labeled FP has the correct empirical saturation point.

It is well known that there are three-nucleon forces[30] which we
denote by V_{ijk}. Particularly the two-pion exchange $V_{ijk}^{2\pi}$ is
theoretically well established for a long time[31], and it can provide the
additional attraction necessary to obtain the correct binding energy of
^{3}H and ^{4}He.[32] However, if one takes only the attractive $V_{ijk}^{2\pi}$, NM gets
overbound and also has too large an equilibrium density. Hence it is
necessary to have other terms in V_{ijk} to obtain reasonable NM. Our
phenomenological approach[23] has been to consider a model of V_{ijk}:

$$V_{ijk} = V_{ijk}^{2\pi} + V_{ijk}^{R} \qquad (3.2)$$

in which V_{ijk}^{R} is a spin-isospin independent interaction of a reasonable
range, and to adjust the strengths of $V_{ijk}^{2\pi}$ and V_{ijk}^{R} to obtain the
experimental binding energies of ^{3}H and ^{4}He and reasonable NM properties
in variational calculations. The obvious problems with this approach
are that the variational calculations are not exact, and the assumed
form of V_{ijk}^{R} , which in principle should contain all the components of
V_{ijk} other than the two-pion exchange $V_{ijk}^{2\pi}$, may be too simplistic.
Better calculational methods are being developed, and attempts to
calculate the binding energies more nuclei like ^{6}Li, ^{8}He, ^{16}O etc. are
underway to address these problems.

The latest model of the type (3.2) is numbered VII; it gives the
experimental ^{3}H and ^{4}He binding energies[23], when used with Argonne or

Urbana potentials, and NM results[19] shown in fig. 2. Note that just by itself the Urbana potential gives lower energies for NM than the Argonne (fig. 1) because of its weaker tensor force. However, the Argonne v_{ij} + V_{ijk} (VII) gives lower energies than Urbana v_{ij} + V_{ijk} (VII) because the $v_{ijk}^{2\pi}$ can better exploit the stronger tensor correlations induced by Argonne v_{ij}.

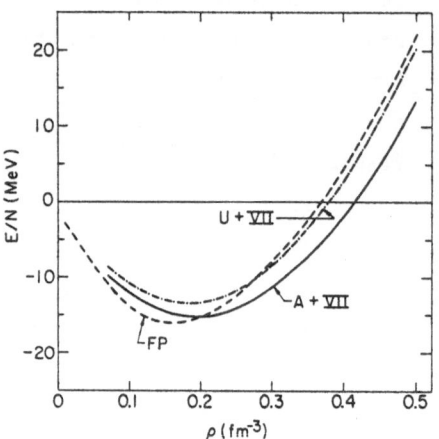

Fig. 2. The energy (per nucleon) of cold nuclear matter. The full and dash-dot curves give results of variational calculations with Argonne and Urbana v_{ij} + V_{ijk} (model VII). The dashed curve (FP) gives the results of variational calculations with the LP Hamiltonian.

The V_{ijk}^{R} has a large effect on the equation of state at higher densities. There is no direct experimental confirmation of this term, though it appears that neutron stars calculated with V_{ijk}(VII) are in better agreement[19] with the little observational data we have than those calculated without the V_{ijk}. Arguments suggesting that the parameters of V_{ijk} model VII are not unrealistic have been given in ref. 33. In particular the contribution of V_{ijk}^{R} in model VII to the energy of the triton is 0.7 MeV, in agreement with the dispersive correction to the two-pion exchange NN interaction with NΔ intermediate states. The latter has been calculated by Sauer[34] with Faddeev calculations including NNΔ intermediate states. Unfortunately neither Argonne or Urbana v_{ij} + V_{ijk}(VII) models give the empirical NM energy (-16 MeV) or density (ρ_0 = 0.16 fm^{-3}).

One can develop even more phenomenological models that are constructed to explain the equilibrium properties of NM. In the model developed by Lagaris and Pandharipande[35] (LP) the Urbana model of v_{ij} is expressed as a sum of pion exchange v_{ij}^{π}, two-pion exchange v_{ij}^{I} and short range v_{ij}^{S} parts:

$$v_{ij} = v_{ij}^{\pi} + v_{ij}^{I} + v_{ij}^{S}. \tag{3.3}$$

The v_{ijk}^R essentially makes the attractive v_{ij}^I weaker as the density increases, while the $v_{ijk}^{2\pi}$ makes v_{ij}^π density dependent. LP assume that the energy of NM (per nucleon) is given by:

$$E(\rho) = E[v_{ij}(\rho)] + \gamma_2 \rho^2 \exp(-\gamma_3 \rho) \ (3-2 \ (\tfrac{N-Z}{A})^2), \qquad (3.4)$$

where $E[v_{ij}(\rho)]$ is the energy obtained with the density dependent interaction:

$$v_{ij}(\rho) = v_{ij}^\pi + v_{ij}^I \exp \ (-\gamma_1 \rho) + v_{ij}^S, \qquad (3.5)$$

and the γ_2 term represents the contributions of all the other many-body interactions. The parameters γ_1, γ_2 and γ_3 are varied to obtain $E_0 = -16$ MeV, $\rho_0 = 0.16 \text{fm}^{-3}$ and K=240 MeV for NM. The LP interaction is used by Friedman and Pandharipande[36] (FP) to study properties of hot and cold NM and neutron matter. Their results are also shown in figs. 1 and 2; they are correct, by construction, near ρ_0, and are similar to those of Urbana $v_{ij} + V_{ijk}$(VII) at high densities.

Unfortunately the LP model cannot be used to study the light ^{3}H and ^{4}He nuclei. In this respect the models with explicit three-body interactions are superior. The main problem with these models is the lack of enough attraction at lower densities of $\sim 0.1 \text{fm}^{-3}$ (fig. 2). Thus it appears that the long standing problem of nuclear forces is still not satisfactorily resolved.

The two-pion exchange interaction v_{ij}^I has contributions from NΔ and $\Delta\Delta$ intermediate states[14], and the v_{ijk}^R was introduced to represent changes in it due to the presence of a third nucleon. However, it has been suggested that relativistic effects[37-40] could also generate a three-body force similar to v_{ijk}^R. In fact some of the recent relativistic Dirac-Brueckner calculations[41,42] give a NM $E(\rho)$ similar to that obtained by FP.

The compressibility K of NM is not well known. The values of K extracted from the observed breathing mode energies of nuclei range from ~ 220 Mev[43] to ~ 300 MeV.[44] However it has been suggested that there may be large uncertainties in these extractions.[45]

IV. HOT NUCLEAR MATTER

In principle the properties of hot matter can be obtained from a chosen nuclear Hamiltonian using methods reviewed in Section II. Lejeune et al.[9] have attempted to calculate them starting from the Paris potential using Brueckner G-matrix, while FP[36] use the LP Hamiltonian

and the variational method. Both the calculations are fairly simplistic
and they give qualitatively similar results. We will first review the
results obtained by FP and then discuss some of the problems in their
theory.

FP have calculated the free energy of nuclear matter at T=0, 5,
10, 15 and 20 MeV up to a density of 2fm^{-3}, and that of neutron matter
at T=0, 3, 6, 10, 13, 16 and 20 MeV at $\rho \leq 0.83 \text{ fm}^{-3}$. It is observed
that at $\rho > 0.04 \text{ fm}^{-3}$ the pair correlation operator F_{ij} has negligible
temperature dependence at T<20 MeV. The $\rho < 0.04 \text{ fm}^{-3}$ low density region
will be discussed separately. The quasi-particle spectrum is
approximated by

$$e(k,\rho,T) = \hbar^2 k^2/2m^*(\rho,T) + \text{constant};\tag{4.1}$$

the constant is irrelevant since it can be absorbed in an effective
chemical potential. In ref. 7 this effective-mass approximation was
found to be quite adequate for simple variational calculations of $F(\rho,T)$.
The pair correlation operator $F_{ij}(\rho)$ (which should not be confused with
the free energy $F(\rho,T)$) is determined at each density by minimizing the
ground state energy $E(\rho) = F(\rho,T=0)$, since the entropy is zero at T=0.
The $m^*(\rho,T)$ is varied to minimize the free energy $F(\rho,T)$; it has a
significant density dependence as shown in fig. 3, and a rather small
variation with temperature at T<20 MeV.

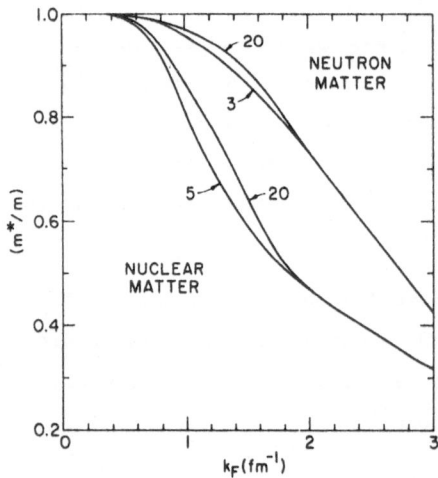

Fig. 3. The effective mass
$m^*(k_F,T)$ in nuclear matter at
T=5 and 20 MeV and neutron
matter at T=3 and 20 MeV
calculated variationally by
FP.

The effective mass of nucleons in nuclear matter is less than
their physical mass because the exchange force is more attractive in all
the realistic models of N-N interaction. Thus nucleons with small
momenta feel more attraction from nuclear forces than those with large

momenta as illustrated in fig. 4. In general the effective mass $m^*(k,\rho,T)$ gives the velocity of particles of momentum $\hbar k$ in matter at ρ,T:

$$\text{velocity } (k,\rho,T) = \frac{\hbar k}{m^*(k,\rho,T)} = \frac{1}{\hbar} \frac{d}{dk} e(k,\rho,T). \qquad (4.2)$$

It plays an important role in the dynamics of heavy-ion reactions[47], and in the energy dependence of the real part of the optical potential in NM (section VI).

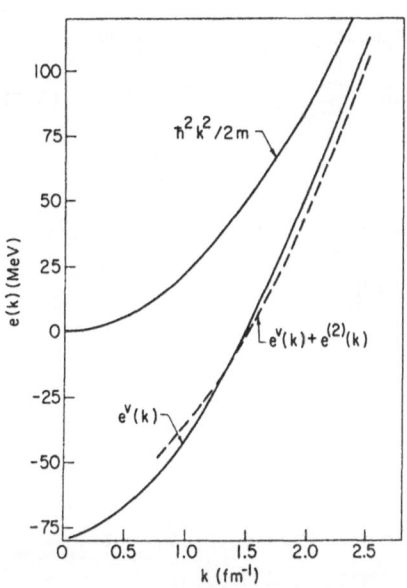

Fig. 4. The single particle energy e(k) of quasi-particles in cold nuclear matter at $\rho=\rho_o$. The results of variational calculations are shown by the full curve $e^V(k)$, and the dashed curve includes effects of the coupling of two-particle one hole or two-hole one particle states. The e(k) of free-nucleons is also shown for comparison.

Quasi-particle states having energies within $\sim\pi T$ of the Fermi energy $e(k_F) \equiv e(k_F,\rho,T=0)$ are important in determining the thermal properties of matter. Thus the $m^*(\rho,T)$ in the variational calculations corresponds to an average value of $m^*(k{\sim}k_F,\rho,T)$ in an interval over which $|e(k,\rho,T) - e(k_F)|{\sim}\pi T$. It decreases with ρ because the interaction contributions increase, and it increases with T because at large temperatures states with large k, which have $m^*{\sim}m$, become more important.

FP calculate $F(\rho,T)$ only at 120 (168) values of ρ and T for nuclear (neutron) matter. Thus it is necessary to interpolate between their results to obtain values of thermodynamic variables at desired ρ_n, ρ_p and T, where ρ_n and ρ_p are neutron and proton densities and $\rho=\rho_n+\rho_p$. A much better procedure developed by Ravenhall is to use a Skyrme Hamiltonian (SH) fitted to the FP results to obtain properties of hot NM. In SH the effective masses $m^*_{n,p}$ depend only upon ρ_n and ρ_p, and thus the small variations in m^* with T (fig. 3) are neglected.

The SH assume that the energy density $\mathcal{H}$ of uniform nuclear matter having $\nabla \rho_n = \nabla \rho_p = 0$ is a function of the densities ρ_n and ρ_p and a linear function of the kinetic densities τ_n and τ_p:

$$\tau_x = \frac{2}{(2\pi)^3} \int d^3k \ k^2 \ n_x \ (k,\rho,T); \ x = n \ or \ p, \tag{4.3}$$

$$\mathcal{H} (\tau_n, \tau_p, \rho_n, \rho_p) = \hbar^2 \tau_n / 2m_n^* (\rho, \rho_n) + \hbar^2 \tau_p / 2m_\rho^* (\rho, \rho_\rho)$$

$$+ g(\rho_n, \rho_p) . \tag{4.4}$$

Since the quasi-particle energies of this Hamiltonian are given by

$$e_x (k, \rho, \rho_x) = \hbar^2 k^2 / 2m_x^* (\rho, \rho_x) + constant \tag{4.5}$$

it is simple to calculate its $F(\rho_n, \rho_p, T)$.

The Ravenhall-Skyrme Hamiltonian (RSH) has the form:

$$m_x^* (\rho, \rho_x) = m / \left\{ 1 + \frac{2m}{\hbar^2} (p_3 \rho + p_5 \rho_x) \ e^{-p_4 \rho} \right\}, \tag{4.6}$$

$$g(\rho_n, \rho_p) = -\rho^2 \left[p_1 e^{-p_6 \rho} + p_2 (1 - e^{-p_6 \rho}) + (\frac{p_{10}}{\rho} + p_{11}) \ e^{-(p_9 \rho)^2} \right]$$

$$- \frac{1}{4} (\rho_n - \rho_p)^2 \left[p_7 \ e^{-p_6 \rho} + p_8 (1 - e^{-p_6 \rho}) + (\frac{p_{12}}{\rho} + p_{13}) \ e^{-(p_9 \rho)^2} \right] . \tag{4.7}$$

The parameters p_{1-13} are obtained by fitting the FP $F(\rho_n, \rho_p, T)$ and are given in Table I.

Table I. Parameters of Ravenhall-Skyrme Hamiltonian.

n	p_n	n	p_n
1	339 MeV fm^3	8	2316 MeV fm^3
2	-1054 MeV fm^3	9	6.50 fm^3
3	89.8 MeV fm^5	10	1.78 MeV
4	0.457 fm^3	11	52.0 MeV fm^3
5	-59.0 MeV fm^5	12	-5.50 MeV
6	0.284 fm^3	13	-197 MeV fm^3
7	-543 MeV fm^3		

Skyrme Hamiltonians have been extensively used in the literature to study nuclear structure in the mean field theory. The more common parameterization has the form[47]

$$m_x^* (\rho, \rho_x) = m / \left\{ 1 + \frac{1}{4} (t_1 + t_2) \rho + \frac{1}{8} (t_2 - t_1) \rho_x \right\}, \tag{4.8}$$

$$g(\rho_n, \rho_p) = \frac{1}{2} t_0 \left[(1 + \frac{1}{2} x_0) \rho^2 - (x_0 + \frac{1}{2}) (\rho_n^2 + \rho_p^2) \right]$$

$$+ \frac{1}{4} t_3 \left[\rho_n \rho_p + \frac{\lambda}{4} (\rho_n - \rho_p)^2 \right] \rho^\alpha. \tag{4.9}$$

and the parameters t_{0-3}, x_0, λ and α are determined by fitting nuclear binding energies, density distributions and energies of giant resonances.

Some of the conventional SH, SkM[47] for example, give rather reasonable values of $F(\rho_n, \rho_p, T)$ at $\rho \sim \rho_0$. However, they generally have unphysical behaviors at $\rho \gg \rho_0$ where their m^* decreases as $1/\rho$ (eq. 4.8), and the $g(\rho_n, \rho_p)$ increases as $\rho^{2+\alpha}$. The $e^{-p_i \rho}$ factors in the RSH ensure that it reproduces the FP $F(\rho, T)$ up to $\rho = 2 \text{fm}^{-3}$. Of course the FP results may not at all be reliable at such high densities, but that is a different issue.

The p_{10}/ρ and p_{12}/ρ terms in RSH are associated with the rather specific behavior of FP energies at low density. Since only pair correlations are considered in these variational calculations, the FP nuclear matter turns into a gas of deuterons at low density. The interaction energy density of this gas has a term proportional to ρ which is simulated by the p_{10}/ρ and p_{12}/ρ terms. We will discuss this term later; however, if in some applications it is desirable to remove it because there is no time for the deuterons or heavier clusters to form, the simplest way is to replace the p_{10}/ρ and p_{12}/ρ in eq. 4.7 by p_{10}/ρ_0 and p_{12}/ρ_0. This substitution has little effect on the equation of state at $\rho \gtrsim \rho_0$.

The pressure is obtained by differentiating the $F(\rho, T)$ (eq. 2.7) and the resulting isotherms $P(\rho, T)$ are shown in fig. 5. They have the typical van-der Waal's shape. At temperatures above the critical temperature $T_C = 17$ MeV the $P(\rho, T)$ increases monotonically with ρ and matter is stable at all densities. Below T_C there exists a density range $\rho_g(T) < \rho < \rho_\ell(T)$ in which matter separates into gas at density $\rho_g(T)$ and liquid at $\rho_\ell(T)$. The $\rho_g(T)$ and $\rho_\ell(T)$ are found by standard Maxwell construction, and their loci form the liquid-gas coexistence curve shown in fig. 6.

Only in the region where $\partial P/\partial \rho |_T$ is negative is the matter unstable against small isothermal density fluctuations. This region is bounded at a given T by the densities $\rho_g'(T)$ and $\rho_\ell'(T)$ at which $\partial P/\partial \rho |_T = 0$. The

$\rho_g'(T)$ and $\rho_\ell'(T)$ are called isothermal spinodal points, and their loci is shown in fig. 6 as the isothermal spinodal curve. Matter at ρ,T in the region in between the isothermal spinodal curve and the coexistence curve is metastable, i.e. it is stable against small fluctuations, but will phase separate via nucleation in the presence of large fluctuations. This is the region in which supercooled liquid or gas can exist, and cloud and bubble chambers operate.

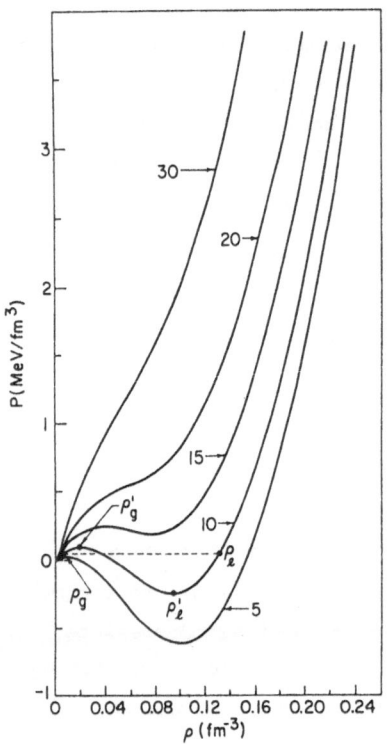

Fig. 5. The isotherms $P(\rho,T)$ of hot nuclear matter labeled with T. The coexistence points ρ_g and ρ_ℓ and the spinodal points ρ_g' and ρ_ℓ' are shown for T=10.

Fig. 6 also shows the curve formed by the loci of points having $P(\rho,T) = 0$. This curve starts at ρ_0, T=0 and has a maximum temperature, which we denote by T_m that is less than T_c. It is obvious that $P(\rho,T>T_m) > 0$ at all densities other than $\rho=0$. Consider a drop of liquid matter, large enough so that the pressure due to surface tension can be neglected. Such a drop can exist in vacuum only if it has zero pressure; otherwise it will expand (if P>0) or contract (P<0). Hence it is possible to have finite liquid drops (or nuclei to the extent that they can be treated as liquid drops) cooling by evaporation in vacuum only at $T<T_m$. Liquid at temperatures $T>Tm$ can exist only in equilibrium with its vapor which produces the pressure, or in dynamically unstable

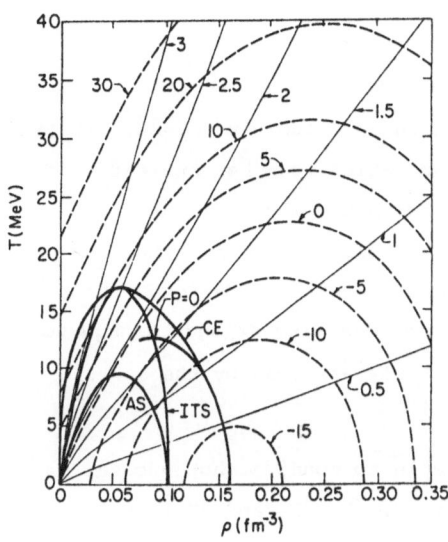

Fig. 6. The phase diagram of nuclear matter as obtained from the calculations of FP. The thick lines show the coexistence (CE), the P=0 and the isothermal (ITS) and adiabatic (AS) spinodal curves. The thin and dashed lines show constant entropy and internal energy contours.

conditions. We note that the $P(\rho, T_m)$ isotherm has P=0 at the spinodal point $\rho'_\ell (T_m)$, and hence the isothermal spinodal passes thru the maximum of the P=0 curve.

The contours for equal energy $U(\rho, T)$ and entropy $S(\rho, T)$ are also shown on fig. 6. The phase-diagrams obtained with the RSH for matter having proton fraction x=0.5 $(\rho_n = \rho_p)$ and x=0.4 $(\rho_n = 1.5 \rho_p)$ and with the SKM interaction[47] for x=0.5 are compared in fig. 7. The differences between these phase-diagrams are not too large.

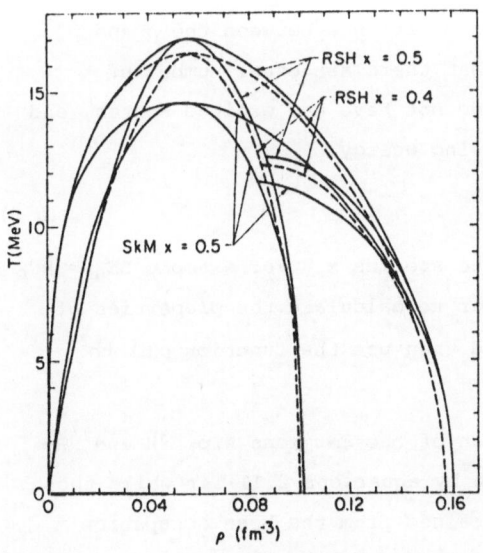

Fig. 7. The phase diagram of nuclear matter at proton fraction x=0.5 and 0.4 as obtained with the Ravenhall-Skyrme Hamiltonian compared with that obtained from the Skyrme M-Hamiltonian. The coexistence, isothermal spinodal and the P=0 curves are shown.

V. HOT NUCLEAR GAS

Matter at $T<T_c$ and $\rho<\rho_g(T)$ can be called hot nuclear gas. It is stable against fluctuations, and will expand and cool in vacuum. It can be created in heavy-ion reactions by evaporation from hot nuclear liquid, or directly by matter expanding adiabatically at large entropies. For example, matter having $S>4$ will not get into the region of instabilities (fig. 6) and expand to become hot nuclear gas. Under some conditions[48] matter with entropies $\gtrsim 2$ can also expand through to the gaseous phase without being significantly affected by the instabilities.

At very low densities one may be able to consider hot nuclear gas as composed of n, p, and the light nuclei d, ^3H, ^3He and ^4He; nuclei heavier than ^4He are not expected to contribute significantly. Such an approach is taken by Pratt, Siemens and Usmani[49] (PSU), at $\rho \lesssim 0.003$ fm^{-3} or $\lesssim$ 1/50th the density of nuclear matter. At low enough densities, and high enough temperatures, it is possible to neglect the effects of interactions between these nuclei.

From the chemical equilibrium between the nuclei in hot nuclear gas one obtains:

$$\mu(p) = \mu(n) = \mu, \tag{5.1}$$

$$\mu(d) = 2\mu, \tag{5.2}$$

$$\mu(^3H) = \mu(^3He) = 3\mu, \tag{5.3}$$

$$\mu(^4He) = 4\mu, \tag{5.4}$$

where $\mu(x)$ is the chemical potential of the species x, and for simplicity we have neglected the mass difference between the n and p, and between ^3H and ^3He and assumed that there are equal number of neutrons and protons. These nuclei do not have any excited states, and hence can be treated as particles having energy:

$$e_x (k) = \frac{\hbar^2 k^2}{2m_x} - BE_x, \tag{5.5}$$

where BE_x is the binding energy of the species x. For example $BE_n = BE_p$ = 0, BE_d = 2.2 MeV etc. It is simpler to calculate the properties of hot gas as a function of μ and T, and then use the function $\rho(\mu)$ to obtain them at the desired ρ, T.

At a chosen μ, T the contribution of the Fermions n, p, ^3H and ^3He to the ρ, S and U of the gas are given by equations 2.11-14; while those of the Bosons d and ^4He should be obtained from the Bose occupation probabilities[1]:

$$\langle n_x(k) \rangle = (e^{\beta(e_x(k) - \mu(x))} - 1)^{-1}. \tag{5.6}$$

The entropy of the Bose gas is given by:

$$\frac{S_x}{V} = - \int \frac{d^3k}{(2\pi)^3}$$

$$\{\langle n_x(k) \rangle \ln \langle n_x(k) \rangle - (1 + \langle n_x(k) \rangle) \ln (1 + \langle n_x(k) \rangle)\}, \tag{5.7}$$

and the density and energy by equations 2.12 and 2.14 respectively. The contributions of n, p, ^{3}H and ^{3}He have to be multiplied by the spin degeneracy 2, and that of d by 3.

PSU also estimate the effects of interactions between these particles directly from their scattering phase shifts. They are not too large; the pressure of matter at $\rho = 0.003$ fm^{-3} is reduced by ~10% by the interactions. The typical composition, S and U obtained by PSU are given in Table II.

Table II. Properties of hot nuclear gas (from PSU).

ρ Nucleons fm^3	T MEV	$\frac{N(d)}{N(p)}$	$\frac{N(^3H + {}^3He)}{N(p)}$	$\frac{N(^4He)}{N(p)}$	S per Nucleon	F MeV per Nucleon
0.00325	7.8	0.535	0.186	0.082	4.12	−24.8
0.00195	6.4	0.471	0.159	0.105	4.27	−21.4
0.00113	4.4	0.444	0.169	0.390	3.88	−15.1

It should be stressed here that this is strictly a high temperature formalism, and should not be used at low temperatures where the effects of Fermi and Bose statistics are pronounced. At $\rho \sim 10^{-3}$ fm^{-3}, if one goes to low enough temperatures, the interaction effects will become important. Consider for example nuclear matter without any d, ^{3}H, ^{3}He and ^{4}He clusters for simplicity. The low density expansion of its energy is a well known[50] series in powers of $k_F a$, where a is the scattering length. The $^1S_0$ scattering length is ~20 fm, so interactions dominate the energy of cold nuclear matter at densities as low as $k_F \sim$ 1/20 fm^{-1} or $\rho \sim 10^{-5}$ fm^{-3}. At $\rho \gtrsim 10^{-5}$ fm^{-3} the interaction energy becomes comparable to or larger than the kinetic energy $3\hbar^2 k^2_F/10$ m. At high temperatures however, U is of order T, and the approach is valid if $T \gg T_F \equiv \hbar^2 k^2_F/2m$. The limitation of this approach to high temperatures may not be too restrictive because matter at T=0 is unstable at densities $\gtrsim 10^{-5}$fm^{-3}.

Within the variational theory FP consider matter in the range $0.004 < \rho < 0.04$ fm^{-3} as hot nuclear gas. Its free energy is minimized by varying both the pair correlation operator F_{ij} (ρ,T) and $m^*(\rho,T)$ at chosen values of ρ,T. Since only pair correlations are considered the calculation contains only the effects of deuteron clustering. These become very important when both ρ and T are small. At $T = 0$ and 5 MeV, $0.004 < \rho < 0.04$ fm^{-3} and at $T=10$ MeV, $0.01 < \rho < 0.04$ fm^{-3} matter is unstable, as can be seen from the phase diagram shown in fig. 6. At these ρ,T values the hypernetted chain equations become unstable as the range of correlations F_{ij} is increased, before a minimum is obtained. However the free energy is not very sensitive to the range of F_{ij} before the equations cease to have solutions, and thus can be crudely estimated. This estimate in the limit $\rho \to 0$, $T \to 0$ is -1.1 MeV/nucleon, which is that of a dilute gas of deuterons.

At $T=10$ MeV, $\rho < 0.005$ fm^{-3} and at $T = 15$ MeV, $\rho < 0.025$ fm^{-3} matter is stable, and in principle one should consider the effect of ^{3}H, ^{3}He and ^{4}He-clusters on its free energy. The effect of ^{4}He-clusters has been estimated using a Skyrme interaction and the Saha equation[51]. It is not too large in this ρ,T range as can be seen from table III, and the FP results seem to be reasonable. Nevertheless developing a more reliable treatment of the composition, entropy and energy of hot nuclear gas over the entire ρ,T region where it is stable is an interesting and challenging task.

Table III. Free-energy of hot nuclear gas in different approximations.
Free-Energy MeV per Nucleon.

ρ Nucleons fm^3	T MEV	No Interactions	Skyrme No α	Skyrme with α	FP
0.00432	15	−53.0	−54.7	−54.8	−55.7
0.00844	15	−42.8	−45.9	−46.4	−48.5
0.00432	10	−29.1	−30.8	−31.7	−32.3
0.00844	10	−22.2	−25.4	−27.5	−28.1

VI. NUCLEAR MATTER AT LOW TEMPERATURES

Thermal properties of Fermi liquids at low temperature are entirely determined by the spectrum $e(k \sim k_F, \rho, T=0)$ of quasi-particles having momenta close to k_F in magnitude. This spectrum is related to the real part of the optical potential $U(e,\rho)$ (not to be confused with the internal energy $U(\rho,T)$)

$$e(k,\rho,T=0) = \frac{\hbar^2 k^2}{2m} + U(e,\rho).$$ (6.1)

We use functions $e(k)$ and $U(e)$ and suppress their density and temperature dependence for brevity. The energy dependent effective mass $m^*(e)$ is given by:

$$\frac{\hbar^2 k(e)}{m^*(e)} = \frac{de(k)}{dk} = \frac{\hbar^2 k(e)}{m} + \frac{dU}{dk} = \frac{\hbar^2 k(e)}{m} + \frac{dU}{de}\frac{\hbar^2 k(e)}{m^*(e)},$$ (6.2)

$$\therefore m^*(e) = m\left(1 - \frac{dU(e)}{de}\right).$$ (6.3)

It was recognized a long time ago that $m^*(e)$ has an enhancement over a small energy interval at the Fermi energy $e_F = e(k_F)$. Generally, states within an energy interval $e_F \pm \pi T$ are important in determining the thermal properties of matter at temperature T. Thus at small T their energies are determined by $m^*(e_F)$ while at larger values of T the average value m^*_{av} over a wide energy interval is more relevant. Since the $m^*(e_F) > m^*_{av}$ the thermal properties have a structure at small T. This structure is very pronounced in the specific heat of atomic liquid ^{3}He where $m^*(e_F)$ is several times m^*_{av} [52,53]; it was pointed out in the NM context by Gerry Brown, who was studying the effects of the relatively small (~30%) difference between the $m^*(e_F)$ and m^*_{av} in nuclear matter.

The enhancement of $m^*(e)$ at e_F is due to the coupling of particle (hole) states with momenta close to k_F with the two-particle one-hole (two-hole one-particle) states, that is not well described either in simple Brueckner calculations or in FP variational calculations. It can be studied in correlated basis perturbation theory[54]. The potential $U(e)$ has been calculated for nuclear matter in the simple variational theory[55], in which there is no enhancement of $m^*(e_F)$, and in second order of perturbation theory using correlated states[56]. The results of these calculations were shown in fig. 4. The second order calculations give the enhancement, and are in reasonable agreement with the empirical data on optical potentials at equilibrium density.[56] The $m^*(e,\rho)$ obtained from the variational and second order calculations is shown in fig. 8. Thus, when these corrections are included in the FP calculations we should expect the specific heat at low temperatures to increase by ~30% at $\rho \sim \rho_o$.

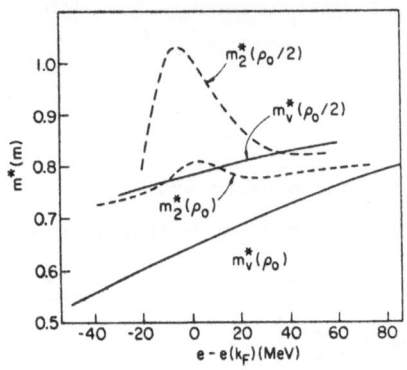

Fig. 8. The energy dependence of the effective mass $m^*(e)$ in cold nuclear matter at density $\rho_0/2$ and ρ_0. The full and dashed lines show results of variational calculations without and with second order corrections.

Neither the perturbative nor the orthogonality corrections[57] to the variational theory of hot nuclear matter have been studied yet. Thus one does not have a quantitative estimate of the errors in the many-body aspects of the FP calculation. It will be most worthwhile to repeat that calculation including these corrections as has been done at zero temperature[56].

The effect of the three-nucleon interactions (TNI) on the spectrum $e(k,\rho,T)$ is also very crudely treated by FP. In their framework a part of the TNI is approximated by a density-dependent two-nucleon interaction, while the contribution of the rest is approximated by a function of ρ alone (eq. 3.4). Thus its contribution to the $m^*(e)$ is totally neglected. Wiringa[58] has recently calculated the optical potential $U(k,\rho)$ in the simple variational theory with the Urbana and Argonne models of v_{ij} and model VII of V_{ijk}. The Argonne v_{ij} + V_{ijk}(VII) gives the largest $m^*(e,\rho)$, while the FP gives the smallest. The differences between the models are small at $\rho < 0.2$ fm^{-3}, but significant at $\rho > 0.3$ fm^{-3} (fig. 9). Thus it seems necessary to include the TNI in the calculation of the thermal properties of matter at high density.

VII. GROWTH OF FLUCTUATIONS IN NUCLEAR MATTER

Matter at ρ,T inside the spinodal curve (fig. 6) is unstable, and small density fluctuations in it would grow and cause it to separate into liquid and gas. The rate of growth of these fluctuations is of interest in studies of fragmentation reactions.[48] Since matter is unstable in this region its free energy has to be defined within perturbation theory as follows. Add the weakest possible interaction δv_{ij}, necessary to stabilize matter at the chosen ρ,T, and let $F_S(\rho,T)$ be the free-energy for the Hamiltonian $H_S=H+\delta v_{ij}$. The $F(\rho,T)$ for "uniform matter" in the unstable region is then defined as:

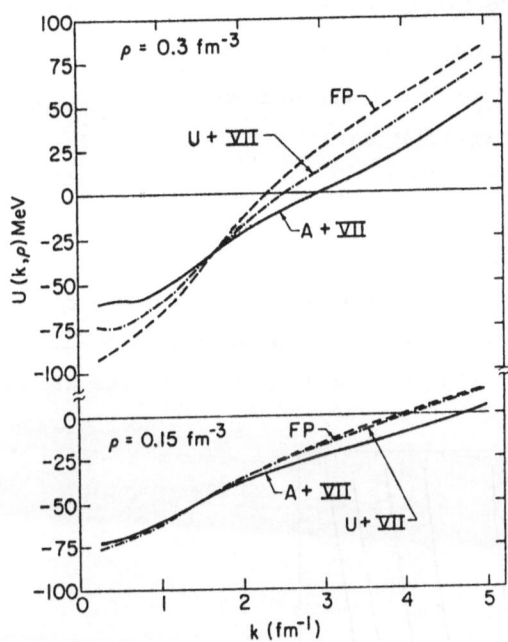

Fig. 9. The single particle potential calculated variationally by Wiringa at $\rho = 0.15$ and 0.3 fm^{-3} with the FP, Argonne and Urbana $v_{ij} + V_{ijk}$ (VII).

$$F(\rho,T) = F_S(\rho,T) - \langle \delta v_{ij} \rangle (\rho,T).\tag{7.1}$$

This procedure is meaningful only if $F(\rho,T)$ does not depend too sensitively on the chosen form of δv_{ij}, and it is found to be useful in classical mechanics.[59,60] The FP calculations do not allow for many-body clusters, and hence are blind to the basic instability of this region. They may provide a reasonable approximation to the $F(\rho,T)$ of "uniform matter" in the unstable region.

Pethick and Ravenhall[61] (PR) have studied the growth rates of density fluctuations in uniform matter in the unstable region. Consider a fluctuation in momentum space:

$$\delta \rho(r,t) = \delta \rho(q,\omega) \, e^{i(\vec{q}\cdot\vec{r}-\omega t)}.\tag{7.2}$$

Its frequency ω generally depends upon λq where λ is the mean-free-path of particles in matter.[62] In the hydrodynamic limit, $\lambda q \ll 1$, one obtains the simple dispersion relation:

$$\omega^2 = c_s^2 q^2,\tag{7.3}$$

$$c_s^2 = \frac{1}{m}\frac{\partial P}{\partial \rho}\bigg|_{S},\tag{7.4}$$

the latter evaluated at constant entropy. When $\partial P/\partial \rho$ is negative, ω is imaginary and fluctuations grow with the rate $\Gamma = -i\omega$. The adiabats $P(\rho,S)$ are shown in fig. 10; their spinodal points are different from

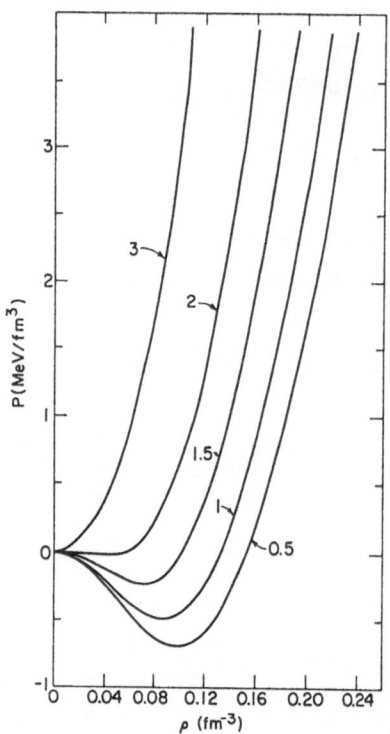

Fig. 10. The adiabats $P(\rho,S)$ of nuclear matter.

those of the isotherms (fig. 5) when T>0, and their loci is shown as the adiabatic spinodal curve in fig. 6 . In classical simulations of heavy-ion collisions,[63] fragmentation is seen to occur inside the region bounded by the adiabatic spinodal, because in these collisions $\lambda <<$ the size of the system, and hence $\lambda q <<1$.

At T<10MeV and $\rho <\rho_0$ the λ in nuclear matter is expected to be $\gtrsim$ 6fm which is comparable to the nuclear radii. Thus the above hydrodynamic limit may not be relevant for real (against classical) heavy-ion collisions. At T=0 PR consider the other limit $\lambda q >>1$ of collisionless sound. Its frequency is obtained from solutions of Landau's kinetic equation and becomes imaginary when c_s^2 becomes negative. The growth rate of collisionless fluctuations is a little smaller than that of fluctuations in the hydrodynamic limit.

The differences between hydrodynamic and collisionless sounds have been illustrated by PR assuming that $m^*=m$ for simplicity. In this case

$$c_s^2 = \frac{1}{3} (1+F_0) V_F^2, \tag{7.5}$$

where V_F is the Fermi velocity $\hbar k_F/m$ and F_0 the Landau interaction parameter. The frequency ω and the growth rate Γ calculated in the long wave-length limits are compared in fig. 11. The F_0 varies between ~ -2 and 0 as NM density goes from 0.04 to 0.16 fm^{-3}. The collisionless sound is Landau damped (i.e. has negative Γ) in the interval $F_0 = -1$ to 0. At finite temperatures the effect of collisions should be included in the Landau kinetic equation[62], and in studies of fragmentation of hot nuclei by growth of fluctuations the effects of finite values of q also need to be considered.[48]

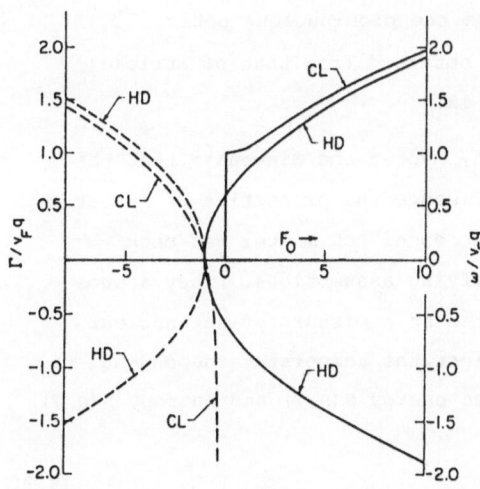

Fig. 11. The frequencies (full lines) and growth rates (dashed lines) of density fluctuations of cold nuclear matter in the collisionless and hydrodynamic limits.

VIII. MATTER AT HIGH TEMPERATURES

At high temperatures the non-nucleonic degrees of freedom are excited and contribute to the properties of matter. The two excitations that appear to be most relevant in the T~50MeV range are pion production and N→Δ (nucleon hole-Δ particle) excitations. The spectrum of noninteracting pions is given by:

$$\omega_F(k) = (\mu^2 + k^2)^{1/2}, \tag{8.1}$$

where μ is the pion mass (~0.7 fm^{-1}), and that of N→Δ excitations is:

$$\omega_R(k) = m_\Delta - m + \frac{k^2}{2m_\Delta} + \delta(\rho), \tag{8.2}$$

where, for the sake of simplicity we neglect the momentum of the nucleon hole, m_Δ is the mass of Δ (~1232 MeV) and its width is also neglected, and $\delta(\rho)$ is the difference between the interaction energies of a nucleon and a Δ in nuclear matter at density ρ. It may be more appropriate to consider $\delta(\rho, k)$ since the interaction energy of Δ could depend upon its momentum, or equivalently use an effective mass m_Δ^* instead of the bare mass m_Δ in eq. (8.2).

The spectra $\omega_F(k)$ and $\omega_R(k)$ are shown in fig. 12 for δ= 50 MeV. At small k the pion excitation has lower energy, while at large k the N→Δ excitation has lower energy. The two modes are coupled with the strong πNΔ coupling and they naturally mix. The coupled low energy excitation, recently named pisobar[45], is essentially a pion at small k, and N→Δ excitation at large k and a mixture of the two at intermediate values of k. The calculation of the spectrum $\omega(k\rho)$ of this excitation is discussed often in the literature[64, 65] and also in this volume[66], and the results obtained in ref. 65 (denoted by FPU) are shown in fig. 12. The $\omega(k, \rho_0)$ can be estimated from the pion-nucleus optical potential at small k, and the values obtained from that of Stricker, McManus and Carr[67] are shown in fig. 12.

It was pointed out by Mishustin, Myhrer and Siemens[68] that the pisobar excitations will strongly influence the properties of matter at T~50 MeV. Their effect on the properties of hot matter has been estimated by FPU by making two simplifying assumptions. They assume that hot hadronic matter can be treated as a mixture of hot nuclear matter and pisobar gas, and also neglect the temperature dependence of the pisobar spectrum $\omega(k, \rho)$. The free energy $F(\rho, T)$ and entropy $S(\rho, T)$ of hot hadronic matter is given by:

$$F(\rho, T) = F_N(\rho, T) + F_{PB}(\rho, T), \tag{8.3}$$

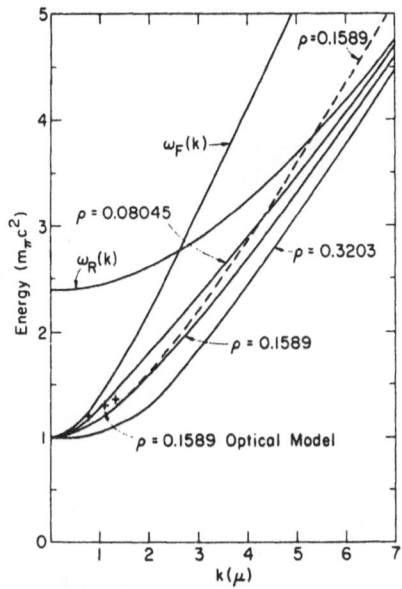

Fig. 12. The spectrum of free pions $\omega_F(k)$, $N \rightarrow \Delta$ excitations $\omega_R(k)$ and pisobars in cold nuclear matter. The results of a crude calculation of the pisobar spectrum are shown by a dashed line.

$$S(\rho,T) = S_N(\rho,T) + S_{PB}(\rho,T),\qquad(8.4)$$

The $F_N(\rho,T)$ and $S_N(\rho,T)$ are calculated using the variational method of FP, and $F_{PB}(\rho,T)$ and $S_{PB}(\rho,T)$ are the free-energy and entropy of pisobar gas at chemical potential $\mu_{PB}=0$, since pisobars can be freely created or absorbed.

The typical results obtained by FPU are shown in fig. 13. The total excitation energy $E^*(\rho,T)$ in this figure is defined as:

$$E^*(\rho,T) = \langle H \rangle(\rho,T) + 16\text{ MeV} = U(\rho,T) + 16\text{ MeV}\qquad(8.5)$$

where the 16 MeV comes from the binding energy of nuclear matter. E_π is just the energy of pisobars per nucleon, and $R_{\pi N}$ is the ratio of pisobars to nucleons. The pisobars are created at temperatures of ~30 MeV, and appear to dominate the energy and entropy of hadronic matter at $T \gtrsim 100$ MeV.

Due to their lower excitation energy pisobars contribute a lot more to the properties of hot hadronic matter than free pions would have. For example, if the spectrum $\omega_F(k)$ is used to calculate the number of pions per nucleon in hot matter we obtain the approximate relation[69]

$$R_{F\pi N} = 0.65\ \frac{\rho_0}{\rho}\ \left(\frac{T}{m_\pi}\right)^4.\qquad(8.6)$$

The $R_{F\pi N} \ll R_{\pi N}$; for example at $T = m_\pi/2$ and $\rho = \rho_0/2$, ρ_0 and $2\rho_0$ $R_{F\pi N}$ is 0.07, 0.035 and 0.0175 respectively while $R_{\pi N}$ is 0.25, 0.18 and 0.13. Thus it is very necessary to take into account the $\pi N\Delta$ coupling in calculating the properties of matter at $T>30$ MeV.

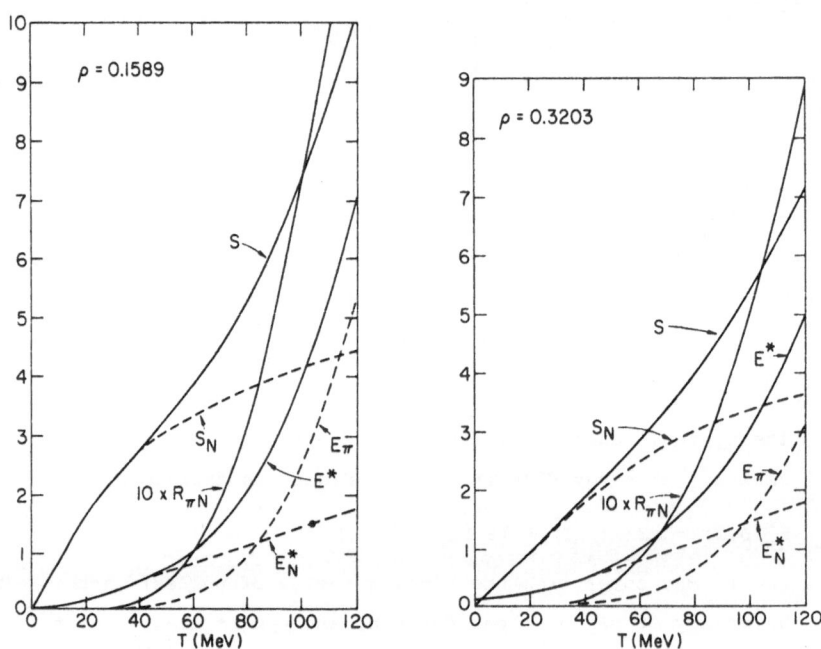

Fig. 13. The properties of very hot nuclear matter at densities ρ_o and $2\rho_o$, as calculated by FPU. The vertical scale gives entropies per nucleon, $10 \times R_{\pi N}$ and the excitation energies in units of 100 MeV/nucleon.

The approximations in the FPU calculations are probably reasonable at lower temperatures where $R_{\pi N}$ is small. Their calculation neglects the interactions between the pisobars, which would be important when $R_{\pi N}$ ~ 1. At T ~ 100 MeV heavier mesons and higher energy baryon resonances may get excited before matter turns into a quark-gluon plasma. At zero baryon density the transition to quark-gluon plasma is expected to occur at T ~ 155 MeV.[11] We may expect this transition to occur at $T \lesssim 155$ MeV at densities ~ρ_0. At some high density matter will become a deconfined quark liquid even at T=0.

ACKNOWLEDGEMENT

This work was supported by the U.S. National Science Foundation under Grant PHY84-15064.

REFERENCES

1. R. P. Feynman, "Statistical Mechanics", Benjamin-Cummings (1982).
2. J. W. Negele and H. Orland, "Quantum Many-Particle Systems", Addison-Wesley (1988).
3. D. M. Ceperley and E. L. Pollock, Phys. Rev. Lett. 56, 351 (1986).
4. E. L. Pollock and D. M. Ceperley, Phys. Rev. B30, 2555 (1984).
5. J. Carlson, Phys. Rev. C36, 2026 (1987); C38, 1879 (1988).
6. C. Mahaux - in this volume.
7. K. E. Schmidt and V. R. Pandharipande, Phys. Lett. B87, 11 (1979).
8. D. Pines and P. Nozières, "Theory of Quantum Liquids I" Benjamin (1966), G. Baym and C. J. Pethick in "The Physics of Liquid and Solid Helium" Part II, Ed: K. H. Bennemann and J. B. Ketterson, Wiley (1978).
9. A. Lejeune et al., Nucl. Phys. A453, 189 (1986).
10. V. R. Pandharipande and R. B. Wiringa, Rev. Mod. Phys. 51 831 (1979).
11. S. Gottlieb et al., Phys. Rev. Lett. 59, 1513 (1987), M. Grady, et al., Phys. Lett. B200, 149 (1988).
12. M. Lacombe et al., Phys. Rev. C21 861 (1980).
13. I. E. Lagaris and V. R. Pandharipande, Nucl. Phys. A359, 331 (1981).
14. R. B. Wiringa, R. A. Smith and T. L. Ainsworth, Phys. Rev. C19, 1207 (1984).
15. R. Machleidt, K. Holinde and Ch. Elster, Phys. Repts. 149, 1 (1987).
16. M. M. Nagels, T. A. Rijken and J. J. DeSwart, Phys. Rev. D17, 768 (1978).
17. P. L. Chung et al., Phys. Rev. C37, 2000 (1988).
18. E. M. Nyman and D. O. Riska, Phys. Rev. Lett. 57, 3007 (1986).
19. R. B. Wiringa, V. Fiks and A. Fabrocini, Phys. Rev. C38, 1010 (1988).
20. C. R. Chen et al., Phys. Rev. C31, 2266 (1985). J. L. Friar, B. F. Gibson and G. L. Payne, Phys. Rev. C37, 2869 (1988).
21. T. Sasakawa, Nucl. Phys. A463, 327c (1987).
22. S. Ishikawa and T. Sasakawa, Phys. Rev. C36, 2037 (1987).
23. R. Schiavilla, V. R. Pandharipande and R. B. Wiringa, Nucl. Phys. A449, 219 (1986).
24. R. B. Wiringa, Private Communication (1988).
25. J. A. Tjon, Phys. Rev. Lett. 40, 1239 (1978).
26. H. Kümmel, K. H. Lührmann and J. G. Zabolitsky, Phys. Reports C36, 1 (1978).

27. M. H. Kalos et al., Phys. Rev. B24, 115 (1981).
28. J. Carlson and R. M. Panoff, Private Communication (1988).
29. B. D. Day and R. B. Wiringa, Phys. Rev. C32, 1057 (1985).
30. B. F. Gibson and B. H. J. McKellar, Few Body Syst. 3, 143 (1988).
31. I. Fujita and H. Miyazawa, Prog. Theo. Phys. 17, 360 (1957).
32. Lecture notes in Physics vol. 260. Ed. B. L. Berman and B. F. Gibson (1986).
33. V. R. Pandharipande, Lect. Notes in Phys. 260, 59 (1986).
34. P. V. Sauer, Lect. Notes in Physics, 260, 107 (1986).
35. I. E. Lagaris and V. R. Pandharipande, Nucl. Phys. A359, 349 (1981).
36. B. A. Friedman and V. R. Pandharipande, Nucl. Phys. A361, 502 (1981).
37. B. Serot, in this volume.
38. R. Malfliet, in this volume.
39. B. D. Keister and R. B. Wiringa, Phys. Lett. B173, 5 (1986).
40. T. L. Ainsworth, et al., Nucl. Phys. A464, 740 (1987).
41. C. J. Horowitz and B. D. Serot, Nucl. Phys. A464, 613 (1987).
42. R. Malfliet, Nucl. Phys. A488, 721c (1988).
43. J. P. Blaizot, Phys. Reports 64, 171 (1980).
44. M. M. Sharma, et al., Phys. Rev. 38C, 2562 (1988).
45. G. E. Brown, Nucl. Phys. A488, 689C (1988).
46. G. F. Bertsch and S. DasGupta, Phys. Rept.160, 189 (1988).
47. H. Krivine, J. Treiner and O. Bohigas, Nucl. Phys. A336, 155 (1980).
48. H. Heiselberg, C. J. Pethick and D. G Ravenhall, Phys. Rev. Lett. 61, 818 (1988).
49. S. Pratt, P. Siemens and Q. N. Usmani, Phys. Lett. 189, 1 (1987).
50. A. L. Fetter and J. D. Walecka, Quantum Theory of Many-Particle Systems, McGraw-Hill, 1971.
51. D. Q. Lamb, et al., Nucl. Phys. A360, 459, 1981.
52. G. E. Brown, C. J. Pethick and A. Zaringhalam, J. Low Temp. Phys. 48k, 349 (1982).
53. S. Fantoni, V. R. Pandharipande and K. E. Schmidt, Phys. Rev. Lett. 48, 878 (1982).
54. E. Feenberg, Theory of Quantum Liquids, Academic Press, 1969.
55. B. A. Friedman and V. R. Pandhairpande, Phys. Lett. 100B, 205 (1981).
56. S. Fantoni, B. L. Friman and V. R. Pandharipande, Nucl. Phys. A399, 51 (1983).
57. S. Fantoni and V. R. Pandharipande, Phys. Rev. C37, 1697 (1988)
58. R. B. Wiringa, Phys. Rev. C38, 2967 (1988).
59. D. Levesque and L. Verlet, Phys. Rev. 182, 307 (1969).
60. A. Vicentini, G. Jacucci and V. R. Pandharipande, Phys. Rev. C31, 1783 (1985).
61. C. J. Pethick and D. G. Ravenhall, Ann. of Phys. 183, 131 (1988).
62. P. J. Siemens, in this volume.
63. T. J. Schlagel and V. R. Pandharipande, Phys. Rev. C36, 162 (1987).
64. A. B Migdal, Rev. Mod. Phys. 50, 107 (1978).
65. B. Friedman, V. R. Pandharipande and Q. N. Usmani, Nucl. Phys. A372, 483 (1981).
66. M. Thies, in this volume.
67. K. Stricker, H. McManus and J. A. Carr, Phys. Rev. C19, 929 (1979).
68. I. M. Mishustin, F. Myhrer and P. J. Siemens, Phys. Lett. 95B, 361 (1980).
69. M. Sobel et al., Nucl. Phys. A251, 502 (1975).

INTRODUCTION TO

NON-RELATIVISTIC TRANSPORT THEORIES [1]

C. Grégoire[2][3]

Physics Department, State University of New York at
Stony Brook, Stony Brook, New-York 11794-3800, USA

ABSTRACT

We give an introduction to non-relativistic transport theories. In the first part, we show how kinetic equations can be derived from the BBGKY hierarchy. Maxwell models are briefly discussed. The hydrodynamical regime and the Chapman-Enskog solution for classical systems are reviewed. Applications to noble gases are given. We give an outline of the Boltzmann-Langevin method for dealing with fluctuations. The semi-classical description of nuclear systems is discussed by comparing TDHF simulations to Vlasov calculations. The second part is devoted to quantal non-relativistic systems. It is based on the Green's function formalism. Special emphasis is given to the time-ordering method proposed by Keldysh. It is shown that a generalized Boltzmann equation can be obtained via smoothness arguments. This equation reduces to the semi-classical Landau-Vlasov equation in the quasi-particle approximation and the T-matrix approximation. Examples like the impurity scattering problem and Fermi liquids are given. It is argued that for nuclear systems as those encountered in heavy-ion collisions above 200-300 MeV/nucleon a proper account of the quantal effects in the collisional regime is required.

INTRODUCTION

In these lectures, we are concerned with classical and quantal transport theories in a non-relativistic framework. We have tried to show how various theoretical methods have been used so far for studying transport properties of gases, liquids and metals. A presentation of some aspects of a relativistic quantum theory will be found in

---

[1]Lectures presented during Les Houches Winter School in theoretical nuclear physics, February 1989

[2]Work supported in part by US DOE Grant DE-FG02-88 ER40388 and by NATO

[3]Permanent address: GANIL,BP5027,14021,Caen,Cedex,France

Dr.P.Siemens'lectures. Since our aim is more to discuss textbook methods than to illustrate their applications to nuclear physics, very little is given as far as these applications are concerned. Nevertheless, two messages should be extracted from this overview of non-relativistic transport theories: first, the semi-classical version the time dependent Hartree-Fock method, namely the Vlasov equation, is a very good approximation for dealing with moderate energy heavy-ion collisions; the quantal behavior of the mean-field is rather unimportant. Secondly, and this is less well known, quantal features, such as off-shell effects, turn out to be crucial for the approach to equilibrium in heavy-ion collisions above 200 MeV per nucleon. In other words, in the regime where collisions dominate over the mean field a generalization of the classical Boltzmann theory is required in order to include quantal effects. Obviously it becomes more and more important when pion degrees of freedom are involved because absorption and emission of pions render the self-energies highly ω - dependent. The first part is devoted to the classical transport theory. The hierarchy of equations for reduced distribution functions is derived. The truncation schemes are discussed and Vlasov and Boltzmann equations are derived. Solvable problems are mentioned. Hydrodynamics and the Chapman-Enskog method for determining transport coefficients are sketched. Results for noble gases are reported. A treatment of fluctuations via the Boltzmann-Langevin equation is refered to. Applications of the classical theory to nuclear systems are considered by comparing TDHF solutions to Vlasov solutions for nuclear slab collisions.

The second part is devoted to the non-relativistic quantal transport theory. The Kadanoff-Baym equations for the propagators are given. It is shown that a time-ordering along Keldysh contours allows one to write these equations as generalized Dyson equations. A transport Boltzmann-like equation is obtained via smoothness arguments(gradient expansion). The central role of the spectral function is emphasized. It is shown that, in the framework of the T-matrix approximation, the quasi-particle limit allows one to write a Landau-Vlasov equation with a Uehling-Uhlenbeck collision integral. Applications to the impurity scattering problem in metals, Fermi liquids, and nucleus-nucleus collisions are indicated.

A.TRANSPORT THEORY FOR CLASSICAL SYSTEMS

A.1.Outline of kinetic theories

A.1.1.Reduced densities and BBGKY hierarchy

We consider a system of N classical particles whose positions are $\vec{r}_1, \cdots, \vec{r}_N$ and whose momenta are $\vec{p}_1, \cdots, \vec{p}_N$. The density of particles in phase-space is given by the distribution function $D(x_1, \cdots, x_N)$, where $x_i \equiv (\vec{r}_i, \vec{p}_i)$. If one is only interested in observables which depend on k particles, it is convenient to define a reduced density or reduced distribution function $f^{(k)}$. The reduced density is proportional to the integral of D performed over $x_{k+1}, \cdots, x_N$. We will denote this integration by $\overset{\mathrm{Tr}}{(k+1...N)}$. In order to keep $\overset{\mathrm{Tr}}{(1...N)} D = 1$, $f^{(k)}$ is defined by:

$$f^{(k)} = \frac{N!}{(N-k)!} \underset{(k+1...N)}{\mathbf{Tr}} \mathbf{D}. \tag{1}$$

The expectation value of any k-body observable A_k reads:

$$\langle A_k \rangle = \underset{(1...N)}{\mathbf{Tr}} A_k \mathbf{D} = \frac{1}{k!} \underset{(1...k)}{\mathbf{Tr}} f^{(k)} A_k. \tag{2}$$

The Bogoliubov-Born-Green-Kirkwood-Yvon (BBGKY) hierarchy is the set of equations of motion for the reduced densities (or distribution functions)[1] . Their derivation can be summarized into three steps : *i)* start with the classical Liouville equation for the full many-body distribution function (the Von Neumann equation in quantum mechanics), *ii)* project the equation by means of the **Tr** operator, *iii)* reorder various terms. The Liouville equation reads:

$$\frac{\partial}{\partial t} \mathbf{D} = \mathcal{L} \mathbf{D}, \tag{3}$$

where $\mathcal{L}$ is the Liouvillien operator defined by: $\mathcal{L}\mathbf{D} = \{\mathcal{H}, \mathbf{D}\}$; the $\{ \, , \, \}$ symbol is the usual Poisson bracket:

$$\{a, b\} = \sum_{i=1}^{N} \left(\frac{\partial a}{\partial q_i} \frac{\partial b}{\partial p_i} - \frac{\partial a}{\partial p_i} \frac{\partial b}{\partial q_i} \right).$$

The Hamiltonian $\mathcal{H}$ being:

$$\mathcal{H} = \sum_{i=1}^{N} K_i + \sum_{i \leq j=1}^{N} V_{ij}, \tag{4}$$

where $K_i = \vec{p_i}\,^2/2m$ is the classical kinetic energy, we decompose the Liouvillien into two parts:

$$\mathcal{L}\mathbf{D} = \left(\sum_{i=1}^{N} \mathcal{L}_i^0 + \sum_{i \leq j=1}^{N} \mathcal{L}_{ij} \right) \mathbf{D}, \tag{5}$$

where $\mathcal{L}_i^0 = \{K_i, \}$ and $\mathcal{L}_{ij} = \{V_{ij}, \}$. The projected Liouville equation is then:

$$\frac{d}{dt} f^{(k)} = \frac{N!}{(N-k)!} \underset{(k+1...N)}{\mathbf{Tr}} \left[\sum_{i=1}^{N} \mathcal{L}_i^0 + \sum_{i \leq j=1}^{N} \mathcal{L}_{ij} \right] \mathbf{D}. \tag{6}$$

From the property $\frac{d}{dt} \overset{\text{Tr}}{(1...N)} D = 0$, it is easy to show that:

$$\overset{\text{Tr}}{j} \mathcal{L}_j^0 D = \overset{\text{Tr}}{jk} \mathcal{L}_{jk} D = 0. \tag{7}$$

These relations and the fact that the summation from $k+1$ to n gives $(N-k) \overset{\text{Tr}}{(k+1...N)}$ $\mathcal{L}_{i,k+1}$ in the interaction term of the projected Liouville equation, allows us to write the equation for the reduced density $f^{(k)}$:

$$\frac{d}{dt} f^{(k)} = \sum_{i=1}^{k} \mathcal{L}_i^0 f^{(k)} + \sum_{j=1}^{k} \sum_{i \leq j} \mathcal{L}_{ij} f^{(k)} + \sum_{i=1}^{k} \overset{\text{Tr}}{(k+1)} \mathcal{L}_{i,k+1} f^{(k+1)}. \tag{8}$$

This equation defines a hierarchy for k ranging from 1 to N, since the equation for $f^{(k)}$ is coupled to the equation for $f^{(k+1)}$, the equation for $f^{(k+1)}$ to the equation for $f^{(k+2)}$ and so on. In other words, it is a *not-closed* set of equations for the reduced densities. The first two equations of this BBGKY hierarchy are:

$$\mathbf{k = 1} \qquad \partial_t f^{(1)} = \mathcal{L}_1^0 f^{(1)} + \overset{\text{Tr}}{(2)} \mathcal{L}_{12} f^{(2)} \tag{9}$$

$$\mathbf{k = 2} \qquad \partial_t f^{(2)} = (\mathcal{L}_1^0 + \mathcal{L}_2^0) f^{(2)} + \mathcal{L}_{12} f^{(2)} + \overset{\text{Tr}}{(3)} (\mathcal{L}_{13} + \mathcal{L}_{23}) f^{(3)}, \tag{10}$$

with obvious notations. In the following sections, we shall consider these two equations as a starting point for the derivation of the Vlasov and of the Boltzmann equations. Since this hierarchy is not-closed, the main purpose of any kinetic theory is to achieve a truncation of the hierarchy depending on the physical situation. Similarly, in relativistic quantal theories, one can derive the so-called Martin-Schwinger hierarchy of equations for the n-point Green's functions[2]. The reduction of this hierarchy to a closed set of equations for the one and two particle propagation is, *in fact*, the main problem to be solved in order to build a theory describing transport properties.

A.1.2. The Vlasov equation

The most drastic truncation of the BBGKY hierarchy is obtained by assuming the absence of any correlation. In such a case, the two-body distribution function reads:

$$f^{(2)}(x_1, x_2) = f^{(1)}(x_1) \cdot f^{(1)}(x_2). \tag{11}$$

Under this assumption, equation (9) becomes:

$$\partial_t f^{(1)} = \{\frac{\vec{p}_1^{\,2}}{2m}, f^{(1)}\} + \{U^{eff}(f^{(1)}), f^{(1)}\}. \tag{12}$$

where the effective selfconsistent potential is:

$$U^{eff}(f^{(1)}) = \int d\vec{r}_2 d\vec{p}_2 V_{12} f^{(1)}(\vec{r}_2, \vec{p}_2). \tag{13}$$

136

Equation (12) is called the Vlasov equation. The mean field defined by equation (13) is obtained by a summation over all the other particles. The self-consistency renders this equation highly non-linear. In spite of the drastic assumption (11), the Vlasov equation is still conserving (the conserving properties are an absolute requirement for deciding if a truncation scheme is acceptable or not): mass conservation, i.e. a continuity equation, momentum conservation and total energy conservation can be checked directly by multiplying the Vlasov equation by $\vec{p}^{\,l}$ with $l = 0, 1, 2$ and then integrating by parts over $d\vec{p}$. Solutions of this equation for colliding nuclear slabs are given in section A.3.2. with a discussion about its limits of applicability in nuclear physics. The Vlasov equation is the basic equation in plasma physics [3] (with the long range Coulomb interaction) and also in astrophysics for globular cluster evolution under the effect of gravitational forces [4].

A.1.3. The Boltzmann equation

For dilute gases, it is possible to neglect the three-body terms in the interaction,i.e. to truncate the second equation of the BBGKY hierarchy:

$$\partial_t f^{(2)} = (\mathcal{L}_1^0 + \mathcal{L}_2^0 + \mathcal{L}_{12}) f^{(2)}. \tag{14}$$

In order to derive an equation for $f^{(1)}$, we shall make two further assumptions:
i) The two-body distribution factorizes at the initial time t_0. This initial time should be understood as the time between two successive interactions of two particles. This initial factorization relies on a randomization assumption known as the molecular chaos ansatz or the Boltzmann Stosszahlansatz. It introduces irreversibility in the equations and leads to the H-theorem.
ii) The two-body distribution is a functional of the one-body distribution function. In other words, one shall time-smooth $f^{(2)}$. As a matter of fact, $f^{(2)}$ varies more rapidly than $f^{(1)}$. Assuming such a functional dependence means that one neglects the rapid time variations of $f^{(2)}$. Only at the level of $f^{(1)}$ will time variations be retained [5].
The assumption *ii)* allows one to express the time derivative of $f^{(2)}$ as a function of the time derivative of $f^{(1)}$:

$$\partial_t f^{(2)} = \frac{\delta f^{(2)}}{\delta f^{(1)}} \partial_t f^{(1)}. \tag{15}$$

Introducing the time evolution operator Δ_t we obtain:

$$\partial_t f^{(2)} = \frac{\delta f^{(2)}}{\delta f^{(1)}} \partial_t (\Delta_t f^{(1)}(t_0)) = \frac{\partial}{\partial \tau} f^{(2)}(x_1, x_2, f^{(1)}(\tau)), \tag{16}$$

where $\tau = t - t_0$. On the other hand, the stosszahlansatz reads:

$$f^{(2)}(x_1, x_2, t_0) = f^{(1)}(x_1, t_0) . f^{(1)}(x_2, t_0). \tag{17}$$

From (16) and (17), one deduces:

$$f^{(2)}(x_1, x_2, t) = f^{(1)}(X_1(t)) f^{(1)}(X_2(t)), \qquad (18)$$

with $X_1(t_0) = x_1(t_0)$ and $X_2(t_0) = x_2(t_0)$. Equation (18) differs from equation (11), because $X_1(t)$ and $X_2(t)$ evolve in time according to their initial conditions. In other words, in order to get expressions for them, one should consider the boundary conditions at t_0 where they are equal to x_1 resp. x_2. A consequence is:

$$\partial_t f^{(2)} = \frac{\vec{P}_1}{m} \frac{\delta f^{(1)}}{\delta \vec{R}_1} f^{(1)}(\vec{R}_2, \vec{P}_2) + \frac{\vec{P}_2}{m} \frac{\delta f^{(1)}}{\delta \vec{R}_2} f^{(1)}(\vec{R}_1, \vec{P}_1), \qquad (19)$$

since the evolution operator Δ_t is deduced from $\mathcal{L}_1^0$. The quantity $\mathcal{L}_{12} f^{(2)}$ can be obtained from equation (14) and used in the first equation of the BBGKY hierarchy, namely equation (9). In the frame of the particle 1, the free propagation terms of equation (14) read:

$$-(\mathcal{L}_1^0 + \mathcal{L}_2^0) = \frac{(\vec{p}_2 - \vec{p}_1)}{m} \nabla_{\vec{r}} + \frac{\vec{p}_1}{m} \nabla_{\vec{r}_1}, \qquad (20)$$

with $\vec{r} = \vec{r}_2 - \vec{r}_1$. Using the equations (9), (14), (19) and (20) and neglecting the terms in $\nabla_{\vec{R}_1}, \nabla_{\vec{R}_2}, \nabla_{\vec{r}_1}$, which vanish for a homogeneous gas:

$$\partial_t f^{(1)} = \mathcal{L}_1^0 f^{(1)} + \int d2 \frac{(\vec{p}_2 - \vec{p}_1)}{m} . \nabla_{\vec{r}} f^{(1)}(X_1(t)) f^{(1)}(X_2(t)). \qquad (21)$$

By choosing the z-axis along the $(\vec{p}_2 - \vec{p}_1)$ direction and with $\vec{r}_2 \equiv (b, \varphi, z)$ in cylindrical coordinates (b is the impact parameter of the two-body collision):

$$\partial_t f^{(1)} = \mathcal{L}_1^0 f^{(1)} + \int d\vec{p}_2 \int d\varphi \int b\,db \int dz \frac{|\vec{p}_2 - \vec{p}_1|}{m} \frac{\partial}{\partial z} [f^{(1)}(X_1(t)) f^{(1)}(X_2(t))]. \qquad (22)$$

The integration over dz should be performed from $-\infty$, before the two-body collision (momenta are $\vec{p}_1$ and $\vec{p}_2$ as they are for t_0), to $+\infty$ after the scattering (momenta are $\vec{p}_1{}'$ and $\vec{p}_2{}'$):

$$\partial_t f^{(1)} + \frac{\vec{p}_1}{m} \frac{\partial f^{(1)}}{\partial \vec{r}_1} = 2\pi \int d\vec{p}_2 \int b\,db \frac{|\vec{p}_2 - \vec{p}_1|}{m} (f^{(1)}(\vec{p}_1{}') f^{(1)}(\vec{p}_2{}') - f^{(1)}(\vec{p}_1) f^{(1)}(\vec{p}_2)). \qquad (23)$$

This equation is the classical Boltzmann equation. It was derived by assuming the stosszahlansatz and also the functional dependence of $f^{(2)}$ on $f^{(1)}$. This latter assumption can be viewed as a time smoothing procedure.

Kirkwood [6] has proposed another derivation of the Boltzmann equation in which time-smoothing arguments play a major role. With the introduction of time-smoothed distribution functions:

$$\bar{f}^{(i)}(t) = \frac{1}{\tau} \int_0^\tau d\eta \, f^{(i)}(t + \eta) \quad i = 1, 2, ..., \tag{24}$$

the first equation of the BBGKY hierarchy becomes:

$$\partial_t \bar{f}^{(1)} + \frac{\vec{p}_1}{m} \frac{\partial}{\partial \vec{r}_1} \bar{f}^{(1)} - \frac{\partial}{\partial \vec{p}_1} \int d2 \frac{\partial V_{12}}{\partial \vec{r}_1} \bar{f}^{(2)} = 0. \tag{25}$$

Assuming that spatial and time variations of $\bar{f}^{(2)}$ can be neglected between two successive collisions, the total time derivative of $\bar{f}^{(2)}$ is approximated by:

$$\frac{d\bar{f}^{(2)}}{dt} \approx \left(\frac{\partial \vec{p}_1}{\partial t} \frac{\partial}{\partial \vec{p}_1} + \frac{\partial \vec{p}_2}{\partial t} \frac{\partial}{\partial \vec{p}_2} \right) \bar{f}^{(2)}. \tag{26}$$

As a consequence one obtains:

$$
\begin{aligned}
-\partial/\partial \vec{p}_1 \textstyle\int d2 \, \partial V_{12}/\partial \vec{p}_1 \, \bar{f}^{(2)} &= \textstyle\int d2 \, d\bar{f}^{(2)}/dt \\
&= \textstyle\int b \, db \int d\varphi \int d\vec{p}_2 \, \tau \mid \vec{p}_1 - \vec{p}_2 \mid /m \times \left((f^{(2)}(t + \tau) - f^{(2)}(t)) \right) /\tau).
\end{aligned} \tag{27}
$$

In the last expression, cylindrical coordinates have been used and the integration about the z-axis gives $\tau \mid \vec{p}_1 - \vec{p}_2 \mid /m$. The smoothing parameter τ should be taken larger than the typical collision time. Then, it is easy to derive the Boltzmann equation by formulating the molecular chaos assumption:

$$f^{(2)}(t + \tau) - f^{(2)}(t) = f^{(1)}(\vec{p}_1{}') f^{(1)}(\vec{p}_2{}') - f^{(1)}(\vec{p}_1) f^{(1)}(\vec{p}_2). \tag{28}$$

This derivation shows that time-smoothing accounts for the residual interactions. As a matter of fact, such a truncation of high frequencies in TDHF has been proposed as a method to go beyond the mean field [7]. The main difficulties are the same as in the classical problem, namely: the smoothing parameter τ is not known and the resulting equations are not necessarily conserving. In the classical case, the resulting equation is conserving only if we impose momentum and energy conservation between time t and the time $t + \tau$.

A.2. Kinetic theory of gases

Dilute gases are typical examples of systems for which the assumptions used for deriving the Boltzmann equation are fulfilled. Since the methods developed for characterizing transport properties of gases are quite general, it is worthwhile to recall them. First of all, solvable models will be mentioned; secondly, the hydrodynamical limit and the Chapman-Enskog method will be discussed; thirdly, an extension of the Boltzmann equation for incorporating fluctuations will be presented.

A.2.1. Solvable problems

In general, the Boltzmann equation is not solvable analytically. Numerical methods are required. Some of these methods are discussed in G. Welke's lectures given in this school. To have at hand special cases for which the Boltzmann equation can be solved exactly is obviously of great interest in order to test these numerical algorithms. They provide also some insight into the relaxation processes. Let us consider an interaction potential between particles proportional to r^{-s}. If the dimensionality of the space is D, it turns out that the cross-section is proportional to the relative momentum to the power $2(D-1)/s$. Since the Boltzmann collision integral is:

$$I_{coll} = \int d\vec{p}_2 \frac{|\vec{p}_2 - \vec{p}_1|}{m} \frac{d\sigma}{d\Omega} d\Omega (f(\vec{p}_1 \,')f(\vec{p}_2 \,') - f(\vec{p}_1)f(\vec{p}_2)), \tag{29}$$

it is clear that the term $|\vec{p}_1 - \vec{p}_2| \, d\sigma/d\Omega$ becomes constant if $(D-1)/s = 1/2$. This allows one to find an analytical solution of the Boltzmann equation. For $D = 2$, and 3, these models are known as the Tjon-Wu and the Krook-Wu models respectively. For such "Maxwell" models, it is found that the approach toward the equilibrium solution is very slow at large velocities and non-uniform. Moreover, this approach is strongly dependent on the initial conditions. In fact, analytical solutions have been found only for a class of initial conditions. For a general discussion of these relaxation processes see ref[8]. Since analytical solutions of the Boltzmann equation are available only for very specific situations such as in the Maxwell models, it is quite natural to try to characterize the one-body distribution function by a moment expansion .

A.2.2. Moment expansion and hydrodynamics

For convenience, the distribution function will be given in velocity space (instead of momentum): $f \equiv f(\vec{r}, \vec{v}, t)$. Let us then define macroscopic quantities as the first moments of f in the velocity variable $\vec{v}$:

$$\mathbf{k = 0}: \quad \rho(\vec{r}, t) = \int f^{(1)}(\vec{r}, \vec{v}, t) d\vec{v} \tag{30}$$

$$\mathbf{k = 1}: \quad \vec{u}(\vec{r}, t) = \frac{1}{\rho(\vec{r}, t)} \int \vec{v} f^{(1)}(\vec{r}, \vec{v}, t) d\vec{v} \tag{31}$$

$$\mathbf{k = 2}: \quad \frac{3}{2} T(\vec{r}, t) = \frac{1}{\rho(\vec{r}, t)} \int \frac{m}{2} (\vec{v} - \vec{u})^2 f^{(1)}(\vec{r}, \vec{v}, t) d\vec{v}. \tag{32}$$

These moments are the local density, the mean velocity and the temperature (the Boltzmann constant has been taken equal to unity) respectively. In order to get the equations of motion for these local quantities, it is sufficient to multiply the Boltzmann equation (23) for $f^{(1)}$ by $\vec{v}^{\,k}$ and to integrate over $\vec{v}$. In fact, the symmetry in the integration arguments allows one to write for any velocity dependent operator Q:

$$\int d\vec{v}\, Q\, I_{coll} = \frac{1}{2}\int \left(Q(\vec{v}\,') + Q(\vec{v}_1\,') - Q(\vec{v}) - Q(\vec{v}_1)\right) f(\vec{v}) f(\vec{v}_1) g \frac{d\sigma}{d\Omega} d\Omega d\vec{v}_1 d\vec{v}, \quad (33)$$

where g is the relative velocity and $d\sigma/d\Omega$ the differential cross section. Then it is clear that for $Q \equiv \vec{v}^{\,k}$ with $k = 0, 1, 2$ the collision term vanishes since mass, momentum and energy are conserved. The equations for $\rho, \vec{u}$ and T are derived only from the kinetic term in the Boltzmann equation. These equations are the hydrodynamical equations involving the pressure tensor, the heat flux and the rate-of-strain tensor. In the limit of local equilibrium, the pressure tensor is diagonal, the heat flux is zero and the equations reduce to the Euler equations for non-viscous hydrodynamics. For non-equilibrium situations, local density, mean velocity and temperatures are still determined by hydrodynamical equations. It does not mean that higher order moments $k = 3, 4, 5 \ldots$ are vanishing; in fact, the zero, first and second order moments only determine a gaussian distribution corresponding to the Maxwell equilibrium.

For rather small deviations from equilibrium, the pressure tensor and the heat flux can be deduced from the equilibrium values via the viscosity coefficients η, κ (shear and bulk viscosity) and the thermal conductivity λ. The coefficients η and κ are defined by:

$$\Pi_{\alpha\beta} = m \int d\vec{v}\, U_\alpha U_\beta f^{(1)} \quad (34)$$

$$\Pi_{\alpha\beta} = \{P + (\frac{2}{3}\eta - \kappa)\nabla\vec{u}\}\delta_{\alpha\beta} - 2\eta D_{\alpha\beta}, \quad (35)$$

where $\vec{U} = \vec{v} - \vec{u}$ is the velocity in the local frame and

$$D_{\alpha\beta} = \frac{1}{2}(\frac{\partial u_\alpha}{\partial r_\beta} + \frac{\partial u_\beta}{\partial r_\alpha}),$$

is the rate-of-strain tensor. The heat conductivity λ is defined by:

$$\vec{J} = -\lambda\nabla T, \quad (36)$$

where $\vec{J} = \int d\vec{v}\frac{1}{2}mU^2\vec{U}f^{(1)}$ is the heat flux. A method for deriving these transport coefficients is to estimate deviations of the distribution function from the non-viscous hydrodynamics. These deviations can be characterized by the non-uniformity of the local quantities (density, mean velocity and temperature). The method is known in the literature as the Chapman-Enskog method. It is an iterative procedure, the distribution being expanded in the non-uniformity parameters about the equilibrium solution. If the Boltzmann equation is written schematically as:

$$D_t(f) = I(ff),\tag{37}$$

the k-th order Chapman-Enskog solution reads:

$$D_t(f_{(k-1)}) = I(f_{(k)}f_{(k)}),\tag{38}$$

f_0 being the equilibrium solution.
In the local frame the total time derivative D_t is:

$$\frac{D}{Dt}f(\vec{r},\vec{U},t) + \vec{U}\frac{\partial f}{\partial \vec{r}} - \frac{D\vec{u}}{Dt}\frac{\partial f}{\partial \vec{U}} - \vec{U}\frac{\partial f}{\partial \vec{U}} \otimes \frac{\partial \vec{u}}{\partial \vec{r}}.\tag{39}$$

If $f_1 = f_0(1 + \Phi)$, the first order solution satisfies:

$$f_0\{\frac{m}{T}[\vec{U}.\vec{U} - \frac{1}{3}U^2\mathbf{1}] \otimes \frac{\partial \vec{u}}{\partial \vec{r}} + [\frac{mU^2}{2T} - \frac{5}{2}]\vec{U}.\frac{\partial LnT}{\partial \vec{r}}\} = I(\Phi).\tag{40}$$

The reader will find easily that for the so-called relaxation time ansatz:

$$I(ff) = \frac{(f - f_0)}{\tau},\tag{41}$$

(which is an economical way to bypass the collision integral in order to describe the approach to the statistical equilibrium), the shear viscosity is:

$$\eta = \frac{1}{15}\int d\vec{U} f_0\tau\frac{(mU^2)^2}{T},\tag{42}$$

and the thermal conductivity is:

$$\lambda = \frac{1}{3}\int d\vec{U} f_0\tau\frac{mU^2}{2T}U^2[\frac{mU^2}{2T} - \frac{5}{2}].\tag{43}$$

Both η and λ are proportional to the relaxation time τ. In general the Φ solution to the Boltzmann equation is obtained by solving an eigenvalue problem $I(\Phi) = \beta\Phi$. For Maxwell gases or realistic Lennard-Jones potential, analytical solutions can be found. They give the well-known Chapman relation between shear viscosity and thermal conductivity:

$$\frac{\lambda}{\eta} = \frac{5}{2}C_v,\tag{44}$$

where C_v is the heat capacity.

142

For gases, the effective molecular interaction has a long range attractive component and a short range repulsive one. Its parametrization leads to the Lennard-Jones potential:

$$V(r) = 4\epsilon[(\frac{\sigma}{r})^{12} - (\frac{\sigma}{r})^6], \tag{45}$$

where σ describes the size of the repulsive core and $-\epsilon$ the depth of the potential. The success of this Chapman-Enskog method and therefore of the classical kinetic theory, is due to its ability to reproduce the experimental shear viscosity, heat conductivity and their temperature variation with a unique set of parameters for the Lennard-Jones potential. For instance, $\epsilon/k = 124K$ and $\sigma = 3.418\text{Å}$ for Argon, $\epsilon/k = 229K$ and $\sigma = 4.055\text{Å}$ for Xenon can be determined from the experimental shear viscosity at two different temperatures. It is found that the experimental temperature variation of the viscosity follows the Chapman-Enskog solution in the entire pressure-temperature range of the gaseous phase. The heat conductivity is also described consistently [9].

A.2.3. Fluctuations of the distribution function

The definition eq(1) for the one-body reduced density $f^{(1)}$ is equivalent to a coarse-graining in phase space for the distribution function. As a matter of fact, the integrations defined by the Tr operator automatically perform this coarse-graining by selecting the elementary integration volumes. One can also interpret $f^{(1)}$ as an ensemble averaging of a distribution function f over some initial distribution in phase space. In other words, f is related to $f^{(1)}$ by $f^{(1)} = \langle f \rangle$. Fluctuations are then just the departure of f from its average value. The evolution equation for f reads:

$$D_t f = I_{coll}(f) + F(\vec{r}, \vec{v}, t). \tag{46}$$

The quantity F is a Langevin force which fulfills the condition:

$$\langle F(\vec{r}, \vec{v}, t) \rangle = 0 \tag{47}$$

It is clear that for a linearized collision term, the Boltzmann equation for $f^{(1)}$ is obtained again after averaging. As a matter of fact, it can be shown that this equation (called the Boltzmann-Langevin equation [10]) is also valid for a non-linearized collision term. It is convenient to introduce a correlation function Γ through:

$$\langle F(\vec{r}_1, \vec{v}_1, t_1) F(\vec{r}_2, \vec{v}_2, t_2) \rangle = 2\Gamma(\vec{r}_1, \vec{v}_1, \vec{r}_2, \vec{v}_2)\delta(t_1 - t_2). \tag{48}$$

The interesting feature of this approach lies in the fact that the correlation function Γ is entirely determined by the collision term of the Boltzmann equation for the smoothed distribution $f^{(1)}$, as soon as the fluctuations are negligible at low densities. As a matter of fact, the Fourier transform with respect to the spatial coordinate $\vec{r}$ of Γ reads:

$$2\Gamma(\vec{k}_1, \vec{v}_1, \vec{k}_2, \vec{v}_2) = 2\delta(\vec{k}_1 + \vec{k}_2)f_0(\vec{v}_2)^{-1}I_{coll}(\vec{k}_1, \vec{v}_1)\delta(\vec{v}_1 - \vec{v}_2). \tag{49}$$

It allows one to calculate the fluctuations of hydrodynamical variables: for example, $\langle \Pi_{\alpha\beta}(\vec{r}_1, t_1) \Pi_{\mu\nu}(\vec{r}_2, t_2) \rangle$. This correlation function for the pressure tensor is proportional to the Chapman-Enskog viscosity coefficient multiplied by the temperature. On the other hand, the Langevin term introduces an extra-term in the Navier-Stokes equations, giving the fluctuating hydrodynamic Landau-Lifshitz equations[11]. This method has been proposed very recently within the framework of nuclear collisions as a possible extension of currently used transport theories[12].

A.3.Application to nuclear systems

A.3.1.Semi-classical approximation

For quantal systems, such as nuclear systems, the N-body density operator is: $D = \sum_\lambda |\Phi_\lambda\rangle p_\lambda \langle\Phi_\lambda|$. A BBGKY hierarchy is straightforward to derive, using the same procedure as for classical systems (section A.1.1.). The Liouville equation becomes the Von Neumann equation and the Poisson brackets become commutators. In the absence of correlations, the first equation of the hierarchy reads:

$$i\hbar \frac{d}{dt}\rho^{(1)} = L_1^0 \rho^{(1)} + \underset{(2)}{\mathrm{Tr}}\, L_{12}\rho_1^{(1)}\rho_2^{(1)}, \tag{50}$$

i.e. the TDHF equation with the Hartree-Fock hamiltonian h:

$$i\hbar \frac{d}{dt}\rho^{(1)} = [\mathrm{h}, \rho^{(1)}]. \tag{51}$$

In order to have a semi-classical expression, it is natural to write the operator in a mixed representation $\vec{r}$ and $\vec{p}$. This is achieved by a Wigner transformation, i.e. a Fourier transform over the non-locality. For the one-body density $\rho^{(1)}$ the Wigner tranform reads (the upper index 1 is dropped):

$$f(\vec{r}, \vec{p}) = \frac{1}{(2\pi\hbar)^{\frac{3}{2}}} \int d\vec{s}\, \exp(\frac{i\vec{p}\vec{s}}{\hbar})\, \rho(\vec{r} + \frac{\vec{s}}{2}, \vec{r} - \frac{\vec{s}}{2}) \tag{52}$$

The Wigner transform of the TDHF equation reads:

$$\frac{df}{dt} = \frac{2}{\hbar}\mathrm{h}_w \sin(\frac{\hbar}{2}\overset{\leftrightarrow}{\Lambda})\, f \tag{53}$$

The operator $\overset{\leftrightarrow}{\Lambda}$ means $\overset{\leftarrow}{\nabla}_{\vec{r}} \cdot \overset{\rightarrow}{\nabla}_{\vec{p}} - \overset{\leftarrow}{\nabla}_{\vec{p}} \cdot \overset{\rightarrow}{\nabla}_{\vec{r}}$ where leftarrows act to the left and rightarrows to the right. The semi-classical approximation is obtained by truncating the sine expansion in eq(53) to lowest order in $\hbar$. In the lowest order one obtains:

$$\frac{df}{dt} = \{\mathrm{h}_w, f\} \tag{54}$$

which is the Vlasov equation with { , } denoting the Poisson brackets. The self-consistent Hartree-Fock hamiltonian can be written in Wigner representation:

$$h_w(\vec{r}, \vec{p}) = \frac{\vec{p}^{\,2}}{2m} + U^{eff}[f(\vec{r}, \vec{p})]. \tag{55}$$

Therefore eq(54) is identical to eq(12-13). The only difference arises from the fact that the Wigner transform f is, in general, *not* definite positive, at variance with a classical distribution function.

A.3.2.Comparison between TDHF and Vlasov solutions for nuclear slabs

Approximate solutions of the Vlasov equation can be obtained either by computing the time evolution of phase space cells (Eulerian method) or by describing the distribution function as an ensemble of classical pseudo-particles (Lagrangian method). It can be shown that the Lagrangian method provides very reliable results in the sense of weak convergency: the approximate solution gives expectation values of one-body observables with the required accuracy [13].

For nuclear collective modes, a comparison between solutions of the Vlasov equation and TDHF solutions shows that the quantal effects of TDHF are unimportant. More precisely, as far as quantum tunneling on the one hand and vibrations with multipolarity higher than the octupole on the other hand are negligible, the semi-classical Vlasov equation is an excellent approximation to TDHF. This can be easily seen for slabs, i.e. semi-infinite pieces of nuclear matter. The figure 1 (from ref [14]) shows the Vlasov and TDHF responses of a slab of thickness $\mathcal{A} = 1.4 fm^{-2}$ to a compressional velocity field given by:

$$v(z) = \alpha \frac{2\hbar}{m} z > \tag{56}$$

The quantity α gives the strength of the velocity field and z is the distance along the axis perpendicular to the slab surface. For α up to .04 fm^{-2}, no differences can be seen. Some deviations between the two responses can be observed very far from the linear regime for a rather long period of time after the initial excitation. Another example is given by the time evolution of the relative distance $d(t) = 2 \int dz \rho(z, t) \mid z \mid /\mathcal{A}$ for two identical colliding slabs. The figure 2 (from ref[15]) shows the results for relative kinetic energies equal to 3.5 MeV/u and 0.5 MeV/u respectively. Even at this low energy, it turns out that the semi-classical solution is in excellent agreement with the TDHF result. For completeness, one should mention that the two reported examples have been obtained for the so-called BKN nuclear effective interaction which accounts for many nuclear properties [16].

A.3.3.Comment about the Uehling-Uhlenbeck integral

The fact that the Vlasov equation seems to reproduce main features of TDHF leads one to use semi-classical dynamics also for the residual interaction. Roughly speaking,

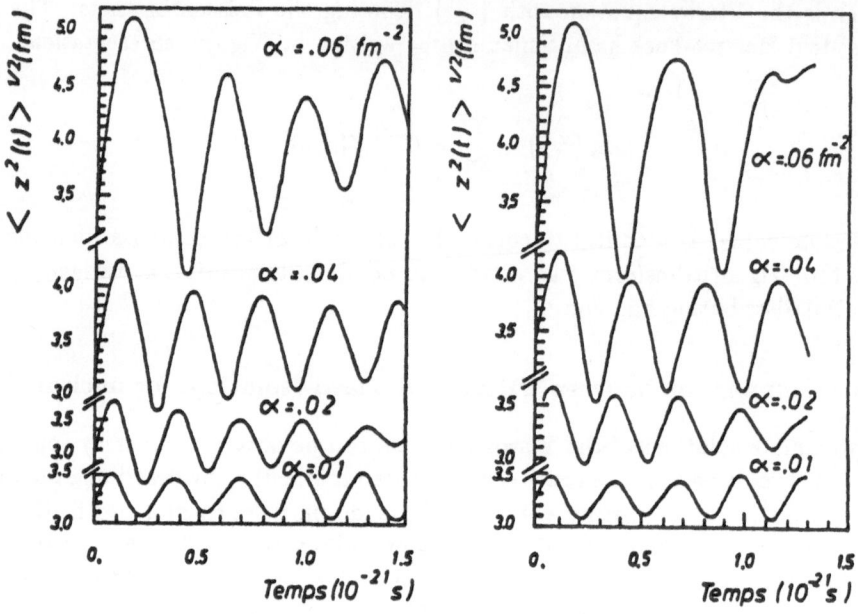

Figure 1

Vlasov(left) and TDHF(right) response of a compressed slab. The mean square thickness is displayed versus time for various strengths of the velocity field.

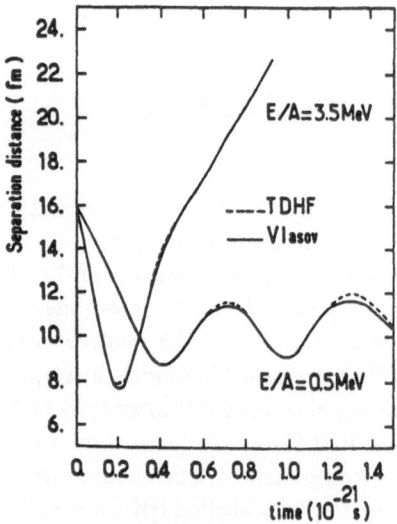

Figure 2

Time evolution of the separation distance between two colliding slabs. Vlasov and TDHF results are compared for two relative energies.

the mean field approach (TDHF or Vlasov) is valid only near the zero temperature Fermi equilibrium. The other extreme limit is the very-far from equilibrium or also the very high temperature situation. The mean field approach is unlikely to be a good description in these latter dynamical situations. We are more faced with a Boltzmann gas. In fact, du e to the quantal statistics, final states in a two-body collisions could be Pauli blocked and the collision integral should be rephrased as:

$$I_{coll} = \frac{g}{2\pi^3}\frac{h^3}{m^2}\int d\vec{k}_2 d\vec{k}_3 d\vec{k}_4 \delta(\epsilon_1 + \epsilon_2 - \epsilon_3 - \epsilon_4)\delta(\vec{k}_1 + \vec{k}_2 - \vec{k}_3 - \vec{k}_4)$$

$$\frac{d\sigma}{d\Omega}[(1 - f_1)(1 - f_2)f_3 f_4 - (1 - f_3)(1 - f_4)f_1 f_2] \qquad (57)$$

The $(1 - f)$ factors are the Pauli blocking factors in the final states, propounded first by Nordheim(1928) [17] and derived by Uehling and Uhlenbeck in 1933 [18]. This collision term conserves energy and momentum. The equilibrium distribution obeys fermion statistics. The non-linearity of the gain term is directly obtained. For all the previous reasons, it has been found reasonable to introduce this collision term on the right hand side of the Vlasov equation. At low energy, the mean field will dominate, whereas, at high energy the Uehling-Uhlenbeck integral becomes more important. Nevertheless, such an equation should be derived from a firm theoretical background. Furthermore, if the Vlasov equation is a reasonable approximation to TDHF, it is not guaranteed that the semi-classical Uehling-Uhlenbeck term is adequate. It is in fact not the case, as soon as the energy becomes high. Let us consider nucleus-nucleus collisions at 300 MeV/u. With a nucleon-nucleon cross section equal to 40 mb, the time between nucleon-nucleon collisions at normal density is of the order of 3 fm/c. Consequently, an energy uncertainty of 70-100 MeV is expected. In other words, the picture of classical colliding nucleons cannot be correct and off-shell effects become crucial. Obviously this feature is strongly enhanced for energies where pionic degrees of freedom are known to be present, i.e. above 400MeV/u. In other words, the Vlasov-Uehling-Uhlenbeck approach could be justified at first glance from a few MeV above the Coulomb barrier and up to 200 MeV/u. However it is very likely not valid outside these limits. As a matter of fact, it is one of the aims of the second part of these lectures to see which kind of approximations have to be done in order to derive a quantal kinetic equation.

B.TRANSPORT THEORY FOR NON-RELATIVISTIC QUANTUM SYSTEMS

B.1.Formalism

The classical phase-space representation with no explicit energy dependence in the distribution function is not sufficient to account for off-shell effects. The propagation of quantal particles are generally described in terms of Green's functions, as introduced in any elementary textbook on the many-body problem (see,for instance ref [19]). The Wigner representation of the Green's functions provides us with this required energy dependence. It will allow one to transform the equations of motion into a kinetic equation formulation. The price one has to pay in order to recover a Vlasov-Uehling-Uhlenbeck equation becomes clear: a quasi-particle approximation and also a T-matrix approximation for the local density. The first approximation is the basis for the Landau theory of liquid Helium. The second approximation (done in Brueckner calculations) removes divergences arising from the hard core of the nucleon-nucleon interaction. The second part of these notes is devoted mainly to a derivation of a kinetic equation, a discussion of the required assumptions and to examples of applications.

B.1.1. Green's functions and Wigner representation

We shall use the notation $X = (\vec{r}, t)$. A field is defined as $\psi(X) = \sum_q \psi_q(X) c_q$ where $\psi_q(X)$ is the single particle wave function with quantum numbers q, and c_q is the annihilation operator. The operators are in Heisenberg representation: $\psi(X) = \exp(i \mathcal{H} t / \hbar) \psi(\vec{r}) \exp(-i \mathcal{H} t / \hbar)$. The fields fulfill equal time commutation relations, namely (upper signs correspond to bosons whereas lower signs correspond to fermions):

$$
\left\{
\begin{array}{lcl}
\psi(\vec{r}, t)\psi(\vec{r}\,', t) & \mp & \psi(\vec{r}\,', t)\psi(\vec{r}, t) \quad = \quad 0 \\
\psi^\dagger(\vec{r}, t)\psi^\dagger(\vec{r}\,', t) & \mp & \psi^\dagger(\vec{r}\,', t)\psi^\dagger(\vec{r}, t) \quad = \quad 0 \\
\psi(\vec{r}, t)\psi^\dagger(\vec{r}\,', t) & \mp & \psi^\dagger(\vec{r}\,', t)\psi(\vec{r}, t) \quad = \quad \delta(\vec{r} - \vec{r}\,').
\end{array}
\right.
$$

The time ordering operator T tells us what is the time arrow. Acting on a vacuum state $| \Phi_0 \rangle$, the creation and annihilation operators are ordered from right to left with increasing times, each permutation giving a sign which corresponds to the relevant statistics:

$$
T(\psi(\vec{r}, t)\psi^\dagger(\vec{r}\,', t')) = \left\{
\begin{array}{ll}
\psi(\vec{r}, t)\psi^\dagger(\vec{r}\,', t') & \text{if } t > t' \\
\mp \psi^\dagger(\vec{r}\,', t')\psi(\vec{r}, t) & \text{if } t < t'.
\end{array}
\right.
$$

To describe the propagation of a particle in a medium one considers a process which involves the destruction of this particle at time t and position $\vec{r}$ and its creation at another time t' and position $\vec{r}\,'$. For a single Heisenberg ground state $| \Phi_0 \rangle$, the one-particle Green's function (or propagator) is defined as:

$$
iG(X, X') = \frac{\langle \Phi_0 \mid T(\psi(X)\psi^\dagger(X')) \mid \Phi_0 \rangle}{\langle \Phi_0 \mid \Phi_0 \rangle}. \tag{58}
$$

For a statistical ensemble the expectation value becomes $\langle \, . \, \rangle = Tr(\rho \, . \,)/Tr\rho$ and the propagator is:

$$
iG(X, X') = \langle T(\psi(X)\psi^\dagger(X')) \rangle. \tag{59}
$$

From the propagator, one can deduce equilibrium one-body observables, currents and energy spectrum. For instance, the density is given by:

$$\langle n(\vec{r}) \rangle = -i \lim_{t \to t'} G(\vec{r}, t, \vec{r}, t').$$ (60)

The adiabatic (Gell Mann-Low) and the Wick theorems and the linked cluster expansion theorem allow one to derive Feynman rules for perturbation theory. The description of systems at finite temperature is achieved by considering imaginary times $t = i\beta$, where β is the inverse of temperature. The expectation values are the canonical ensemble average values. It is straightforward to show that:

$$\begin{cases} G(\vec{r}, \tau, \vec{r}\,', 0) = \pm G(\vec{r}, \tau, \vec{r}\,', \beta) & \text{for } \tau < \beta \\ G(\vec{r}, 0, \vec{r}\,', \tau') = \pm G(\vec{r}, \beta, \vec{r}\,', \tau') & \text{for } \tau' < \beta. \end{cases}$$

In these relations, the Hamiltonian for the Heisenberg representation has been taken equal to $\mathcal{H} - \mu N$. For fermions (bosons) the Green's function is antiperiodic (periodic) over a β interval. In both cases, the periodicity is achieved over a 2β interval along the imaginary time axis. The periodicity properties can be used in order to decompose the Green's functions in a Fourier series. Only odd (even) multiples of π/β (called the Matsubara frequency in the literature) are found with non-zero Fourier coefficients for fermions (bosons). The Feynman rules can be extended for this finite temperature case and the thermodynamical properties at equilibrium deduced from the Matsubara decomposition.

For *non-equilibrium* situations, the theorems of the perturbation theory are not valid anymore. For instance, in the equilibrium perturbation theory, the adiabatic theorem allows one to express a state at a time $t \to +\infty$ as identical to the state at time $t \to -\infty$ modulo a phase factor. In non-equilibrium problems, this property is lost because the processes are irreversible. Since the time reversibility is not fulfilled anymore, the time ordering should be treated independently for increasing and decreasing times. Instead of *one single* time ordered Green's function, we will need the knowledge of *two* Green's functions, one chronological g^c and one anti-chronological g^a. Shortening the notation $X_1 = (\vec{r}_1, t_1)$ to 1, their definitions are:

$$g^c(1, 1') = \theta(t_1 - t_1')g^>(1, 1') + \theta(t_1' - t_1)g^<(1, 1')$$ (61)

$$g^a(1, 1') = \theta(t_1' - t_1)g^>(1, 1') + \theta(t_1 - t_1')g^<(1, 1'),$$ (62)

where :

$$\pm ig^<(1, 1') = \langle \psi^\dagger(1')\psi(1) \rangle$$ (63)

$$ig^>(1, 1') = \langle \psi(1)\psi^\dagger(1') \rangle,$$ (64)

and θ is the Heaviside function. The usual advanced (retarded) Green's functions g^- (g^+) are:

$$g^-(1, 1') = g^<(1, 1') - g^a(1, 1') = g^c(1, 1') - g^>(1, 1')$$ (65)

$$g^+(1, 1') = g^>(1, 1') - g^a(1, 1') = g^c(1, 1') - g^<(1, 1').$$ (66)

As seen in section A.3.1., a Wigner transformation allows one to design a classical representation of a quantal operator, by Fourier integration over the non-locality. Since there is now an explicit dependance on time, one should also consider the non-locality in this variable. The Wigner transform of an operator $A(\vec{r}_1, t_1, \vec{r}_2, t_2)$ is defined as:

$$A_W(\vec{R}, T, \vec{p}, \omega) = \frac{1}{(2\pi)^4} \int d\vec{s} \int dt \; \exp[\frac{-i}{\hbar}(\vec{p}\vec{s} - \omega t)] A(\vec{r}_1, t_1, \vec{r}_2, t_2). \tag{67}$$

In this expression $\vec{R} = (\vec{r}_1 + \vec{r}_2)/2$, $T = (t_1 + t_2)/2$, $\vec{s} = \vec{r}_2 - \vec{r}_1$ and $t = t_2 - t_1$. The Wigner transform of the Green's function describing particle propagation $g_W^{\lessgtr}(\vec{R}, T, \vec{p}, \omega)$ is our quantal distribution function. The off-shell effects are contained via the ω dependence. The classical counterpart $f_{cl}(\vec{R}, T, \vec{p})$ is obtained by integration over ω (one drops the w subscript):

$$f_{cl}(\vec{R}, T, \vec{p}) = \frac{1}{2\pi} \int d\omega (\pm i) g^{<}(\vec{R}, T, \vec{p}, \omega) \tag{68}$$

The Wigner transformation will be used to establish a quantal kinetic theory (section B.2). In the two following sections, we shall keep the original space-time representation.

B.1.2. Kadanoff and Baym equations

The equation of motion for the Green's function eq(59) should be derived from the Von-Neumann equation for the fields. In the Heisenberg representation, the time evolution of an operator $A_H(t)$ is given by:

$$i\partial_t A_H(t) = [A_H(t), \mathcal{H}]. \tag{69}$$

For particle fields $\psi(X)$, the Hamiltonian reads:

$$\mathcal{H} = \int d\vec{r}\psi^\dagger(X)\frac{\nabla^2}{2m}\psi(X) + \frac{1}{2}\int\int d\vec{r}d\vec{r}\,'\psi^\dagger(\vec{r}, t)\psi^\dagger(\vec{r}\,', t)v(|\,\vec{r} - \vec{r}\,'\,|)\psi(\vec{r}\,', t)\psi(\vec{r}, t) \tag{70}$$

Forming an expression for the Green's functions and using the relation:

$$\partial_{t_1}\langle T(\psi(1)\psi^\dagger(1'))\rangle = T\langle \partial_{t_1}\psi(1)\psi^\dagger(1')\rangle + \delta(t_1 - t_1')\delta(\vec{r}_1 - \vec{r}_1\,'), \tag{71}$$

one obtains two equations, one for the time derivative with respect to the first argument of the Green's function and another one for the time derivative with respect to the second time argument:

$$(i\partial_{t_1} + \frac{\nabla_1^2}{2m})G(X_1, X_1') = \delta(t_1 - t_1')\delta(\vec{r}_1 - \vec{r}_1\,') \pm i\int d2v(1,2)G_2(1,2,1',2^+) \tag{72}$$

$$(-i\partial_{t_1'} + \frac{\nabla_{1'}^2}{2m})G(X_1, X_1') = \delta(t_1 - t_1')\delta(\vec{r}_1 - \vec{r}_1\,') \pm i\int d2v(1,2)G_2(1,2^-,1',2). \tag{73}$$

In these expressions $v(1,2) = v(|\vec{r}_1 - \vec{r}_2|)\delta(t_1 - t_2)$ and G_2 is the two-body Green's function:

$$G_2(1,2,1',2') = \langle T(\psi^\dagger(1')\psi^\dagger(2')\psi(1)\psi(2))\rangle. \qquad (74)$$

The limit 2^+ (2^-) means $\lim(t_2' - t_2) \to 0^+$ (0^-). It turns out that the propagator G is coupled to the two-particle Green's function. These equations have the same structure as the first equation of the BBGKY hierarchy but with time variables included. The integration $d2$ over the second particle indicates that the propagation of the particle 1 is governed by its interaction (via v) with the other particles in the medium.

A convenient description of this propagation is given by the *self-energy* Σ of the particle 1:

$$\Sigma(1,1') = \pm i \int d1'' d2 v(1,2) G_2(1,2,1'',2^+) G^{-1}(1'',1'), \qquad (75)$$

with G^{-1} defined by:

$$\int d1'' G^{-1}(1,1'') G(1'',1') = \delta(1-1'). \qquad (76)$$

The perturbation theory allows one to calculate the self-energy, the first order approximation being nothing but the Hartree and the Hartree-Fock approximations. In the Hartree approximation $G_2(1,2,1',2') = G(1,1')G(2,2')$. In the Hartree-Fock approximation, the exchange contribution to the self energy is $iv(1,1')g^<(1,1')$. The second order approximation introduces correlation and polarization terms etc...
The equation of motion eq(72-73) for the propagator reads:

$$(i\partial_{t_1} + \frac{\nabla_1^2}{2m})G(1,1') = \delta(1-1') + \int d1'' \Sigma(1,1'') G(1'',1') \qquad (77)$$

$$(-i\partial_{t_1'} + \frac{\nabla_{1'}^2}{2m})G(1,1') = \delta(1-1') + \int d1'' G(1,1'') \Sigma(1'',1'). \qquad (78)$$

This set of equations are the Kadanoff and Baym equations. They are our starting point for the derivation of a quantal kinetic theory. From these equations, we should derive equations for the chronological and antichronological Green's function in order to deal with irreversible processes.

B.1.3. Generalized Dyson equations

Let us define a time ordering contour as follows: the upper branch of the contour goes from $-\infty$ up to $+\infty$ and is described chronologically; the lower branch of the contour goes from $+\infty$ to $-\infty$ and is described anti-chronologically. In the complex plane

151

where real axis corresponds to real times and imaginary axis to finite temperatures (imaginary time), this contour (Keldysh contour [20]) can be drawn along the real axis with a rightarrow for the upper branch and a leftarrow for the lower branch. The Green's function has two time arguments t_1 and t'_1. In order to specify it along the Keldysh contour ,one must modify the definition of the time ordering T, by defining a time-ordering along the contour:

$$\begin{cases} G(1,1') &= g^c \quad \text{if } t_1 \text{ and } t'_1 \text{ are on the upper branch} \\ G(1,1') &= g^a \quad \text{if } t_1 \text{ and } t'_1 \text{ are on the lower branch} \\ G(1,1') &= g^< \quad \text{if } t_1 \text{ is on the upper branch and } t'_1 \text{ on the lower one} \\ G(1,1') &= g^> \quad \text{if } t_1 \text{ is on the lower branch and } t'_1 \text{ on the upper one.} \end{cases}$$

The expectation values characterizing G in eq(59) can by defined as the averages over all possible initial conditions at time $t \to -\infty$.

For completeness, one should mention that any choice of contour, which allows one to define what these expectation values are, and which achieves the time ordering, is possible. For instance, in the original Kadanoff-Baym derivation [21], the contour was chosen with boundaries along the imaginary axis. Hence, the expectation values are the equilibrium thermal expectation values of the grand canonical ensemble. The periodicity along the imaginary axis and some analytical continuation properties allow one to write the relevant equations along the real axis. The complication of analytical continuation is not needed if one considers only the Keldysh contour, keeping in mind the meaning of the expectation values. This formulation is well adapted to situations like nucleus-nucleus collisions where it is easy to define an ensemble of initial conditions for which the nuclei are far apart in the entrance channel. It was used very recently for a relativistic formulation of the kinetic equations [22].

The Kadanoff-Baym equations can be rewritten along the Keldysh contour by performing the integration $\int \Sigma G$ along it. Schematically:

$$SG = \delta + \oint \Sigma G. \tag{79}$$

On the upper branch the δ function is $\delta(1-1')$. On the lower branch, it is $-\delta(1-1')$. For time arguments on different branches, it is zero. If one gathers the four Green's functions in a matrix:

$$\mathbf{G} = \begin{bmatrix} g^c & -g^< \\ g^> & -g^a \end{bmatrix},$$

the corresponding four equations of motion along the Keldysh contour read (the integral being implicit):

$$\mathbf{G}_0^{-1}\mathbf{G} = \delta + \Sigma\mathbf{G}, \tag{80}$$

with:

$$G_0^{-1} = \begin{bmatrix} S & 0 \\ 0 & -S \end{bmatrix} \quad \delta = \begin{bmatrix} 1 & 0 \\ 0 & -1 \end{bmatrix} \quad \Sigma = \begin{bmatrix} \Sigma^c & -\Sigma^< \\ \Sigma^> & -\Sigma^a \end{bmatrix}.$$

G_0 is the free propagation Green's function matrix. The equation of motion along the Keldysh contour can be recognized as a generalized Dyson equation for the 2×2 matrices:

$$G = G_0 + G_0 \Sigma G. \tag{81}$$

It has been shown by Keldysh and Craig [20,23] that the Feynman diagrammatic analysis of the perturbation theory can be extended to these matrices. In other words, the expression of the Green's functions along Keldysh contours allows one to reduce the non-equilibrium case to a form similar to the equilibrium problem.

B.2.Quantal kinetic theory

B.2.1.Quantum Boltzmann equation

The equation of motion for the retarded Green's function g^+ is obtained by subtraction of the equations for $g^>$ and g^a. From eq(66) and eq(80) one gets:

$$S g^+ = \delta(1 - 1') + \int \Sigma(1, 1'') g^+(1'', 1'), \tag{82}$$

the summation being performed over repeated indexes. Therefore the Wigner transform of g^+ is:

$$g^+(\vec{R}, T, \vec{p}, \omega) = [\omega - \frac{p^2}{2m} - \Sigma^+(\vec{R}, T, \vec{p}, \omega)]^{-1}. \tag{83}$$

It is clear from this expression that Σ^+ plays the role of an effective optical potential. We will take advantage of this physical interpretation to derive a kinetic equation from the Kadanoff- Baym equations (in the generalized Dyson equation formulation). In the space-time representation, one can write the equation for $g^>$ as:

$$S g^> = \int \Sigma^+ g^> + \int \Sigma^> g^-, \tag{84}$$

i.e. more schematically and using the eq(82):

$$(g^+)^{-1} g^> = \Sigma^> g^-, \tag{85}$$

and similarly for the adjoint equation:

$$g^{\lessgtr} (g^-)^{-1} = g^+ \Sigma^{\lessgtr}. \qquad (86)$$

As we did for the TDHF equation in section A.3.1., we will Wigner transform the equations, the Wigner transformation being now performed in space-time. Let $\overleftrightarrow{\Lambda}$ be:

$$\overleftrightarrow{\Lambda} = \frac{\overleftarrow{\partial}}{\partial \omega} \frac{\overrightarrow{\partial}}{\partial T} - \frac{\overleftarrow{\partial}}{\partial T} \frac{\overrightarrow{\partial}}{\partial \omega} + \frac{\overleftarrow{\partial}}{\partial \vec{r}} \frac{\overrightarrow{\partial}}{\partial \vec{p}} - \frac{\overleftarrow{\partial}}{\partial \vec{p}} \frac{\overrightarrow{\partial}}{\partial \vec{r}}. \qquad (87)$$

After Wigner transformation, the previous equations eq(85-86) reads:

$$(g^+)^{-1}(\vec{R}, T, \vec{p}, \omega) \, \exp(\frac{i}{\hbar} \frac{\overleftrightarrow{\Lambda}}{2}) \, g^{\lessgtr}(\vec{R}, T, \vec{p}, \omega)$$

$$= \Sigma^{\lessgtr}(\vec{R}, T, \vec{p}, \omega) \, \exp(\frac{i}{\hbar} \frac{\overleftrightarrow{\Lambda}}{2}) \, g^-(\vec{R}, T, \vec{p}, \omega), \qquad (88)$$

and a similar expression for the second equation. Just as TDHF yields the Vlasov equation, a kinetic equation is derived via *a smoothness argument*. Terms of higher order than two in $\overleftrightarrow{\Lambda}$ are neglected. This approximation is justified either if the self-energy Σ is peaked in $(\vec{R}, T)$ and spread in $(\vec{p}, \omega)$, and the Green's function is peaked in $(\vec{p}, \omega)$, and spread in $(\vec{R}, T)$ or vice-versa. In physical problems, it is the first case which is relevant, the self-energy being not too broad in momentum and the Green's functions being rather sharply defined in energy-momentum. Nevertheless, the conditions of applicability of the truncation of the gradient expansion should be checked for every specific situation. A subtraction of the two dynamical equations directly gives, after first order truncation:

$$\{(Reg^+)^{-1}, ig^<\} - \{i\Sigma^<, Reg^+\} = \Sigma^> g^< - \Sigma^< g^>, \qquad (89)$$

where the generalized Poisson bracket is now defined by our space-time $\overleftrightarrow{\Lambda}$ operator as : $\{A, B\} = A \, \overleftrightarrow{\Lambda} \, B$. The real part of $(g^+)^{-1}$ is, as seen previously, equal to $\omega - p^2/2m - Re\Sigma^+$. This dynamical equation is a transport equation as can be seen by introducing the so-called spectral function.
The spectral function A is defined by:

$$A(\vec{R}, T, \vec{p}, \omega) = i(g^+ - g^-) = i(g^> - g^<). \qquad (90)$$

If $-ig^< = Af$ represents the probability amplitude to find a fermion with $\vec{R}, t, \vec{p}, \omega$ then $ig^> = A(1-f)$ represents the probability amplitude to find an available quantum state. The spectral function is a Lorentzian function:

$$A = \frac{\Gamma}{[\omega - p^2/2m - Re\Sigma^+]^2 + \Gamma^2/4},$$
(91)

with $\Gamma = -2\ Im\Sigma^+$. For a uniform medium where $\Sigma \equiv \Sigma(\vec{p})$, Γ turns out to be the decay rate of fermion states since $|\ g^<(\vec{p}, t)\ |^2 \propto \exp(-\Gamma(\vec{p})t)$. If σ is the cross-section associated to the two-body processes responsible for this decay, Γ is proportional to $\rho\sigma\langle v\rangle$. The spectral function contains all the information about the propagation through the real part of Σ^+ and also about the decay through the imaginary part of Σ^+. For dilute gases, i.e. in the limit $\Sigma^+ = 0$, it becomes merely $2\pi\delta(\omega - p^2/2m)$, which corresponds to the free propagation.

The dynamical equation (89) for $g^<$ can be transformed, after some elementary algebra, into a *Generalized Boltzmann Equation*:

$$\underbrace{\{\omega - \frac{p^2}{2m} - Re\Sigma^+, ig^<\}}_{MEAN\ FIELD} - \overbrace{\{i\Sigma^<, Reg^+\}}^{OFF\ SHELL} = \overbrace{\underbrace{-\Gamma\ (ig^<) + (i\Sigma^<)A}_{COLLISIONS}}^{OFF\ SHELL}.$$
(92)

The various contributions have been identified: mean field, collision terms and off-shell components. In order to be explicit, one should determine the self-energies. In section B.3.1. and B.3.3., we shall give examples where the self energy is given by an interaction potential and the Born approximation. In nuclear physics, it is not possible to make such an approximation because of the hard core of the nucleon-nucleon interaction. The method proposed for the nuclear matter problem is to perform a summation of a class of diagrams in order to determine the two-body Green's functions and hence the self-energies. Since this problem has been treated in other lectures during the school, we will just recall the salient features which are relevant for our purpose.

B.2.2. T-matrix approximation

In Brueckner calculations for nuclear matter, the two-body Green's functions are determined by summation of ladder diagrams. These diagrams include direct successive two-body interactions between particles (above the Fermi sea) and exchange terms. The scattering amplitude is obtained by a T matrix which fulfills an integral equation:

$$T = v + vggT.$$
(93)

The self-energy Σ along Keldysh contours can be obtained from the corresponding four matrix-components of the Keldysh $\mathbf{T}$ matrix:

$$\mathbf{T} = \begin{bmatrix} T^c & -T^< \\ T^> & -T^a \end{bmatrix}.$$

It reads:

$$\Sigma(1,1') = \oint d2 \oint d2' iG(2',2)[\langle 1,2 \mid \mathbf{T} \mid 2'1' \rangle - \langle 1,2 \mid \mathbf{T} \mid 1'2' \rangle]. \tag{94}$$

The T matrix fulfills an optical theorem, which can be deduced very easily by writing the integral equation (93) for the T matrix in terms of Keldysh matrices. One finds:

$$T^{\lessgtr} = T^{+} \mathcal{G}^{\lessgtr} T^{-}, \tag{95}$$

where $\langle \vec{r}_1, \vec{r}_2 \mid \mathcal{G} \mid \vec{r}_1\,', \vec{r}_2\,' \rangle = iG(1,1')G(\vec{r}_2, t_1, \vec{r}_2\,', t_1')$. For *a uniform medium or in the local density approximation*, one obtains an expression for the self energy. The Wigner transform for two-particle functions, like the T matrix, is defined by considering first the non-locality between the two ingoing or outgoing particles and then the non-locality between the centers of mass of these in and outgoing particles. The result can be expressed explicitly by using the spectral function:

$$-i\Sigma^{<}(\vec{p},\omega) = \int \frac{d\vec{p}_1 d\omega_1}{(2\pi)^4} \int \frac{d\vec{p}\,'d\omega'}{(2\pi)^4} \int \frac{d\vec{p}\,'_1 d\omega'_1}{(2\pi)^4} \delta(\vec{p}+\vec{p}_1-\vec{p}\,'-\vec{p}_1\,')\delta(\omega+\omega_1-\omega'-\omega'_1)$$

$$\frac{1}{2}[\langle \frac{\vec{p}-\vec{p}_1}{2} \mid T^{+}(\vec{p}+\vec{p}_1, \omega+\omega_1) \mid \frac{\vec{p}\,'-\vec{p}_1\,'}{2} \rangle - \langle \frac{\vec{p}-\vec{p}_1}{2} \mid T^{+}(\vec{p}+\vec{p}_1, \omega+\omega_1) \mid \frac{\vec{p}_1\,'-\vec{p}\,'}{2} \rangle]^2$$

$$A(\vec{p}_1,\omega_1)[1-f(\vec{p}_1,\omega_1)]A(\vec{p}\,',\omega')f(\vec{p}\,',\omega')A(\vec{p}_1\,',\omega'_1)f(\vec{p}_1\,',\omega'_1). \tag{96}$$

B.2.3. The Landau-Vlasov equation

The Vlasov-Uehling-Uhlenbeck equation (we prefer to call it the Landau-Vlasov equation following the current textbooks [1]) can be deduced from the generalized Boltzmann equation eq(92) under two main assumptions:

i) The quasi-particle limit for the spectral function i.e. a vanishingly small width Γ. The spectral function readily becomes:

$$A(\vec{R},T,\vec{p},\omega) = 2\pi\delta(\omega-\tilde{\omega})Z(\vec{R},T,\vec{p}), \tag{97}$$

where $\tilde{\omega}$ is the on-shell energy, i.e. the solution of :

$$\tilde{\omega} - \frac{p^2}{2m} - \Sigma^{+}(\vec{R},T,\vec{p},\tilde{\omega}) = 0, \tag{98}$$

and where the Z factor is:

$$Z(\vec{R},T,\vec{p}) = 1/[1 - \frac{\partial \Sigma^{+}}{\partial \omega} \mid_{\omega=\tilde{\omega}}]. \tag{99}$$

ii) The T-matrix approximation for the local density.

The classical distribution function f_{cl} is obtained by integration of $g^<$ over ω. The quasi-particle approximation allows one to perform the ω integration of the generalized Boltzmann equation explicitly. The left hand side is the classical Vlasov equation eq(12), the mean field being determined by the real part of the self-energy taken on-shell:

$$U^{eff}(f_{cl}) = Re\ \Sigma^+(\vec{R}, T, \vec{p}, \tilde{\omega}). \tag{100}$$

The right hand side is the Uehling-Uhlenbeck integral eq(57), the effective cross-section being determined by the same T matrix which was used for determining the self-energy. The off-shell collisions are neglected because of the quasi-particle approximation. In a collisional regime where the imaginary part of the self-energy can not be neglected with respect to the real part, the Landau-Vlasov equation should clearly not be used.

The merit of the quantal derivation is to exhibit the fact that mean-field and collision terms should be derived from the same Brueckner T-matrix. The propagation in the medium reduces neither to a Vlasov equation nor to a classical Boltzmann equation, but to a Landau-Vlasov equation where mean-field and collisions are present simultaneously. In particular, no double-counting occurs since the mean field corresponds to the real part of the self-energy and the collisional integral to the imaginary part.

B.3. Illustrations of the quantal kinetic theory

B.3.1. The impurity scattering problem

Let us consider electrons propagating in a metal [24]. These electrons are scattered by local impurities. The impurity scattering modifies the momentum distribution $f(\vec{p})$ of the electrons according to the quantal generalized Boltzmann equation. If n_i is the concentration of impurities and $v(\vec{r})$ the local scattering potential, the Born approximation gives the self-energy :

$$\Sigma(\vec{R}, T, \vec{p}, \omega) = n_i \int \frac{d\vec{p}\,'}{(2\pi)^3} \mid v(\vec{p} - \vec{p}\,') \mid^2 G(\vec{R}, T, \vec{p}\,', \omega), \tag{101}$$

with $v(\vec{p})$ the Fourier transform of $v(\vec{r})$. Assuming that Σ varies slowly, the real part of Σ^+ can be taken to be a constant equal to U. The equation (92) reads:

$$\{\omega - \frac{p^2}{2m} - U, ig^<\} \simeq -\Gamma(ig^<) + (i\Sigma^<)A. \tag{102}$$

In the dilute limit, the spectral function is equal to $2\pi\delta(\omega - p^2/2m - U)$ and the right hand side of eq(102) becomes, after integration over ω,

$$-2\pi n_i \int \frac{d\vec{p}\,'}{(2\pi)^3} \mid v(\vec{p} - \vec{p}\,') \mid^2 \delta(\frac{p^2}{2m} - \frac{p'^2}{2m})[f(\vec{p}) - f(\vec{p}\,')]. \tag{103}$$

This equation is a semi-classical Landau-Vlasov equation, with a self-energy given by the impurity potential. It describes the randomization of the electron momentum orientations. To my knowledge, it is one of the simplest examples of a kinetic equation which can be derived from the quantal Boltzmann equation.

B.3.2. Fermi liquids

It is beyond the scope of these lectures to discuss the physics of Fermi liquids (see, for example, ref[25,26]). Our aim is only to put the kinetic theory of Fermi liquids in perspective of the generalized Boltzmann equation. A Fermi liquid is a highly degenerate system. At low temperature, $T \ll \mu$ (where μ is the Fermi energy), all states below the Fermi energy are occupied, whereas all the states above the Fermi energy are unoccupied. Near the Fermi surface, dispersion relations for Σ^+ allow one to show that the equation (98) for $\tilde{\omega}$, the on-shell energy, has a unique solution near the Fermi surface. This energy is the quasi-particle energy $\epsilon(\vec{R}, T, \vec{p})$. As a matter of fact, at low temperature the quasi-particle approximation can be done near the Fermi surface (for $\omega > \mu$ and $\omega < \mu$ the kinetic equation is automatically fulfilled by trivial stationary solutions). The Green's function $g^<$ reads:

$$-ig^<(\vec{R}, T, \vec{p}, \omega) = 2\pi\delta(\omega - \epsilon)Z(\vec{R}, T, \vec{p})n(\vec{R}, T, \vec{p}), \qquad (104)$$

where $n(\vec{R}, T, \vec{p})$ is the quasi-particle distribution function. The Generalized Boltzmann equation gives, after integration over ω, the so-called Landau-Silin equation for the quasi-particle distribution near the Fermi surface at low temperature:

$$\underbrace{\frac{\partial n}{\partial T} + \nabla_{\vec{p}}\epsilon(\vec{R}, T, \vec{p})\nabla_{\vec{r}}n - \nabla_{\vec{r}}\epsilon\nabla_{\vec{p}}n}_{LANDAU\ EQUATION} = \underbrace{(-i\Sigma^<)(1 - n) - i\Sigma^> n}_{COLLISIONS}. \qquad (105)$$

This equation is conserving. In particular, a continuity equation can be established. On the other hand, the quasi-particle energies contain the interaction energy of the quasi-particle with the medium in which it is propagating. For short range forces and long wave length excitations, the interaction can be chosen local (this gives the Landau theory). For short wave length excitations, non-local potentials could be defined phenomenologically (this gives the extension of the Landau theory propounded by Aldrich and Pines with polarization potentials [27]).

In the Landau theory, the quasi-particle energy reads:

$$\epsilon(\vec{r}, \vec{p}) = \epsilon_0(\vec{p}) + \int d\vec{p}\,'f(\vec{p}, \vec{p}\,')\delta n(\vec{r}, \vec{p}\,'), \qquad (106)$$

where f is the local interaction and δn the deviation from the ground state density. Since we are only interested in momenta located near the Fermi surface, the local interaction f depends only on their angular coordinates. This allows one to decompose this interaction in a Legendre polynomial expansion. For sake of simplicity, we shall just consider linear momenta (spin variables will be omitted). The expansion of f reads:

$$2N(0)f(\vec{p},\vec{p}\,') = \sum_l F_l \; P_l(\cos\theta), \qquad (107)$$

where $N(0)$ is the density of states near the Fermi surface and θ the relative angle between $\vec{p}$ and $\vec{p}\,'$. The F_l are the so-called Landau parameters.

The first limit of the Landau-Silin equation is the collisionless regime (elastic regime). If $U(\vec{r},t) = U.\exp(i(\vec{q}\vec{r}-\omega t))$ is an external field with $|\vec{q}| \ll p_f$, the Fermi momentum, the collisionless kinetic equation reads:

$$(\omega - \vec{q}.\vec{v})\nu + \vec{q}.\vec{v}\int d\vec{p}\,' f(\vec{p},\vec{p}\,')\frac{\partial n_0}{\partial \epsilon_0}\nu = \vec{q}.\vec{v}U. \qquad (108)$$

In this equation, the time dependence of the quasiparticle density is given by:

$$n(\vec{r},\vec{p},t) = n_0(\vec{p}) + \frac{\partial n_0}{\partial \epsilon_0}\nu(\vec{r},\vec{p},t), \qquad (109)$$

and $\nu(\vec{r},\vec{p},t) = \nu(\Omega)\exp(i(\vec{q}\vec{r} - \omega t))$. The ground state density and energy are indicated by the 0 indices. The various modes of vibration can be identified by decomposing the density variations themselves in a Legendre polynomial expansion: $\nu(\Omega) = \sum_l \nu_l \; P_l(\cos\theta)$. The orthogonality relations can be used readily to derive the vibration amplitudes. If one considers only the $l=0$ term and $F_0 \neq 0$, eq(108) reads, after use of eq(107):

$$\nu_0 = \frac{-\Xi(\lambda)}{1 + F_0\Xi(\lambda)}, \qquad (110)$$

with :

$$\Xi(\lambda) = 1 + \lambda\ln[\frac{(\lambda - 1)}{(\lambda + 1)}], \qquad (111)$$

and $\lambda = (\omega/q)/v_f$, the ratio between the phase velocity and the Fermi velocity. If $F_0 > 0$, one gets a pole as soon as the phase velocity associated to the excitation is larger than the Fermi velocity($\lambda > 1$). It gives rise to a collective excitation of the Fermi liquid, which is a collective vibration of the Fermi surface called zero-sound. If λ is slightly smaller than 1, then this oscillation is damped (ν has an imaginary component): this is the Landau damping. The nature of the vibration depends, of course, on the strength of the interaction F_0. If $F_0 \gg 1$, the vibration turns out to be isotropic in momentum space, whereas if $F_0 \ll 1$, it is forward peaked along the $\vec{q}$ direction.

The second limit of the kinetic equation is the hydrodynamical limit where collisions are dominant. This limit is achieved for excitation frequencies much smaller than the frequency of quasi-particle collisions. The derivation is identical to that of the hydrodynamic equations in classical transport theories (section A.2.2). Sound waves (called first sound) are found, with a velocity c given by:

$$c^2 = \frac{v_f^2}{3}(1 + \frac{F_1}{3})(1 + F_0).$$ (112)

The quasi-particle approximation generates a significant simplification of the collision term. It turns out to be a Nordheim-Uehling-Uhlenbeck term for the quasi-particle densities. Furthermore, the fact that the momenta are all located near the Fermi surface allows for a complete decoupling between energy and angular variables. The transport coefficients can be explicitly calculated. The thermal conductivity is found to be proportional to the inverse of the temperature, the viscosity to its square. The attenuation α of the sound modes (zero and first sound) can be also be derived readily. For the first sound α is proportional to ω^2/T^2, whereas for the zero sound it is proportional to T^2. This property gave the principle of the experiment exhibiting the presence of the sound regimes in the liquid helium [28].

In the polarization potential theory, which aims to lift the long wave length approximation of the Landau theory, the effective interaction is described by non local restoring forces. The definition of these forces is done via Fourier transformation of configuration space pseudo-potential. The determination of this potential is phenomenological with two requirements (for details see ref[29]). First, the zero sound and Landau parameters should be reproduced for the small $\vec{q}$ limit. Secondly, the potential should have a scalar and a vector component to take into account the polarization of the medium as well as the backflow (or drag current) effect. This approach, originally developed for liquid Helium at non-vanishing $\vec{q}$ values, provides a convenient framework for the description of multipair excitations. It allows one to account for the dispersion relation found in neutron scattering experiments in the $1\mathring{A}$ domain. Very recently, it has been introduced in nuclear physics and a first determination of the polarization potential has been done in all the spin and isospin channels [30].

B.3.3. Pieces of nuclear matter in relative motion

We have seen in section A.3.3. that for bombarding energies above 300 MeV/u, off-shell effects should be very important. The width of the spectral function, which is $\Gamma = -2 \, Im\Sigma^+$, is found larger or equal to 70-100 MeV. Furthermore, the emission and absorption of pions makes Σ very ω-dependent. As a matter of fact, with the contribution of the pisobar at 800 Mev/u, calculated in ref[31], the imaginary part of the self-energy should be larger than 200 MeV at twice normal density. These two essential features should imply a quantal kinetic theory *without* the quasi-particle approximation. In other words, an examination of the width of the spectral function indicates that the so-called VUU-BUU treatments of heavy ion collisions are *inappropriate* above 200 Mev/u (even in their relativistic versions), since the quasi-particle approximation, which is done implicitly, is not valid. A model case studied in ref[32] shows that the quantal behaviors could drastically affect the relaxation towards equilibrium.

In this reference, Danielewicz has studied the problem of two pieces of nuclear matter in relative motion. The self-energy is calculated by neglecting the Hartree-Fock term and by taking into account second order diagrams. The self-energy has no ω-dependence and the effects studied are those given by the two-body collision spreading widths. These second order terms are calculated using the Born approximation for an effective potential:

$$v(\vec{p}) = \pi^{3/2} \mu^3 V_0 \exp(-\frac{1}{4}\mu^2 p^2),$$
(113)

with $\mu = .57fm$ and $| V_0 |= 453 MeV$. This parametrization is supposed to reproduce the hard core part of the nucleon nucleon potential responsible for much of the scattering between 0 and 500 MeV. At 400 MeV/u, it is found that the quantum dynamics given by the Kadanoff-Baym equations proceeds twice as slow as in classical Boltzmann dynamics, as far as the relaxation processes are concerned. The author attributes this fact to particle decay. Nevertheless, this model has some limitation. In particular, it does not take into account the first order term of the self energy (mean field) , whereas it is the mean field which should ensure the stability of the nuclear medium.

In spite of this fact, one can conclude by saying that this model exhibits strong differences between classical and quantal dynamics at 400 MeV/u. It deserves further studies with more realistic description of the medium and the inclusion of pion degrees of freedom.

CONCLUSION

In the book "Concepts and theories of modern physics" written in 1882, the scientific journalist J.B.Stallo discussed the Boltzmann kinetic theory: "....I do not hesitate to declare that the kinetic hypothesis has none of the characteristics of a legitimate physical theory. Its premises are as inadmissible as reasoning upon them is inconclusive. It postulates what it professes to explain....". In fact, we have seen in this introduction to kinetic theories that their developments were numerous, wide and fruitful. They include the quantal version, originally derived by Kadanoff and Baym. These kinetic theories describe a variety of physical phenomenon in molecular and solid state physics. As far as the quantal problem is concerned, very little has been done beyond the quasi-particle approximation. This is a problem if one wishes to analyse results obtained, for instance, in heavy ion collisions. Whereas this approximation could be legitimate in nucleus-nucleus collisions below 200 Mev/u and above a few MeV above the Coulomb barrier, it is unlikely to stay valid outside this range. In fact, the theoretical study of these nuclear systems requires a quantal treatment of the collisional regime.

Acknowledgements: I am indebted to R.Malfliet for numerous discussions on transport theories. He introduced me to the Keldysh contour method which makes the quantal theory so elegant. It is my pleasure to attribute part of these lectures to him. S.Ayik, N.L.Balazs, T.Kuo, M.Prakash, E.Suraud, G.Welke are acknowledged for enlightening discussions. Thanks are due to D.Pines for communication of his recent work on the polarization potential theory. The author is very grateful to G.E.Brown and the Nuclear Theory Group of the University of New-York at Stony-Brook for the warm hospitality and financial support. I would like to express also my gratitude to my colleagues M.Pi, B.Remaud, P.Schuck, F.Sebille and E.Suraud for a very fruitful collaboration over many years, which was devoted solving the nuclear Landau-Vlasov equation in order to simulate intermediate-energy heavy ion collisions. Finally, I would like to thank M.Prakash and G.Welke for a careful reading of the manuscript. Last, but not least, the editors, especially H.Flocard, are acknowledged for their patience during the preparation of the final version of these lecture notes.

REFERENCES

[1] R.Balescu, *Equilibrium and non-equilibrium statistical mechanics*, J.Wiley (1975)

[2] P.C.Martin and J.Schwinger, *Phys.Rev.115*(1959)1352

[3] A.A.Vlasov, *Many particle theory and its application to plasma*, Gordon and Breach, New-York(1961)

[4] A.P.Lightman and S.L.W.McMillan, *Rev.Mod.Phys.50*(1978)437

[5] N.N.Bogoliubov, *Studies in Statistical Mechanics*, deBoer ed., North-Holland, Amsterdam(1952)

[6] J.G.Kirkwood, *J.Chem.Phys.15*(1947)72 and R.L.Liboff, *Introduction to the theory of kinetic equations*, R.Krieger Publ, New-York(1979)

[7] R.Balian and M.Veneroni, *Ann.Phys.135*(1981)270

[8] M.Ernst, *Phys.Rep.78*(1981)1

[9] in J.O.Hirschfelder, Ch.F.Curtiss, R.B.Bird, *Molecular theory of gases and liquids*, Wiley ed. (1954)

[10] M.Bixon and R.Zwanzig, *Phys.Rev.187*(1969)267

[11] L.D.Landau, E.M.Lifshitz, *Fluid Mechanics*, Pergamon, London(1959) Ch.17

[12] S.Ayik and C.Gregoire, *Phys.Lett.212B*(1988)269

[13] P.A.Raviart and J.M.Thomas, *Introduction a l'analyse numerique des equations aux derivees partielles*, Masson, Paris(1983)

[14] L.Vinet, *Ph.D.Dissertation*, Orsay(1986), unpublished

[15] C.Gregoire, B.Remaud, F.Sebille, L.Vinet, Y.Raffray, *Nucl.Phys.A465* (1987)317

[16] P.Bonche, S.E.Koonin, J.W.Negele, *Phys.Rev.C13*(1976)1226

[17] L.W.Nordheim, *Proc.Roy.Soc.A119*(1928)689

[18] E.A.Uehling and G.e.Uhlenbeck, *Phys.Rev.43*(1933)552

[19] A.L.Fetter and J.D.Walecka, *Quantum theory of many-particle systems*, McGraw-Hill (1971)

[20] L.V.Keldysh, *Sov.Phys.JETP 20*(1965)1018

[21] L.P.Kadanoff and G.Baym, *Quantum statistical mechanics*, Benjamin, New-York(1962)

[22] W.Botermans and R.Malfliet, *Phys.Lett.215B*(1988)617 and W.Botermans, Ph.D.Dissertation, Groningen(1989)

[23] R.A.Craig, *J.Math.Phys.9*(1968)605

[24] J.Rammer and H.Smith, *Rev.Mod.Phys.58*(1986)323

[25] G.Baym and C.Pethick, in *Physics of liquid and solid Helium*, Interscience monographs and texts in physics and astronomy, Wiley (1979) Vol.29, Part II.

[26]D.Pines and P.Nozieres, *The theory of quantum liquids*, Benjamin (1966)

[27]C.H.Aldrich III and D.Pines, *J.Low.Temp.Phys.25*(1976)677

[28]W.R.Abel, A.C.Anderson, J.C.Wheatley, *Phys.Rev.Lett.17*(1966)74

[29]D.Pines, in *Highlights of condensed matter theory*, Proc.89th Int.School of Physics, Soc.Italiana di Fisica(1985)p.580

[30]D.Pines, K.F.Quader, J.Wambach, *Nucl.Phys.A477*(1988)365

[31]G.E.Brown, E.Oset, M.V.Vacas, W.Weise, *preprint Stony-Brook*(1988)

[32]P.Danielewicz, *Ann.Phys.152*(1984)239

RELATIVISTIC TRANSPORT THEORY OF FLUCTUATING FIELDS
FOR HADRONS [*]

Philip J. Siemens

Oregon State University
Corvallis, OR 97331, U.S.A.

ABSTRACT

We analyze the physics of relativistic nuclear collisions, and demonstrate that an adequate treatment of pions must include the quantum time-energy uncertainty principle and non-sequential 3-body collisions. We apply relativistic quantum field theory to obtain exact equations determining the time evolution of hadronic fields and their fluctuations in terms of the effective interactions describing scattering in matter. These equations relate the main physical observables to the sought-after properties of nuclear matter, and involve only these quantities and the corresponding properties of free-space 1- and 2-body processes. We show how to regularize the singularities by employing the internal structure of the mesons, while maintaining causality and unitarity. We show that, in a very good approximation, the dynamic quantities reduce to functions of 8 variables – those of the Boltzmann equation, supplemented by the energy of the hadron. Our method appears capable of deducing the properties of hot dense nuclear matter from data already measured in experiments on the collisions of heavy nuclei.

---

[*] The material covered by these lectures is presented in a paper by P.J. Siemens, M. Soyeur, G.D. White, L.J. Lantto and K.T.R. Davies with the above title and abstract, to be published in Phys. Rev. **C40** (1989).

PION ABSORPTION IN NUCLEI

Michael Thies

Institut für Theoretische Physik, Universität Erlangen-Nürnberg
Glückstraße 6, D - 8520 Erlangen, FRG

INTRODUCTION

The nucleon-nucleon interaction becomes inelastic at $\sim$ 300 MeV lab energy, signalling the onset of pi-meson production. Heavy ion collisions at energies of several 100 MeV per nucleon will therefore inevitably be accompanied by the appearance of pions. The issue of pion absorption in nuclei[1] is relevant for these processes in two ways: First, the pion production mechanism is closely related to that of the inverse reaction, pion absorption. Secondly, some of the produced pions will be re-absorbed by the nuclei involved, an effect which has to be accounted for when analyzing the data.

In these two lectures, I should like to review what has been learned about pion absorption in nuclei during the last decade at the meson factories LAMPF (Los Alamos), PSI (formerly SIN, Villigen) and TRIUMF (Vancouver). In spite of considerable experimental effort and an impressive amount of good data, I have to admit that the physical picture of pion absorption which has emerged so far is not yet very clear. Some of the simplest conceivable questions (e. g. "How many nucleons are involved in pion absorption in nuclei?", ref. [2]) are still very much debated. Whether these difficulties just reflect the complexity of the final nuclear states at high excitation energies or point to some interesting, new phenomenon remains to be seen.

EXPERIMENTAL PRIMER

Pion absorption on a free nucleon is kinematically forbidden. This is evident if one thinks of the inverse reaction, decay of a nucleon (at rest) into pion and nucleon. The momentum mismatch is

$$\Delta p = \sqrt{2Mw} - k = 0(500 MeV/c) \tag{1}$$

where M, ω, k denote the nucleon mass, pion energy and momentum respectively. Since $\Delta p >> p_F$, we may safely ignore one-nucleon absorption in nuclei as well. The simplest allowed absorption reaction is therefore

$$\pi^+ d \rightarrow pp \tag{2}$$

The total cross section is shown in Fig. 1. It has a strong enhancement at an energy close to the Δ-resonance in πN-scattering. The peak value (12 mb) represents about

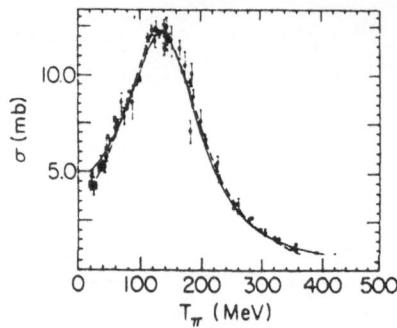

FIGURE 1
Total cross section for the reaction $\pi^+ d \to pp$, Ref. 1.

10 % of the πd reaction cross section. Hence, pion absorption is a relatively weak process in the deuteron with a clear dynamical signature, pointing to a participation of the Δ- resonance. When studying pion absorption on more complex nuclei, one quickly discovers two things: The deuteron is not very typical as far as the strength is concerned (absorption plays a much more prominent role in heavy nuclei), and it is by no means obvious that the basic mechanism is the same in both cases. Let me briefly remind you of the type of experimental information which is available nowadays:

Total Absorption Cross Sections

Is there a way of determining total absorption cross sections without summing over all the complicated final states? There is indeed such a short-cut, but it involves 3 steps:

1. A standard transmission experiment to obtain the total π-nucleus cross section, σ_{tot}.

2. An inclusive pion scattering experiment (π, π') in which the angular distribution of all elastically - or inelastically scattered pions is measured. Integration yields $\sigma_{scatt} = \sigma_{el} + \sigma_{inel}$.

3. A measurement (or, in practice, estimate) of the total single-charge- exchange cross section $(\pi^\pm, \pi^0)$, σ_{cex}.

The total absorption cross section is then obtained by subtraction:

$$\sigma_{abs} = \sigma_{tot} - \sigma_{scatt} - \sigma_{cex} \tag{3}$$

Since one relies on independent absolute cross section measurements and has to subtract large numbers, the accuracy of this method is not higher than 20 - 30 % at present. This method has been extensively applied by Ashery et al.[1], [3]. In Fig. 2, the A-dependence of various integrated π-nucleus cross sections is shown (165 MeV). Pion absorption and inelastic scattering are of competitive size in heavy nuclei, in contrast to what we saw in the deuteron. The energy dependence of σ_{abs} shown in Fig. 3 is still reminiscent of the Δ-resonance, especially for lighter nuclei.

Single Arm Proton Spectra

The simplest measurement of reaction products consists in determining the spectrum of emitted protons (or neutrons) at a fixed angle. Nucleons above a certain

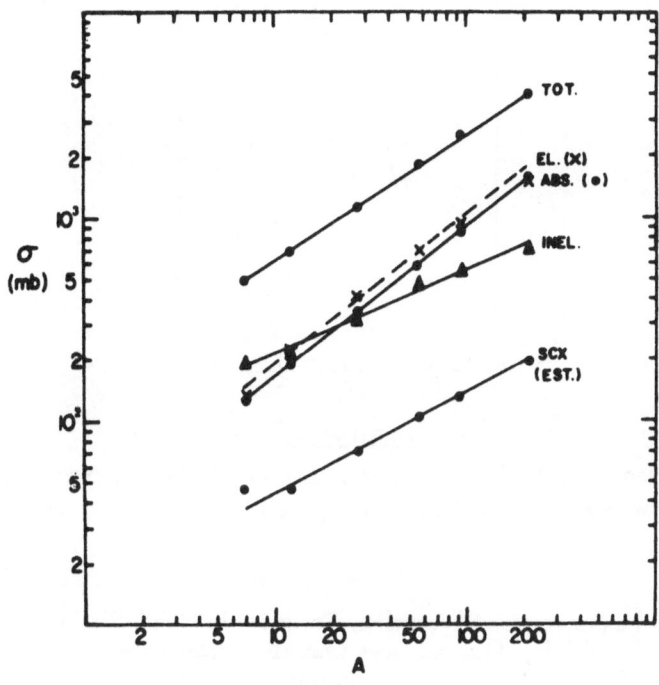

FIGURE 2

A-dependence of π^+-nucleus integrated cross sections at 165 MeV, Ref. 1.

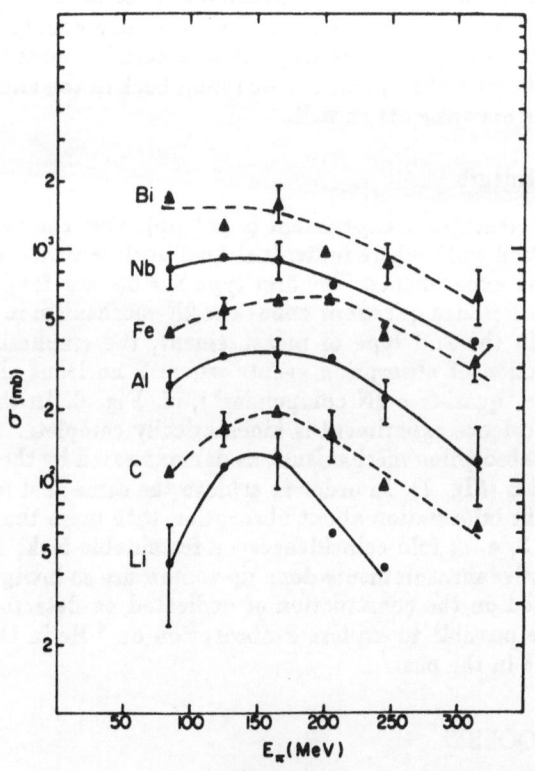

FIGURE 3

Energy-dependence of total π^+-nucleus absorption cross sections, Ref.1.

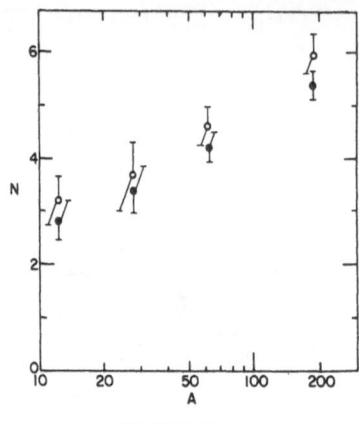

FIGURE 4

Number of nucleons involved in pion absorption at 220 MeV versus A, as obtained from rapidity analysis of inclusive proton spectra, Ref. 2. Open circles: π^-, Full circles: π^+.

energy can only be due to pion absorption. For pions in flight, the data are amenable to a "rapidity analysis" [2]. One assumes that the protons are emitted isotropically in the cm-frame of a cluster of n participating nucleons, and extracts n from the data. The values obtained by the Argonne group (Fig. 4) were unexpectedly high and have been taken as evidence for "multi-nucleon absorption mechanisms" [4]. However, this interpretation has also been criticized because the isotropy assumption would certainly not hold for the 2N-mechanism, and because the purely kinematical signature of such an inclusive measurement does not give enough clues about the underlying dynamics. The general concept that a certain, measurable number of nucleons is involved in pion absorption can be found back in the analyses of virtually all recent coincidence experiments as well.

Coincidence Measurements

The prototype coincidence experiment is (π^+,pp). One can further distinguish between high-resolution work where individual final nuclear states are resolved (Fig. 5), and more inclusive experiments. The first type has become feasible only recently and allows to address detailed questions about the 2N-mechanism in nuclei or nuclear structure aspects. In the 2nd type of measurement, the emphasis is typically on determining the fraction of absorption events where 2 nucleons share most of the available energy (the "quasi-free 2N component"), cf. Fig. 6. In the special case of ^{3}He, a two-fold coincidence experiment is kinematically complete. One can literally map out the various absorption mechanisms, as demonstrated by the Basel-Karlsruhe group[5] over the years (Fig. 7). In order to achieve the same goal for heavier nuclei, or to get more specific information about absorption with more than 2 participating nucleons, one needs 3, 4, ... fold coincidences - a formidable task. Nevertheless, the results of more inclusive measurements done up to now are so intriguing that several groups have embarked on the construction of dedicated 4π-detectors[6]. With such devices, it should be possible to explore π-absorption on ^{4}-He in the near future as completely as on ^{3}He in the past.

THEORETICAL TOOLS

I told you before that a pion cannot be absorbed on a single nucleon. This is not really true if the nucleon can be internally excited: It is in fact easy for a nucleon to absorb a 180 MeV pion and turn into a $\Delta\left(S = T = \frac{3}{2}, M_\Delta = 1232 MeV\right)$. However,

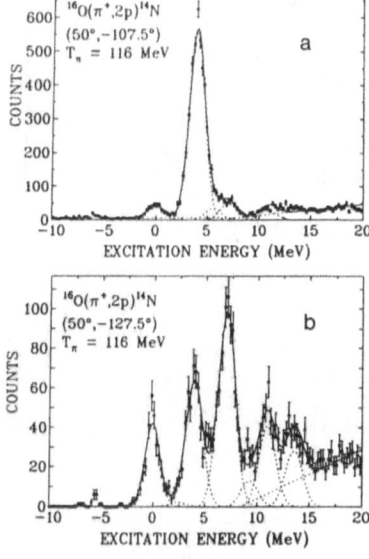

FIGURE 5

Example of good resolution (π^+, pp) coincidence experiment on ^{16}O. The 3.9 MeV L = 0 state is clearly seen at the quasi-free angles (a), whereas L = 2 states at 0,7 and 11 MeV show up at larger recoil momenta (b). From Ref. 8.

since the Δ is short-lived $(\Gamma = 110$ MeV) and can only decay back (strongly) into pion and nucleon, such a process gives rise to pion scattering on a single nucleon (Fig. 8 a).

In a nucleus, the Δ can also be de-excited in an inelastic collision $(\Delta N \rightarrow NN)$, without pion emission. The pion has been absorbed (Fig. 8b). This is the basic physical picture of pion absorption. It explains naturally the resonant energy dependence of the $\pi^+ d \rightarrow pp$ cross-section shown in Fig. 1 (up to a poorly understood downward shift). Moreover, with the one extra assumption that the $\Delta N \rightarrow NN$ transition goes dominantly via the ΔN S-wave, one can predict the angular distribution

$$\frac{d\sigma}{d\Omega} \sim 1 + 3 \cos^2\theta \qquad (4)$$

in good agreement with experiment. From the theoretical point of view, the splitting of the π-absorption mechanism into a $\pi N \rightarrow \Delta$ vertex followed by a $\Delta N \rightarrow NN$ 2-baryon interaction is very important, since it allows to bypass some of the difficult field-theoretic aspects inherent in creation- or annihilation processes.

There is a whole industry of Faddeev 3-body calculations of the $\pi^+ d \rightarrow pp$ reaction, often in the context of the more ambitious program to describe $NN \rightarrow NN$, $NN \rightarrow NN\pi$ and $NN \rightarrow \pi d$ in a unified way[7]. Without going into any detail, it is probably fair to say that one cannot predict the magnitude of pion absorption or production on 2 nucleons at present, using as only input NN and πN 2-body interactions. It seems more productive to turn things around and try to use the data to pin down the $NN \leftrightarrow NN$, $NN \leftrightarrow N\Delta$ and $N\Delta \leftrightarrow N\Delta$ potentials, in the same spirit as one uses low energy NN phaseshifts to construct NN potentials (Paris, Bonn etc.). In detail, the problem is unfortunately even more complicated since not all of pion absorption and production has to go via the Δ-resonance.

Next, let me discuss absorption on 2 nucleons bound in a nucleus. In the "quasi-deuteron model", one assumes that the residual nucleus acts as a spectator, aside from possible initial or final state interactions. Both phenomenological and microscopic

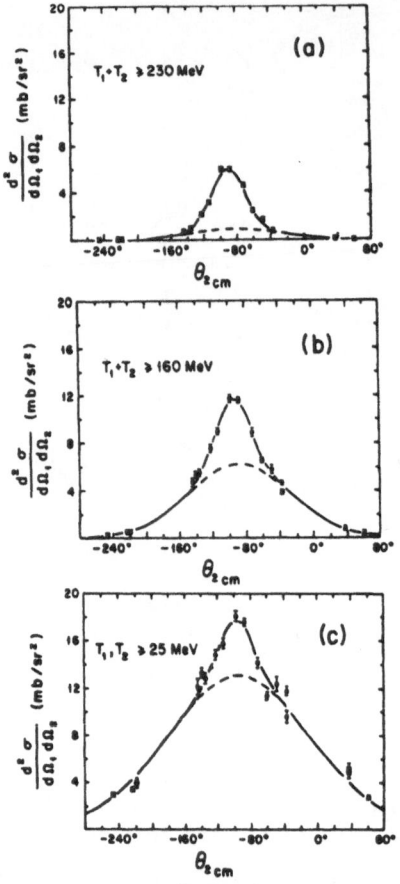

FIGURE 6

Angular correlation for low resolution $^{58}Ni(\pi^+, 2p)$ coincidence reaction at 160 MeV, for 3 different energy cuts. The area under the narrow Gaussian is taken as a measure of the quasi-free 2N-absorption component, Ref. 25.

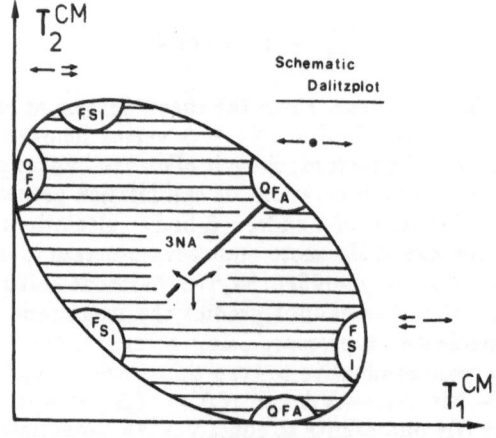

FIGURE 7

Schematic illustration of different π absorption mechanisms in ^3He (from H. J. Weyer). QFA: quasi-free 2N-absorption, FSI: final state interaction peaks, 3NA: 3 Nucleon share the available energy and momentum.

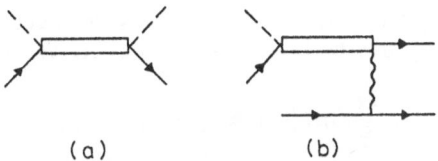

FIGURE 8

a) πN-scattering via the Δ resonance.

b) Basic π-absorption mechanism on 2 nucleons, involving the $\Delta N \to NN$ transition.

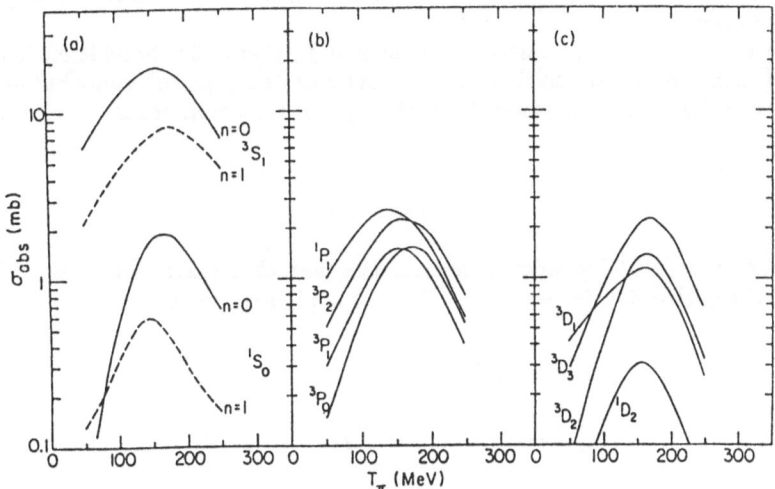

FIGURE 9

Energy dependence of total π-absorption cross section on nucleon pairs with various quantum numbers, Ref. 9. Harmonic oscillator radial wavefunctions have been used.

applications exist[8], [9], [10]. For the latter ones, it is clearly necessary to resolve the $\pi^+ d \to pp$ process into smaller building blocks before transporting it into the nucleus, simply because the size and quantum numbers of pairs in nuclei can be different from those of a physical deuteron. The two-step process via the Δ discussed above provides just such a decomposition. By way of illustration, let me show you some model calculations studying the dependence of the absorption cross section on the quantum-numbers of the absorbing pair (Fig. 9). Here, a separable two-baryon interaction fitted to NN phaseshifts and inelasticities has been used as input[9]. The $^3S_1$ "deuteron" sticks out by one order of magnitude as compared to all other pairs available in a nucleus like ^{16}O. Inspection of the partial wave decomposition of these results reveals that the Δ-N p-wave interaction is dominant in all pairs except $^3S_1$, as a result of a selection rule (T = 1 pairs) or the centrifugal barrier (T = 0, l ≠ 0 pairs).

We now turn to the total absorption cross sections. Can one predict them in some elegant way? Let me first remind you of an easier problem where the solution is well known, namely the calculation of the total reaction cross section (σ_{re}) via the optical potential. The unitarity relation of the hadron-nucleus T-matrix reads

$$T - T^+ = T^+(G_0 - G_0^+)T + (1 + T^+ G_0^+)(U - U^+)(1 + G_0 T) \tag{5}$$

It follows directly from the Lippmann Schwinger equation

$$T = U + U G_0 T \tag{6}$$

where U is the optical potential, G_0 the free Green's function. Taking a diagonal matrix-element of eq. (5), one finds the familiar optical theorem

$$\frac{4\pi}{k} Im f(0^0) = \sigma_{el} + \sigma_{re} = \sigma_{tot} \tag{7}$$

as well as the following closed expression for σ_{re}:

$$\sigma_{re} = -\frac{2\omega}{k} < \psi^{(+)} \mid ImU \mid \psi^{(+)} > \tag{8}$$

Hence, one can predict σ_{re} on the basis of elastic scattering, even if U is only known phenomenologically.

For pion absorption, we need to go one step further and subdivide σ_{re} into "scattering" (mainly quasi-free nucleon knock-out) and "absorption" contributions. Well below the Δ resonance, this is still possible, given the corresponding decomposition of U:

$$\begin{aligned} U &= U_{sc} + U_{abs} \\ U_{sc} &= i\varrho, \ U_{abs} \sim \varrho^2 \end{aligned} \tag{9}$$

The different density-dependence reflects the fact that scattering (absorption) involves 1 (2) nucleon(s), respectively. Then, it is plausible to set

$$\begin{aligned} \sigma_{re} &= \sigma_{sc} + \sigma_{abs} \\ \sigma_{sc} &= -\frac{2\omega}{k} < \psi^{(+)} \mid ImU_{sc} \mid \psi^{(+)} > \\ \sigma_{abs} &= -\frac{2\omega}{k} < \psi^{(+)} \mid ImU_{abs} \mid \psi^{(+)} > \end{aligned} \tag{10}$$

This recipe works rather well at low energies[11]. At higher energies, it becomes increasingly questionable, as remarked by Masutani and Yazaki [12]: The decomposition (9) of U only characterizes the first process responsible for the loss of flux out of the elastic channel. The scattered pions contained in Im U_{sc} may still be absorbed at a later stage. One can try and go deeper into the ramifications of possible multi-step processes and recover σ_{abs} as a sum of absorption after 0, 1, 2, ... quasi-free scatterings, but many more approximations have to be invoked in order to keep the calculation tractable.

The Δ-hole model[13] which has extensively been used in the analysis of π-nucleus reactions near the Δ-resonance is plagued by similar problems. The naive prediction of σ_{abs} in the same spirit as eqs. (9-10) agrees remarkably well with the experimental data, but should be taken with a grain of salt above resonance (Fig. 10). In between the quasi-deuteron mechanism as the simplest conceivable reaction on the one hand and the total absorption cross section as the most global observable on the other hand, there is a wide range of potentially interesting phenomena and measurements, but a theoretical desert. The only tool which we have at our hands to deal with processes involving 3 or more nucleons are intra-nuclear cascade (INC) calculations. They have been applied in various forms, e. g. transport equations[14], Monte Carlo simulations[15]. An obvious drawback of such INC calculations is the fact that we are not in a regime where a classical description can be justified. This has prevented wide acceptance of such techniques for π-nucleus reactions and left the comparison between theory and experiment somewhat inconclusive. In my opinion, Monte Carlo integration techniques have been most useful in those cases where they were restricted to kinematics (and possibly Fermi motion), i. e. phasespace calculations. Here, they yield at least a resonably well-defined background against which some experimental findings may acquire more contrast.

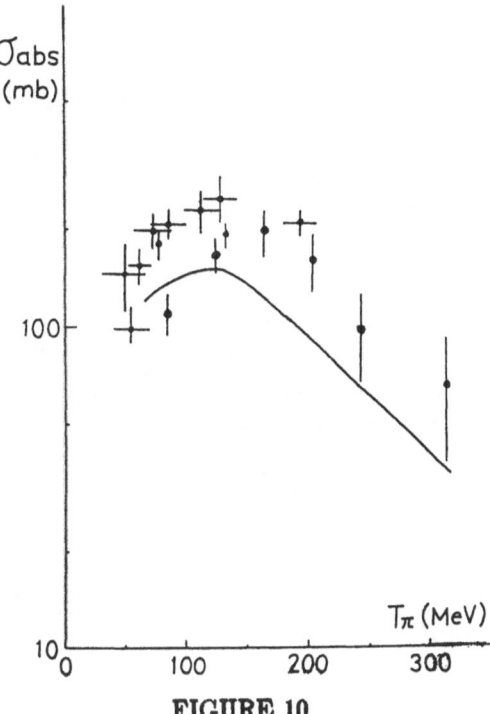

FIGURE 10

$\pi^+ - {}^{13}C$ total absorption cross section, Ref. 26. The data are compared to the prediction of the Δ-hole model.

FEW BODY SYSTEMS

If one accepts the hypothesis that a small number of nucleons only are directly involved in pion absorption in nuclei, it is particularly important to understand first absorption on the lightest systems, the deuteron, ^{3}He and ^{4}He. In the present section we focus on the transition from the deuteron to ^{3}He and ask the questions: How does 2N-absorption evolve with increasing density? How big is absorption of a π^- on a pp-pair, as compared to a np-pair? Is there evidence for events where all 3 nucleons in ^{3}He share the pion energy?

As mentioned before, a two-fold coincidence experiment with a ^{3}He-target [16] determines the final state completely and allows to disentangle several mechanisms (Fig. 7). Quasi-free absorption on a np-pair is the strongest contribution and shows practically the same characteristics as in the deuteron (Fig. 11). The 10 % increase in strength is less than expected by naive pair counting (1.5 ${}^3S_1$ pairs in ^{3}He), perhaps as a result of some shadowing. In any case, a drastic dependence of 2N-absorption on the nuclear density seems to be ruled out. The quasi-free character of the (π^+,pp) reaction is confirmed nicely by the spectator momentum distribution, Fig. 12.

By means of the (π^-,pn) reaction, it is possible to focus on quasi-free absorption on ${}^1S_0(T = 1)$ pairs [1], [17]. The cross section is lower than for ${}^3S_1$ $(T = 0)$ pairs by an order of magnitude throughout the Δ-region (Fig. 13). Table 1 is a reminder of the lowest few partial waves which can contribute to this particular reaction. The sequential process via $\Delta N \rightarrow NN$ requires a ΔN p-wave interaction which is predicted to be very weak[9]. On the other hand, the asymmetry of the angular distribution (c. f. Fig. 11) indicates interference between odd and even NN partial waves, i. e. an admixture of absorption leading to final $T_f = 0$ pairs where the ΔN intermediate state is forbidden. At 65 MeV, a partial wave analysis of the data by Piasetzky et al.[18] yields

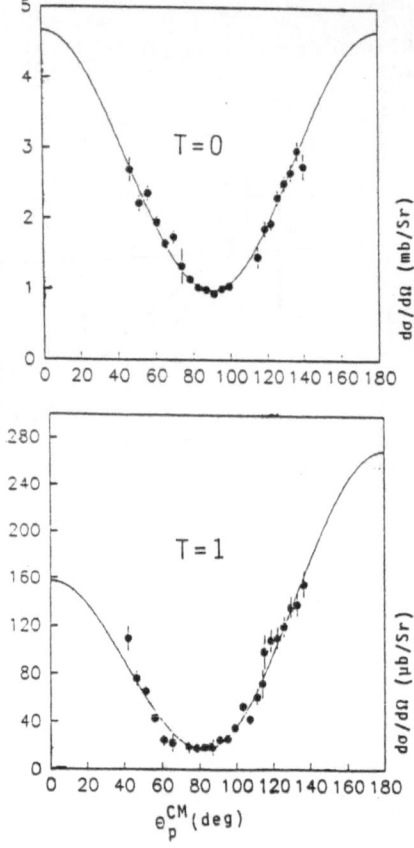

FIGURE 11

Differential cross section for absorption on nucleon pairs with $T = 0$ and $T = 1$ in ^{3}He at 121 MeV pion energy, Ref. 17.

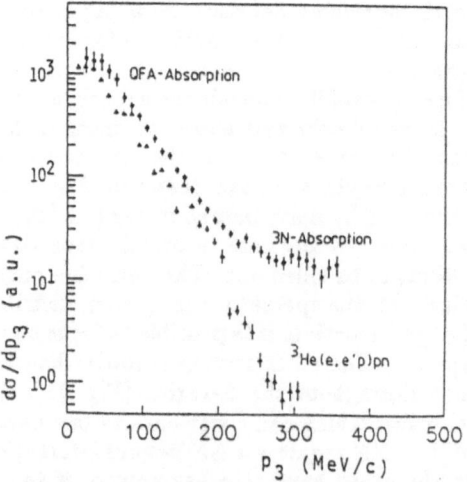

FIGURE 12

Recoil momentum distribution for $^3He(\pi^+, pp)p$ at 121 MeV, compared with momentum distribution obtained with the (e,e'p) reaction, Ref. 17.

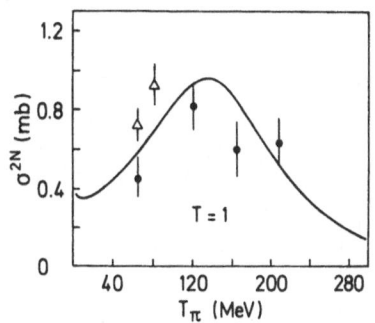

FIGURE 13

Energy dependence of total $T = 1$ pair absorption in ³He. The curve is the $\pi^+ d \to pp$ cross section divided by 12, Ref. 17.

TABLE 1

l_π	ΔN	NN
0	³P_0	³$P_0(T = 1)$
1	-	³S_1, ³$D_1(T = 0)$

$$\frac{\sigma(T_f = 0)}{\sigma(T_f = 0) + \sigma(T_f = 1)} = 0.92 \tag{11}$$

Hence, for absorption on $T = 1$ pairs, non-resonant absorption is 10 times more important than the $\Delta N \to NN$ driven process, at least at low energies. This may explain why the predictions of ref. 9 for the (π^-,np) reaction on ³He - based solely on the Δ-mechanism - strongly underestimated the data.

Whatever the non-Δ-absorption mechanism is, the important message here is that absorption on $T = 1$ pairs is not likely to be very important in complex nuclei. This might be considered as a posteriori justification of the quasi-deuteron model in its most naive form where only ³D_1-pairs are kept.

What about absorption mechanisms involving all 3 nucleons? The Basel-Karlsruhe group has identified final state interaction peaks which I don't want to discuss because they are specific for few-body systems. Beyond that, they have found some background consistent with 3-body phasespace, assuming incoherence[19]. This background adds up to ~ 25 % of σ_{abs} at 120 MeV and has the remarkable property that

$$\sigma(\pi^+, ppp) \approx \sigma(\pi^-, pnn) \tag{12}$$

Such an isospin dependence seems to rule out a sequential process (of the type quasi-free scattering followed by absorption), as one can see from the following comparison:

$$\pi^+ p \to \quad \pi^+ p \qquad (strong)$$
$$\pi^+(np) \to pp \qquad (strong)$$

versus

$$\pi^- p \to \quad \pi^- p \qquad (weak)$$
$$\pi^-(np) \to nn \qquad (strong) \tag{13}$$

or

$$\pi^- n \to \quad \pi^- n \qquad (strong)$$
$$\pi^-(pp) \to np \qquad (weak)$$

The processes indicated as weak are suppressed by one order of magnitude typically.

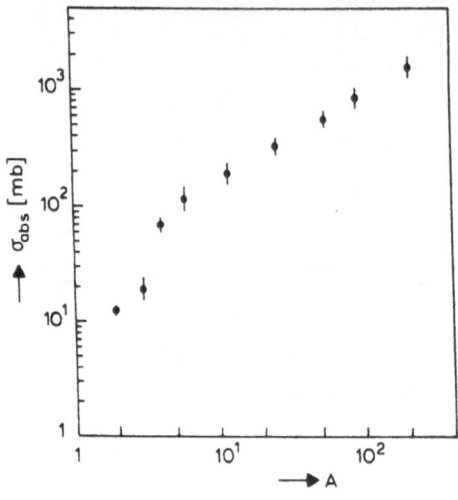

FIGURE 14

A-dependence of total π-nucleus absorption cross section at 165 MeV, Ref. 27.

Collecting all the results, the total $\pi-^3$He absorption cross section at 120 MeV amounts to

$$\sigma_{abs}^{tot} = \underset{(T=0)_{qf}}{14mb} + \underset{(T=1)_{qf}}{1mb} + \underset{(3N)}{5mb} = 20mb \tag{14}$$

Thus, the transition from the deuteron to ^{3}He is rather smooth. The total absorption cross section scales approximately with the number of $^3S_1$ pairs. Part of this strength is not found in the quasi-free kinematics region, but in the background, and has been attributed to "3N-absorption". Absorption on $T=1$ pairs is interesting in its own right as a window on non-Δ absorption mechanisms, but is too weak to play a significant role for the bulk of pion absorption in nuclei.

THE ^{4}HE-PUZZLE

^{4}He is the lightest nucleus for which the total absorption cross section has been obtained via the subtraction method. As shown in Fig. 14, the result nicely extrapolates the values for heavier nuclei, assuming approximately a $\sim A^{2/3}$ dependence. However, there appears to be a gap between ^{3}He and ^{4}He, as if a new absorption channel would open up at A = 4. A related anomaly has been observed in inclusive (π,π') scattering: At 100 MeV for instance, the Seattle group [20] finds the ratios

$$\frac{^3He}{d}(\pi^+,\pi^{+'}) = 2$$

$$\frac{^4He}{^3He}(\pi^+,\pi^{+'}) = .67 \tag{15}$$

i. e. less π^+ scattering from ^{4}He than from ^{3}He! The magnitude of σ_{abs} for ^{4}He is consistent with the Δ- hole analysis based on elastic scattering (Fig. 15). Unfortunately, such an approach cannot be applied to ^{3}He, in practice, so that this does not give any clue concerning the difference between the He-isotopes. We have also included in Fig. 15 the result of a microscopic calculation of σ_{abs}, using a resonant 2N-absorption mechanism[9]. As a result of the increased density, the maximum absorption cross section goes up to 20 mb per $^3S_1$ pair. However, because of shadowing effects, the final prediction is lower than 3 x 20 mb (3 $^3S_1$-pairs in ^{4}He) and misses about 1/2 of the measured cross section.

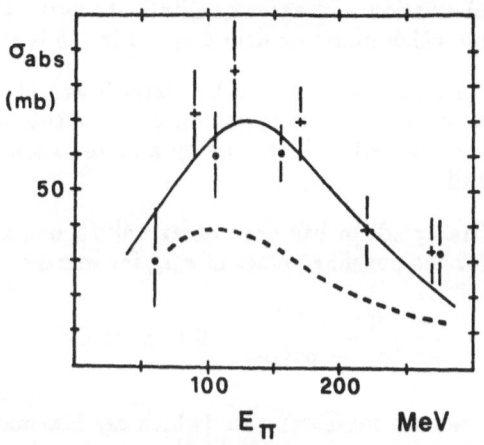

FIGURE 15

Total $\pi - {}^4He$ absorption cross section, Ref. 28. Solid curve: Δ-hole model, dashed curve: microscopic prediction, Ref. 9, based on quasi-deuteron mechanism.

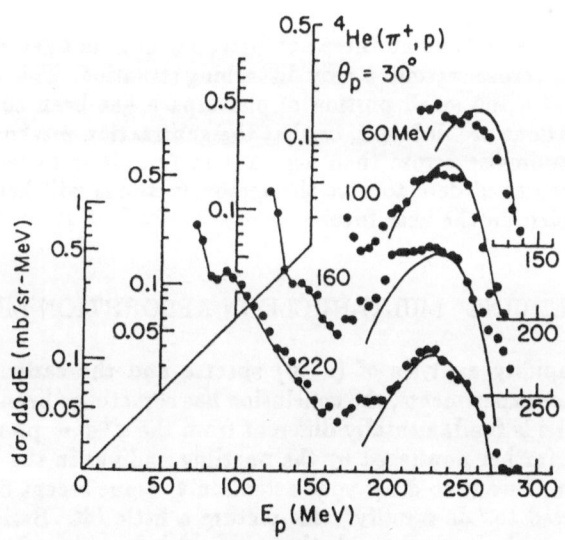

FIGURE 16

${}^4He(\pi^+,p)$ reaction at several incident pion energies. The curves correspond to the quasi-free 2N-absorption, Ref. 9.

These findings suggest strongly the presence of a sizeable absorption mode involving more than 2, perhaps even 4 nucleons. Therefore, let us take a closer look at what is known about the reaction products in the case of the α-particle. We proceed by order of increasing number of detected particles:

1. Inclusive single arm proton spectra (π^+,p) exhibit a peak where expected from quasi-free 2N-absorption and agree qualitatively with theory (Fig. 16). The same calculation which underpredicts σ_{abs} in Fig. 15 is shown.

2. Two-fold coincidence measurements by Backenstoss et al[17] show again an angular distribution for the absorption on np pairs consistent with the $\pi^+d \rightarrow$ pp reaction. The strength is increased by a factor 1.4 only. Absorption on T = 1 pairs is small.

3. The reaction ^{4}He (π^+,dp)p has been observed[21], but at a level too weak to be significant for the puzzling values of σ_{abs}, for instance

$$\frac{\pi^+ \, ^4He \rightarrow dp(p)}{\pi^+ \, ^4He \rightarrow pp(pn)} = 0.16 \text{ at } 120 MeV, \, 113° \tag{16}$$

4. Three-fold coincidence measurements (which are kinematically complete) ^{4}He (π^+,ppp)n and ^{4}He (π^+,ppn)p at 120 MeV support a picture where one nucleon acts as spectator, whereas the other 3 share energy and momentum according to phasespace[22]. In addition, final state interaction effects are noted. The total 3N-contribution extrapolated from these data is rather small though:

$$\sigma_{ppp}^{Q3N} = 2.1 \pm 0.5 mb, \; \sigma_{npp}^{Q3N} = 4.4 \pm 1.3 mb \tag{17}$$

For absorption on 4 nucleons, a value of 0.5 mb only is reported.

Obviously, if one adds up all these numbers, one does not get anywhere close to the total absorption cross section, a very disturbing situation. This discrepancy may indicate either that a too small portion of phasespace has been covered so far and that the extrapolations are deficient, or that the subtraction method of determining σ_{abs} has larger systematic errors than assumed so far. It is to be hoped that the planned large acceptance detectors at the meson factories will help to clarify this experimental problem in the near future.

IS THERE A "GENUINE" MULTI-NUCLEON ABSORPTION MECHANISM?

From the rapidity analysis of (π^+,p) spectra and the earlier low- resolution (π^+,pp) coincidence experiments, the conclusion has sometimes been drawn that pion absorption in nuclei is fundamentally different from the $\pi^+d \rightarrow$ pp reaction[4]. Such speculations were further nourished by the puzzling findings in the Helium-isotopes. In this last section, I want to draw your attention to some recent data which in my opinion have started to "de-mystify" this picture a little bit. Besides, I would like to address the theoretical question whether one can define properly what one means by "multi-nucleon absorption mechanism", as opposed to two-nucleon absorption augmented by initial- or final state interactions. Specifically, let me discuss the following 3 experiments:

1. $^{16}O(\pi^+, pp)$ at 115 MeV (Schumacher et al[8]).

   In this most complete two-fold coincidence measurement on a heavier nucleus done so far, the authors have extracted the two-nucleon component of σ_{abs}. The result is shown in table 2, together with estimates of FSI effects. The corrected numbers take into account the fact that 2N-absorption strength can

TABLE 2

E_x(MeV)	$\sigma(\pi^+, 2p)$ [mb]	σ/σ_{abs}	corrected (FSI)
≤ 20	38 ± 5	$19 \pm 4\%$	$\sim 50\%$
≤ 50	78 ± 16	$38 \pm 10\%$	$> 50\%$

be redistributed in phasespace due to NN collisions, and are based on DWBA-calculations. In contrast to earlier claims, they find that more than 50 % of the total absorption cross section of 200 mb can be accounted for. Note however that this percentage might decrease at higher energies.

2. $^{12}C(\pi^+, ppp)$ at 130, 180 and 228 MeV (Tacik et al.[23]).

The authors find that the triple coincidence data can very well be reproduced by assuming that absorption takes place on a 3 N cluster according to phasespace, with the rest of the nucleons, acting as spectator. They exclude a similar 4N process (Fig. 17). Extrapolation of the measured cross section with the help of (Monte Carlo) phase space calculations yields a fraction of 4.6 % at 130, 11.1 % at 180 and 18.6 % at 228 MeV of (π^+, ppp) out of σ_{abs}. If one would assume equal absorption probability on ppn, pnn and nnn clusters, one would arrive at numbers which are not unreasonable as compared to 1).

3. $^{12}C(\pi^+, ppp)$ and $^{13}C(\pi^+, ppd)$ at 65 MeV (R. Hamers et al.[24]).

The contribution of 2N-absorption is found to account for $\sim \frac{3}{4}$ of σ_{abs}. Fits to the data using phasespace and quasifree absorption on 3 resp. 4 nucleons are satisfactory at low resp. high excitation energies of the residual nucleus. Deuterons were found at an unexpected high rate: The ratio of (pd) to (ppn) is approximately 0.6.

If these trends are confirmed by the next generation of multi-coincidence measurements, it appears that 2N ("quasi-deuteron") absorption may still be the dominant contribution in the Δ-region. However, some discrepancies among different data still have to be resolved. The α-particle may play a key role in clarifying the absorption mechanism in the future.

Let me now turn to a theoretical inventory of "multi-nucleon absorption mechanisms". I think that the most economic language here is not that of Feynman diagrams, but rather a decomposition of the Hilbert space into certain subspaces. As in the theory of pre-equilibrium reactions, we shall use the number of holes as ordering principle. Three types of subspaces will be important:

1. Pion Channels: P_n π np - nh $(n \geq 0)$

2. Delta channels: D_n Δ(n-1)p-nh $(n \geq 1)$

3. Absorption channels: A_n np-nh $(n \geq 2)$

The subscript n always refers to the number of holes. Furthermore, let us assume that the Hamiltonian H only contains 2-body interactions, namely a $\pi N \Delta$-vertex and 2-baryon-potentials:

$$H = H_0 + H_1 + H_2 + H_3 + H_4 \tag{18}$$

H_0 is the kinetic energy, $H_1 - H_4$ are given schematically in Fig. 18. A little thought shows that the H_i connect (dominantly) the following subspaces:

$$
\begin{aligned}
&H_1: \quad P_n \leftrightarrow D_{n+1}(n \geq 0), P_n \leftrightarrow D_n(n \geq 1) \\
&H_2: \quad P_n \leftrightarrow P_{n+1}(n \geq 1), D_n \leftrightarrow D_{n+1}, A_n \leftrightarrow A_{n+1}(n \geq 2) \\
&H_3: \quad D_n \leftrightarrow A_{n+1}(n \geq 1) \\
&H_4: \quad D_n \leftrightarrow D_{n+1}(n \geq 1)
\end{aligned} \tag{19}
$$

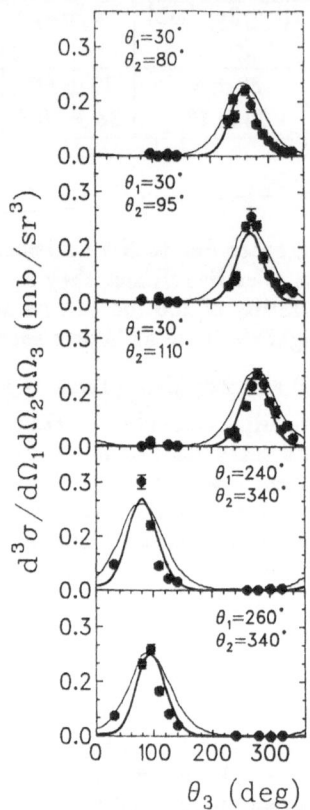

FIGURE 17

3-fold angular correlations in the reaction $^{12}C(\pi^+, ppp)X$ at 180 MeV, Ref. 23. Fat curves: quasi-free 3N absorption, thin curves: 4N absorption.

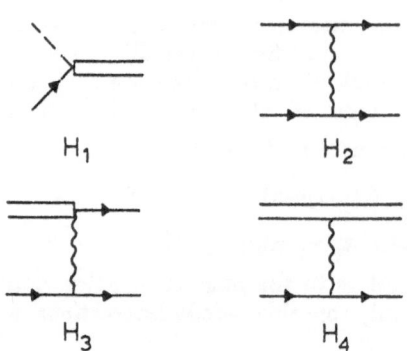

FIGURE 18

2-body interactions considered in the model Hamiltonian, eq. (18). H_1: $\pi N\Delta$ vertex, $H_2 - H_4$: 2-baryon potentials.

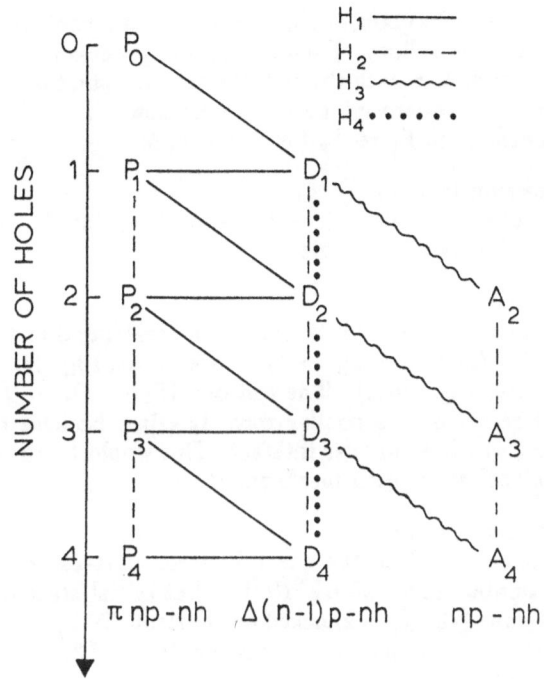

FIGURE 19
Schematic illustration of eq. (19) (see text).

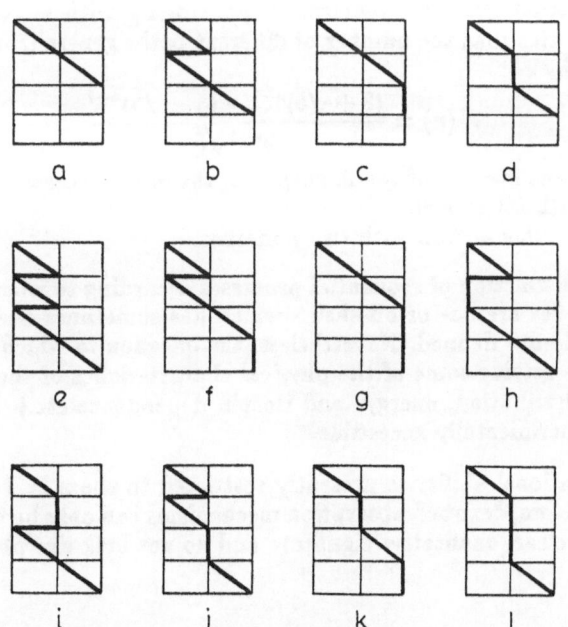

FIGURE 20
Classification of absorption processes according to the path taken through diagram (19) (see text).

This is illustrated in the diagram of Fig. 19, our main book-keeping device. The diagram has to be used as follows: Each absorption process leading to the emission of n nucleons corresponds to a path through the diagram starting from P_0 and ending at A_n, with D_1 acting as "doorway" space. To see how this works, let us enumerate all the (shortest) paths from P_0 to A_n for n = 2, 3, 4:

1. 2-nucleon-emission (n = 2)
   There is only one possibility, $P_0 \rightarrow D_1 \rightarrow A_2$ (Fig. 20 a). This corresponds to the pure quasi-deuteron mechanism.

2. 3-nucleon emission (n = 3)
   There are 3 possible processes, two of which correspond to the quasi- deuteron mechanism with ISI ($P_0 \rightarrow D_1 \rightarrow P_1 \rightarrow D_2 \rightarrow A_3$, Fig. 20 b) and FSI ($P_0 \rightarrow D_1 \rightarrow A_2 \rightarrow A_3$, Fig. 20 c). The 3rd one ($P_0 \rightarrow D_1 \rightarrow D_2 \rightarrow A_3$, Fig. 20 d) cannot be resolved in a similar way: It arises because the Δ has a finite lifetime during which it can also interact. This would be the obvious candidate for defining a "3N-absorption mechanism".

3. 4-nucleon emission (n = 4)
   There are 5 paths in which the quasi-deuteron pattern ($P_n \rightarrow D_{n+1} \rightarrow A_{n+2}$) remains identifiable, Figs. 20 e-i. Notice that initial state interactions can be mediated by pions (20 e,f), a knocked out nucleon (20 h) or the Δ (20 i). Then, there are 2 diagrams which can be interpreted as "3N-absorption" (Fig. 20 d) with ISI (Fig. 20 j) or FSI (Fig. 20 k). Finally, there is one truly new "4N-absorption mechanism", Fig. 20 l.

The classification of absorption mechanisms emerging from such a model is now clear: At each n, there is only one type of process which cannot be decomposed into absorption on less then n nucleons plus ISI or FSI. Because of the short lifetime of the Δ, it seems very unlikely that these processes play any role for n> 3. The number of sequential processes increases exponentially with n. One can show by simple combinatorics that the number of different paths generalizing those shown in Fig. 20 is given by

$$N(n) = \frac{(3 + \sqrt{5})^{n-1} - (3 - \sqrt{5})^{n-1}}{2^{n-1}\sqrt{5}} \qquad (20)$$

In the limit that the lifetime of the Δ goes to 0, the only processes which survive are 2N-absorption with ISI or FSI.

Let me close this section with two remarks:

1. A further distinction of sequential processes according to whether the intermediate particles are on- or off-shell - which has sometimes been attempted - is theoretically ill - defined. Nevertheless, the question to which extent a sequential process retains some of the physical characteristics of the individual steps (angular distribution, energy- and isospin dependence etc.) is legitimate and may be experimentally accessible.

2. Our calculational ability is presently restricted to the n = 1 level in diagram (19). Therefore, "exotic" absorption mechanisms can only be discovered if they have some clear, qualitative signature and do not look like phasespace.

CONCLUSIONS

Whereas the experimental information about pion absorption in nuclei has increased tremendously over the last decade, you will have noticed that theory is way behind. The principal difficulty of disentangling the "trivial", kinematical aspects from the dynamics is presumably familiar to this audience. What is badly needed are new ideas and calculational methods which would allow to evaluate directly the more inclusive quantities. If such techniques were available, they would undoubtedly have an impact on the understanding of other deep inelastic reactions - for instance heavy ion collisions - as well.

REFERENCES

[1] D. A. Ashery and J. P. Schiffer, Ann. Rev. Nucl. Part. Sci. 1986, 36: 207

[2] R. D. McKeown et al., Phys. Rev. Lett. 44 (1980) 1033

[3] D. A. Ashery et al., Phys. Rev. C 23 (1981) 2173

[4] J. P. Schiffer, Comm. Nucl. Part. Phys. 10 (1981) 243; 14 (1985) 15

[5] G. Backenstoss, Workshop on Pion-Few Nucleon Reactions and Dibaryons, Hannover, May 10-12, 1984

[6] G. Smith and U. Sennhauser, Workshop on Pion-Nucleus Interactions, SIN, Aug. 4-7, 1987

[7] M. P. Locher, LAMPF Workshop on Pion-Nucleus Physics, Los Alamos, Aug. 17-20, 1987

[8] R. A. Schumacher et al., Phys. Rev. C 38 (1988) 2205

[9] K. Ohta, M. Thies and T.-S. H. Lee, Ann. of Phys. (N.Y.) 163 (1985) 420

[10] M. Gouweloos and M. Thies, Phys. Rev. C 35 (1987) 631

[11] K. Stricker, H. McManus and J. A. Carr, Phys. Rev. C 19 (1979) 929

[12] K. Masutani and K. Yazaki, Nucl. Phys. A 407 (1983) 309

[13] F. Lenz and E. J. Moniz, Comm. Nucl. Part. Phys. 9 (1980) 101

[14] V. Girija and D. S. Koltun, Phys. Rev. Lett. 52 (1984) 1397

[15] L. L. Salcedo et al., to be published

[16] H. J. Weyer, Helvetica Physica Acta 60 (1987) 667

[17] H. Ullrich et al, XI-th European Conf. on Few Body Physics, Fontefraud, 31.8. - 5.9.1987

[18] E. Piasetzky et al, Phys. Rev. Lett. 57 (1986) 2135

[19] G. Backenstoss et al., Phys. Rev. Lett. 55 (1985) 2782

[20] M. A. Khandaker, PhD. Thesis, Univ. of Washington (1986)

[21] G. Backenstoss et al., Phys. Rev. Lett. 59 (1987) 767

[22] G. Backenstoss et al., PSI-preprint PR-88-04

[23] R. Tacik et al., to be published

[24] R. Hamers, PhD. Thesis, Free University of Amsterdam (1989)

[25] W. J. Burger et al., Phys. Rev. Lett. 57 (1986) 58

[26] K. Yazaki, Proc. of the Symposium on Delta-Nucleus Dynamics, May 2-4, 1983, Argonne National Laboratory, p. 297

[27] M. Thies, Proc. 21st LAMPF Users Group Meeting, 1987, LA-11248-C, p. 18

[28] M. Baumgartner et al., Nucl. Phys. A 399 (1983) 451

FORMATION AND DECAY OF HOT NUCLEI: THE EXPERIMENTAL SITUATION

Daniel Guerreau

GANIL
BP 5027, 14021 Caen Cedex, France

1. INTRODUCTION

With the achievement of new facilities in the 80's providing us with heavy ion beams well above the coulomb barrier, a unique opportunity was offered to the experimentalists to produce and study nuclear matter under extreme conditions[1-2]. Effectively, in the energy range 20-100 MeV/u, on which we will concentrate in these lectures, it appeared very rapidly that excited nuclei could be formed at rather high temperatures.

The first experiments were started at the CERN-SC[3] and continued then in many laboratories as GANIL, GSI, MSU, HMI, Texas AM, and to some extent also at SATURNE. The study of the formation and decay of hot nuclei is now one of the major effort pursued in this intermediate energy domain. One should notice that a similar effort has been undertaken at the same time, on the theoretical side.

How to define such nuclei that are commonly labelled "hot nuclei" ? First of all, the word nucleus means that, at least for a while, a self bound system has been formed. Furthermore, as we are going to define a temperature T, this implies for the hot system to be in thermal equilibrium. It is well known that pairing effects are disappearing very rapidly around T = 1 MeV as that shell effects are washed out at T = 3 MeV. In the following, we then chose quite arbitrarily to consider a nucleus as beeing hot when its temperature exceeds 3 MeV. The interest in producing hot nuclei is not only relevant for nuclear physics but also for astrophysics[4-5]. In both cases, one is interested in better characterizing the nuclear equation of state. The ultimate gool of these studies is then to determine the highest temperature that a nucleus can sustain without breaking into its constituants. The most simple (and naive) definition for this critical temperature is what could be called the boiling point of nuclear matter that is when the excitation energy equals the binding energy, i.e. 8 MeV per nucleon. In the Fermi gas model, with a level density parameter taken to be equal to A/8, this would correspond to 8 MeV temperature. How to reach such high values ? What are the limitations which may prevent from heating enough the nucleus ? In addition to the intrinsic properties of hot nuclei, does the dynamics play an important role ? What happens to the nucleus above this critical temperature ? Besides this crucial aspect of the existence of a limiting temperature, the production of hot nuclei opens up also a large field for studying intrinsic properties as a function of temperature (surface energies, coulomb barriers, fission barriers, level densities, collective excitations ...).

These lectures are intended to give an overview of the experimental status in this "hot" domain. Other lectures are more focused on the theoretical point of view. This paper will be arranged as follows : In a first part, the conceptual problems one might have to face will be introduced. A long chapter will be then devoted to the different experimental methods used so far in order to characterize the hot nucleus. We shall then discuss what can be learned from the study of deexcitation of nuclei at high T. The fourth part will be focused on the experimental evidences fro the existence of limiting temperatures. Finally, a brief discussion will follow related to possible clues for the onset of nuclear instabilities at this critical temperature.

2. CHARACTERISTIC TIMES - ONSET OF CONCEPTUAL PROBLEMS

As it will be discussed in section 3, a large number of the experimental techniques used for signing the production of a hot nucleus are based on the observation of its statistical decay. Furthermore, some criteria have to be defined to conclude that a thermal equilibrium has been reached. Before starting to perform such experiments and to further analyze them, one must then be aware of the applicability of the thermodynamical concepts at temperatures as high as 5 or 6 MeV[6]. First of all, let me recall that the main assumption for a statistical analysis is a microscopic equilibrium. In other words, it means that all states with excitation energy E^* are equally populated. The system may then be described within a few macroscopic parameters. In fact, it is more convenient to express the density of states as a function of the nuclear temperature, that is to replace the microcanonical description by a canonical ensemble. T is then related to the level density ρ through the relation :

$$\frac{1}{T} = \frac{d}{dE^*} \ln \rho \, (E^*)$$

Besides the necessary achievement of thermodynamical equilibrium the principle of detailed balance from which emission probability cross sections are derived implies a long life time for the compound system. The key point is then to insure that the life time is larger than the relaxation time !

A lower limit for the relaxation time is obviously the time of a light signal through the nucleus, i.e. 2 R/c where R is the nucleus radius, c the velocity of light. For a heavy system of A = 200, 2R/c $\simeq 6.10^{-23}$ sec. However recent dynamical calculations using the Landau-Vlasov equation indicate that the thermalization time for a heavy system is reached rather shortly after 10^{-22}sec[7]. Let us assume now that an equilibrated nucleus has been formed and consider its statistical decay life time[8,9]. Fig. 1 shows the calculated decay time for neutron evaporation from a ^{208}Pb nucleus. It examplifies the very sharp decrease of life-time with increasing temperature of the system. For T = 5 MeV, the lifetime is close to 10^{-22} sec, i.e. very comparable to the equilibrium time, if not the collision time. Consequently one can reasonably imagine that for still higher temperatures, the characteristic time for energy dissipation and the decay-time can start to overlap. It would then imply a dependance of the de-excitation on the mode of formation. This introduces quite naturally the role of the dynamics in the process, through, for instance, the increasing effect of non-equilibrium processes.

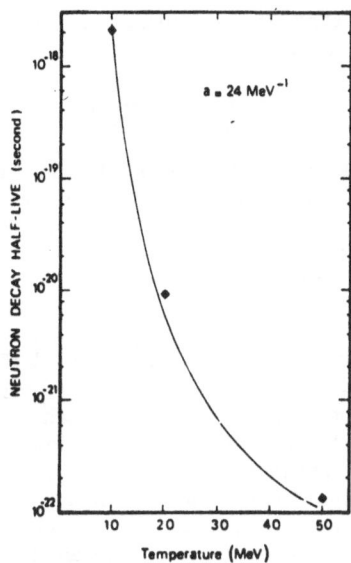

Fig. 1. Neutron decay half-life of a ^{208} Pb nucleus as a function of its temperature (ref. 9).

One sees now clearly the impossibility to decouple the probability of formation of a compound system from the study of its properties. However, we shall see later on that, usually, the experimentalist asserts that an equilibrated system is formed by checking if the experimental observables can be accounted for by the statistical model.

Anyhow, this will be a rather crude method, as if one admits that the intrinsic degrees of freedom can be equilibrated in a very short time-scale, it is definitely not the case for all collective modes. Another assumption of the statistical model is the hypothesis of equal life-time for the different decay channels. It will be discussed in more details in section 4 but an example can be given when observing again fig. 1. At T = 5 MeV the neutron decay time τ_n is 10^{-22} sec, to be compared with the prescission life-time[10], whose estimate is close to 10^{-20} sec, i.e. 100 times larger than τ_n ! One may then expect large consequences for the decay of very heavy systems (see section 4).

The last point to be stressed in this section is connected with the relationship between the excitation energy and the temperature. Usually, one experiment does not provide at the same time the temperature T and the excitation energy E^* and one is led to use a model to connect these two quantities. This is commonly done by using the degenerated Fermi gas model where one gets :

$$E^* = aT^2 - T \qquad (1)$$

where a is the level density parameter.

However, due to the fluctuations of the fundamental thermodynamical quantities, both E^* and T cannot be defined exactly[11-14]. As pointed out recently by Feshbach[14], statistical mechanics tells us that energy and temperature are complementary variable related through the uncertainty relation :

$$\Delta E \cdot \Delta(1/T) \simeq 1$$

where ΔE is related to the energy spread of the populated states.

It can be shown that the temperature fluctuation can be expressed as a function of excitation energy E^*

$$\frac{\Delta T}{T} = \frac{2}{\sqrt{AT}} = \left(\frac{2}{AE^*}\right)^{1/4} \qquad (2)$$

The use of the temperature concept must then be taken with care for small systems. For a nucleus with A = 40 excited at 130 MeV, the temperature cannot be defined within 15 %. One can notice that the validity of this concept becomes much more reliable for massive systems and at high excitation energy.

3. EXPERIMENTAL METHODS : CHARACTERIZATION OF HOT NUCLEI

Ideally, the aim would be to get an as complete as possible identification of such nuclei, i.e. a determination of the mass, excitation energy, temperature, angular momentum, shape ... Unfortunately, the experimentalist is unable to get a simultaneous measurement of all these quantities. Moreover, he doesn't have at his disposal a nuclear thermometer or a calorimeter with automatic reading ! All experiments are concerning indirect measurements of mainly the excitation energy deposit or the temperature. The results must then be taken with caution keeping in mind all the hypothesis which have been used in the analysis. This is the aim of this section to explain in detail these different experimental methods together with their advantages and limitations.

3.1. Which projectile?

Most of the studies performed so far in order to form and study hot nuclei have been using heavy ion induced reactions. These reactions have obvious advantages as for instance the large stopping power due to the number of nucleons involved. With such heavy projectiles, a quite large linear momentum can be transferred at very moderate incident energies (< 50 MeV/u >). Fusion like processes are known to dominate for central and intermediate impact parameters. Finally, heavy ions provide us with a quite unique tool to study dynamical effects, excitation of collective modes, angular momentum dependance ...

The situation is somewhat more complex with light projectiles (p, α-particles)[15-16]. One needs to use quite energetic beams to produce a substantial dissipation. This is obviously not an ideal situation and exclusive experiments have shown that one gets very broad energy distributions making difficult a clear identification of the formation of a well defined hot nucleus. However, compressional effects, excitations of collectives degrees are surely strongly minimized.

Another alternative is to use antiprotons[17]. A few experiments are presently undertaken at LEAR which seem to be rather promising[18,19]. Intranuclear cascade calculations[20-21] predict that half of the five pions created at the surface of the nucleus may interact with the target and that in some cases, a hot nucleus is thermal equilibrium can be formed. For example, they indicate the possibility to reach very high temperatures (T $\sim$ 6 MeV) for a Mo nucleus for about 5 % of the events. Annihilation of $\bar{p}$ at rest could then be the best way to produce hot nuclei without compression, nor angular momentum. Such measurements might be worth to be compared with heavy ion induced reactions. In the rest of the paper, we shall mainly concentrate on heavy ion reduced reactions.

3.2. From complete to incomplete fusion

Intuitively, one would expect to get the highest energy dissipation for the most central collisions. At low energy (E < 10 MeV/u), the reaction is indeed dominated by the fusion process or strongly damped collisions and has been extensively studied with a variety of projectiles at different incident energies[22-24]. When increasing the bombarding energy, a quite large number of fast particles having almost the beam velocity are observed which are very likely emitted during the early stage of the collision. The consequence will be a decrease of the available linear momentum in the entrance channel for fusion. Complete fusion processes will then be strongly reduced and the so called incomplete fusion or massive transfer process will take place. One should note that this fast particle emission phenomenon has been first observed almost 30 years ago by Britt and Quinton[25]. This emission has been discussed in terms of hot spot, preequilibrium process or Fermi Jets[26-28].

A simple interpretation of incomplete fusion can be given within the framework of a Fermi gas[29]. Let us define V_L, the relative velocity associated with the light fragment in the entrance channel (assuming we are using an asymmetric system $A_L + A_H$)

$$V_L = \frac{A_H}{A_L + A_H} V_{rel} \tag{3}$$

where A_H and A_L are the masses of the heavy and light fragment respectively, and V_{rel} the relative velocity in the entrance channel expressed by

$$V_{rel} = \left[2 \left(E_{cm} - V_B \right)/\mu \right]^{1/2} \tag{4}$$

where V_B is the coulomb barrier, μ the reduced mass.

The energy threshold for the onset of incomplete fusion can be obtained if one adds collinearly the Fermi velocity V_F of a nucleon m_o with V_L

$$\frac{1}{2} m_o \left(V_F + V_L \right)^2 > \varepsilon_F + E_S \tag{5}$$

ε_F is the Fermi Energy and E_s the separation energy of this nucleon. The threshold corresponds approximately to $V_L \simeq 0.06$ c.

From relation (5), one sees also that the probability for one nucleon of the heavy fragment A_H to escape will be much smaller. This has been nicely demonstrated by Morgenstern et al.[29]. Consequently, this legitimates the hypothesis of a "massive" transfer where part of the projectile may fuse with all nucleons of the target. However, this assumption may not be valid any more at too high bombarding energies (and is not for symmetric systems). An analysis of existing data shows that the fraction of the initial linear momentum P_I transferred to the target is quite independant of the nature of the projectile. Anticipating on the discussion of section 3.5, an "universal" energy dependance of this momentum transfer is shown in fig. 2. (This was first pointed out by Viola[30]). The onset of incomplete fusion corresponds roughly to a bombarding energy of 10 MeV/u.

3.3. Recoil velocity measurements

A simple experimental signature for the onset of incomplete momentum transfer during the fusion process may be found in the

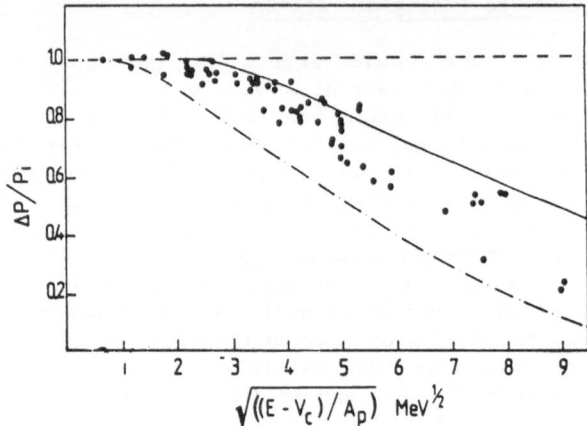

Fig. 2. Average proportion of the initial linear momentum transferred to the target versus the square root of the incident energy per nucleon above the coulomb barrier. Dots are experimental points. Dashed curve is calculated assuming a momentum transfer <p> = 180 MeV/c per nucleon. Dashed and solid curves correspond to calculations[31] (see section 3.5)

measurement of the recoil velocity V_R of the fusion nucleus. The deviation from the center of mass velocity will tell us how far the fusion has been completed.

The first method concerns the direct measurement of the recoil velocity of the evaporation residues in the case of light systems which are known to mainly deexcite by light particle evaporation[32-39]. This can be done by the standard energy-time of flight method or by using radiochemical techniques. The second method is using reverse kinematics reactions where one looks at complex fragment evaporation. The third one has to deal with the observation of the folding angle distribution between fission fragments and is well appropriated for very heavy systems.

3.3.1. Measurements of heavy residues

As, to the first order, the initial recoil velocity v_R of the fused nucleus is not modified by the evaporation process, the recoil velocity measurements of the evaporation residues should provide us with a reliable value of V_R. Two examples are shown in fig. 3 - 4. Fig. 3 is a mass-velocity contour plot for the system Ar + ^{124}Sn at 27 MeV/u[40]. The evaporation residue bump is well separated from the other components (i.e. peripheral collisions and fission products) around A = 115 and with a velocity close to 70 % of the center of mass velocity. Velocity spectra of these heavy products are presented in fig. 3 for the system Ar + Ag at 27 and 35 MeV/u[33]. The shape of these spectra calls for two remarks :

i) the residue velocity may exceed the center of mass velocity. We shall see later on that it can be clearly attributed to fluctuations in the evaporation process. This might already indicate that only the most probable velocity can be used to determine the recoil velocity of the system.

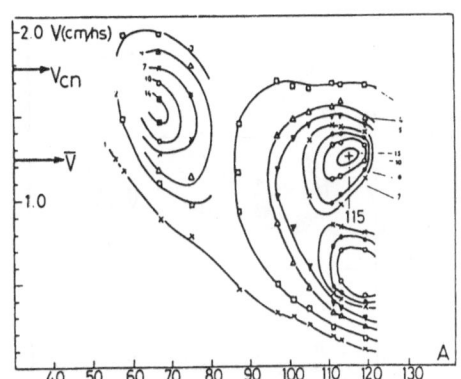

Fig. 3. Mass velocity plot of residual products for the system Ar + ^{124}Sn at 27 MeV/u. The fusion products at high velocity exhibit a well defined bump around A = 115 (Ref. 40).

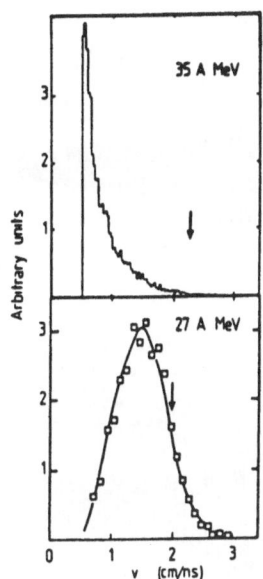

Fig. 4. Evaporation residue velocity distribution for the system Ar+Ag at 27 and 35 MeV/u. The arrows indicate the center of mass velocity[33].

ii) the almost gaussian-like distribution observed at 27 MeV/u has been replaced at 35 MeV/u by a continuous increase of the cross section down to the velocity corresponding to the experimental threshold. It appears then difficult, if not dangerous, to define a most probable value for V_R, as this shape seems to sign for a continuous evolution of momentum transfers. Clearly, more exclusive experiment would be needed before to draw any conclusion. It should be noted that the same quite sudden change around 30 MeV/u in the velocity distribution has been seen in other systems as Ar + Al[34].

3.3.2. Complex fragment evaporation

It is a somewhat trivial effect that cluster evaporation (emission of fragments with Z > 2) should become rather probable for large excitation energies as the influence of the binding energy will be progressively washed out. This process will be discussed in more details in section 4.6. We shall just demonstrate here that it can be successfully used to determine V_R. This way has been investigated by several authors using reverse kinematics[41-46]. The use of reverse kinematics gives an obvious advantage as the source velocity of the complex fragments is only slightly lower than the beam velocity. The locus of all velocity vectors of the emitted fragments is then expected to be a circle centered at the tip of the velocity vector characterizing the source. Fig. 5-6 are examples of experimental results obtained in the systems ^{86}Kr + C, Al at 35 MeV/u and ^{139}La + C at 18 MeV/u. At least for the low Z's in fig. 5, one identifies clearly the two branches cor-

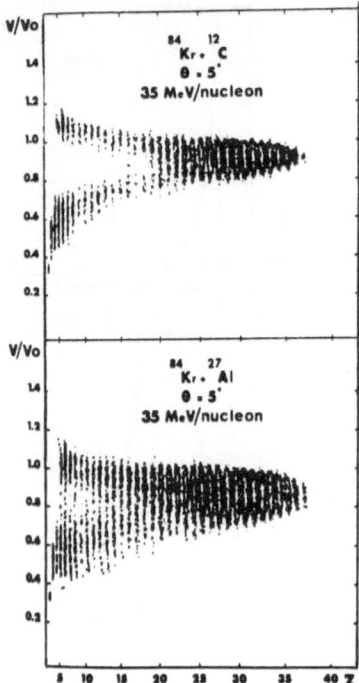

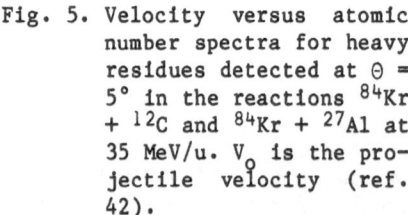

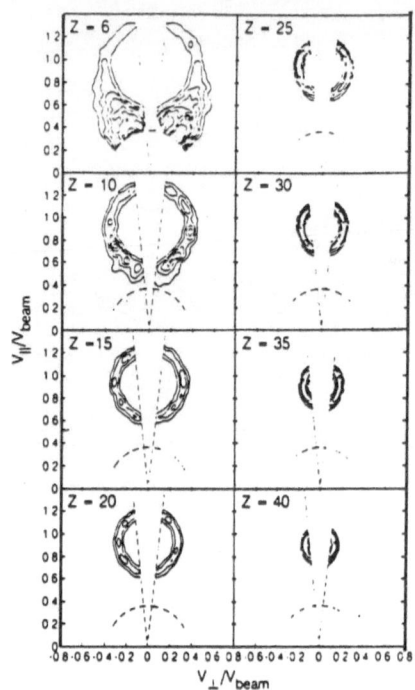

Fig. 5. Velocity versus atomic number spectra for heavy residues detected at $\Theta = 5°$ in the reactions ^{84}Kr + ^{12}C and ^{84}Kr + ^{27}Al at 35 MeV/u. V_0 is the projectile velocity (ref. 42).

Fig. 6. Contour plots of invariant cross sections in the $V_\parallel - V_\perp$ plane for various Z values detected in the reaction 18 MeV/u ^{139}La + ^{12}C. Beam direction is vertical (ref. 46).

responding to the two expected kinematical solutions. Contour plots of invariant cross sections in fig. 6 show very nicely the coulomb circles expected from a statistical evaporation of an equilibrated system. The extracted source velocities have been shown to be independant of the observed Z and the binary nature of the reaction signed by coincidences between the heavy projectile residues and the emitted fragments. It appears clearly that for very assymmetric systems, the main source of complex fragment is the equilibrated fusion-like system. The situation may become more complicated for more symmetric systems or at higher energies where non equilibrated sources may contribute significantly[47].

3.3.3. Fission fragment angular correlation techniques

For very heavy systems which are known to undergo very easily fission, the schematic diagram in fig. 7 shows clearly how the measurement of the coincident fission fragments allows in principle to deduce the recoil velocity V_R of the fissioning system. It can be easily demonstrated[48] that in case of a longitudinal momentum transfer, V_R can be expressed as :

$$V_R = \frac{V_F}{2} \sqrt{\frac{-TK}{(1-KT/4)}}$$

$$\text{where } K = \frac{1}{tg\ \Theta_1} + \frac{1}{tg\ \Theta_2} \qquad (6)$$

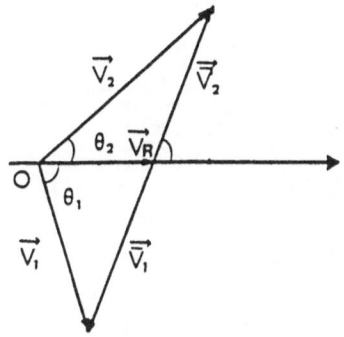

Fig. 7. Velocity diagram for the emission of two fission fragments from a recoiling source.

Θ_1 and Θ_2 correspond to the emission angle of the fission fragments with respect to the beam direction.

$$T = tg\left(\Theta_1 + \Theta_2\right)$$

$$V_F = \sqrt{2 <E_K>/A}$$

(7)

where $<E_K>$ is the average total kinetic energy of fission fragments in the center of mass frame which can be deduced from the Viola systematics[50].

Expression (6) is then totally valid only for pure symmetric fission.

Since the pionnering work of Sikkeland,[49], such experiments have been performed by many groups[51-63]. Two beautiful examples[57,59] are shown in Fig. 8-9 for the systems $^{40}Ar + Th$ and $^{14}N + U$. For these very heavy systems, the distributions are supposed to be representative of the total reaction cross section σ_R as fission is expected to be the only decay channel, even for very peripheral collisions. Let us mention already that this statement might be questionable for high incident energies. For moderate incident energies, the folding angle distribution shows clearly 2 peaks which can be easily identified as due to peripheral collisions $\left(\Theta_{fold} \sim 180° \rightarrow V_R \approx 0\right)$ and central collisions $\left(\Theta_{fold}\text{ much smaller than }180°\right)$. For both examples, the position of the central collision bump seems to saturate above 30 MeV/u, an indication for the recoil velocity of the composite system to reach a limiting value. At higher bombarding energies (E > 35 MeV/u), this peak either becomes very broad for the N + U case, or disappears completely for heavier projectiles as in the case of Ar + Th.

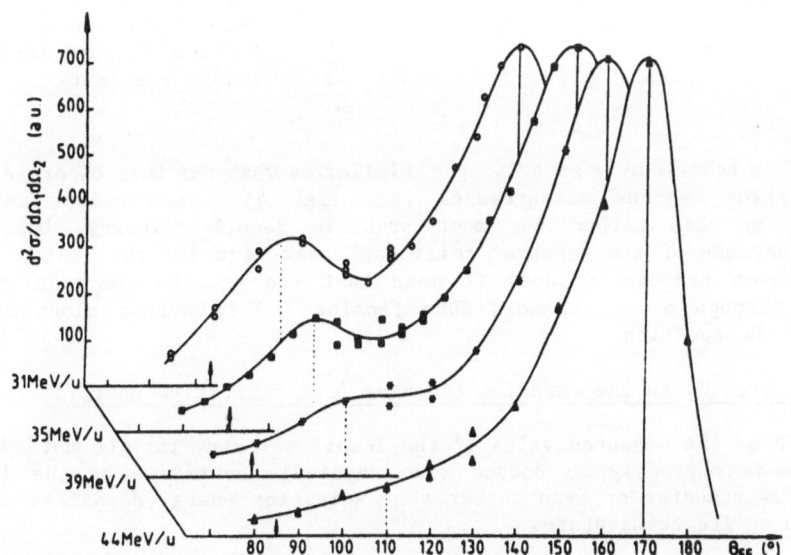

Fig. 8. Folding angle distributions for the fission fragments in the reaction Ar + ^{232}Th at 4 incident energies (31 to 44 MeV/u)[59].

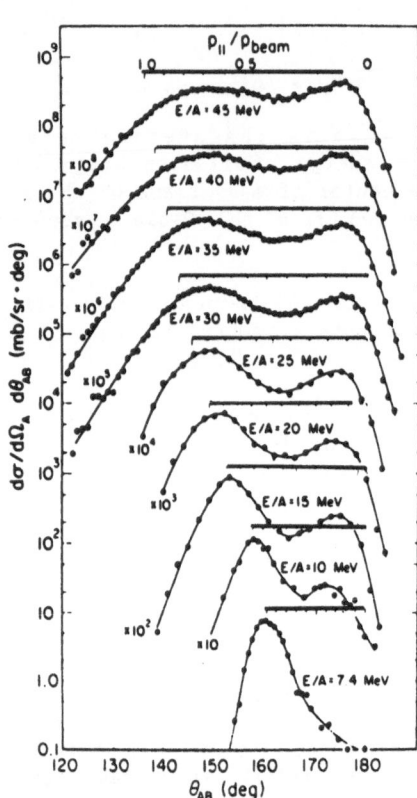

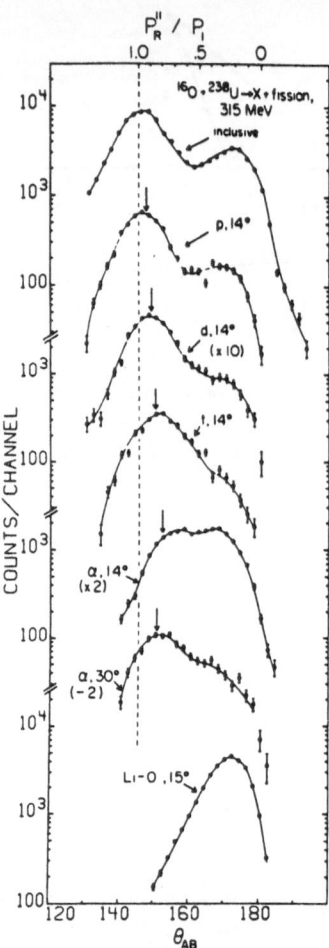

Fig. 9. Same as Fig. 8 for the system N + ^{238}U in the energy range 7.4 - 45 MeV/u (ref. 57).

Fig. 10. Folding angle distributions of fission fragments measured inclusively and in coincidence with light charged particles[64]. For more details, see text.

This behaviour is surprisingly similar to what has been observed with evaporation residue measurements (see fig. 4) : one cannot anymore define by this method the most probable recoil velocity. Does the disappearance of the "central collision" peak sign for the vanishing of the fusion process or does it mean that the excited compound system decay through other channels than fission ? This method alone cannot answer the question.

3.3.4. From V_R to the momentum transfer (and excitation energy)

From the measured value of the recoil velocity for the hot system, one needs obviously to deduce more physical quantities as the linear momentum transfer or even better the excitation energy deposited in the system or its temperature.

Unfortunately, the transformations are model dependant. To deduce the fraction ρ of the initial angular momentum which is effectively transferred, it is generally assumed an incomplete fusion model implying the following conditions :

i) only a part of the projectile fuses with the whole target.

ii) the non-fusing nucleons of the projectile are emitted with the beam velocity at 0°. They can be considered as spectators in the reaction.

iii) the recoil velocity is not changed on the average by the evaporation process.

This hypothesis of incomplete fusion (or massive transfer) has been already discussed in section 3.2. Fig. 10 illustrates the validity of such a model, at least for not too large bombarding energies (20 MeV/u $^{16}O + ^{238}U$)[64]. It displays the folding angle distributions for inclusive fission events (only the two fission fragments detected) as well as in coincidence with p,d,t,α and fragments with Z > 3. Arrows in the figure indicate the position where the recoil momentum is equal to the difference between the beam momentum and the momentum carried away by the emitted particle. They almost nicely coincide with the maximum of the distribution. More massive is the emitted particle, (with about the beam velocity), smaller is the folding angle. There is a clear relationship between the mass of the particle flighing away and the momentum transferred in the fusing nucleus.

Assuming this massive transfer process leads to the following expression for ρ

$$\rho = \frac{A_T \ R}{A_T + (1-R) \ A_p} \tag{8}$$

where A_p, A_T are the projectile and target mass respectively and R the ratio of the recoil velocity V_R to the center of mass velocity V_{cm}. This expression is valid only for normal kinematics ($A_p < A_T$). In case of reverse kinematics ($A_p > A_T$), the following expression may be used :

$$\rho = \frac{1}{R} \ (1 + A_p/A_c) - \frac{A_p}{A_t} \tag{9}$$

Another expression can be easily calculated, the excitation energy per nucleon of the system

$$E^*/A = \left(V_p - V_R\right) * V_R/2 \tag{10}$$

This relation implies that the spectator nucleons of the projectile keep their initial velocity and that the spectators of the target, if any, are at rest. (The Q value is neglected). It is also implicit that the transferred nucleons from the projectile have the average initial momentum per nucleon, an assumption which may not be justified theoretically.

As mentionned already, for very asymmetric systems, one can assume a massive transfer process where the missing momentum is carried away at 0° by particles having the beam velocity. Using expression 8, the absolute excitation energy can thus be deduced :

$$E^* = \frac{A_T}{A_T + \rho A_p} \ \rho \ E_p + Q \tag{11}$$

where E_p is the projectile energy, Q the Q value. In the frame of the Fermi gas, the temperature can be obtained from the relation

$$E^* = a \ T^2 - T \simeq \frac{A}{8} \ T^2 - T \tag{12}$$

where a is the level density parameter. However, this would mean that the total excitation energy E^* has been transferred into heat, which might not be the case as one could have to take into account the energy transfer into collective modes as compression or rotation.

A compilation of existing data concerning the evolution of the linear momentum transfer with the projectile bombarding energy shows a very striking feature. It appears clearly (see fig. 2) that all the data follow more or less an universal curve. Leray[65] has shown that the proportion ρ of the linear momentum transferred to the target can be parametrized by the expression :

$$\rho = 1 \qquad \text{for } v_{rel}/c \leqslant 0.1$$
$$\rho = -1.904 * v_{rel} + 1.19 \text{ for } v_{rel}/c \geqslant 0.1 \qquad (13)$$

where v_{rel} is the relative velocity in the entrance channel. It follows that, on the average, the momentum transfer per incident nucleon is, to the first order, a constant value. (The dashed curve in fig. 2 has been calculated assuming 180 MeV/c per incident nucleon). This trend cannot be reproduced by calculations taking into account only the influence of the mean field (dot dashed curve), a much better agreement being obtained when adding the two body dissipation (solid curve)[31].

This universal value close to 180 MeV/c per nucleon is now confirmed by a large variety of projectiles from Li up to Kr as demonstrated by fig. 11. This value, close to the Fermi momentum strongly suggests that nucleon-nucleon collisions should play a major role in the momentum transfer process and that all nucleons of the projectile have the same probability to induce such n-n collisions[66]. It also implies that the mean free path of the incident nucleons in the target is projectile mass independant. Finally it proves that heavy projectiles are much more efficient than the light ones to induce high momentum transfers. For example, in the reaction 35 MeV/u ^{86}Kr + ^{232}Th, a momentum transfer as high as 13 GeV/c has been observed[63].

This technique has been used very often and seems to indicate a saturation of the momentum transfer above 30 MeV/u for a wide variety of projectiles ranging from N to Kr. It would imply, in the frame of a massive transfer process, a linear increase of the total excitation energy deposition with increasing bombarding energy. However one should be very cautious before to draw definite conclusions :
i) The existence of a transverse momentum is usually neglected in the estimate of V_R.
ii) One is expecting, when increasing the incident energy, an enhancement of non-equilibrium processes in the early stage of the collision (particle or cluster emission). This might affect, perhaps strongly, the position of the so-called "central collision" bump in the folding angle distribution as shown clearly in fig. 12. This picture displays the folding angle distributions of fission fragments in the reaction N + Th at 35 MeV/u in coincidence with intermediate mass fragments (IMF) detected in the backward and in the forward direction[67]. A comparison with the inclusive measurement demonstrates the strong kinematic shift induced by IMF emission. This will then strongly distord the inclusive data as the forward emitted clusters display much broader spectra than do the backward ones.
iii) One clearly observes a complete disappearance of the "central collision" peak when increasing the bombarding energy. In that case, a determination of the most probable momentum transfer is no more meaningful and more exclusive experiments are needed.

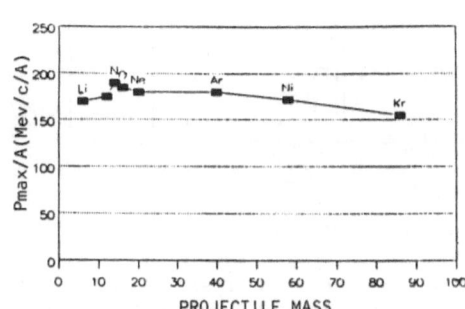

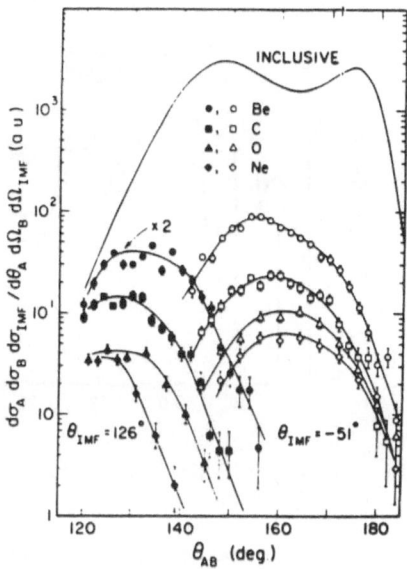

Fig. 11. Most probable linear momentum transfer per projectile nucleon, versus the projectile mass.

Fig. 12. Folding angle distributions for fission fragments for the reaction N + Th at 35 MeV/u (inclusive and in coincidence with IMF detected forwards and backwards[67]

3.4. **Emission of light particles. Is it a good thermometer?**

The properties of evaporated light particles by the hot composite nucleus may definitely shed some light on the temperature of the emitter. Their energy spectrum can be predicted by the statistical theory. If one describes the system within a canonical representation where the nucleus is supposed to be placed in a heat bath, the density of states may be expressed as a function of the temperature T and one is led to the following expression for the energy spectrum of the particle :

$$P(E)dE = (E-E_c)/T^{2} * \exp-(E-E_c)/T \ dE \qquad (14)$$

where E_c is the coulomb barrier of the emitted particle.

The observation of a Maxwellian spectrum is then the first condition for identifying evaporated particles, the second being an isotropic emission in the center of mass of the emitting system. However, the crucial requirement is to be clearly sure that one is selecting unambiguously the source responsible for the emission.

Therefore, it appears very dangerous to extract temperatures from purely inclusive spectra. Moreover, T in expression (14) is the instantaneous temperature. If the detected particles are emitted all along the cooling chain, one is then observing a spectrum resulting from a superposition of several spectra, each of them being defined by a different temperature. Fitting such spectra using several moving sources will not provide any meaningfull temperature information. The only

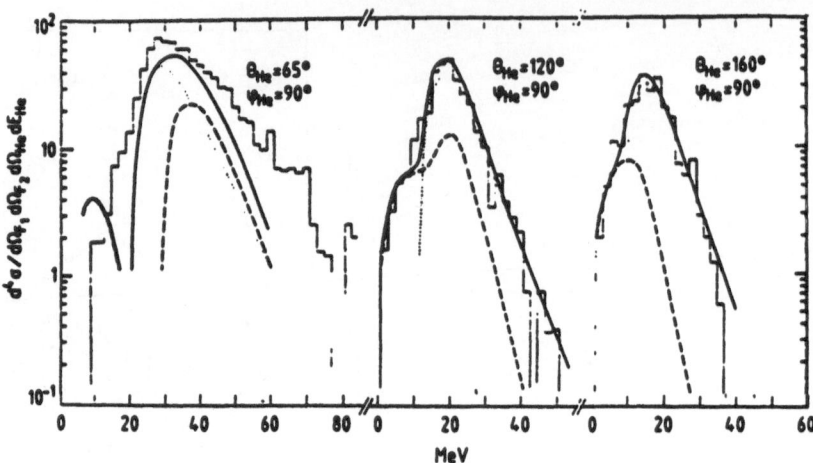

Fig. 13. Experimental α-particle spectra (histograms) in the reaction
27 A.MeV Ar + ^{238}U. Large momentum transfers are selected by
triggering on small folding angles for the fission fragments.
Calculated spectra have been obtained by the Monte Carlo simulation code GANES : evaporation contribution from the composite
system (dotted), the fully accelerated fission fragments
(dashed). The sum is represented by solid lines[48].

reliable experiments are those where the particles are detected in
coincidence with the emitting nuclei. Such experiments are concerning
charged particles or neutrons detected in coincidence, for instance with
the two correlated fission fragments or evaporation residues[3,10,68,71].
One example is given in figure 13 for the reaction 27 MeV/u ^{40}Ar + ^{238}U.
Alpha-particle spectra have been measured for large momentum transfers
by triggering on the two fission fragments detected in coincidence at a
small correlation angle. This angle was giving the recoil velocity of
the emitting nucleus. A complete kinematic Monte Carlo simulation
code[72] has been used and compared with experiment. It clearly signs for
the evaporative nature of the particles in the backward direction and
indicates that these alpha particles were emitted by the composite
system before fission (the calculated evaporative component is indicated
by dotted lines in the figure). Moreover, the angular distribution was
found to be isotropic in this backward hemisphere. (It should be noted
that the inclusive spectrum measured at 160° is quite identical to the
one observed in this coincidence experiment. Looking at very backward
angles seems then to be a very good filter to select central
collisions). It appears then very clear that the thermalization of the
system has been reached and a temperature close to 4 MeV was deduced
from the spectrum analysis. Let us remind that at such a temperature,
the decay life time is estimated to be about 2.10^{-22}sec (see fig. 1).
This would then imply a shorter thermalization time, an assertion which
is confirmed by recent dynamical calculations (see section 5).

The observation of the forward emitted light particles associated
with large momentum transfer provides us also with a better knowledge of
the early stage of the process. For instance, the emission of protons
having the beam velocity has been interpreted as resulting from a
transparency effect, the nucleons of the projectile going through the
target without any interaction[70].

3.5. Multiplicity measurements of evaporated light particles : a good calorimeter

It has been shown before that the identification of the evaporation residues or fission fragments produced in the most central collisions would not be sufficient to determine with a good accuracy the excitation energy. The ideal experiment would be to combine this type of measurement with the identification of all decay channels contributing to the decay process of the hot nucleus. This includes the observation of neutrons, light charged particles, complex fragments and γ-rays. Such a complete measurement has not been undertaken so far, but several groups have concentrated their effort on one or several of these decay channels[10,68,69,71,73,77].

One series of experiment is concerning the study of Ar induced reactions on heavy targets (Au or Th)[76,77]. For such very heavy systems, neutron evaporation is expected to carry away a large part of the excitation energy. Using a 4 π neutron detector essentially sensitive to low E neutrons, i.e. to neutrons evaporated by the slow moving source (the recoiling hot nucleus), the neutron multiplicity has been measured in coincidence with the fission fragments in the reaction Ar + Th at 3 bombarding energies 27,35 and 44 MeV/u. As examplified in fig. 14, the average neutron multiplicity (M_N) exhibit a very strong dependance with linear momentum transfer. However, for small values of the folding angle, it saturates at a somewhat constant value whatever the bombarding energy is $\langle M_N \rangle \simeq 20$, a value not corrected for the efficiency of the detector. The real value is close to 34 neutrons). This saturation clearly indicates that the width of the central collision bump is only due to the combined effects of particle evaporation, velocity and mass dispersion of the fission fragments. The observation of events at very small folding angle by no means indicates the possible existence of some complete fusion events. The observed saturation of neutron emission is not compensated by an increase of the light charged particle (LCP) multiplicity, the same behaviour beeing observed for these other decay channels. From the observed multiplicities for neutrons, and LCP at backward angles, an estimate of the total excitation energy deposit in the system was deduced assuming an energy reduction of 14 and 20 MeV per neutron and LCP respectively. The emission of heavy clusters has been neglected, first because the multiplicity is small, but mainly because their emission does not pump almost any excitation energy due to the large positive Q value. The results are plotted in fig. 15 together with

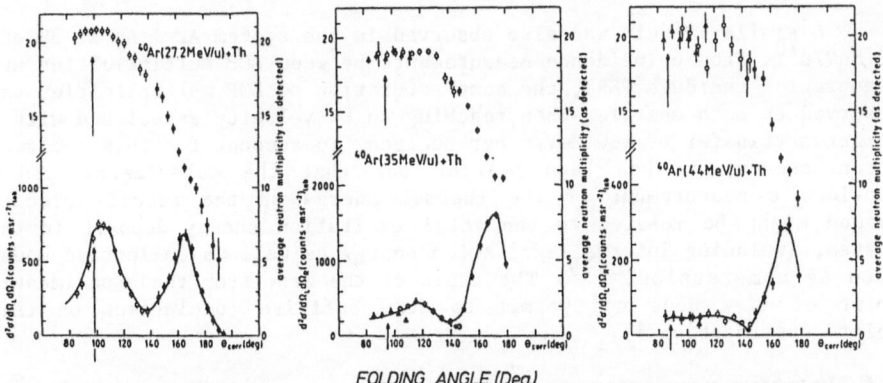

FOLDING ANGLE (Deg.)

Fig. 14. Folding angle distributions (bottom) of the fission fragments and associated average neutron multiplicities (not corrected for detector efficiency) for ^{40}Ar + Th at 27,35 and 44 MeV/u[77].

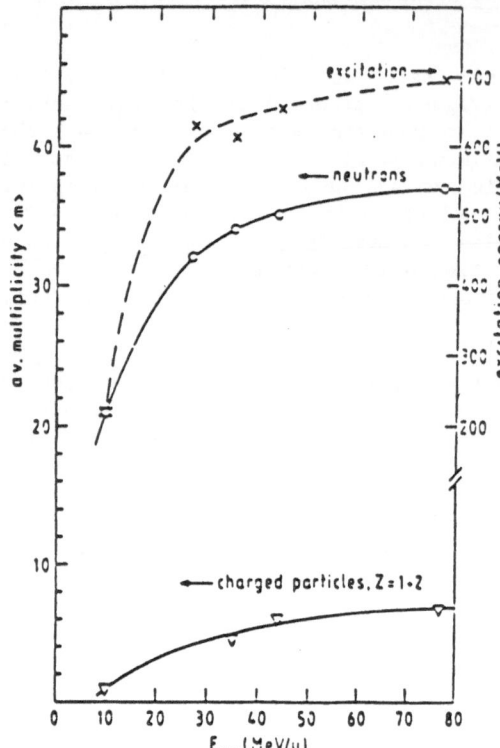

Fig. 15. Excitation function for the system ^{40}Ar + ^{232}Th. Evolution of the total number of neutrons (corrected for detector efficiency) as well as evaporated charged particles associated with the most central collisions (left scale). Estimated excitation energy deposit (right scale)[76,104].

a recent result obtained at 77 MeV/u. Within the experimental uncertainties, the excitation energy remains constant in this wide energy range. (E^* ≃ 650 MeV). Assuming a level density parameter equal to A/8, this corresponds to a temperature of 4.5 MeV.

This result contradicts the conclusions deduced from folding angle measurements on the same system[59], a saturation of the linear momentum transfer associated with an increase of the excitation energy deposit of 200 MeV between 31 and 44 MeV/u. This might be another indication for this method to become irrelevant for very large values of the missing momentum. It should be stressed that the agreement between the two methods is very satisfactory when the momentum transfer is larger than 75 % of the initial one. This consistency was nicely checked at low energy for the system Ne + U where complete fusion was still the dominant process[74].

A similar result was also observed in the system Ar + Ag at 39 and 60 MeV/u[70]. From coincidence measurements between LCP multiplicities and evaporation residues (ER), the same saturation of LCP multiplicities was observed at both energies when reaching an ER velocity associated with a momentum transfer of 180 MeV/c per nucleon. To account for this apparent inconsistency, it has been pointed out that the calorimetry method provides a measurement of the thermal energy as the recoil velocity method might be related to the total excitation energy deposit in the system, including internal excitation energy as well as collective modes such as compression[70,79]. The state of the art, from the experimental point of view does not permit to draw definite conclusions on that point. (see section 5).

3.6. <u>Relative population of excited states : a good thermometer?</u>

Assuming the hot system to be in thermal equilibrium, in the frame of the Fermi gas, one can calculate the decay width ratios for the evaporation of two different clusters :

$$\frac{P_1}{P_2} = \frac{g_1\mu_1}{g_2\mu_2} \frac{E^*-E_1}{E^*-E_2} \frac{a_2}{a_1} \exp\left[2\left[a_1(E^*-E_1)\right]^{1/2} - 2\left[a_2(E^*-E_2)\right]^{1/2}\right] \tag{15}$$

where E^* is the excitation energy of the emitter, $E_{1,2}$ the energy removed by particle 1 and 2,, g and μ the spin degeneracy and reduced mass, a the level density parameter.

If clusters 1 and 2 are identical but in different excited states, expression 15 can be easily simplified[13] providing $E^* \gg E_1$:

$$R = \frac{P_1}{P_2} \simeq \exp\frac{\Delta E}{T} \tag{16}$$

where $\Delta E = E_2-E_1$ is the difference between the internal energies for the two excited states of the same cluster. One gets then within the Fermi gas model the Boltzmann factor law. The knowledge of this ratio can thus be a quite powerful way to deduce the temperature of the emitting system. The method has been applied in two different situations. If the two states are particle stable R can be calculated through the detection of the clusters and decay γ-rays[80,83]. If both clusters are particle unstable (highly excited levels) the decay particles can be detected in coincidence and their kinetic energy measured in order to properly identify the quantum state[84-91].

Two main conditions have to be fullfilled in order to make this method reliable :
i) one has to be sure that the emitting system is in thermal equilibrium
ii) the observed products have to be primary products. If the observed level has been partially fed up by a previous decay of higher lying states or more massive clusters decaying by particle emission, the apparent measured temperature may be wrong (side feeding effect).

As far as the first point is concerned, most of these experiments were unable to sign for a thermal equilibrium. It is then an a priori assumption. As for the side feeding, this problem has been discussed in ref. 85. It appears that its contribution is all the more important since the difference E_2-E_1 is small. This is the case for the γ-ray method for the identification of low lying states. In that case, as soon as the temperature of the emitter exceeds 2 MeV, the sequential feeding becomes dominant and leads to an underestimation of the temperature. The method is then not applicable to the study of very hot nuclei.

The only accurate method for high T is then the particle decay method. An example of such a measurement is shown in fig. 16a and is concerning the α-d correlation[86]. In this plot, displaying the coincidence yield versus the relative kinetic energy of the particles, the two peaks corresponding to the 2.19 and 4.3 MeV levels of ^{6}Li are clearly visible. One should notice that the spectrum has been obtained after substraction of the "background" contribution of sequential emission. A detailed analysis of the procedure may be found elswhere[84]. Fig. 16b shows the apparent temperatures obtained for the system ^{40}Ar + ^{197}Au at 60 MeV/u. The largest temperatures are observed for wide apart quantum states, i.e. for cases where the side-feeding is assumed to be minimized. The results are in good agreement with the predictions of the quantum statistical model of Hahn and Stöcker,[92] where the emitting source is supposed to be in thermal and chemical equilibrium. Side feeding corrections are also taken into account in the calculation. The main finding is that temperatures close to 5 MeV are deduced from this method, in very good agreement with the one obtained with the calorimetry method on a very similar system (see section 3.5).

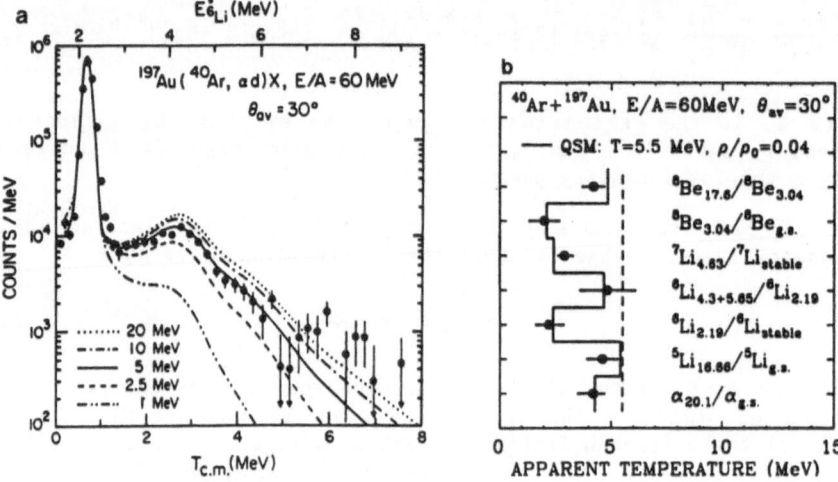

Fig. 16 (a) Decay of a ^{6}Li cluster. $T_{c.m.}$ is the relative kinetic
energy of the outgoing particles [86].

(b) Apparent emission temperatures. The histogram shows the
results of a quantum statistical calculation [92].

3.7. **What about deep inelastic collisions ?**

Up to now, central collisions and the so called incomplete fusion
process have been considered to be the only source of production of hot
nuclei. Several experimental facts indicate that one has to seriously

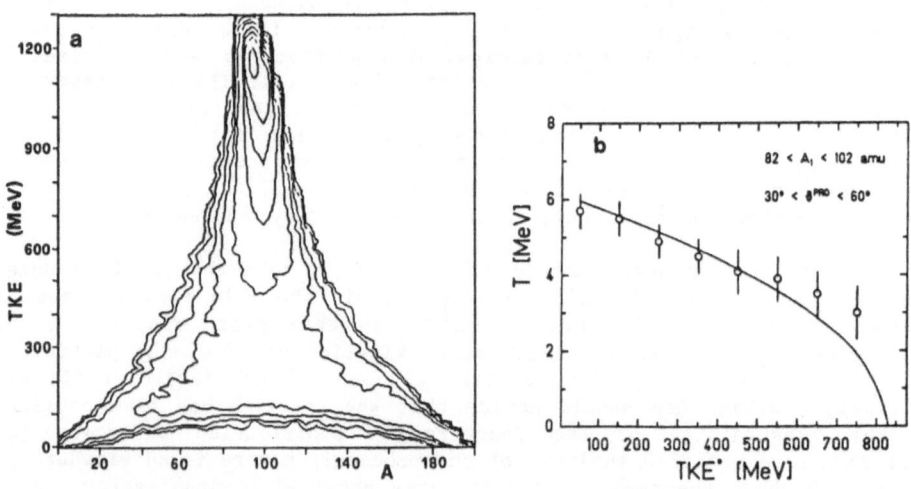

Fig. 17 (a) Diffusion plot for the system ^{100}Mo + ^{100}Mo at 23.7
MeV/u [96].

(b) System Mo + Mo at 18.2 MeV/u. Slope parameter T deduced from
fits of the proton energy spectra as a function of the total
kinetic energy of the fragments (points). The solid line
shows the calculated T from the measured total kinetic
energy loss [94].

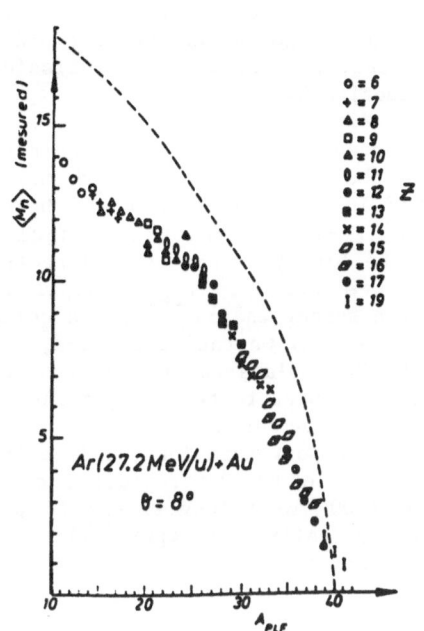

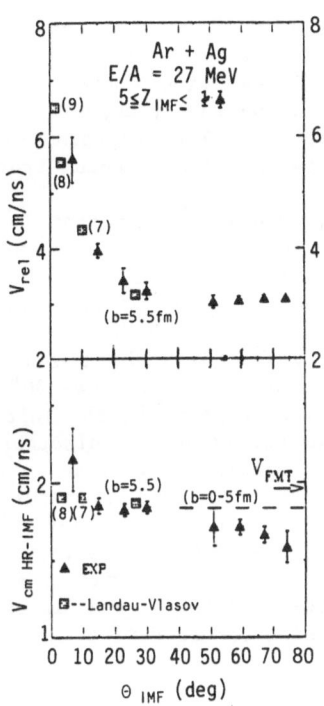

Fig. 18. Average neutron multiplicities (not corrected for
the efficiency) associa-
ted with projectile-like
fragments detected near
the grazing angle[100].

Fig. 19. System Ar + Ag at 27
MeV/u. Relative velo-
city of the fragments
and of the system be-
fore splitting as a
function of the lab
angle of the light
fragments. Comparison
with Landau-Vlasov
calculations[101].

consider the large effect of the mean field in peripheral reactions and
intermediate impact parameters[93-99]. These experiments remind in many
aspects of the well know deep inelastic collisions (DIC) at lower
energy. For instance, fig. 17 shows a diffusion plot for binary events
observed in the reaction 23.7 MeV/u ^{100}Mo + ^{100}Mo. Events correlated
with lowest total kinetic energies (TKE) can be identified as due to a
completely damped reaction[96]. Such reactions have obviously the
advantage to produce fragments with a huge range of excitation energies
in a single experiment and then to offer the possibility of studying
their decay properties as a function of this variable. For the same
system, at 18 MeV/u, the temperature of the dinuclear complex computed
from the fragment TKE is in good agreement with the one deduced from the
slope of the coincident proton spectra and temperatures as high as 6 MeV
are effectively reached for the most dissipative collisions[94] (fig.
17b).

The observation of the associated neutron multiplicities to a
given projectile like fragment detected at the grazing angle in the
reaction Ar + Au at 27 MeV/u shed also some light on the occurence of

highly dissipative processes[100] (fig. 18). The light fragments (Z < 10), emitted with a velocity close to 80 % of the beam velocity are associated with large neutron multiplicities signing for a high excitation energy deposit in the target like fragment, of the same order as the one observed when triggering on central collisions (i.e. fission following large momentum transfer, see section 3.5).

Would it mean that DIC are still occuring at such high incident bombarding energies ? Recent results on the system Ar + Ag at 27 MeV/u where light fragments have been detected in coincidence with the heavy residues seem to support this idea[97]. Fig. 19 shows the angular evolution of both the c.m. velocity of the binary system as well as the heavy residue-light fragment relative velocity V_{rel}. The rapid saturation of V_{rel} above 30° is given as a strong indication for a quite complete damping of the initial linear momentum. On the other hand, the large out of plane anisotropy signs for the coplanarity of the process and a multiplicity close to the unity with respect to the heavy residues is found, a good signature for the occurence of a binary process. Dynamical Landau Vlasov calculations[101] support this idea, indicating that intermediate impact parameters give rise to the formation of a dinuclear complex with a sticking time (300 fm/c) long enough to get orbiting effects and incompletely damped collisions (preequilibrium emission occurs during the first stage of the collision).

3.8. Hard photon production

It is well established that statistical γ-ray emission from a system in thermal equilibrium may compete with light particle decay channels. The observation of high energy γ-rays could then be a good probe for the temperature of the system. It has an obvious advantage, namely it is not distorted by coulomb effects and no kinematical shift is expected.

In the system Mo + Mo at 19 MeV/u, the emission of γ-rays has been carefully studied in cases where the exit channel was properly identified (the two coincident fragments)[102]. As mentionned before, a DIC process was identified in which two very hot fragments in thermal equilibrium are produced decaying essentially by sequential particle emission. We have then to deal here with two compound systems which have to decay also by statistical γ-ray emission. The exclusive photon energy spectra are well reproduced using statistical model calculations. The exponential tail at high energy can be well fitted by a Boltzman distribution :

$$f(E_\gamma) \; \alpha \; E_\gamma^2 \; \exp \left(-E_\gamma/T\right) \qquad (17)$$

The temperatures deduced from such fits match nicely those deduced from the measured TKE losses. Furthermore, the integrated multiplicities are almost perfectly reproduced by a statistical model calculation. Temperature as high as 5 MeV can be deduced in good agreement with the one obtained from the charged particle energy spectra.

It should be mentioned that the Mo + Mo system is the only one where a pure statistical γ-ray emission was found without any ambiguity. Extensive work has been indeed devoted to the study of the interplay between mean field effects and the two body dissipation through the observation of hard photons emitted in heavy ion reactions[103]. The experimental data are found to be consistent with γ-ray production from an incoherent nucleon-nucleon bremstrahlung process during the very early stage of the collision. (This is discussed by E. Gross at this school).

The apparent contradiction with the Mo + Mo system may be solved by considering that most of the experiments have been performed with much faster projectiles and quite asymmetric systems where the statistical process, although still there, is strongly mixed-up with direct processes. This method for measuring temperatures is then probably restricted to rather low energy reactions (< 20 MeV/u).

4. NUCLEAR PROPERTIES OF HOT NUCLEI

As shown in the previous chapter, there are now clear experimental evidences on the possibility to form hot nuclei in thermal equilibrium at temperatures ranging between 3 and 6 MeV. We also know that it is quite reasonable to assume that both the pairing and shell effects have completely disappeared above 3 MeV. But how do nuclei behave at higher T, how do they decay ? More generally what is the evolution of the deexcitation properties with increasing temperature ? As mentionned several times before, one should be very cautious in applying the standard statistical model to very high temperatures and high angular momenta. The incomplete fusion process probably leads not only to a rather broad initial mass distribution of the system but also to large initial fluctuations in the energy deposit. Moreover, it is very likely that, if a thermal equilibrium has been reached, it does not concern the collective degrees. Finally, the measured spectra are often averaged over a long evaporation chain indicating that the measured values have then to be considered as weighted average value over time and/or deexcitation chain. Drawing precise conclusions from the experiments on nuclear properties at high T such as 5 MeV is then out of range of the present experimental possibilities. However, from the series of experiments which were connected with decay studies of hot nuclei, quite interesting features can be stressed which will be discussed in this chapter.

4.1. Particle-fission competition

For very heavy systems as those mentionned before (Ar + U, Th) one knows from low-energy experiments that fission of the compound nucleus strongly competes with particle emission.

Starting again from the statistical theory, and if for the sake of simplicity, one restricts the competition between fission and neutron emission, one is led to the following expression for the ratio of emission widths for fission and neutron decay[105] :

$$\frac{P_f}{P_n} = k(E^*) \, \exp \frac{2 \, \sqrt{a_f(E^*-B_f)}}{2 \, \sqrt{a_n(E^*-S_n)}} \tag{18}$$

where B_f is the fission barrier, S_n the neutron binding energy. a_f and a_n are the level density parameter for the saddle point shape and for the spherical nucleus respectively.

The evolution of a with temperature will be discussed in 4.2. Let us first concentrate on the fission barriers. The evolution of B_f at high temperature has been discussed in several theoretical papers[106-110]. An example is shown in fig 20 taken from ref. 108. It shows the T evolution of B_f for ^{208}Pb and ^{240}Pu calculated using an extended Thomas-Fermi density variationnal method. The fission barrier has completely vanished at 5 MeV temperature. This reflects the continuous decrease of surface and curvature energies with increasing temperature. From this simple argument, i.e. $B_f \rightarrow 0$, one would then expect the first chance fission to become the dominant, if not the unique decay channel.

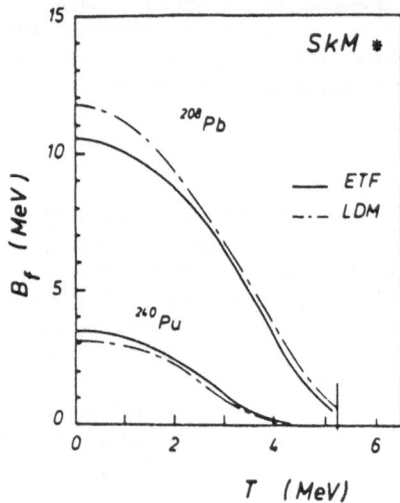

Fig. 20. Temperature evolution of the fission barrier obtained by semi-classical ETF calculations (solid lines). Dashed dotted lines are liquid drop model barriers[108].

This is clearly not what is observed experimentally. Fig. 21 nicely illustrates that the opposite effect is seen[36]. For the system Ar + Ho, it appears that the probability for the compound system to fission which is about 95 % at low energy vanished down to 70 % at 27 MeV/u. Several reasons may be found to account for this a priori surprising result. At once, the onset of incomplete fusion at around 8 MeV/u above the coulomb barrier will tend to lower the Z of the fusion nucleus and consequently increase its fission barrier as compared to a complete fusion. However, the main explanation may be searched in the foundations of the statistical model. This model implies that an equilibrium has been reached and that the decay times for the different channels can be neglected. What is the situation at high temperature ? If thermalization of the system may occur within 10^{-22} sec, this is probably not the case for the collective degrees of freedom. It appears clearly that dynamical effects have to be introduced in the decay process which are not included in the standard statistical model.

The experiments support this statement as they demonstrate that most of the light evaporated particles are emitted before fission[10,68,111,113]. Fig. 22 is a typical example showing iso-contours of invariant

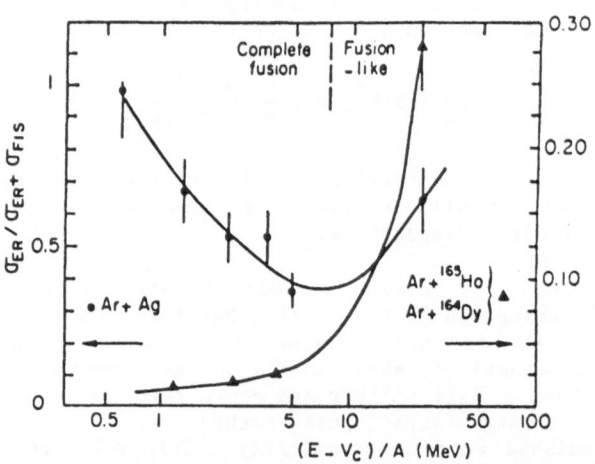

Fig. 21. Evolution of the evaporation residue cross section over the fusion like cross section with incident energy[36].

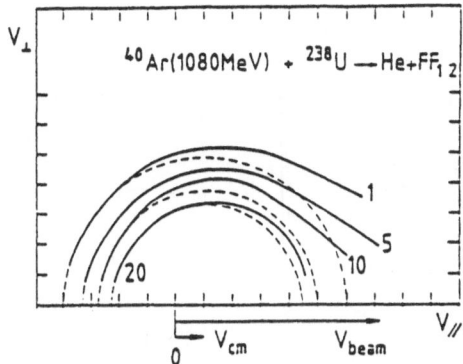

Fig. 22. Iso contours of invariant cross sections for α-particles in the velocity plane measured in coincidence with fission following large momentum transfer (solid lines). Dashed lines correspond to an isotropic emission from a source moving with the c.m. velocity (ref. 68).

cross sections for α-particles detected in coincidence with 2 fission fragments following large momentum transfer[68]. (The system under study is here 27 MeV/u ^{40}Ar + ^{238}U). At backward angle, an isotropic emission from the recoiling composite system is clearly identified. In other words, the spectra are identical whatever the fission fragment angle is [112], which means that the particles are not evaporated from the fission fragments. Other experiments dealing with neutron emission, have permitted a precise measurement of the prefission and post-fission neutron numbers. It appears that the number of neutrons emitted after fission is rather insensitive to the initial temperature of the hot nucleus[10]. From the knowledge of the prescission neutron number, an estimate of the prescission lifetime τ_{presc} can be deduced from the relation :

$$\tau_{presc} = \sum_{i=1}^{M_n^{presc}} \frac{\hbar}{\Gamma_{n,i}} \qquad (19)$$

where $\Gamma_{n,i}$ is the neutron decay width. The est. time is a few times 10^{-20}sec. From fig. 23 which represents the calculated compound nucleus lifetimes as a function of the emitted neutrons, one sees that, for the ^{186}Pt compound nucleus excited at 500 MeV, more than 15 neutrons can be emitted before the scission[10] (in addition to charged particles). If most of the initial thermal energy is dissipated in a very short time through light particle evaporation it may even happen that after a long evaporation chain, the fission will be totally hindered.

Such results are well explained in dynamical models where the collective motion of the nucleons towards the scission point is supposed to be a slow process as compared to one evaporation step[111,114-116]. The phenomenon is described as a diffusion process using a Fokker Planck equation. The coupling between intrinsic and collective degrees is followed as a function of time and the probability for particle evaporation calculated for each time interval. The fission width is then strongly reduced in the early stage of the decay as compared to the static calculations (fig. 24). It should be stressed that such

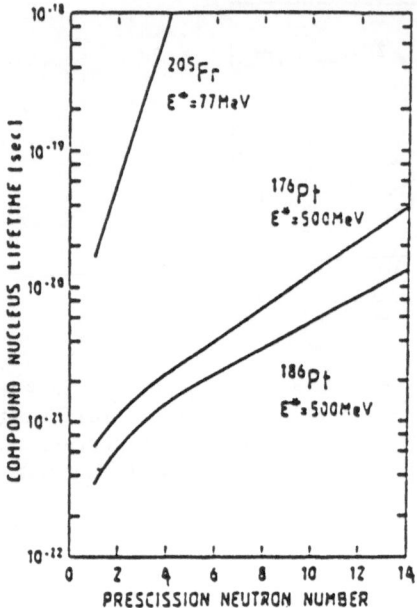

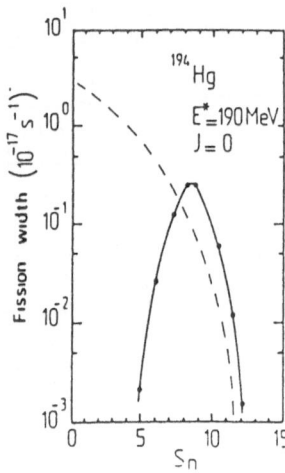

Fig. 23. Calculated compound-nucleus lifetimes as a
function of the emitted
neutrons for the nuclei
^{205}Fr, ^{176}Pt and ^{186}Pt
excited at 77, 500, 500
MeV respectively (ref.
10).

Fig. 24. Evolution of the fission
width as a function of
the number of neutrons
evaporated. Dashed curve : standard statistical model ; solid curve : dynamical calculation (ref. 115).

calculations introduce quite naturally a limitation in excitation energy
transfer as particle emission during the thermalization time is
introduced in the model.

4.2. Evolution of level densities

One of the most important parameter used in the statistical model
is the so called level density parameter a which, in the Fermi gas model
is related to the temperature T and the excitation energy E^* by the well
known relation :

$$E^* = a\ T^2 - T \qquad (20)$$

At low excitation energies, the value of a has been experimentally
determined and is close to A/8[117].

More precisely, still in the frame of the Fermi gas, a is defined by:

$$a = \frac{1}{6}\ \pi^2\ g(\varepsilon_F) = \frac{1}{4}\ \pi^2\ \frac{2m^*}{h^2}\ \frac{A}{k_F^2} \qquad (21)$$

where g is the density of single particle states at the Fermi energy ε_F,
m^* the effective mass and k_F the Fermi momentum.

The experimental study of the evolution of the level density
parameter with increasing temperature is not an easy task. However, a
few attempts have been made in the study of the systems Ne + Ho[71] and N
+ Sm[69,118]. The principle of the method is rather simple but its appli-

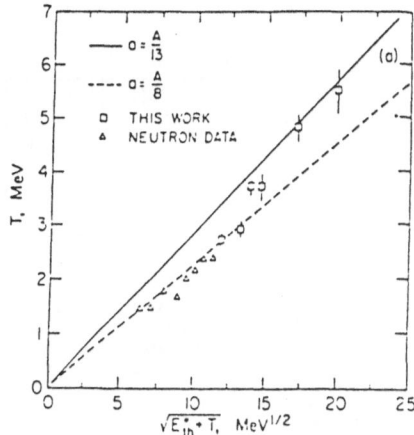

Fig. 25. The extracted relation ship between the temperature and the excitation energy seems to indicate a transition for the level density parameter value from A/8 to A/13 when increasing the excitation energy (ref. 69).

cation not straightforward. It is based on an independant determination of both T and E*, a being immediately deduced from expression 20.

Energy spectra of neutrons or charged particles are measured in coincidence with evaporation residues (ER). The thermal energy is then deduced from the ER recoil velocities. The temperatures are obtained from the slopes of the associated energy spectra. However, the apparent temperature extracted from the spectra are weighted average temperatures over the whole deexcitation chain. A quite complex analysis has to be made in order to extract the temperature for first chance emission. According to fig. 25, and for a compound system with A ~ 160, the results seem to be consistent with level density parameter increasing from A/8 to A/13 in the excitation energy range 100-400 MeV [i.e. from 1.5 to 5.5 MeV temperature].

The temperature dependance of a has been investigated in several theoretical papers. Hartree Fock calculations[119] as well as Thomas-Fermi calculations[120] show a very little change of a with T in the range 1 to 5 MeV. More recently, other authors have looked for the temperature dependence of the effective mass in finite nuclei and the correlation effects[121-123]. These results are in qualitative agreement with the data. However, one should note that the experimental value of A/8 found at low excitation energy is usually not reproduced by the calculation. On the other hand, as already mentionned, more precise experiments are needed before to draw definite conclusions.

4.3. Collective excitations built on excited nuclear states

Recent experimental studies of γ-decay of giant resonances built on very excited states have pointed out the interest of such studies for improving our understanding of the behaviour of hot nuclei. The very dominant resonance which has been presently observed is the giant dipole resonance (GDR), i.e. a counter phase vibration of neutrons and protons in the nucleus. Its study provides a rather unique tool to follow the relaxation of collective degrees in a very hot nucleus[124].

To illustrate this, it appears quite instructive to concentrate ourselves on results concerning Sn nuclei for which a rather large set of data is available[125-128]. Typical γ-ray spectra obtained when observing the decay of Sn nuclei excited at moderate energies (from 60 to 130 MeV) are shown in fig. 26a. Besides the high E exponential tail attributed to statistical γ-rays (see 3.8), the bump located at E_γ ~ 15 MeV can be easily attributed to the GDR component.

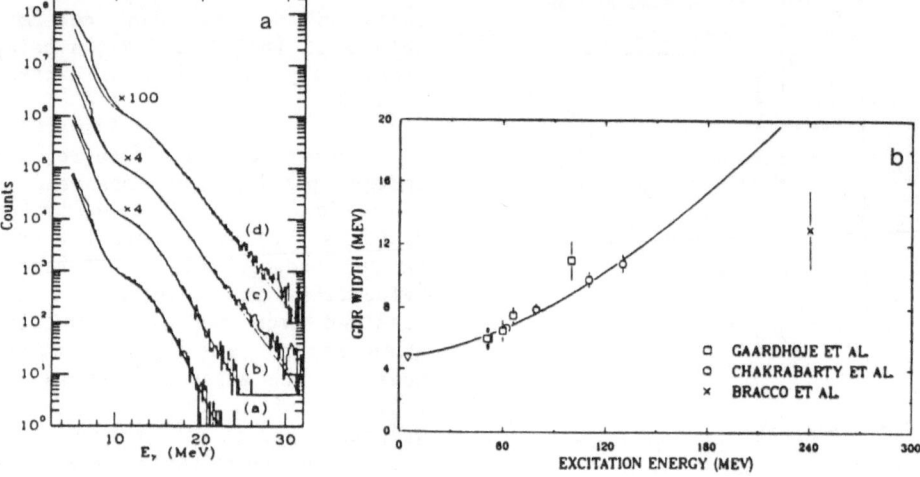

Fig. 26. Excitation of giant resonances in Sn nuclei
    a) Measured gamma ray spectra for Sn nuclei at excitation
    energies E^* = 60, 80, 110, 130 MeV (ref. 127).
    b) Evolution of the GDR width built on excited states in Sn
    isotopes. Figure from ref. 128. Data from ref. 125, 127, 129.

While the centroid of the GDR does not move with excitation
energy, there is a clear broadening of the width Γ_{GDR}[127]. Two combined
effects have been discussed to account for the increase of Γ_{GDR}[128]. The
first one is a thermal effect which is expected to follow a square root
dependance with T in a hot Fermi-gas :

$$\Gamma = \frac{1}{2} \sqrt{\frac{T}{k}}$$

But the major one seems to be connected with spin effects which should
play a very important role as the shape of these Sn isotopes is expected
to move from a spherical shape at I = 0ħ to oblate shapes for I ⩾ 40 ħ.
However this effect should saturate for angular momenta corresponding to
the vanishing of the fission barrier. (For the example under discussion
here, this corresponds to $E^* \sim$ 100 MeV). Above this value, one should
then observe the variation due the T. This seems to be the case, at
least qualitatively, as indicated in fig. 26b which summarizes the data
obtained for these Sn isotopes. The last point at E^* = 240 MeV has been
obtained in the reaction 15 MeV/u Ar + ^{70}Ge.

Nevertheless, results obtained on the same system at 24 MeV/u[126]
do not seem to follow this trend. Statistical calculations are not able
anymore to account for the observed γ-ray spectra associated with
central collisions. The data indicate a rather strong inhibition of the
γ decay for these very high excitation energies (E^* ⩾ 250 MeV) together
with a reduction of the width of the GDR. More recent results obtained
for the system 44 MeV/u ^{40}Ar + ^{158}Gd show an even stronger reduction of
the GDR strengh[130].

No clear explanation can be put forward. Does it mean that collective
nuclear motion is progressively lost for temperature exceeding 4 MeV ?
Does one observe here only a real statistical γ-ray emission from a

nucleus in thermal equilibrium ? Again, it is quite clear that the time scale for equilibrium has to be taken into account. It could be really to short to allow a complete equilibrium of collective modes. An explanation has been suggested by Broglia, the existence of a motionnal narrowing, which assumes that the damping width of the GDR in very hot systems is essentially due to the coupling of this resonance to thermal fluctuations of the nuclear surface. Anyhow, the study of the GDR damping at high T seems to be a very promising way to study hot nuclear systems and determine to what extent we do have a thermal equilibrium and at which temperature could we observe a change from an ordered to a chaotic behaviour.

4.4. Emission barriers, deformation, angular momenta

Some attempts to deduce information on deformation and angular momentum have been done detecting either γ-rays[71] or charged particles[48]. Information on the emission barriers has been obtained also by fitting the experimental energy spectra[48,71,132]. There is no doubt that a reduction of the barrier for α-particles is observed as compared to what one would expect from pure angular momentum effects or thermal fluctuations[71]. Even larger reduction of the proton emission barrier has been found. They might be explained by the existence of a hot and very diffuse nuclear surface which allows evaporation before its relaxation[132,133]. This lowering of the emission barriers has been already observed on many systems even at moderate temperature (T = 2-3 MeV) and was interpreted by Alexander et al.[134] as an indication of large distorsions of the emitters. On the other hand, there are serious evidences for the occurence of large shape fluctuations. For instance, in the reaction 27 MeV/u Ar + U, large deformations of the hot system have been deduced from the analysis of the α-particle spectra. (Deformations as large as a factor of two for the ratio between the major axis have been observed[48]).

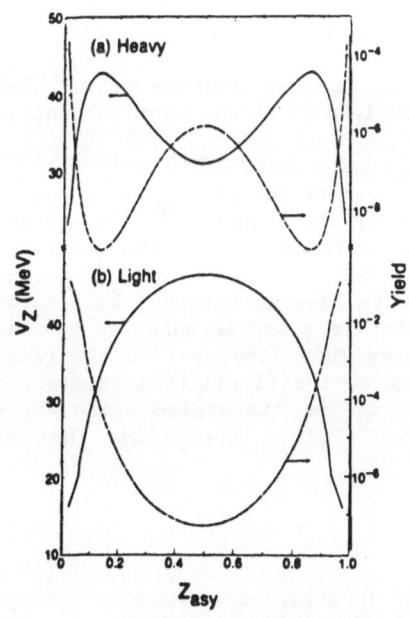

Fig. 27. Ridge line potentials (solid curves) and expected yields (dashed curves) versus the mass asymmetry coordinate for a heavy and light nucleus respectively (ref. 46).

4.5. Complex fragment evaporation

As already mentionned in 3.3.2., with increasing excitation energy deposited in the system, the phase space available for complex fragment evaporation will become much larger as compared to the low energy situation. The emission width ratio for two different decay channels is given by expression 15. One sees easily that, despite the fact that the statistical factor $g_1 m_1$ is obviously in favour of the cluster emission, the high coulomb barrier will strongly inhibit this channel at low temperature but its influ-

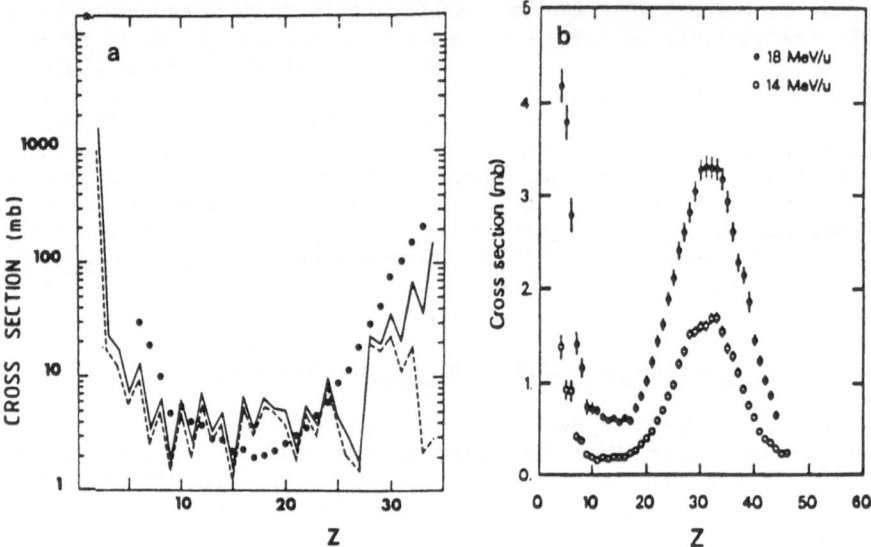

Fig. 28. Total experimental yields as a function of the fragment Z-value
(points) for : a) 35 MeV/u Kr + C. The solid line shows the
prediction of the statistical model (ref. 42) ; b) 14,18 MeV/u
La + C (ref. 136).

ence will be washed out at very high excitation energies. The problem of
cluster emission has been first discussed in 1960 by Ericson[6]. More
recently it has been treated by Friedman[135] within the standard
statistical model. An interesting result concerns the production of
complex fragments which is predicted to reach a maximum near 5 MeV
temperature, as higher temperatures will lead to a dominant sequential
decay of the primary fragments.

Another approach has been developped by Moretto[136] and recently
applied to the analysis of experiments by two groups[41-46]. It assumes
that fission and evaporation are the two extremes of the same statistical process. Moretto has considered a cut in the potential energy
surface along the mass asymmetry coordinate passing through the fission
saddle point in such a way that each point is a saddle point as soon as
one freezes the mass asymmetry coordinate. This line, called the ridge
line, can have different shapes according to the fissility parameter of
the system (see fig. 27). On the basis of the transition state theory[136], the expected emission width for a given charge may then be
expressed by :

$$\Gamma_Z \; \alpha \; e^{-\,V(Z)/T_Z}$$

For light systems the charge distribution is expected to have an U shape
(very asymmetric fission) and for heavy systems, one should observe,
besides the ordinary fission, two wings associated with light particle
emission and the complementary heavy fragment. Typical experimental
results[42,136] follow this expected trend as indicated in fig. 28 for the
system 35 MeV/u ^{84}Kr + C and 18 MeV/u La + C. Another interesting
feature of the calculation is that it predicts a transition in the shape
of the E spectra from a Maxwellian shape for light particles (p,α) to a

gaussian one for heavier products (B,C) as for standard fission. This trend has been effectively observed[137].

Such an identification of a sequential evaporative process in intermediate energy heavy ions is very important as it demonstrates one has to look for "conventional" decays before to search for new exotic processes as multifragmentation. It is quite probable that such an evaporation mechanism is not the only process responsible for the production of the so-called intermediate mass fragments (IMF) but it has to be properly quantified. (See chapter 6 for a discussion concerning the possibility of multifragment evaporation).

5. EVIDENCE FOR THE EXISTENCE OF A LIMITING TEMPERATURE

5.1. The present status

Besides the studies of nuclear properties of hot nuclei shortly reviewed in section 4, a very fondamental question is related to the possible thermal energy that a nucleus can sustain. In other words, how hot can be a self bound system in thermal equilibrium ?

The first naive answer which occured to us has been to define this limit as the average binding energy of the nucleus (i.e. between 7.5 and 8.5 MeV/u). Obviously above this limit, any nucleus will disintegrate into its main constituents. Having this simple idea in mind, the existing results can be reviewed. Among the quite large amount of experimental data existing on the market, we have tried to extract a very limited number of results showing the <u>highest</u> excitation energy per nucleon (E^*/A) reached so far for a given mass of the composite system. It has been deliberately chosen to select only a few results complying with severe criteria which ensures a rather good confidence in the measured excitation energy. Three types of measurements have been selected :
i) T measurements from relative population of excited states : only experiments where side feeding effects clearly play a minor role have been selected.
ii) Recoil velocity measurements : the requirement was to observe a well defined maximum in the recoil velocity distribution of the heavy residues or in the correlation function between the two fission fragments. In these cases, the excitation energy per nucleon has been extracted from the most probable value of the recoil velocity. (It has been pointed out that the inexistence of a maximum may lead to very questionable value of the excitation energy deposit).
iii) Excitation energy determination by means of the counting of particle multiplicities : the requirement was the measurement of both neutron and charged particle emission.

These highest values of E^*/A over the mass range 60-300 are displayed in fig. 29. These upper values are evolving from 6.5 MeV/u for the light systems down to about 3 MeV/u for the heaviest systems and are thus far below the corresponding binding energies of the system. Assuming a = A/8, this would correspond to a temperature decreasing from almost 7 MeV down to 4.6 MeV for the heavy systems. Let us now examine how do these results compare with the theoretical predictions. It is not our purpose to compare the advantages and drawbacks of the different theories. More detailed descriptions of the existing models will be found in Ngô and Gregoire's lectures.

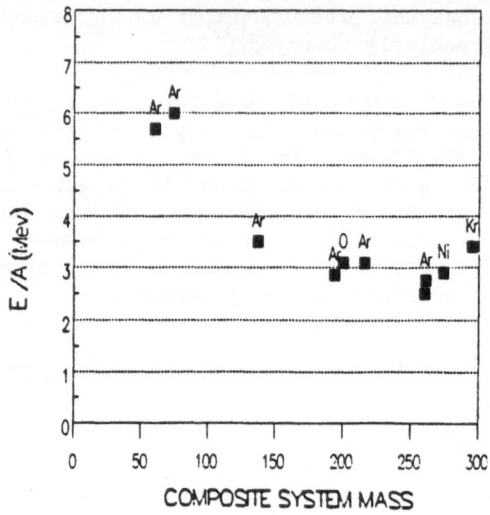

Fig. 29. Systematics of the highest excitation energy per nucleon
reached so far in heavy ion induced experiments for a given
mass of the composite system. The labels above the points are
referring to the projectiles used for the experiments.

5.2. **Static approaches**

Theorists have first approached the problem from a static point of
view. The first calculations have been undertaken for infinite nuclear
matter[138]. The equation of state (pressure versus density) which is
obtained is similar to that of a Van der Waals gas. Nuclear matter may
exist in liquid and vapor phases, the critical temperature for the phase
transition beeing close to 18 MeV.

Hartree Fock calculations as well as semiclassical methods have
been developped to solve the problem of the single particle states in
the continuum[8,119]. A consistent treatment can be achieved only when
considering the hot nucleus in equilibrium with a surrounding nucleon
vapor. Bonche et al,[8]. have applied the substraction procedure to
determine the properties of the hot nucleus without vapor. Other
calculations have to deal with Thomas Fermi method[139] and the hot
liquid-drop model[140]. Contrary to infinite nuclear matter calculations,
the limiting temperature corresponding to the instability is found to be
much smaller than 18 MeV. The instability of the finite real nucleus is
indeed largely influenced by the balance between the surface tension and
the coulomb repulsion. This induces a significant dependence of T_{lim} on
the nuclear charge as well as on the charge to mass ratio (fig. 30). An
example of the predicted limiting T for masses along the β-stability
line is shown in fig. 31. This has been calculated within the hot liquid
drop model[140]. Coulomb repulsion forces clearly diminish the value T_{lim}
for very heavy systems. The solid curve, using a soft nuclear EOS with
an incompressibility modulus K_∞ = 210 MeV, seems to be in rather good
agreement with the actual data.

Auger et al[34] have estimated the projectile bombarding energy at
which the limiting temperature should be reached for various effective
masses of the composite nucleus. This has been computed for Ar induced

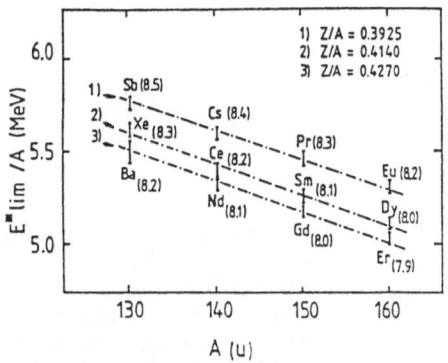

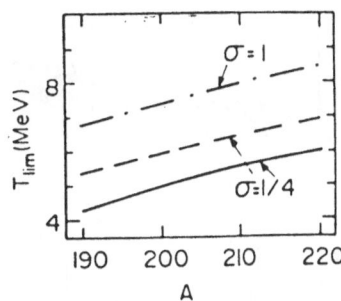

Fig. 30. Isospin and mass effect on the limiting excitation energy and (temperature) Ref. 140.

Fig. 31. Evolution of the limiting temperature as a function of the mass along the β-stability line. Full and dashed curves are calculated using a soft EOS and different surface tensions. Dashed dotted line is obtained with a stiff EOS. Ref. 140.

reactions by coupling the results of the momentum transfer systematic with the values of T_{lim} deduced by Levit and Bonche,[140]. The results, drawn in fig. 32 predict that, above a composite mass of 60, the limit is reached for an almost constant bombarding velocity close to 30 MeV/u. Surprisingly, the highest excitation energies reached fo far (see fig. 29) have been mainly obtained with an Ar projectile at this energy.

Does it mean that one has a complete comprehensive picture of the problem ? It might be premature to conclude as there are still two open questions :
i) What's happening with Ar projectiles above 30 MeV/u ?
ii) Is there just no way to reach higher temperatures, for instance, with heavier beams ?
As far as the first question is concerned, an enlightning result is the systematic study of the excitation energy deposit in the energy range 27-77 MeV/u for the system Ar + ^{232}Th. The results are summarized in fig. 15 (see §3.5). Within the experimental uncertainties, one may reasonably consider that a saturation around 650 MeV is reached already at 27 MeV/u. This is a clear signature that we cannot deposit more thermal energy in the nucleus, at least with an Ar projectile. However is that really the maximum energy that a nucleus can sustain ? This refers directly to the second question. Recent results with heavier projectiles such as ^{86}Kr indicate clearly that more energy can be deposited[63,141]. They are concerning linear momentum transfer studies as well as neutron multiplicity measurements. For instance[141], the neutron multiplicities are 50 % larger using 35 MeV/u ^{86}Kr instead of 35 MeV/u ^{40}Ar. The observation of a saturation of the E deposit for a given projectile, together with an increase of this value for heavier projectiles cannot be explained without considering the influence of the dynamics.

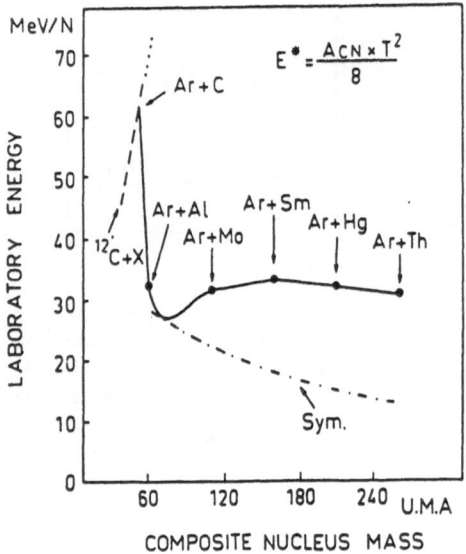

Fig. 32. Expected bombarding energy at which the limiting temperature is reached for a given composite nucleus mass. The full and dashed lines correspond to Ar and C induced reactions, the dot-dashed line to symmetric systems (ref.34).

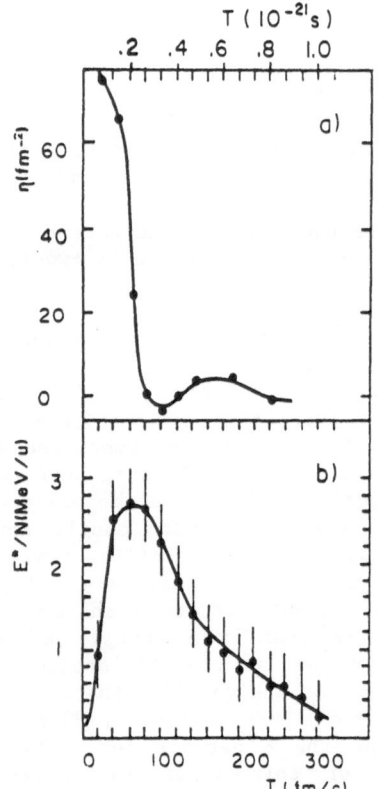

Fig. 33. Dynamical calculation (Landau-Vlasov) performed for the system Ar + U for central collision (ref. 7)
a) anisotropy η of the momentum distribution of nucleons in the composite system.
b) excitation energy per nucleon of the composite system.

5.3. Dynamical approaches

It is not our aim to review the different existing models ; let us just mention that the dynamical mean field approach has been used through the TDHF method,[143] time dependant Thomas Fermi[144], Monte Carlo Hartree-Fock[145] and the semi classical TDHF (the Vlasov equation)[146]. Nevertheless, one promising way is to use semi classical kinetic equations which allow to take into account the balance between mean field and individual collisions[147,148] and to follow the complete evolution of the system.

Two reasons may explain the observed saturation of the thermal energy with increasing Ar bombarding energy. First, particle emission at the very early stage of the collision should play an important role as the decay time of such a very hot system is becoming comparable to the collision time. Moreover, the possible excitation of collective modes which might be strongly enhanced in heavy ion collisions could strongly limit the thermal energy. Recent dynamical calculations (Landau-Vlasov) have shed additional light on this problem. An example is shown in fig. 33 for the system Ar + U at 27 MeV/u. This results from a Landau Vlasov simulation performed for an head-on collision. It indicates that a complete thermalization occurs in a very short time, 80 fm/c ($\sim 2 \ 10^{-22}$sec). At that time, the anisotropy of the nucleon momentum distribution indeed tends to zero. At this point, a maximum energy per nucleon of 2.6 MeV has been reached, and the fused system in thermal equilibrium starts to evaporate. One should notice that the agreement with the experiment is quite satisfactory. However, the calculation exhibits a very interesting feature, namely that the collective modes are far from being relaxed so rapidly. This emerges clearly from the study of the system Ar + Au at 60 MeV/u[90,147]. Even at this high energy, the thermal energy of the system remains saturated at 650 MeV. The explanation is found in the balance between thermal and collective energy. Densities up to 1.3 ρ_0 are obtained and the coupling between preequilibrium emission and monopole vibrations contributes to remove a large part of the total available intrinsic excitation energy.

There are still many open questions concerning the limiting temperatures. For instance, a precise study of non equilibrated particle emission with increasing bombarding energy is strongly needed (n,p,α, clusters). Up to now, very few results are available.

On the other hand, it is very likely that the limit in the nucleus heating has still not been reached. Extrapolating our present knowledge on linear momentum systematics for much more massive projectiles, would indicate that a Au + Au collision at 30 MeV/u could heat up the system to about 7 MeV temperature. This is clearly emphasized from the existing systematics of excitation energy deposits for heavy ion projectiles (from N to Kr) impinging a heavy target at 30 MeV/u (fig. 34). This will be a challenge for the upgraded GANIL facility before the end of 1989.

6. CLUES FOR THE APPEARANCE OF NEW PROCESS AT THE CRITICAL TEMPERATURE

The question of the fate of a very hot nucleus near and above the critical temperature may provide information on the knowledge of the nuclear equation of state. In our energy domain, one is considering rather low densities and several authors have considered the time evolution of this very hot nuclear system, namely its cooling and expansion. Above some critical temperature, the possible onset of a new

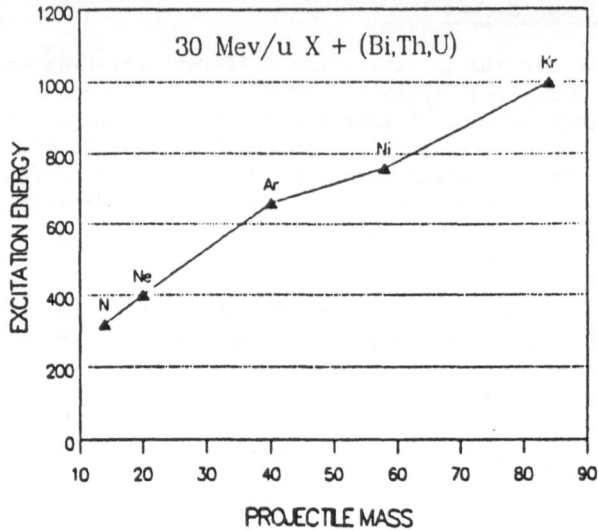

Fig. 34. Experimental systematics of the excitation energy deposit for heavy ion projectiles impinging a heavy target at 30 MeV/u.

process where a nucleus breaks into clusters, has been investigated using various theoretical approaches[149-151]. In the following, the general denomination for such a process will be "multigragment emission". As C. Ngô's lectures are devoted to this subject, we shall only here briefly comment on this point.

6.1. <u>Does the fusion process vanish above the critical bombarding energy?</u>

Many authors claim that the fusion process vanishes completely above some critical bombarding energy, let us say for instance around 35 MeV/u in Ar induced reactions. Moreover this statement is used to assert the appearance of a new process. The argument which is put forward is the vanishing of the maximum in the recoil velocity distribution of the heavy residues (see for instance the reaction Ar + Ag[70]). For very heavy systems, it is the complete disappearance of the so called central collision peak in the folding angle distribution of the fission fragments. In the following, we would like to stress that the above observations do not sign automatically for the extinction of the fusion process. To illustrate this, fig. 35 displays again the correlation function of the two fission fragments measured in the reaction Ar + Th from 31 up to 44 MeV/u[59] together with recent measurements of the inclusive neutron multiplicities for the same reaction in the energy range 27-77 MeV/u[76,152]. The comparison is highly instructive. In the same way the folding angle is representative (in principle) of the transferred linear momentum, the neutron multiplicity spectrum reflects directly the energy dissipation in the entrance channel. Two peaks are clearly visible in this spectrum in the whole energy range which may evidently be associated with peripheral collisions (soft process) and more central collisions (highly dissipative process). As mentionned already, the average neutron multiplicity remains more or less constant above 27 MeV/u, a signature of a saturation of the thermal energy deposit in the system. Furthermore, the cross section corresponding to the highly dissipative collisions remains also constant, with a value close to 2 barn (i.e. about 50 % of the total reaction cross section).

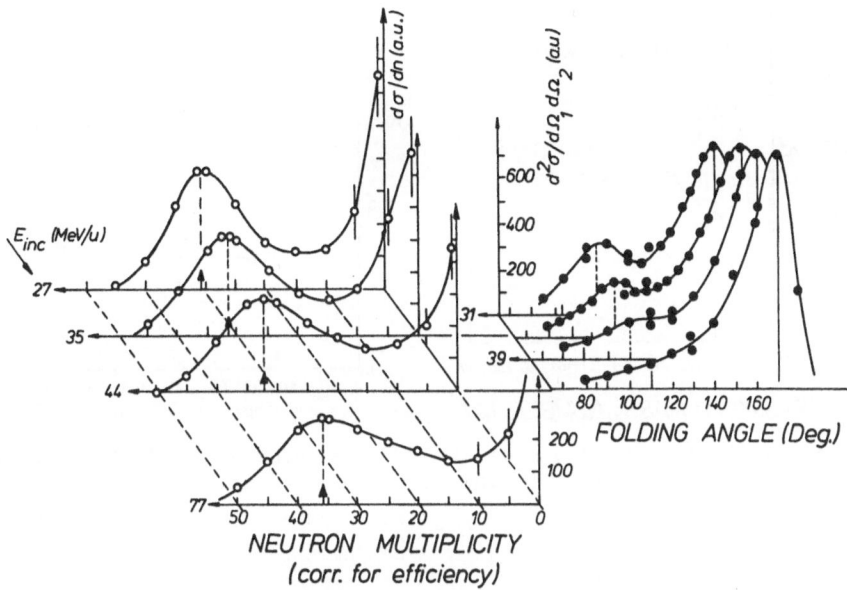

Fig. 35. Neutron multiplicity (ref. 76) and folding angle distributions (ref. 59) from the reaction ^{40}Ar + Th within the incident energy range 27 to 77 MeV/u.

The disappearance of the high linear momentum transfer peak in the folding angle distributions has then to be connected with the decay properties of the composite system (vanishing of binary fission ?).

One may then question ourselves about the repartition of the cross section corresponding to highly dissipative collisions which does not seem to contribute much to the fission channel. A partial answer has been given by radiochemical measurements. A quite violent transition is observed between 32 and 44 MeV/u as far as the shape of the mass distribution for the target fragment production is concerned (fig. 36) [153,154]. A central bump around A = 130-140 has build up at 44 MeV/u which did not exist at all at 32 MeV/u. At the same time, the fission cross section has vanished by 40 %. This has been confirmed recently by an on line experiment at 44 MeV/u[155] on the same system where equal cross sections (~ 1 barn) for the production of heavy residues as well as fission fragments have been found and are both associated with the same large neutron multiplicities. One explanation for the non observation of the 1 barn fission cross section for central collisions in the folding angle distribution may be explained for instance by the existence of a third body which will destroy not only the clear relationship between the LMT and the folding angle but also possibly the coplanarity. As far as the increase of the heavy residue cross section is concerned, it could be accounted for by a copious increase of preequilibrium emission, large enough to sufficiently lower the mass of the hot system down to a mass region where the fission barrier becomes significantly high. However, the emission of a few complex fragments (IMF) will be highly efficient to increase the fission barrier. Trockel et al.,[78] have reported a multiplicity of 1.5 IMF for the reaction Ar + Au at only 30 MeV/u.

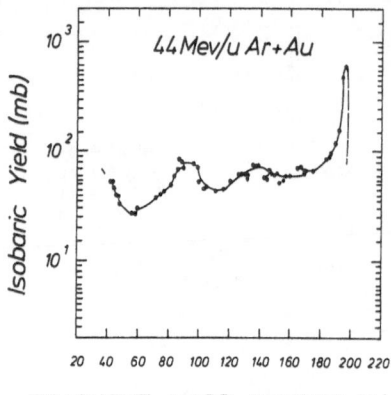
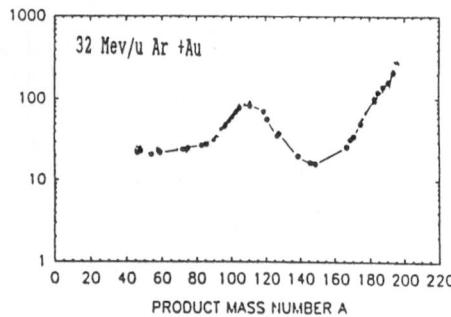

FRAGMENT MASS NUMBER (A)

Fig. 36. Radiochemical measurements of the mass distribution for the residual products in the reaction Ar + Au at 32 and 44 MeV/u. (ref. 153-154).

To end with the rather long digression on the fate of the fusion process, our feeling is that one cannot claim easily that fusion types of reaction have vanished above 35 MeV/u. For us, the observation of a fission-like product or an heavy residue associated with the emission of more than 35-40 light particles is at least very similar to a fusion-evaporation process ! However, one should admit that a dramatic change has occured in the way the distribution of the cross section among fission and heavy residues has been operated (for the most dissipative collisions). Whether or not this unambiguous observation can be related to the onset of multifragment emission is still on open question.

6.2. **Clues for the onset of a multifragment process?**

There are presently several experiments strongly focussed on the search for the onset of multifragmentation emission[78,156,157]. All of them strongly suggest that the onset of IMF multiplicities larger than one appears to become non negligeable above 30-35 MeV/u (for rather heavy projectiles as Ar and Kr). To illustrate this point, an excitation function for the multiple IMF production in the system Ar + Al is shown in fig. 37[156].

However, a real multifragmentation of the system is still a very rare phenomenon at these energies, although it has been observed in emulsion experiments[158]. One should probably speak in terms of multifragment emission without prejudging of the process. Some experiments indicate clearly that all the nucleons are not concerned with this emission[157]. As a matter of fact, IMF multiplicities larger than the unity do not sign automatically the reaction mechanism. For instance Moretto has shown that a pure process of sequential binary decay that is completely controlled by the statistical model at each step of the deexcitation might well explain high IMF multiplicities at very high excitation energies (nuclear comminution)[46]. This pure sequential decay process has to be opposed to the multigragmentation process where the system breaks into several pieces within a very short time (prompt emission). There are needs for experiments that could unambiguously differentiate a sequential from a prompt process. The results seem presently to be quite contradictory, sequential in the

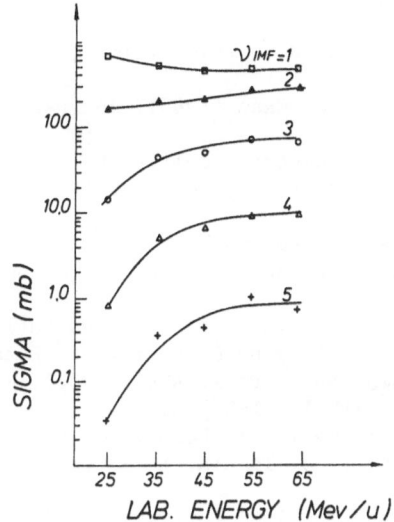

$$\nu_{IMF=1}$$

SIGMA (mb)

LAB. ENERGY (Mev/u)

Fig. 37. Excitation functions for the reaction Ar + Al corresponding to 1 up to 5 detected intermediate mass fragments (ref. 156).

reaction ^{16}O + Ag, Au at 84 MeV/u,[159] prompt for the reaction Kr + Au at 43 MeV/u[157]. However this might be not the right signature as the decay life time for IMF evaporation may become so short that the definition itself of the sequentiality will become obsolete as there will be strong interaction between two successive decays.

A clear experimental signature of the onset of multifragmentation has still to be established. Further experiments in this field will, without any doubt, strongly contribute to our knowledge of the different modes of energy storage in a nucleus.

7. CONCLUSIONS

These lectures were intended to present an overview of the very active field of "hot nuclei". Heavy ion induced reactions in the intermediate energy range 20-100 MeV/u appear to be very efficient for the formation of nuclei at very high temperature. However, the determination of the momentum transfer, the excitation energy or the temperature is not straightforward from the experiment. The different available techniques must be used very cautiously. For instance, it may well be that recoil velocity measurements become inadequate when the linear momentum transfer falls below 70 % of the initial one. However quite reliable experiments have now established without any ambiguity that nuclear systems can be formed in thermal equilibrium up to temperatures as high as 5 or 6 MeV.

We learned already a great deal of the investigation of nuclear properties with increasing temperatures, as the evolution of level densities, the dominance of dynamical effects in the competition between fission and evaporation, the increase of cluster evaporation and the excitation of the GDR build on excited nuclear states.

The concept of a limiting temperature (and limiting excitation energy deposit) is now an experimental fact. It really widens the area of our research on the behaviour of nuclear matter under extreme conditions : onset of large dynamical effects, excitation of compression modes, possible existence of a transition towards multifragmentation. There are still open and fascinating questions as the evolution of both compression and thermal energies, the timescale for thermalization, the role played by non-equilibrium processes ... This is clearly a field where theorists and experimentalists need to combine their forces.

The apparent limitations in the temperature which have been reported so far probably do not reflect the intrinsic properties of the nuclei but are mostly related to the dynamics of the entrance channel. One has then still probably not reached the maximum excitation energy that a heavy nucleus can sustain. Further experiments with very heavy beams, such as Uranium, are thus very promising.

References

1. Several recent Conferences have been focused on this field. The reader may find very valuable information in :
a) Proceedings of the HICOFED Conference, Caen (1986). Journal Phys. (France) 47 (1986) Colloque C4.
b) Proceedings of the Symposium on Central Collisions and fragmentation Processes (Denver, 1987) edited by C.K. Gelbke, Nucl. Physics A471 (1987)
c) Proceedings of the TEXAS A.M. Symposium on hot nuclei, College Station (1987). Edited by S. Shlomo, R.P. Schmitt and J.B. Natowitz, World Scientific.
d) Proceedings of the Third International Conference on Nucleus-Nucleus Collisions (Saint-Malo, 1988) edited by C. Détraz, C. Estève, C. Grégoire, D. Guerreau and B. Tamain. Nucl. Phys. A488 (1988).
2. C. Grégoire and B. Tamain, Ann. Phys. Fr. 11 (1986) 323.
3. S. Song, M.F. Rivet, R. Bimbot, B. Borderie, I. Forest, J. Galin, D. Gardès, B. Gatty, M. Lefort, H. Oeschler, B. Tamain, X. Tarrago, Phys. Lett. 130B (1983) 14.
4. H.A. Bethe, G.E. Brown, J. Applegate and J.M. Lattimer, Nucl. Phys. A324 (1979) 487.
5. S.L. Shapiro, S.A. Teukolsky : Blackholes, White dwarfs and neutron stars, New York, J. Wiley and Sons (1983).
6. T. Ericson, Adv. Phys. 36 (1960) 425.
7. C. Gregoire, J.N. De, D. Jacquet, B. Remaud, F. Sébille, P. Schuck and L. Vinet, Proceedings of the XXV International Winter Meeting on Nuclear Physics, Bormio (1987).
8. P. Bonche, S. Levit and D. Vautherin, Nucl. Phys. A436 (1985) 265.
9. H. Delagrange, C. Grégoire, F. Scheuter and Y. Abe, Z. Phys. A323 (1986) 437.
10. D. Hilscher, D.J. Hinde and H. Rossner, Proceedings of the Texas A.M. Symposium on hot Nuclei, College Station (1987), World Scientific, p. 193.
11. L.D. Landau and F.M. Lifshits, Statistical Physics, Pergamon Press, part 1, p. 333.
12. V.S. Stavinskii, Soviet Journal of Particles and Nuclei 4 (1973) 417.
13. B. Tamain, 9ème Session d'Etudes Biennales de Physique Nucléaire (Aussois, 1987). LYCEN 8702 - C12.
14. H. Fesbach, Physics Today, November 1987, p. 9.
15. F. Saint-Laurent, M. Conjeaud, R. Dayras, S. Harar, H. Oeschler and C. Volant, Phys. Lett. 110B (1982) 372 ; Nucl. Phys. A422 (1984) 307.
16. C. Klotz-Engmann, H. Oeschler, E. Kankeleit, Y. Cassagnou, M. Conjeaud, R. Dayras, S. Harar, M. Mostefai, R. Legrain, A.C. Pollaco and C. Volant, Phys. Lett. B187 (1987) 245.
17. J. Rafelski, Phys. Lett. 91B (1980) 281.
18. H. Machner, S.A. Jun, G. Riepe, D. Protic, H. Daniel, T. Von Egidy, F.J. Hartmann, P. Hofmann, W. Kanert, W. Markiel, H. Plendl, K. Ziock, R. Marshall, J.J. Reidy, Proceedings of the International Workshop on Gross Properties of nuclei and nuclear interactions XVI, Hirschegg (1988) p. 226.
19. E.F. Moser, H. Daniel, T. Von Egidy, F.J. Hartmann, W. Kanert, G. Schmidt, M. Nicholas and J.J. Reidy, Phys. Lett. B179 (1986) 25.
20. J. Cugnon, J. Vandermeulen, Nucl. Phys. A445 (1985) 717.
21. J. Cugnon, P. Jasselette and J. Vandermeulen, Nucl. Phys. A470 (1987) 558.
22. M. Lefort, Heavy Ion Collisions, vol. 2, edited by R. Bock, North Holland (1980).
23. M. Lefort and C. Ngô, Ann. Phys. (Paris) 3 (1978) 5.
24. W.U. Schröder and J.R. Huizenga, Treatise on Heavy Ion Science (ed. D.A. Bromley), vol. 2, 1984.

25. H. Britt and A. Quinton, Phys. Rev. C124 (1961) 877.

26. G.D. Harp and J.M. Miller, Phys. Rev. C3 (1987) 1847.

27. M. Blann, Phys. Rev. C3 (1985) 295 ; ibid C31 (1985) 1245.

28. J.P. Bondorf, J.N. De, G. Fai and A.O.T. Karvinen, Nucl. Phys. A333 (1980) 285.

29. H. Morgenstern, W. Bohne, W. Galster, K. Grabisch and A. Kyanowski, Phys. Rev. Lett. 52 (1984) 1104.

30. V.E. Viola, B.B. Back, K.L. Wolf, T.C. Awes, C.K. Gelbke and H. Breuer, Phys. Rev. C26 (1982) 178.

31. C. Grégoire and F. Scheuter, Phys. Lett. 146B (1984) 21.

32. G. Auger, D. Jouan, E. Plagnol, F. Pougheon, F. Naulin, H. Doubre and C. Grégoire, Z. Phys. A321 (1985) 243.

33. G. Bizard, R. Brou, H. Doubre, A. Drouet, F. Guilbault, F. Hanappe, J.M. Harasse, J.L. Laville, C. Lebrun, A. Oubahadou, J.P. Patry, J. Péter, G. Ployard, J.C. Steckmeyer and B. Tamain, Z. Phys. A323 (1986) 459.

34. G. Auger, E. Plagnol, D. Jouan, C. Guet, D. Heuer, M. Maurel, N. Nifenecker, C. Ristori, F. Schussler, H. Doubre, C. Grégoire, Phys. Lett. 169B (1986) 161.

35. C. Cerruti, D. Guinet, S. Chiodelli, A. Demeyer and K. Zaïd, S. Leray, P. Lhenoret, C. Mazur, C. Ngô and M. Ribrag, Nucl. Phys. A453 (1986) 175.

36. M.F. Rivet, B. Borderie, H. Gauvin, D. Gardès, C. Cabot, F. Hanappe and J. Péter, Phys. Rev. C34 (1986) 1282.

37. T. Batsch, J. Blachot, Q. Chen, J. Crançon, M. Fatyga, A. Gizon, J. Jastrzebski, H. Karwowski, W. Kurcewicz, A. Llères, T. Miroz, L. Pienkowski, P.P. Singh, S.E. Vigdor and I. Zychor, Phys. Lett. 189B (1987) 287.

38. A. Fahli, J.P. Coffin, G. Guillaume, B. Heusch, F. Jundt, F. Rami, P. Wagner, P. Fintz, A.J. Cole, S. Kow and Y. Schutz, Phys. Rev. C34 (1986) 16.

39. M. Gonin, J.P. Coffin, G. Guillaume, F. Jundt, P. Wagner, P. Fintz, B. Heusch, A. Malki, A. Fahli, S. Kox, F. Merchez and J. Mistretta, Phys. Rev. C38 (1988) 135.

40. J. Blachot, J. Crançon, B. de Goncourt, A. Gizon, A. Llères, H. Nifenecker, Proceedings of the XXIII Intern. Winter Meeting on Nuclear Physics (Bormio, 1985) p. 598.

41. W. Mittig, A. Cunsolo, A. Foti, J.P. Wieleczko, F. Auger, B. Berthier, J.M. Pascaud, J. Québert and E. Plagnol, Phys. Lett. 154B (1985) 259.

42. F. Auger, B. Berthier, A. Cunsolo, A. Foti, W. Mittig, J.M. Pascaud, E. Plagnol, J. Québert and J.P. Wieleczko, Phys. Rev. C35 (1987) 190.

43. R.J. Charity, M.A. McMahan, D.R. Bowman, Z.H. Liu, R.J. McDonald, G.J. Wozniak, L.G. Moretto, S. Bradley, W.L. Kehoe, A.C. Mignerey, M.N. Namboodiri, Phys. Rev. Lett. 56 (1986) 1354.

44. D.R. Bowman, W.L. Kehoe, R.J. Charity, M.A. McMahan, A. Moroni, A. Bracco, S. Bradley, I. Iori, R.J. McDonald, A.C. Mignerey, L.G. Moretto, M.N. Namboodiri and G.J. Wozniak, Phys. Lett. 189 (1987) 282.

45. R.J. Charity, D.R. Bowman, Z.H. Liu, R.J. McDonald, M.A. McMahan, G.J. Wozniak, L.G. Moretto, S. Bradley, W.L. Kehoe and A.C. Mignerey, Nucl. Phys. A476 (1988) 516.

46. L.G. Moretto and G.J. Wozniak, Nucl. Phys. A488 (1988) 337c.

47. A.C. Mignerey, Proceedings of the Texas A.M. Symposium on hot Nuclei, College Station (1987) p. 406, World Scientific.

48. D. Jacquet, Thèse d'Etat (Orsay), 1987, unpublished.

49. T. Sikkeland, E.L. Haines and V.E. Viola, Jr., Phys. Rev. 125 (1962) 1350.

50. V.E. Viola, K. Kwiatkowski and M. Walker, Phys. Rev. C31 (1985) 1550.

51. V.E. Viola, Jr., B.B. Back, K.L. Wolf, T.C. Awes, C.K. Gelbke, H. Breuer, Phys. Rev. C26 (1982) 178.

52. J. Galin, H. Oeschler, S. Song, B. Borderie, M.F. Rivet, I. Forest, R. Bimbot, D. Gardès, B. Gatty, H. Guillemot, M. Lefort, B. Tamain and X. Tarrago, Phys. Rev. Lett. **48** (1982) 1787.

53. G. La Rana, G. Nebbia, E. Tomasi, C. Ngô, X.S. Chen, S. Leray, P. Lhenoret, R. Lucas, C. Mazur, M. Ribrag, C. Cerruti, S. Chiodelli, A. Demeyer, D. Guinet, J.L. Charvet, M. Morjean, A. Peghaire, Y. Pranal, L. Sinopoli, J. Uzureau and R. De Swiniarski, Nucl. Phys. **A407** (1983) 233.

54. S. Leray, G. Nebbia, C. Grégoire, G. La Rana, P. Lhenoret,C. Mazur, C. Ngô, M. Ribrag, E. Tomasi, S. Chiodelli, J.L. Charvet and C. Lebrun, Nucl. Phys. **A425** (1984) 345.

55. M.B. Tsang, D.R. Klesch, C.B. Chitwood, D.J. Fields, C.K. Gelbke, W.G. Lynch, H. Utsunomiya, K. Kwiatkowski, V.E. Viola Jr. and M. Fatyga, Phys. Lett. **134B** (1984) 169.

56. E.C. Pollacco, M. Conjeaud, S. Harar, C. Volant, Y. Cassagnou, R. Dayras, R. Legrain, M.S. Nguyen, H. Oeschler and F. Saint-Laurent, Phys. Lett. **146B** (1984) 29.

57. M. Fatyga, K. Kwiatkowski, V.E. Viola, C.B. Chitwood, D.J. Fields, C.K. Gelbke, W.G. Lynch, J. Pochodzalla M.B. Tsang, and M. Blann, Phys. Rev. Lett. **55** (1985) 1376.

58. S. Leray, O. Granier, C. Ngô, E. Tomasi, C. Cerruti, P. L'Henoret, R. Lucas, C. Mazur, M. Ribrag, J.L. Charvet, C. Humeau, J.P. Lochard, M. Morjean, Y. Patin, L. Sinopoli, J. Uzureau, D. Guinet, L. Vagneron, A. Peghaire, Z. Phys. **A320** (1985) 533.

59. M. Conjeaud, S. Harar, M. Mostefai, E.C. Pollacco, C. Volant, Y. Cassagnou, R. Dayras, R. Legrain, H. Oeschler and F. Saint-Laurent, Phys. Lett. **159B** (1985) 244.

60. Y. Patin, S. Leray, E. Tomasi, O. Granier, C. Cerruti, J.L. Charvet, S. Chiodelli, A. Demeyer, D. Guinet, C. Humeau, P. Lhenoret, J.P. Lochard, R. Lucas, C. Mazur, M. Morjean, C. C. Ngô, A. Peghaire, M. Ribrag, L. Sinopoli, T. Suomijärvi, J. Uzureau and L. Vagneron, Nucl. Phys. **A457** (1986) 146.

61. G. Bizard, R. Brou, H. Doubre, A. Drouet, F. Guilbault, F. Hanappe, J.L. Laville, C. Lebrun, A. Oubahadou, J.P. Patry, J. Péter, G. Ployart, J.C. Steckmeyer and B. Tamain, Nucl. Phys. **A456** (1986) 173.

62. C. Volant, M. Conjeaud, S. Harar, M. Mostefai, E.C. Pollacco, Y. Cassagnou, R. Dayras, R. Legrain, G. Klotz-Engmann and H. Oeschler, Phys. Lett. **B195** (1987) 22.

63. E.C. Pollaco, Y. Cassagnou, M. Conjeaud, R. Dayras, S. Harar, R. Legrain, J.E. Sauvestre and C. Volant, Nucl. Phys. **A488** (1988) 319c.

64. B.B. Back, K.L. Wolf, A.C. Mignerey, C.K. Gelbke, T.C. Awes, H. Breuer, V.E. Viola, Jr., P. Dyer, Phys. Rev. **C22** (1980) 1927.

65. S. Leray, Journal de Physique, Colloque **C4** (1986) 275.

66. J.B. Natowitz, S. Leray, R. Lucas, C. Ngô, E. Tomasi and C. Volant, Z. Phys. **A325** (1986) 467.

67. M. Fatyga, K. Kwiatkowski, V.E. Viola, W.G. Wilson, M.B. Tsang, J. Pochodzalla, W. Lynch, C.K. Gelbke, D.J. Fields, C.B. Chitwood, Z. Chen and T. Nayak, Phys. Rev. Lett. **58** (1987) 2527.

68. D. Jacquet, J. Galin, B. Borderie, D. Gardès, D. Guerreau, M. Lefort, F. Monnet, M.F. Rivet, X. Tarrago, E. Duek, J.M. Alexander, Phys. Rev. **C32** (1985) 1594 and Phys. Rev. Lett. **53** (1984) 226.

69. D. Hilscher, Nucl. Phys. **A471** (1987) 77c.

70. G. Bizard, R. Brou, P. Eudes, J.L. Laville, J. Natowitz, J.P. Patry, J.C. Steckmeyer, B. Tamain, A. Thiphagne, H. Doubre, A. Peghaire, J. Péter, E. Rosato, J.C. Adloff, A. Kamili, G. Rudolf, F. Scheibling, F. Guilbault, C. Lebrun, F. Hanappe, Ibid ref. 1c, p. 262 ; P. Eudes, Thesis (Caen 1988) LPCC-T-88-01.

71. K. Hagel, D. Fabris, P. Gonthier, H. Ho, Y. Lou, Z. Majka, G.

Mouchaty, M.N. Namboodiri, J.B. Natowitz, G. Nebbia, R.P. Schmitt, G. Viesti, R. Wada and B. Wilkins, Nucl. Phys. **1486** (1988) 429.

72. N.N. Ajitanand, R. Lacey, G.F. Peaslee, E. Duek and J.M. Alexander, Nucl. Instrum. Methods Phys. Res. **A243** (1986) 111.

73. D. Hilscher, H. Rossner, A. Gamp, U. Jahnke, A. Giorni, C. Morand, A. Dauchy, P. Stassi, B. Cheynis, B. Chambon, D. Drain, C. Pastor and G. Petit, Phys. Rev. **C36** (1987) 208.

74. J. Galin, G. Ingold, U. Jahnke, D. Hilscher, M. Lehmann, H. Rossner and E. Schwinn, Z. Phys. **A331** (1988) 63.

75. U. Jahnke, G. Ingold, D. Hilscher, M. Lehmann, E. Schwinn and P. Zank, Phys. Rev. Lett. **57** (1986) 190.

76. U. Jahnke, J.L. Charvet, B. Cramer, H. Doubre, J. Fréhaut, J. Galin, B. Gatty, D. Guerreau, G. Ingold, D. Jacquet, D.X. Jiang, B. Lott, M. Morjean, C. Magnago, Y. Patin, J. Pouthas, Y. Pranal, E. Schwinn, A. Sokolov and J.L. Uzureau, XX Inter. Summer School on Nucl. Phys., Mikolajki, Poland, Sept. 1-11, 1988. To be published.

77. J. Galin, J.L. Charvet, H. Doubre, J. Fréhaut, D. Guerreau, G. Ingold, D. Jacquet, U. Jahnke, D.X. Jiang, B. Lott, C. Magnago M. Morjean, Y. Patin, J. Pouthas, Y. Pranal, J.L. Uzureau, Texas A.M. Symposium on hot nuclei, Texas (1987), World Sci., p. 241.

78. R. Trockel, K.D. Hildenbrand, U. Lynen, W.F. Müller, H.J. Rabe, H. Sann, H. Stelzer, W. Trautmann, R. Wada, E. Eckert, P. Kreutz, A. Küm-Michel, J. Pochodzalla and D. Delte, preprint GSI-88-54 (October 1988).

79. K. Sneppen, D. Cussol and C. Grégoire, Niels Bohr Institute, Preprint NBI-88-52 (1988).

80. D.J. Morrissey, W. Benenson, E. Kashy, C. Bloch, M. Lowe, R.A. Blue, R.M. Ronningen, B. Sherrill, H. Utsunomiya, Phys. Rev. **C32** (1985) 877.

81. K. Siwek-Wilczynska, R.A. Blue, L.H. Harwood, R.M. Ronningen, H. Utsunomiya, J. Wilczynski, D.J. Morrissey, Phys. Rev. **C32** (1985) 1450.

82. D.J. Morrissey, C. Bloch, W. Benenson, E. Kashy, R.A. Blue, R.M. Ronningen, R. Aryaeinejad, Phys. Rev. **C34** (1986) 61.

83. L.G. Sobotka, D.G. Sarantites, H. Puchta, F.A. Dilmanian, M. Jaaskelainen, M.L. Halbert, J.H. Barker, J.R. Beene, R.L. Ferguson, D.C. Hensley, G.R. Young, Phys. Rev. **C34** (1986) 917.

84. H.M. Xu, D.J. Fields, W.G. Lynch, M.B. Tsang, C.K. Gelbke, D. Hahn, M.R. Maier, D.J. Morrissey, J. Pochodzalla, H. Stocker, D.G. Sarantites, L.G. Sobotka, M.L. Halbert, D.G. Hensley, Phys. Lett. **182B** (1986) 155.

85. J. Pochodzalla, W.A. Friedman, C.K. Gelbke, W.G. Lynch, M. Maier, D. Ardouin, H. Delagrange, H. Doubre, C. Grégoire, A. Kyanowski, W. Mittig, A. Péghaire, J. Péter, F. Saint-Laurent, Y.P. Viyogi, B. Zwieglinski, G. Bizard, F. Lefebvres, B. Tamain, J. Québert, Phys. Rev. Lett. **55** (1985) 177 and Phys. Lett. **161B** (1985) 275.

86. J. Pochodzalla, C.K. Gelbke, W.G. Lynch, M. Maier, D. Ardouin, H. Delagrange, H. Doubre, C. Grégoire, A. Kyanowski, W. Mittig, A. Péghaire, J. Péter, F. Saint-Laurent, B. Zwieglinski, G. Bizard, F. Lefebvres, B. Tamain, J. Québert, Y.P. Viyogi, W.A. Friedman, D.H. Boal, Phys. Rev. **C35** (1987) 1685.

87. D. Fox, D.A. Cebra, J. Karn, C. Parks, A. Pradhan, A. Vander Molen, J. van der Plicht, G.D. Wesfall, W.K. Wilson, R.S. Tickle, Phys. Rev. **C38** (1988) 146.

88. Z. Chen and C.K. Gelbke, Preprint MSUCL-662 (1988).

89. C.K. Gelbke, Preprint MSUCL-663 (1988).

90. F. Saint-Laurent, A. Kyanowski, D. Ardouin, H. Delagrange, H. Doubre, C. Grégoire, W. Mittig, A. Péghaire, J. Péter, G. Bizard, F. Lefebvres, B. Tamain, J. Québert, Y.P. Viyogi, J. Pochodzalla, C.K. Gelbke, W. Lynch and M. Maier, Phys. Lett. **B202** (1988) 190.

91. Z. Chen, Phys. Rev. **C36** (1987) 2297.
92. D. Hahn and H. Stöcker, Phys. Rev. **C35** (1987) 1311.
93. A. Olmi, Nucl. Phys. **A471** (1987) 97c.
94. K.D. Hildenbrand ; ibid ref. 13 - C6.
95. A. Olmi, P.R. Maurenzig, A.A. Stefanini, J. Albinski, A. Gobbi, S. Gralla, N. Herrmann, K.D. Hildenbrand, J. Kuzminski, W.F.J. Müller, M. Petrovici, H. Stelzer and J. Toke, Europhys. Lett. 4 (1987) 1121.
96. A. Olmi and R. Freifelder, Nouvelles du Ganil, 23 (1987).
97. B. Borderie, M. Montoya, M.F. Rivet, D. Jouan, C. Cabot, H. Fuchs D. Gardès, H. Gauvin, D. Jacquet, F. Monnet, Phys. Lett. **B205** (1988) 26.
98. R. Vandenbosch, R.C. Connolly, S. Gil, D.D. Leach, T.C. Awes, S. Sorensen and C.Y. Wu, Phys. Rev. **C37** (1988) 1301.
99. J.L. Charvet, G. Duchene, S. Joly, C. Magnago, M. Morjean, Y. Patin, Y. Pranal, L. Sinopoli, J.L. Uzureau, R. Billerey, B. Chambon, A. Chbihi, A. Chevarier, N. Chevarier, D. Drain, C. Pastor, M. Stern and A. Péghaire, Phys. Lett. **B189** (1987) 388.
100. M. Morjean, J. Fréhaut, D. Guerreau, J.L. Charvet, G. Duchêne, H. Doubre, J. Galin, G. Ingold, D. Jacquet, U. Jahnke, D.X. Jiang B. Lott, C. Magnago, Y. Patin, J. Pouthas, Y. Pranal and J.L. Uzureau, Phys. Lett. **B203** (1988) 215.
101. M.F. Rivet, B. Borderie, C. Grégoire, D. Jouan, B. Remaud, Internal Report Orsay, IPNO-DRE-88-17.
102. N. Herrmann, R. Bock, H. Emling, R. Freifelder, A. Gobbi, E. Grosse, K.D. Hildenbrand, R. Kulessa, T. Matulewicz, F. Rami, R.S. Simon, H. Stelzer, J. Wessels, P.R. Maurenzig, A. Olmi, A.A. Stefanini, W. Kühn, V. Metag, R. Novotny, M. Gnirs, D. Pelte, P. Braun-Munzinger and L.G. Moretto, Phys. Rev. Lett. 60 (1988) 1630.
103. V. Metag, Nucl. Phys. **A488** (1988) 483c.
104. G. Ingold et al., to be published.
105. N. Bohr and J.A. Wheeler, Phys. Rev. 56 (1939) 426.
106. M. Brack, C. Guet and H.B. Hakamsson, Phys. Rep. 123 (1985) 275.
107. D. Dalili, J. Nemeth, C. Ngô, Z. Phys. **A321** (1985) 335.
108. X. Campi and S. Stringari, Z. Phys. **A309** (1983) 239.
109. C. Guet, E. Strumberger and M. Brack, Phys. Lett. **B205** (1988) 427.
110. J. Treiner, Nucl. Phys. **A488** (1988) 279c.
111. A. Gavron, A. Gayer, J. Boissevain, H.C. Britt, J.R. Nix, A.J. Sierk, P. Grangé, S. Hassani, H.A. Weidenmüller, J.R. Beene, B. Cheynis, D. Drain, R.L. Ferguson, F.E. Obenshain, F. Plasil, G.R. Young, G.A. Petitt and C. Butler, Phys. Lett. **B176** (1986) 312.
112. S. Song, M.F. Rivet, R. Bimbot, B. Borderie, I. Forest, J. Galin, D. Gardès, B. Gatty, M. Lefort, H. Oeschler, B. Tamain, X. Tarrago, Phys. Rev. Lett. **130B** (1983) 14.
113. D. Jacquet, E. Duek, J.M. Alexander, B. Borderie, J. Galin, D. Guerreau, M. Lefort, F. Monnet, M.F. Rivet, X. Tarrago, Phys. Rev. Lett. 53 (1984) 2226.
114. S. Hassani and P. Grangé, Phys. Lett. **B137** (1984) 281.
115. H. Delagrange, C. Grégoire, F. Scheuter and Y. Abe, Z. Phys. **A323** (1986) 437.
116. C. Grégoire, H. Delagrange, K. Pomorski and K. Dietrich, Z. Phys. **A329** (1988) 497.
117. J.R. Huizenga and L.G. Moretto, Ann. Rev. Nucl. Sci. 22 (1972) 427.
118. G. Nebbia, K.Hagel, D.Fabris, Z.Majka, J.B.Natowitz, R.P.Schmitt, B.Sterling, G.Mouchaty, P.L.Gonthier, B.Wilnins, N.N.Namboodiri, H.Ho, Phys. Lett. **B176** (1986) 20.
119. P. Bonche, S. Levit and D. Vautherin, Nucl. Phys. **A427** (1984) 278.
120. E. Suraud, P. Schuck, Phys. Lett. **164B** (1985) 212.
121. R.W. Hasse and P. Schuck, GSI Preprint 86-20.
122. P.F. Bortignon and C.H. Dasso, Proceeding of the Texas A.M. Symposium on hot nuclei, College Station (1987), World Scientific, p. 210.

123. D. Vautherin and N. Vinh-Mau, Orsay Preprint IPNO-TH-87-03 (1987).

124. First topical meeting on giant resonance excitation in Heavy Ion Collisions, Padova 1987. Nucl. Phys. **A482** (1988).

125. J.J. Gaardhoje, C. Ellegaard, B. Herskind, R.M. Diamond, M.A. Deleplanque, G. Dines, A.O. Macchiavelli and F.S. Stephens, Phys. Rev. Lett. **56** (1986) 1783.

126. J.J. Gaardhoje, A.M. Bruce, J.D. Garett, B. Herskind, M. Maurel, H.N. Nifenecker, J.A. Pinston, P. Perrin, C. Ristori, F. Schussler, A. Bracco and M. Pignanelli, Phys. Rev. Lett. **59** (1987) 1409.

127. D.R. Chakrabarty, S. Sen, M. Thoennessen, N. Alamanos, P. Paul, R. Schicker, J. Stackel and J.J. Gaardhoge, Phys. Rev. **C36** (1987) 1886.

128. J.J. Gaardhoje, Nucl. Phys. **A488** (1988) 261c.

129. A. Bracco, J. Gaardhoje, A.M. Bruce, J.D. Garett, B. Herskind, M. Pignanelli, D. Barneoud, M. Maurel, H. Nifenecker, J.A. Pinston, J. Perrin, C. Ristori and F. Schussler, NBI Preprint 1987. To be published.

130. R. Hingmann, W. Kühn, V. Metag, R. Mühlhans, R. Novotny, A. Ruckelshausen, W. Cassing, H. Emling, R. Kulessa, J. Wollerscheim, B. Haas, J.P. Vivien, A. Boullay, H. Delagrange, H. Doubre, C. Grégoire and Y. Schutz, Phys. Rev. Lett. **58** (1987) 769.

131. R.A. Broglia and W.E. Ormand, Nucl. Phys. **A482** (1988) 141c.

132. R. Lacey, N. Ajitanand, J.M. Alexander, D. De Castro Rizzo, G.F. Peaslee, L.C. Vaz, M. Kaplan, M. Kildir, G. La Rana, D.J. Moses, W.E. Parker, D. Logan, M.S. Zisman, P. de Young and L. Kowalski, Phys. Rev. **C37** (1988) 2541 ; ibid p. 2561.

133. G. Batko and O. Civitarese, Phys. Rev. **C37** (1988) 2647.

134. J.M. Alexander, D. Guerreau and L.C. Vaz, Z. Phys. **A305** (1982) 313.

135. W.A. Friedman and W.G. Lynch, Phys. Rev. **C28** (1983) 16 ; ibid p. 950.

136. L.G. Moretto, Nucl. Phys. **A247** (1975) 211.
     L.G. Moretto and G.J. Wozniak, Preprint LBL-25744 (1988).

137. L.G. Sobotka, M.L. Podgett, G.J. Wozniak, G. Guarino, A.G. Pacheco, L.G. Moretto, Y. Chan, R.G. Stokstad, I. Tserruya and S. Wold, Phys. Rev. Lett. **51** (1983) 2187.

138. G. Sauer, H. Chandra and U. Mosel, Nucl. Phys. **A264** (1976) 221.

139. E. Suraud, Nucl. Phys. **A462** (1987) 109.

140. S. Levit and P. Bonche, Nucl. Phys. **A462** (1987) 109.

141. J. Galin, Nucl. Phys. **A488** (1988) 297c ; see fig. 8.

142. G. Fai and J. Randrup, Nucl. Phys. **A487** (1988) 397.

143. P. Bonche, D. Vautherin, M. Vénéroni, J. de Phys. Suppl. 47, C4 (1986) 339.

144. J. Nemeth, M. Barranco, C. Ngô and E. Tomasi, Z. Phys. **A320** (1985) 691.

145. J. Knoll and B. Strack, Phys. Lett. **149B** (1984) 45.

146. L. Vinet, F. Sébille, C. Grégoire, B. Remaud and P. Schuck, Phys. Lett. **172B** (1986) 17.

147. B. Remaud, C. Grégoire, F. Sébille and P. Schuck, Nucl. Phys. **A488** (1988) 423c and references therein.

148. J. Aichelin, G. Perlert, A. Bohnet, A. Rosenhauer, H. Stöcker and W. Greiner, Nucl. Phys. **A488** (1988) 437c.

149. D.H. Gross, Zhang Xiao Ze, Xu Shu-Yan and Zheng Yu-Ming, Nucl. Phys. **A461** (1987) 641 ; Nucl. Phys. **A461** (1987) 668.

150. J.P. Bondorf, R. Donangelo, I.N. Mishustin, C.J. Pethick, H. Schülz and K. Sneppen, Nucl. Phys. **A443** (1985) 321.

151. C. Ngô, Nucl. Phys. **A488** (1988) 233c.

152. M. Morjean, J.L. Charvet, B. Cramer, H. Doubre, J. Fréhaut, J. Galin, D. Guerreau, G. Ingold, D. Jacquet, U. Jahnke, D.X. Jiang, B. Lott, C. Magnago, Y. Patin, J. Pouthas, Y. Pranal, E. Schwinn, A. Sokolov, Third Intern. Conf. on Nucleus-Nucleus Collisions (Saint-Malo June 1988) Contributed paper B51.

153. W. Loveland, K. Aleklett, L. Sihver, Z. Xu and G.T. Seaborg, Nucl. Phys. **A471** (1987) 175c.

154. K. Aleklett, W. Loveland, L. Sihver, Z. Xu, C. Casey, D.J. Morrissey, J.O. Liljenzin, M. de Saint-Simon, G.T. Seaborg, ibid ref. 152 ; contribution C81.

155. E. Schwinn et al. (to be published).

156. Jin G.M., A. Péghaire, H. Doubre, J. Péter, F. Saint-Laurent, G. Bizard, R. Brou, M. Louvel, J.P. Patry, R. Regimbart, J.C. Steckmeyer, B. Tamain, Y. Cassagnou, R. Legrain, C. Lebrun, E. Rosato, J.C. Jeong, S.M. Lee, T. Nakagawa, Y. Nagashima, M. Ogihara, J. Kasagi, T. Motobayashi, R. Mc. Grath, K. Hagel, ibid ref. 152 ; contribution B19.

157. R. Bougault, F. Delaunay, A. Genoux-Lubain, C. Le Brun, J.F. Lecolley, F. Lefebvres, M. Louvel, J.C. Steckmeyer, J.C. Adloff, B. Bilwes, R. Bilwes, M. Glaser, G. Rudolf, F. Scheibling, L. Stuttge, J.L. Ferrero, Nucl. Phys. **A488** (1988) 255c.

158. B. Jakobsson, G. Jönsson, L. Karlsson, B. Noren, K. Söderström, F. Schussler and E. Monnand, Nucl. Phys. **A488** (1988) 251c.

159. R. Trockel, U.Lynen, J.Pochodzolla, W.Trautmann, N.Brummund, E.Eckert, R.Glasow, K.D.Hildenbrand, K.H.Kampert, W.F.J.Mueller, D.Pelte, H.J.Rabe, H.Sann, R.Santo, H.Stelzer, R.Wada, Phys.Rev.Lett. **59** (1987) 2844.

MULTIGRAMENTATION OF NUCLEI

Christian Ngô

Laboratoire National Saturne
91191 Gif sur Yvette Cedex
France

One uses often heavy ion nuclear reactions to convert all or part of the incident kinetic energy of the projectile (which is provided by the accelerator) into other forms of excitation. This is done with the following questions in mind :
1- How much of the incident kinetic energy can we convert?
2- Is this transfer of energy used to excite collective degrees of freedom or to create incoherent excitations (heat)?
3- Which part of the system is concerned with this energy transfer : the projectile, the target, the composite system, part of them...?

At low bombarding energies ($\leq$ 10 MeV/u) one has identified now rather well the different types of energy transfer which can take place[1]. For instance, we know that grazing collisions are concerned with small energy transfers while there can be a complete transformation of the incident kinetic energy into heat in central collisions (by fusion of the projectile and target). Collective modes like giant resonances are also excited in this energy domain.

Very heavy projectiles are now available at several accelerator facilities. The question of the energy transfer occuring in nuclear reactions is crucial since one of the reasons of using very energetic beams is to form nuclear species as excited as possible. The formation of highly excited and compressed pieces of nuclear matter is an important issue for at least two reasons : first, concerning the possibility of creating new states of nuclear matter (deconfining the quarks for instance[2]). Second, because a subsequent evolution of the system may lead to conditions where new phenomena take place (for example multifragmentation[3] where a self organization of the system takes place).

Heavy ion collisions are very useful to create nuclear systems at high temperatures and densities. The reason to do so can be understood in the schematic diagram presented in fig. 1. On very general arguments one can predict that two new states of nuclear matter may exist :
The first one corresponds to density values smaller than normal and temperatures of the order of a few MeV. In this case, nuclear matter is expected to disassemble into clusters. This process, which is usually called multifragmentation, has received a considerable attention within the last few years. It will be the subject of these lectures.
The second one corresponds to high temperature and density values.

If such conditions could be achieved in a sizeable volume, one expects a deconfinement of the quarks and a quark-gluon plasma could be created[2].

In these lectures we shall deal with some of the problems connected to nuclear multifragmentation and try to give a quick outlook of the present situation as well as of its perspectives. Other aspects will be treated during this school by other lecturers. Due to the large amount of litterature in the field we shall not be able to quote every work on the subject. We apologize for that and hope that the references will be able to complement this short review.

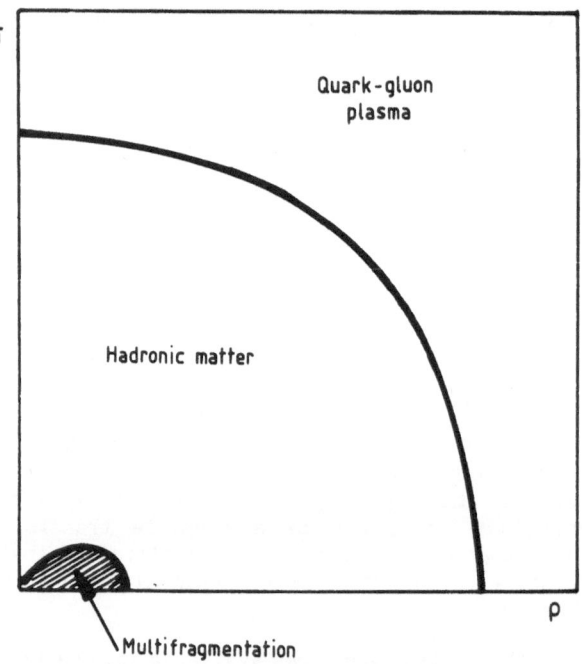

Figure 1. Schematic drawing in the density (ρ) temperature (T) plane of the different phases of nuclear matter which are expected to exist.

1. INTRODUCTION

A high energy proton going through a nucleus can break it up. The same occurs with a heavy ion but, since the energy transfer is more efficient, little less kinetic energy is required. Multifragmentation occurs if enough excitation energy is brought into the system by the projectile. It is an important issue in heavy ion physics to know why and how nuclei break up. In this introduction we would like to demonstrate, using simple physical arguments, that multifragmentation should take place if certain conditions on the temperature and on the density can be obtained.

1.1 Nature likes to save energy

At moderate temperatures (T ≤ 2-3 MeV) a nucleus decays by particle evaporation, γ rays emission and/or fission. The number of emitted particles increases with the temperature and they are emitted more and more quickly. As a consequence, fission, which is a slow process, may be hindered by this fast particle evaporation[4]. At the same time, heavier particles have a larger probability to be emitted[5], especially if the angular momentum of the system is large.

If we extrapolates our knowledge at low excitation energies to higher values, we would expect that a nucleus boils off into single nucleons if it has enough excitation energy. This extrapolation is probably too simple and seems not to be observed in practice. This is not too surprising since nature likes to save energy. This is what occurs if clusters are formed because one gains their binding energy. Consequently, from an energetic point of view, it is more efficient for an excited nucleus to break up into several clusters rather than separate in nucleons.

1.2 Fluctuations of the mean field

Two kinds of interactions exist in the nuclear medium : the long range repulsive Coulomb forces and the short range attractive nuclear forces. At normal densities the nucleus is a homogeneous fluid of nuclear matter. The physics of this object is, to a good approximation, governed by the mean field created by the whole set of nucleons. This system can be well described in terms of a local average density. The nucleon density varies smoothly from one point to another and a nucleon cannot be singled out from the density profile. If, in a nuclear reaction, the average density of the system decreases a lot, then, as we move in the medium, we have regions where there are nucleons, i.e. where the nuclear density is large, and regions where there are no nucleon, i.e. where the nuclear density is zero. For such a situation, it is not very accurate to describe the properties of the system in terms of an average density, i.e. in terms of a mean field, and one expects the many body correlations to become important. At normal densities one says that the fluctuations of the mean field are small while at low densities they are large. In the first case the correlation length extends over the whole system while it is of the order of the size of one cluster in the second case.

To illustrate a little bit more the importance of the correlations between the nucleons let us consider a volume V in which we put A nucleons, neutrons and protons (fig.2)[6]. At normal densities (fig.2a) each nucleon interacts by nuclear forces with all its close neighbours. The physics of the system is dominated by the mean field created by the whole set of nucleons. A nuclear signal can easily propagate through the whole system. If the volume V becomes bigger, as in fig.2b, each nucleon no longer interacts with all its neighbours by nuclear forces because they are of short range. Clusters can be formed if several nucleons are close enough to each other but isolated nucleons can also exist. The physics of each cluster is of course dominated by its own mean field but the mean field of a particular cluster cannot describe properly the physical properties of the whole system. Such a configuration is very unstable because of the Coulomb and centrifugal forces which are repulsive. Consequently, the system breaks up into pieces: one has multi-fragmentation.

The above considerations are based on very general arguments and should apply to any nuclear system at low density. They also apply to macroscopic systems which are dominated by a short range interaction (like the Van der Waals one).

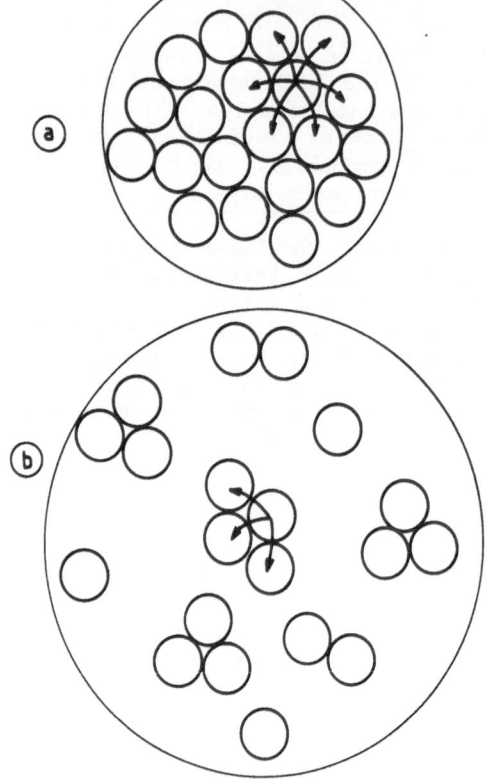

Figure 2. Schematic illustration of the importance of density
effects on the formation of clusters in a nuclear system.

1.3 What happens in central heavy ion collisions at high energy?

In central heavy ion collisions the following scenario probably
happens :

In a first phase, the projectile and the target merge and form a non
equilibrated system. At not too large bombarding energies this system
contains a lot of excitation energy which is not shared among all degrees
of freedom. The system does not like to have so much excitation and
several things happen in order to sustain it. Part of the available
energy is removed by prompt particles emission while the rest is shared
among all the intrinsic degrees of freedom by two body collisions. This
leads to a thermalisation of the system. Furthermore, part of the initial
energy is stored in the compression mode of the system.

After a time, which one can estimate to be of the order 10^{-22}s from
the point where the projectile and the target have merged[7], one is left
with a system which has almost reached thermal equilibrium and is
compressed. This system expands, very likely almost isentropically, and
cools down. The larger the compression and the thermal excitation, the
larger the expansion. Consequently, it is easy to imagine that, above
some certain bombarding energy threshold, one reaches a point where the
fluctuations of the mean field become large. If it is the case, the
system breaks up into clusters.

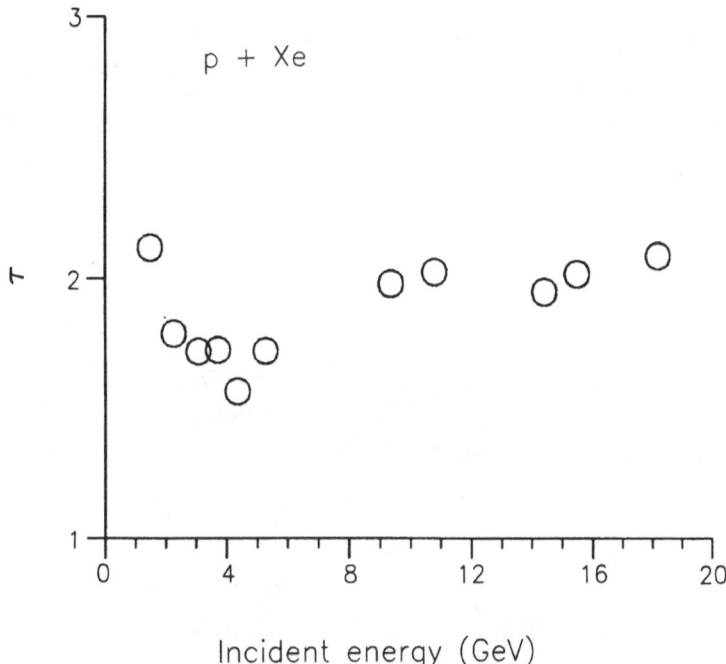

Figure 3. Effective exponent τ as a function of the incident kinetic energy. From Porile et al[11].

Protons induced reactions may proceed a bit differently since the compression of the system, which is essential for an expansion, is probably negligeable. Therefore, protons of relatively high bombarding energy (a few GeV) are required to trigger multifragmentation. At these energies, the incident proton cascades through the target nucleus. This leads to a shower of secondary particles which may either leave the target nucleus, create mesons, or thermalize in the medium. In the end, one is left with a thermally excited system which has a lower density due to the emission of particles. There can also be large density fluctuations due to the fast emission of particles. Consequently, the system may break up. Because of a lack of compression, the bombarding energy required in proton induced reaction in order to have multifragmentation is larger than the total bombarding energy needed in heavy ion collisions.

In conclusion, it seems likely that one gets multifragmentation because a nuclear medium at low density is formed. Since these conditions cannot be made directly, one has to use heavy ion collisions to create a compressed and thermal excited system which subsequently expands and eventually reaches the point of disassembly.

2. EXPERIMENTAL EVIDENCES OF MULTIFRAGMENTATION

There are not so many direct evidences of nuclear multifragmentation. This is because it is a difficult experimental problem due to the fact that several fragments are produced in the exit channel. The more direct indications that nuclei break up in several pieces is given by emulsion experiments[8,9]. However, they do not tell us if the process is sequential or not. Counter experiments are now trying to pin down the main characteristics of this process but we are just at the beginning of very extensive studies. Let us give a brief overview of the subject.

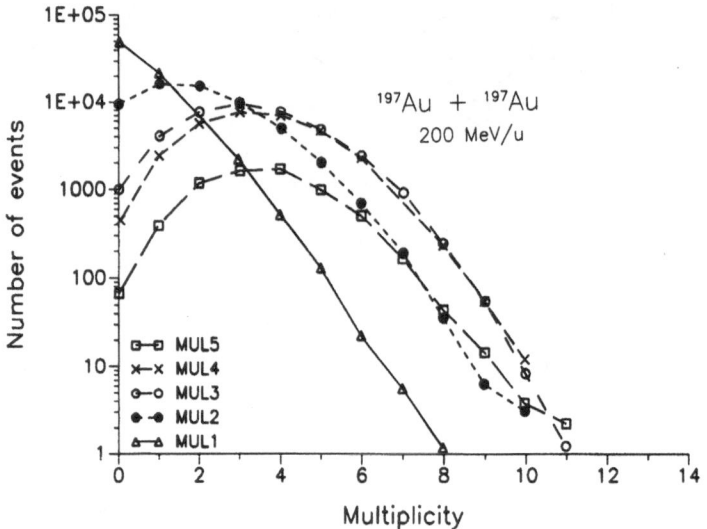

Figure 4. Fragment multiplicity distributions corresponding to different bins in the multiplicity of light particles recorded in a forward wall. From Doss et al[16].

2.1 Proton induced reactions

All the story about nuclear multifragmentation has started with the beautiful experiments made by the Purdue group with very high energy protons[10]. They found that the inclusive mass distribution of intermediate mass fragments produced from the interaction of 50 to 350 GeV protons with Kr and Xe nuclei followed a power law $A^{-\tau}$, where A is the fragment mass and τ and exponent equal to 2.6. This was interpreted as the result of a gas liquid phase transition near the critical point. Although this interpretation is still questionable for systems like nuclei, its great merit was to draw the attention of the physicists about the interesting problem of multifragmentation. As a consequence, it has allowed to better investigate, from the theoretical point of view, the question of cluster formation in a nuclear medium.

More recently, this group, Porile et al[11], has investigated the p+Xe reaction for proton energies ranging from 1 to 20 GeV. The energy spectra of multifragmentation events detected at 48.5° and 131.5° have been measured. Below 6 GeV these authors could identify, from the kinetic energy spectra, two distincts mechanisms. One corresponds to a binary splitting of the system. Its probability decreases as the bombarding energy increases and is negligible above 6 GeV. The second mechanism is multifragmentation. It starts to contribute around 1 GeV and its importance increases until it saturates at about 12 GeV. At 1.6 GeV, it is worth to note that the multifragmentation cross section is only one tenth of its limiting value at high bombarding energy. One can parametrize the inclusive mass distribution of the multifragmentation contribution by a power law. The apparent exponent τ found in the fit depends on the proton kinetic energy as one can see in fig.3. It is minimum around 4 GeV. According to ref.12 this minimum should occur when the excitation energy is comparable to the binding energy of the system.

2.2 Heavy ion reactions

A direct way to see that nuclei break up into several pieces is provided by emulsion experiments[8,9]. Several of them have been performed so far and interesting results have been obtained. Above a certain bombarding energy threshold, which depends upon the system, several tracks associated to medium mass fragments are observed. These events are associated to multifragmentation. The great merit of emulsion experiments is that one can have sometimes a full knowledge of all charged particles emitted during the explosion of the system. This is for example the case in the beautiful experiment performed by Waddington and Freier[9] with 990 MeV/u incident ^{197}Au nuclei.

Many inclusive experiments have been made with heavy ions. The inclusive mass distribution of the so called intermediate mass fragments is not so clear to interpret since one does not know how to separate other contributions than multifragmentation. Nevertheless, attempts to fit the mass distribution have been made. The result of the compilation of some of the earlier data is presented in ref.13 together with those obtained in proton induced experiments. One finds that τ varies from system to system in a range of values between 2-4. At intermediate bombarding energies one should be very careful before saying that the intermediate mass fragments are all coming from multifragmentation. Indeed, the excitation energy of the system is not always large enough for this process to take place and very often one has just binary processes as it has been beautifully examplified in ref.14. In particular one may now question about the values of the exponent τ deduced from the inclusive atomic or mass distributions at rather low bombarding energies.

An exclusive measurement of all the fragments emitted in a multifragmentation process is difficult because they are produced with a rather small kinetic energy in the disassembling system. Furthermore, many light particles are also emitted. A full characterization of a multifragmentation event requires a tracking detector with a large solid angle, a good atomic number and mass resolution and a small energy threshold. Some attempts have been made to show that more than two medium mass fragments are emitted in some collisions involving very heavy projectiles and targets[15]. However, a full measurement of all the characteristics of the products of the reaction is not yet available. So far, the most complete measurements about multifragmentation have been obtained for the 200 MeV/u Au+Au system at the Bevalac with the plastic ball[16]. In this experiment large multiplicities of fragments with $3 \leq Z \leq 10$ have been observed in central collisions. This is illustrated in fig.4 which shows the yield distribution as a function of the multiplicity for different bins of the multiplicity of light particles recorded in a forwards plastic wall. This light particle multiplicity is supposed to be closely related to the impact parameter of the collision : the larger its value, the more central the collision. In fig.4 MUL5 corresponds to the most central collisions while MUL1 corresponds to the most peripheral ones.

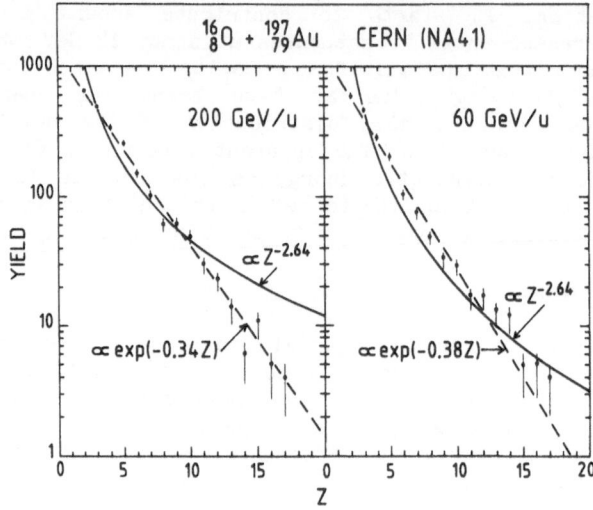

Figure 5. Atomic number distribution obtained by Berthier et al[17].

An interesting problem is to know what happens at ultrarelativistic energies and to compare the results with proton induced experiments. This has been possible with the ^{16}O and ^{32}S beams recently available at CERN. It has been found, in the case of heavy ion collisions, that the inclusive atomic number distribution does not follow a power law dependence at 200 GeV/u. This is illustrated in fig.5 for the ^{16}O + ^{197}Au system at 200 GeV/u measured by Berthier et al at CERN[17]. The measurement performed at 60 GeV/u, with the same experimental set up, is also displayed. The full line corresponds to the power law $Z^{-2.64}$ which fits the proton data results[10]. The dashed curve corresponds to a fit to the heavy ion data. At 60 GeV/U one could say that the proton and heavy ion distributions follow the same kind of power law dependence. This seems not to be the case at 200 GeV/u where an exponential curve fits by far better the atomic number distribution. Similar results have been obtained with a 200 GeV/u ^{32}S beam[18]. Where does the difference between ultrarelativistic projectiles comes from? This is still an open question. One may note that, in proton induced reactions, many impact parameters may contribute to the multifragmentation process while this is no longer the case in a heavy ion collision (central collisions lead to a complete destruction of the target while peripheral collisions do not deposit enough excitation energy).

An indirect indication of the onset of multifragmentation might be found in the linear momentum transfer measurements performed in heavy ion collisions at intermediate bombarding energies[19]. Let us now discuss a little bit this problem. In the case of heavy nuclei the fused nucleus decays essentially by fission. Therefore, measurement of the fission fragments allows to study directly the fusion process. This method is known for a long time[20] and has been extensively used at low bombarding energies to deduce fusion cross sections. At intermediate bombarding energies the transfer of linear momentum from the projectile to the target is no longer complete because light particles are emitted before

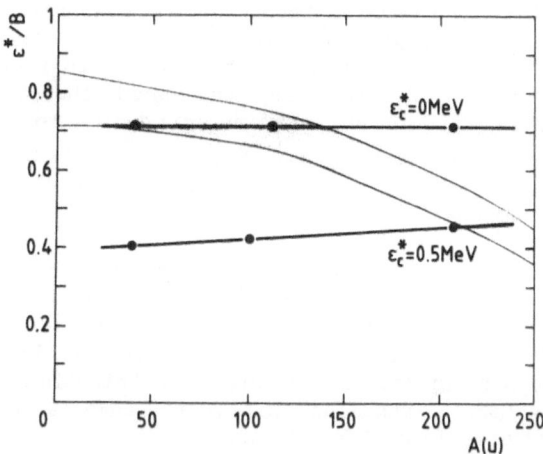

Figure 6. The critical excitation energy per nucleon ε^* above which multifragmentation takes place, divided by the binding energy per nucleon B, is plotted as a function of the mass of the nucleus. The two curves are calculations from ref.[41] corresponding to different initial compressional energies per nucleon. The shaded area corresponds to the region where incomplete fusion disappears as deduced experimentally[19].

the system fuses : one has incomplete linear momentum transfer and the process is called incomplete fusion because only part of the projectile fuses with part of the target. It turns out that, above a certain bombarding energy threshold, incomplete fusion (i.e. the fission products) disappears. This occurs very gradually and depends critically upon the mass asymmetry of the system. For instance, for the Ar+Au, U systems the fusion cross section vanishes around 40-50 MeV/u[21-22] while it is different from zero for the C+AU and U systems at 86 MeV/u[23]. Using reasonnable assumptions one can deduce the amount of excitation energy deposited into the system. Although this quantity depends upon the assumptions which have been made, it gives a good order of magnitude of how much thermal excitation has been deposited into the system. By correlating this quantity with the observation of fusion, one can deduce the maximum energy that can be contained in a fused system. The critical energy deposition above which incomplete fusion disappears is about 3-6 MeV/u, depending on the system under consideration. This is schematically shown in fig.6 where, from a compilation of the data relative to linear momentum measurements, one has plotted the region where the transition occurs as a function of the mass of the compound system (shaded area). One observes that the energy necessary to break up heavy fused nuclei is smaller than for lighter ones.

If incomplete fusion gradually disappears as the bombarding energy increases where goes the cross section?. Furthermore is there really a disappearance of fusion? The answer to this last question is not clear. Indeed, one may imagine that the fused system is so excited that a burst of prompt particles is emitted very quickly removing a non negligeable part of the mass of the system. Consequently, the resulting incomplete fused nucleus might be not heavy enough to subsequently fission with a high probability. This would mean that evaporation residues should be observed. Experiments with emulsions and visual detectors indicates that, indeed, nuclei that could be interpreted as evaporation residues are detected at relatively high bombarding energies[24]. Nevertheless, this does not excludes that one would not have multifragmentation for some events. But, as in the proton case, the onset should probably occur smoothly. At any rate it is an interesting problem to study experimentally.

2.3 What are the relevant observables?

One of the main problems of nuclear reactions where several fragments are produced in the exit channel is to know how to present the experimental data and to correlate the different measured quantities. It is clear that it is insufficient, for a model, to just reproduce the inclusive mass distribution. An important contribution to this problem has been made by Campi[25] assuming that multifragmentation is a critical phenomena. Let us briefly outline the method.

For an infinite system near the critical point several theories predict scaling laws characterized by critical exponents. Let us consider a system which exhibits a critical phenomenon when a variable X reaches a value X_c (for example, X_c might be the critical temperature in the case of a thermal phase transition, or the time t_c in the case of a kinetic process,...). Let us introduce the variable $\varepsilon = X - X_c$ which represents the departure of X with respect to the critical value X_c. The size distribution of the fragments, $P(A, \varepsilon)$ (where A denotes the size of a cluster), close to the critical point is :

$$P(A, \varepsilon) \alpha A^{-\tau} f(\varepsilon \cdot A^{\sigma}) \tag{1}$$

where τ and σ are two critical exponents. At the critical point $f(0) = 1$ and one finds a power law distribution as we discussed aboved. Away from the critical point f decays rapidly and exponentially.

Even if one cannot have a real phase transition for a finite system, there should be a reminiscence of it and this should be seen. This was the starting point of Campi who proposes a method to look at the experimental data, provided they are exclusive. For an infinite system, one usually defines the moments $M_k(\varepsilon)$ of the size of the mass distribution by :

$$M_k(\varepsilon) = \sum_{A \neq A_{max}} A^k P(A, \varepsilon) \tag{2}$$

For k>1 they diverge at the critical point with critical exponents :

$$M_k \alpha \varepsilon^{-\mu_k} \quad \text{and in general} \quad \mu_k = -(\tau - 1 - k)/\sigma \tag{3}$$

Since it is not possible to know experimentally ε in a nuclear system, one cannot evaluate $P(A, \varepsilon)$. To go around this problem one introduces the moments of a given multifragmentation event, M_k^j, as :

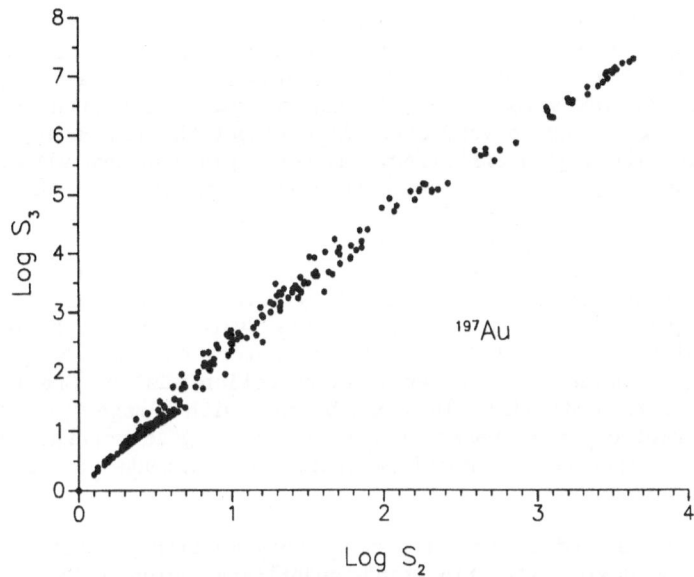

Figure 7. Correlation between Log S_3 and Log S_2 for the emulsion data of Waddington and Freier[9].

$$M_k^j(\varepsilon) = \sum_{A \neq A_{max}} A^k \, m^j(A) \tag{4}$$

where $m^j(A)$ is the number of fragments of size A in the multifragmentation event number j. In eq. 4 the sum excludes the heaviest fragment. From these moments one defines normalized ones :

$$S_k^j = \frac{M_k^j}{M_1^j} \tag{5}$$

For a finite system all these moments remain finite. If there is some reminiscence of a critical behaviour, they should exhibit some particular correlation. For example, if one plots Ln S_3 as a function of Ln S_2 one should have about a linear correlation with a slope equal to $\lambda_{3/2} = 1 + 1/\sigma\mu_2$. For a gas liquid phase transition $\lambda_{3/2} = 2.5$ while $\lambda_{3/2} = 2.22$ for a lattice percolation model. From the emulsion experiments of Waddington and Freier[9] one gets a value close to the percolation number. In fig. 7 we show this correlation for the emulsion events of ref. 9.

2.4 Perspectives

Since we are just at the beginning of serious experimental investigations of nuclear multifragmentation we see that a lot of studies are still needed before a good understanding of the process is reached. The most urgent thing is probably to completely characterize multi-fragmentation events using counter experiments. The main experimental problem is the low velocity of the multifragmentation products in their emitting system which is very often also almost the laboratory system. To detect them with a good efficiency one needs gas counters with a very low energy threshold, i.e. with very thin entrance windows. Since a large solid angle should be covered, this makes the experimental problem rather tricky. To remove this threshold problem one can use inverse kinematics, i.e. bombard a light target with a heavy projectile. In this case the recoiling system moves in the laboratory with a large velocity and there are no longer threshold problems. If this system breaks up the products have almost the same velocity in the laboratory and they are focussed at very forward angles. This makes a 4π detection easier. One should then separate and identify them. This can be done with a magnet located in the forward direction as proposed in ref.26, but a good separation power is required to separate the multifragmentation fragments which have about the same momentum.

Once we succeed to identify completely multifragmentation events we need to determine the important quantities which will be later on compared with theories. In other words what are the relevant physical characteristics of the multifragmentation process? This question has so far not received enough attention. Indeed, one should not only compare the inclusive mass distribution of the fragments which we know to be reproduced by many models. In this context, the approach of Campi[25], based on the moments of the mass of the fragments, is interesting. One may note that previous comparisons between experimental emulsion data and percolation models are encouraging but more should be done. In particular it seems very probable that the nucleation process, during which clusters are formed, is highly non linear and chaotic in the sense that a small change of the initial conditions may lead to quite different final geometries. If it is so, one has no chance to be able to reproduce accurately a given multifragmentation event. However, one has to try to reproduce the characteristics of the system which are not dependent upon this evolution. The correlation between the moments proposed by Campi are perhaps a first step in this direction.

3. STATIC THEORETICAL APPROACHES

3.1 The theoretical problem

A large part of nuclear physics can be understood within the mean field approximation. Even dissipative phenomena like deep inelastic reactions are dominated by the mean field of the system[1]. The residual interactions are just needed to convert part of the incident kinetic energy into heat. Many experimental results obtained at intermediate energies can also be described within the framework of mean field theories improved by including two body collisions. Multifragmentation is a more difficult problem because it involves the many body correlations during the formation of the clusters. We actually know that the conventional mean field descriptions used in nuclear physics should break down at low densities. For macroscopic systems we also know that the mean field approximation, which amounts to replace the operators associated to some macroscopic variables by their mean value, breaks down at the critical point. It can be applied before and after the critical point but not at, nor close to it. We have the same problem for nuclear multifragmentation. This difficult problem cannot be solved without any approximation and this is what has been done during the last few years. We shall now describe some of the results of this interesting area.

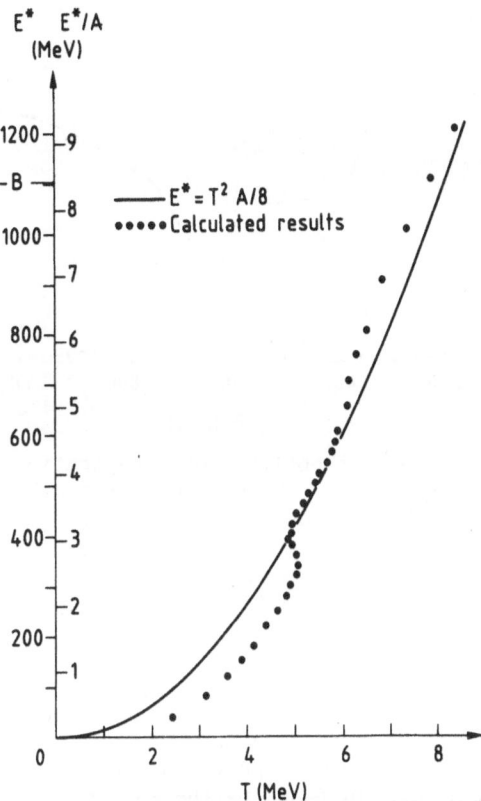

Figure 8. Microcanonical calculation of the temperature T as a function of the excitation energy E* for the [131]Xe nucleus. From Gross et al[32].

3.2 Statistical approaches

The first approaches to multifragmentation are static ones where one tries to see if an excited nucleus can break up or not. These models start with a spherical nucleus in thermal equilibrium. On the basis of statistical mechanics they study if the system can break up into several pieces. These approaches do not consider the whole dynamical evolution of the system from its initial compressed and heated up state until the moment where it breaks up into fragments. They concentrate on the disassembly stage only, as it takes place when the system reaches the break up volume and undergoes multifragmentation. In order to fragment the nucleus should have a volume which is substantially larger than its initial one, therefore the system should have expanded. Since statistical approaches to multifragmentation will be extensively developed by Jacob Bondorf[27] at this school, we shall quickly go through the subject and stress only a few points.

The possibility that nuclear matter can separate in two phases, a gas and a liquid, has been considered and pointed out a long time ago within the framework of several approaches[28]. One of the main implication of these studies are in the field of astrophysics. It has been investigated again more recently with the multifragmentation problem in mind[29]. Critical temperatures of the order of 17-18 MeV are obtained. However, a nucleus is different from nuclear matter : it is finite and

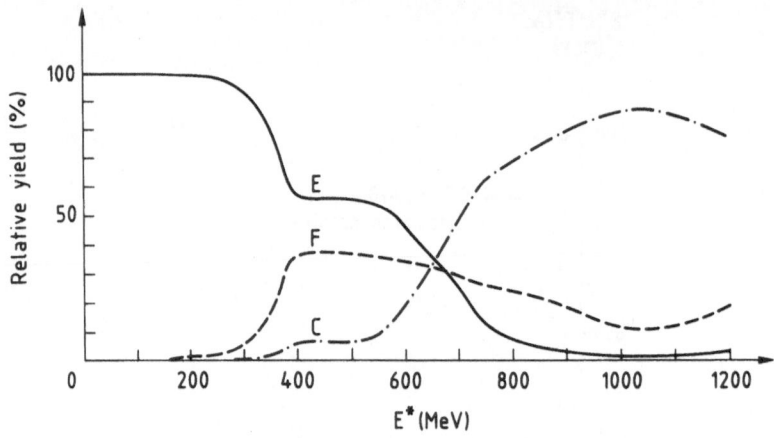

Figure 9. Relative probability of evaporation (E), binary fission (F) and cracking (C) as a function of the excitation energy of the ^{131}Xe nucleus (microcanonical calculation from Gross et al[32]).

charged. As a consequence, the binding energy in nuclear matter, which is about -16 MeV, decreases to -7,-8 MeV in normal nuclei. A similar shift should be expected for the critical temperature and one would expect a value around 10 MeV for a nucleus, provided a gas liquid phase transition exists. This is about what is found by the models[30].

If one uses equilibrium statistical mechanics to describe a nucleus at the breaking point one has to take care that it is a finite isolated system which contains a small number of particles and that long range forces are present in the medium. A proper description of this situation requires the use of the microcanonical ensemble[31,32]. Since a practical use of this ensemble is rather tedious, the first attempts were made in the framework of the grand canonical ensemble (mean number of particles and mean energy fixed)[33] or using the canonical ensemble (mean energy fixed)[34].

In order to illustrate some of the results of statistical approaches we show two examples from Gross et al[32]. In fig.8 the excitation energy per nucleon, E^*/A, is displayed as a function of the temperature T for the ^{131}Xe nucleus. The points are the results of the microcanonical calculation. The full curve is just the Bethe formula. For E^*/A around 3 MeV and T around 5 MeV there is a change in the evolution which could be attributed to a phase transition in a finite system. Another transition, but not so apparent, can be seen at larger E^*/A ($\simeq$ 5 MeV) and T ($\simeq$ 6 MeV) values. Similar results have also been obtained by Bondorf et al using the canonical ensemble[33].

It is interesting to look at the evolution of the number of fragments produced in the de-excitation of a nucleus as a function of the excitation energy E^*. This is illustrated in fig.9 for the ^{131}Xe nucleus by the microcanonical calculation of ref.32. The evaporation like component where there is one fragment of mass A≧10 decreases as E^* increases. First because binary fission increases and second because three or more fragments with a mass A≧10 are produced. This latter

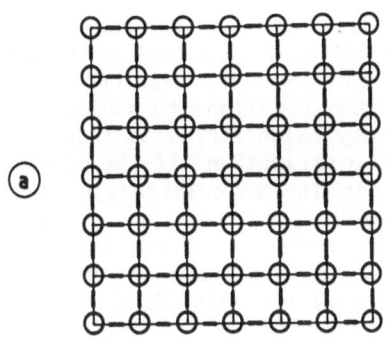

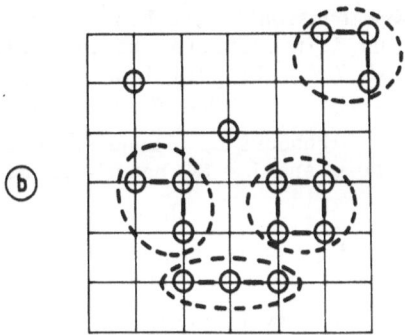

Figure 10. Principle of a site-bond percolation calculation shown on a 7×7 two dimensional cubic lattice.

component , called cracking in ref.32, corresponds to a multifragmentation of the system.

As far as statistical approaches are concerned, instead of assuming that all the available microstates of the system are equiprobable (microcanonical ensemble), one may assume that it is the different break up partitions of the initial system which are equiprobable. This was assumed by Aichelin and Hüfner[35] who tried to describe the multifragmentation of nuclei in a similar way as the shattering of glass. In this approach, the assumption of partitions equiprobability implies that the available microstates are not equiprobables. Nevertheless, the computed mass distributions of the multifragmentation products also agree with the experimental measurements.

3.3 Percolation

Percolation is a method which allows to study long range correlations in a system composed of points linked together according to some given mathematical law[36]. It is one of the simplest statistical approaches which allows to describe critical phenomena. It has been applied successfully to many physical problems like gelation, polymerisation, communication networks, resistor networks, porous media (for oil production or nuclear wastes pollution for example)... It is also used in many other fields to describe various problems like the propagation of forest fires or the spread of diseases. It has been introduced recently in nuclear physics to study the multifragmentation of nuclei[37-38] and the[39] transition from hadronic matter to the quark gluon plasma . The first applications were static ones assuming that multifragmentation is a critical phenomena which can be described by percolation. Given a percolation model one then can find the parameters corresponding to the critical point and thereby calculate, with these values, other properties like the inclusive mass distribution. If the agreement is reasonnable one may then try to connect the percolation parameters to some physical quantities. More refined approaches try to investigate whether the disassembling system is below or above the critical point. In this respect the moment analysis discussed above is useful. Let us, as an introduction, briefly recall the main ideas of a percolation approach.

A percolation system is essentially defined by two ingredients : a collection of points and a rule to connect them. For example, the percolation model used in refs.[40,41] to evaluate the fluctuations of the mean field is a 3-dimensional site-bond percolation model based on a cubic lattice[38]. Let us illustrate the basic principle of the approach with a two dimensional 7×7 cubic lattice. Initially (fig.10a) all the N_0=49 sites of the lattice are occupied by a particle. Each particle is connected to its close neighbours by a bond indicated by a thick line. In the present example there are L_0= 84 bonds. If we draw at random N_0- N particles and break up L_0- L bonds, there remain N occupied sites and L unbroken bonds. This configuration can be described by two parameters : $p = N/N_0$, the concentration of sites, and $q = L/L_0$, the concentration of bonds. These two variables vary between 0 and 1. A cluster is defined as a set of connected points. Fig.10b shows the result of a particular simulation (another one would give another distribution) , where N=15 and L=10. There is one cluster of size 4, 3 clusters of size 3 and 2 isolated particles. Let us denote by N_{max} the size of the largest cluster. It is convenient to introduce a dimensionless parameter $P(p,q) = N_{max}/N_0$. This quantity varies between 0 and 1. It turns out that $P(p,q)$ has a strong variation in the p-q plane. This is illustrated for example for a 5×5×5 cubic lattice in fig.11[38]. When p and q are close to unity a large size cluster is always present. It is called *percolation cluster*. The percolation cluster disappears for small p and q and the size of the largest cluster is small. Systematic simulations in the p-q plane show two regions. The first one corresponds to the case where a percolation cluster is observed while the second one (which, in a nuclear system, we shall associate to the multifragmentation) corresponds to the situation where several small and medium size clusters are obtained. The fluctuations of the mean field are small in the percolation cluster region. They are large in the multifragmentation region. It is worth to note that the transition region between percolation and multifragmentation is smooth for the examples displayed here. This is due to the small number of particles contained in the system. For an infinite system the transition region is very sharp. Furthermore the percolation cluster is infinite and spans throughout all the system.

Several static applications of percolation have been made for proton as well as for heavy ion collisions. Sites, bonds, site-bond percolation models have been used as well as percolation in phase space[37,38]. All these models are able to reproduce reasonnably well the inclusive mass or atomic number distribution associated to multifragmentation as well as to spallation in the case of proton induced reactions. The main problem with these approaches is to make a good connection with the nuclear physical properties of the system. For example in a site-bond percolation model

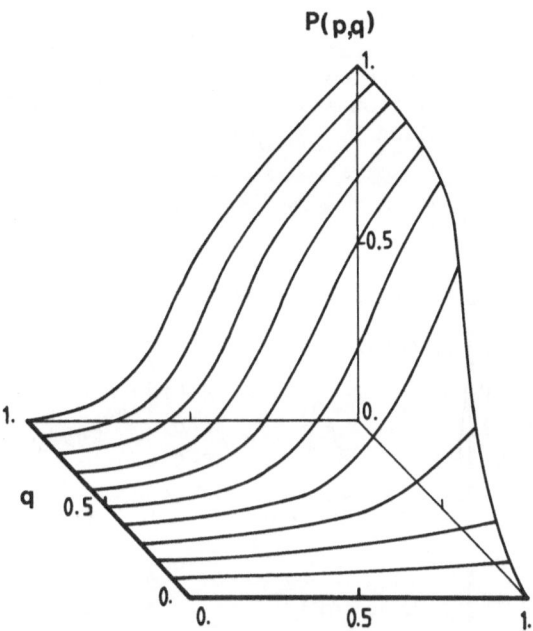

P(p,q)

Figure 11. Dimensionless size of the largest cluster displayed as a function of p and q. from Desbois[38].

what are the p and q parameters related to? We shall now try to give a simple analogy below.

In the evolution of a nuclear system, which we want to describe by a lattice percolation model, the change of the percolation parameters, like p or q in the case of site bond percolation, are expected to depend on the distance between the particles rather than on the occupation of the sites. In this respect the picture is a little bit different from the example described just above. If we want to make an analogy, let us consider the spread of a disease among trees which are distributed according to a lattice in an orchard. Every sick tree can contaminate its nearest neighbours (located at distance d, the lattice spacing) with a probability p. Clearly p is a decreasing function of the distance d. When the system has equilibrated two situations may be observed : if the spacing d is large the disease does not extend over the whole orchard but remains localized in clusters of different sizes. If the spacing is small all the orchard is contaminated and one has a percolation cluster of sick trees which extends over the whole orchard. For cluster formation in a nuclear medium we have about a similar situation. The spacing corresponding to the transition from a percolation cluster situation to multifragmentation depends upon the nuclear interaction distance. The problem is even more complicated for two reasons:
1- Nucleons are not at rest like the trees of the above example. They are moving due to the fermi motion and to their thermal excitation. If we make an analogy with a disease, it would more look like influenza which propagates because peoples are travelling.
2- Because of the strong surface interaction of nuclear objects, one does not expect to get filamented clusters like in macroscopic systems. These clusters restructure either in a single one, or they break up into smaller clusters. This makes the nuclear physics problem complicated.

For the above reason the restructured aggregation model proposed in ref.39 for the quark-gluon plasma and in ref.42 for nuclear fragmentation is interesting because it tries to go a little bit further compared to the simple lattice percolation models. We shall now briefly describe this approach in the next section.

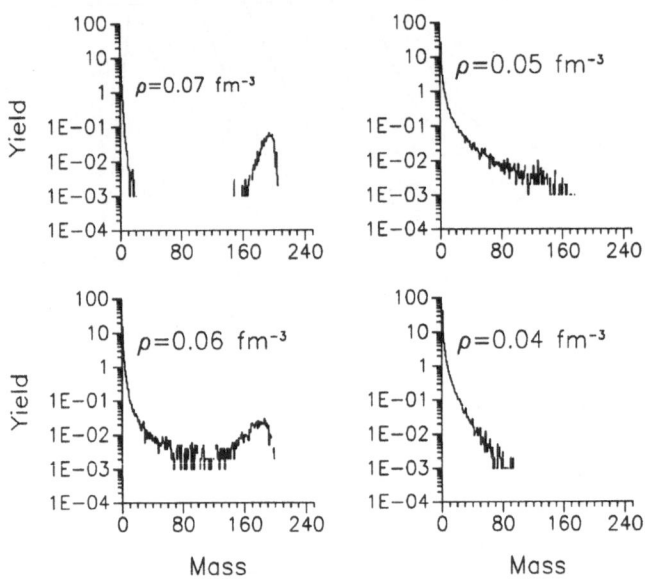

Figure 12. Mass distribution of the clusters after restructured aggregation for the ^{208}Pb nucleus at different density values. From ref.[42].

3.4 Restructured aggregation

The aim of this approach is to try to evaluate the importance of the fluctuations of the mean field and thereby to find when the system becomes unstable with respect to multifragmentation. In this model each nucleon is supposed to be represented by a sharp sphere of radius $\lambda = 1.03$ fm. This value corresponds to about the range of the nuclear forces. For a given configuration of a system of mass A, one draws at random the position of the A nucleons and look if they overlap. Some of them do. In this case the overlapping nucleons are supposed to restructure in a single sphere of radius $R = 1.15\, A_c^{1/3}$, where A_c is the mass of the cluster. After a first iteration one is left with a distribution of clusters and nucleons. Some of these particles may still overlap due to the restructuration. If it is the case, one iterates again the aggregation process and the restructuration until no overlap exists anymore. In the end one is faced with two situations : either there is a percolation cluster or not. In the first case the fluctuations of the mean field are small while they are large in the second case which ends by multifragmentation. This process, done for a particular sampling of the nucleons, is repeated several times in order to generate an ensemble average and study average properties.

The above restructured aggregation model is probably an improvement compared to lattice percolation models. However there are still some physical points of the real situation which are not well described. Among them one may quote the followings :
- Any cluster is restructured to a spherical one. However, if one gets a very filamented cluster it may rather break up in two or more pieces before a restructuration of each of them takes place.
- The model is a kind of black and white approach. If nucleons are close enough to overlap they fuse, otherwise not. Reality is very likely to be smoother and the probability of fusion is expected to vary continuously with the distance.

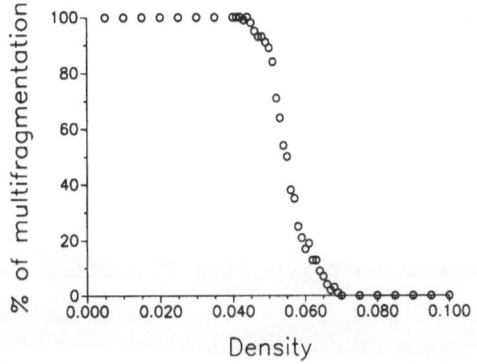

Figure 13. Percentage of multifragmentation (defined for each event when the largest cluster has a mass smaller than half the total mass) as a function of the density of the ^{208}Pb nucleus. From ref.[42].

In order to illustrate the kind of result that one can obtain with a restructured aggregation model, let us present some results of ref. 42 Fig. 12 shows the mass distribution of the products obtained for a ^{208}Pb nucleus at different densities ρ :
- There are two components in the mass distribution for $\rho=0.07$ fm^{-3}: one, peaked around mass 200, corresponds to the percolation cluster. The other component, at smaller mass values, contains the nucleons and the light mass clusters. This mass distribution is typical of a situation in which the fluctuations of the mean field are small.

- For $\rho = 0.04$ fm^{-3} the shape of the mass distribution is completely different. The percolation cluster has disappeared and it remains a component containing light and medium mass fragments only. In this case, the fluctuations of the mean field are large and the system disassembles.
- The two other examples $\rho = 0.06$ fm^{-3} and $\rho = 0.05$ fm^{-3}, are in the transition region where one goes from percolation to multifragmentation.

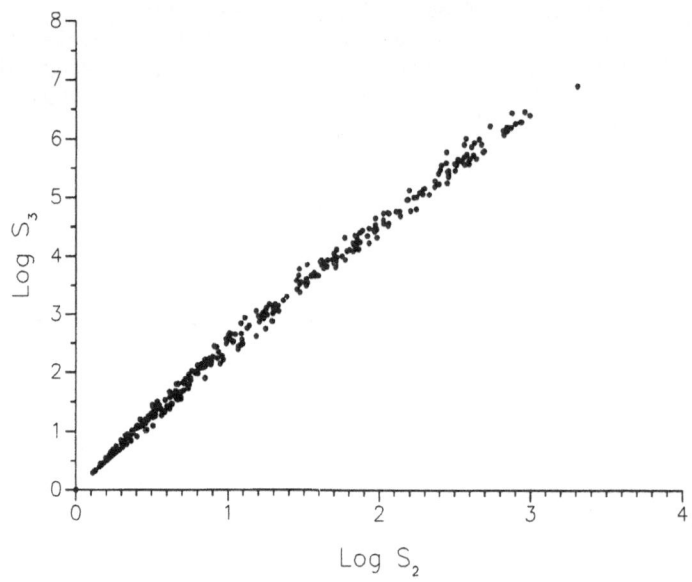

Figure 14. Correlation between Log S_2 and Log S_3 for the ^{208}Pb nucleus. from ref.[42].

Due to the small number of particles present in the system and to its finite size, a given event can lead to a percolation cluster or to multifragmentation with a probability which varies smoothly with the density of the system. For an infinite system the size of the percolation cluster is infinite. When it disappears the mass of the largest cluster drops suddenly. For nuclei, we do not have such an evolution but rather a continuous decrease of the size of the largest cluster. In order to estimate the influence of the finite size of the system it has been assumed in ref.42 that one has a percolation cluster if the size of the largest cluster is larger than half the initial mass of the system. Consequently, for a given event, the system undergoes multifragmentation only if the size of the largest cluster is smaller than half the initial mass of the system. This is arbitrary but one cannot avoid to introduce a convention (which of course could be chosen differently) due to the smoothness of the transition. The multifragmentation probability is deduced by averaging over many different events for each density. The results for the ^{208}Pb are shown in fig.13. Multifragmentation appears for density values between 0.06 and 0.07 fm^{-3} and the multifragmentation probability is equal to unity below $\rho = 0.04$ fm^{-3}. Similar results are obtained for the ^{40}Ca nucleus$_{42}$ but the transition is smoother due to the smaller number of particles.

It is interesting to perform a moment analysis as suggested by Campi and discussed already in section 2.3. The correlation between $\text{Log } S_2$ and $\text{Log } S_3$ is related to the critical exponent τ entering into the inclusive mass distribution of the fragments. It is shown in fig 14 for an ensemble of multifragmentation events at different densities. This correlation is almost linear and, from the slope of the average behaviour of $\text{Log } S_3$ as a function of $\text{Log } S_2$, one finds $\tau = 2.2$ which is very close to the value deduced from emulsion experiments ($\tau=2.2\pm0.1$)[9]. This value is also close to the one obtained in a lattice percolation model of Campi[43]. It is worth noting that the slope of the correlation does not change with the mass of the nucleus[42].

This moment analysis seems to indicate that the restructured aggregation model of ref.42 contains probably the necessary physical ingredients governing nuclear multifragmentation induced in heavy ions collisions.

4. DYNAMICAL THEORETICAL APPROACHES

4.1 Disassembly of a hot and compressed system

What happens to a hot and compressed nucleus? Does it breaks up or not? In this latter case under which conditions does it happens? Such questions are important issues to understand excited nuclei. In a first step of a dynamical theoretical understanding of hot nuclei it is interesting to study what happens to a hot and compressed nucleus in thermal equilibrium.

This question has been first attacked within the framework of mean field theories : Thomas Fermi[44], Landau Vlasov calculations[45], or time dependent Hartree Fock calculations in the case of compression only[46]. It is found that an initial compressed and hot nucleus first expands. Afterwards, its fate depends very much of the initial conditions. If it is not too much excited and heated, it oscillates back and forth in a large scale monopole oscillation. If it is more excited one may create a bubble nucleus (because the models usually keep a spherical symmetry to the system) which probably indicates that the system is unstable. However it is not possible to give the nature of the instability. At any rate these models guess that the system becomes unstable at excitations energy values which are substantially larger than those for which one looks at the fluctuations of the mean field.

If one restricts oneselves to the mean field approximation, one may nevertheless estimate at which point the system becomes unstable with respect to some collective degree of freedom by using the method of stability of dynamical systems. In its simplest form this has been used in ref.42 where a nucleus is described by a simple liquid drop. The monopole instability is investigated during the expansion of the system. Some of the results of the model of ref.42 are displayed in fig.15. There, the stability of the ^{208}Pb nucleus is investigated as a function of the initial temperature T_0. The minimum initial density ρ_0 above which the system becomes unstable is also calculated. The top left part of the fig.15 shows the evolution of the time t_f at which the instability takes place as a function of T_0. The bottom left part shows the temperature T_f

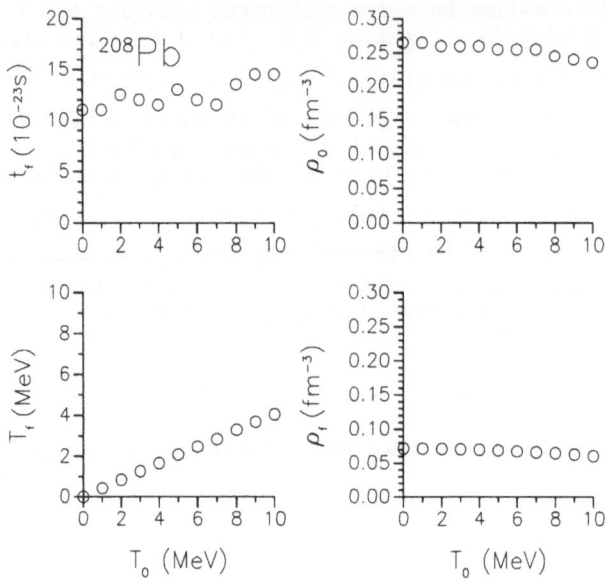

Figure 15. For different initial temperature values T_0 the following quantities are shown : ρ_0, the minimum initial density above which the system becomes dynamically unstable. ρ_f and T_f the density and temperature at the point of instability. t_f the needed for the system to reach this configuration. From ref.[42].

at the instability point. The top right and the bottom right part show respectively the minimum initial density and the density at the break up point as a function of T_0. One sees that t_f increases slightly with increasing temperature. The final temperature T_f increases regularly with T_0 but is always substantially smaller than the initial one. The critical initial density ρ_0 above which the instability takes place remains almost constant as a function of T_0. It is equal to about 0.26-0.27 nucleon/fm^3. The density at the instability point also remains essentially constant as a function of T_0. It is equal to about 0.07 nucleon/fm^3 (a little less than a factor of two smaller than the ground state density). These results show that the initial compression is a more important factor of instability than the initial temperature. This result agrees with those obtained previously in ref.40,41 based on an hydrodynamical expansion coupled to a 3-dimensional site-bond percolation on a cubic lattice.

Even if one can find the conditions under which a system is unstable with respect to multifragmentation, the next thing is to calculate how the fluctuations and instabilities grow up. This is a difficult problem which is presently under investigation[47] and Jacob Bondorf will give more details about it.

The dynamical instability of a hot and compressed nuclear liquid drop with respect to nucleation has been investigated by Schulz et al[48] who have looked under which conditions a gas-liquid phase transition is possible. However, they neglect the surface and Coulomb energies of the drop and their results should be taken as semiquantitative. This problem has also been very recently studied by Friedman[49] who follows the decay of a highly excited nucleus in thermal equilibrium using the Weisskopf detailed balance. He finds that most of the intermediate mass fragments are emitted at a temperature of about 5 MeV which essentially does not depend on the initial temperature of the nucleus. This phenomenon is interpreted as coming from a heating of the residual nucleus due to cluster formation (coalescence heating). This should happen in a low density nuclear medium when the binding energy of the system has decreased a lot.

The above models show that one has to take care of the fluctuations of the mean field at some point. A step further has been made by Nemeth et al[41] assuming that the initial nucleus is in statistical equilibrium and remains spherical during its evolution. These constraints are of course rather drastic but allow to study many of the features of a real situation. The hot and compressed nucleus expands and this process is described by a time dependent Thomas Fermi model[44]. This expansion is assumed to be isentropic, which is probably a very good approximation in regards to cascade calculations performed for heavy ion collisions at higher bombarding energies[50] or by molecular dynamical calculations[51]. The time dependent Thomas Fermi calculation corresponds to irrotational hydrodynamics in which the internal energy is evaluated using the Thomas Fermi approximation. Using a change of variable first introduced by Madelung[52], the hydrodynamical equations for neutrons and protons can be transformed into non linear Schrödinger equations which are solved numerically by methods developed for time dependent Hartree-Fock[53].

At each stage of the expansion, the magnitude of the fluctuations of the mean field are evaluated using a 3-dimensional site-bond percolation model on a cubic lattice[38]. The connection between hydrodynamics and percolation is made through the parameters p and q which are function of time as follows :

$$\rho(t) = \frac{\langle\rho\rangle}{\langle\rho_0\rangle} \qquad \text{and} \qquad q(t) = 1 - \frac{\varepsilon_T^*}{B(t=0)} \qquad (6)$$

where $\langle\rho\rangle$ ($\langle\rho_0\rangle$) are the average densities at time t (t=0) and B(t=0) is the binding energy at time t=0. As long as the fluctuations of the mean field remain small, the mean field approach used to describe the expansion process is perfectly suited. However, it can happen that this is no longer the case. Then, the nucleus breaks up into several pieces. The primary mass distribution of the fragments is obtained by means of the percolation model. The most important results of this model are the followings :
- Compressional energy is more efficient to break up nuclei than thermal energy. This is illustrated for example in fig.16 which shows the phase diagram associated to the Pb nucleus as a function of the ε_T and of ε_C (ε_T^* and ε_C^* are the thermal excitation energy per nucleon and the compressional excitation, respectively). If the system is not compressed

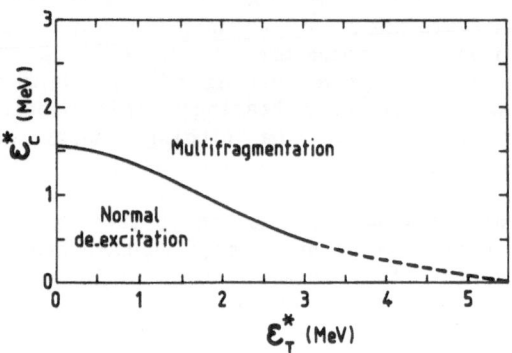

Figure 16. Multifragmentation and normal decay regions calculated for the ^{208}Pb nucleus as a function of ε_T^* and ε_c^* the thermal and compressional excitation per nucleon. From Nemeth et al[41].

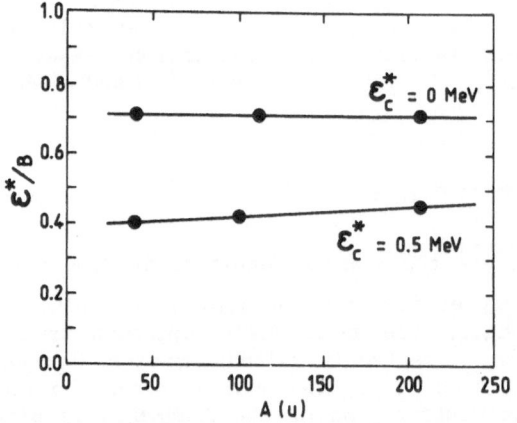

Figure 17. ε^*/B, where B is the binding energy per nucleon of the ground state nucleus and ε^* its total excitation energy per nucleon, is plotted as a function of A for $\varepsilon_c^*=0$ MeV and $\varepsilon_c^*=0.5$ MeV. From Nemeth et al[41].

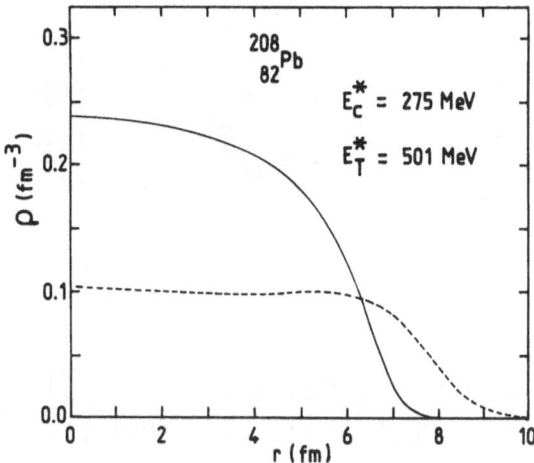

Figure 18. Initial density (full line) and density profile at the multifragmentation point (dashed curve) for typical initial conditions of the ^{208}Pb nucleus. From ref.[6].

(ε_c^*=0) it should have about 5.5 MeV/u of thermal excitation in order to break up. This quantity drops down to about 1 MeV/u of compressional excitation in the extreme case where there is no thermal excitation. Therefore, the more compressed the system, the lower the amount of excitation energy needed to get multifragmentation. This effect can be easily understood if one remembers that thermal excitation corresponds to desorganized energy while compression corresponds to coherent energy in the mode of instability. Another way to look at the results is displayed in fig.17 which shows ε /B (the amount of total excitation energy per nucleon (compressional plus thermal) divided by the binding energy per nucleon) required to break up a nucleus as a function of the mass number for nuclei along the beta stability line. If there is no compression about 70% of the binding energy are needed in order to break up the nucleus. This amount decreases to about 40% if the initial compressional energy is equal to 0.5 MeV/u. In fig.6 the results of this calculation are compared with the limit where fission following incomplete fusion disappears.

- At the point of break up it is interesting to note that the nucleus has not expanded too much. This is illustrated, for a typical example, in fig.18 which shows the initial density profile (full line) as well as the one at the point of multifragmentation instability (dashed line). The central density is still equal to 0.1 nucleon/fm^3 at the point of disassembly. The mean square radius has changed from 4.94 fm, for the initial value, to 6.2 fm at the point of instability. In other words, the volume of the nucleus has doubled. The temperature of the system, which was initially equal to 6 MeV has decreased down to 3.4 MeV at the multifragmentation configuration. This is due to the isentropic expansion. The time needed to go from the initial configuration to the disassembly point is very short : 1-1.5 10^{-22}s.

One may note that the kinetic energy of a multifragmentation product is of the order of the Coulomb repulsion between this fragment and the rest of the system. Actually it has three different contributions : one coming from the Coulomb repulsion, one arising from the thermal motion due to the heat of the system and finally one coming from its the collective expansion. Typically, the Coulomb kinetic energy of a fragment

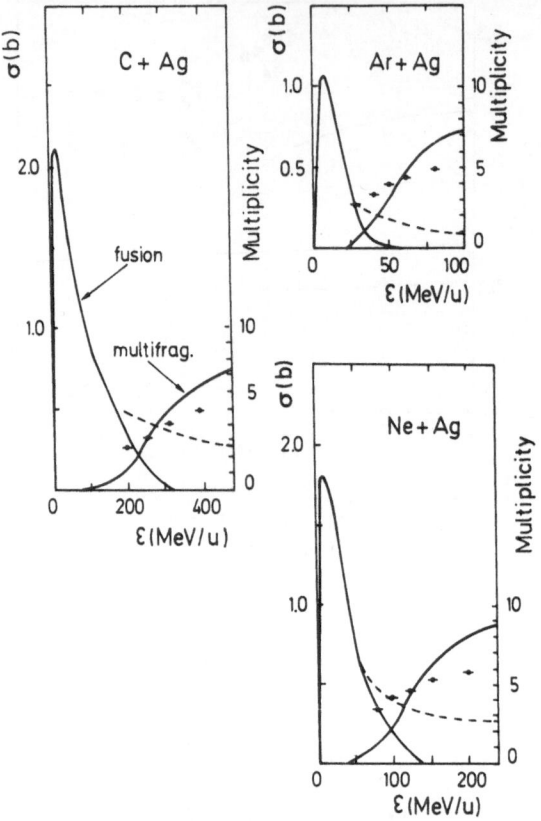

Figure 19. Incomplete fusion and multifragmentation excitation functions for different systems calculated by Cerruti et al[55]. The multiplicity of multifragmentation fragments of mass larger than 4 is also shown. From ref.[55].

emitted at distance r from the center of a nucleus of mass A atomic number Z and radius R is given by[54] :

$$E_{kin} = \frac{Z_f Z e^2}{R^3} r^2 \left[1 - \frac{A_f}{A} \right]^2 \qquad (7)$$

where Z_f and A_f are the atomic number and mass of the emitted fragment.

4.2 Multifragmentation excitation functions

One may use a simple preequilibrium model to describe part of the first phase of the collision between the two heavy ions. This allows to describe the thermalization of the system and the prompt emission of fast particle occurs simultaneously. After this first stage one is left with a hot and compressed system which is treated by the same method as

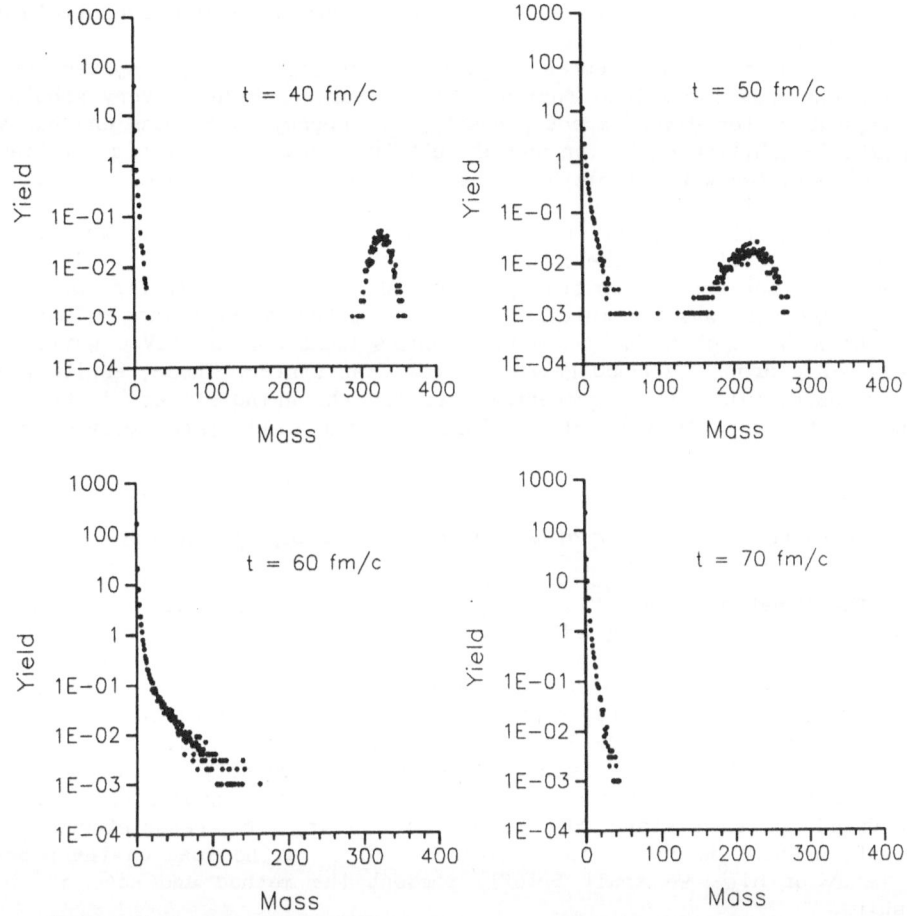

Figure 20. Mass distribution of the fragments obtained at different time steps in the 200 MeV/u Au+Au collision for zero impact parameter. From Leray et al[60].

described in the previous section. This is basically the physics contained in the model of Cerruti et al[7,55] which describes multifragmentation and in particular calculates the associated cross section.

The equilibration of the fused system is described by the preequilibrium model of Harp, Miller and Berne[56]. The projectile is assumed to be trapped in the potential well of the target. Because of that this approach was restricted to asymmetric systems in its first form but has been extended, meanwhile, to symmetric systems[57]. Due to two body collisions the system thermalizes while nucleons are emitted (preequilibrium emission). This emission of prompt particles leads to a substantial cooling of the system and only part of the initial available excitation energy goes into heat. After a time of the order of 10^{-22}s the remaining system has reached thermal equilibrium and expands. Its stability towards multifragmentation is then studied along the lines

described in the preceeding section. Two important results are obtained
with this model :
1- Because of the small number of particles contained in the system, the
transition from incomplete fusion to multifragmentation is very smooth.
Consequently, for given impact parameter and energy conditions one has a
certain probability of fusion and of multifragmentation. These quantities
can be evaluated with simple hypotheses and one can obtain the excitation
functions for incomplete fusion and multifragmentation.
2- The transition region from incomplete fusion to multifragmentation
depends critically upon the initial mass asymmetry of the system. This is
illustrated for example on the C, Ne and Ar + Au systems in fig.19. This
figure also displays the multiplicity of fragments of mass larger than 4.
For the Ar+Au system the transition occurs around 40-50 MeV/u while it
takes place around 200-300 MeV/u for the C+Au system. These results are
in agreement with the experimental data for the Ar+Au system[21,22] but no
experimental data is not yet available for the C+Au system between 200
and 300 MeV/u.

4.3 *Landau Vlasov kinetic equation and restructured aggregation*

The kinetic Landau Vlasov equation of refs.58,59 is rather successful to reproduce many of the experimental data related to one body
observables in a heavy ion collision at medium and high bombarding
energies. This approach cannot describe multifragmentation since it is
based on the mean field approximation only. However, it allows to rather
well reproduce the overall features of a heavy ion collision as long as
the density of the system is not too small. Consequently, it can be used
when the fluctuations of the mean field remain small. These latter can be
evaluated at each stage of the reaction using the restructured aggregation model described in section 3.4. This is the idea of the model
of Leray et al[60]. We shall briefly present the method and some of the
results.

At each time of the evolution of the colliding nuclei, the kinetic
equation gives the Wigner transform of the one body distribution function
of the A-nucleons system. Since the numerical method of ref.59 was used
to solve the Landau Vlasov equation, one nucleon is represented by N_g
gaussians in phase space. One draws at random A positions among $N_g \times A$
gaussian positions and assume that they are the positions of the nucleons
of radius r as defined in section 3.4. The restructured aggregation
process is then applied at each stage of the evolution of the system to
evaluate the magnitude of the fluctuations. If they are large, the Landau
Vlasov calculation is stopped and the system is assumed to break up. If
the fluctuations are small one continues to describe the evolution of the
system by the Landau Vlasov equation. A large number of samplings is done
to make an ensemble average.

This model is applied to the system ^{197}Au + ^{197}Au at 200 MeV/u which
has been measured in ref.16. Fig.20 presents, at different time steps of
a head on collision, the mass distribution of the clusters. At 40 and 50
fm/c most of the nucleons aggregate into a big fragment and there are
just some small mass clusters left indicating that the mean field
approximation is still valid. At 60 and 70 fm/c only small and
intermediate mass fragments are left indicating a complete fragmentation
of the system. The transition is smooth and occurs between 50 and 60
fm/c. This is due to the small number of particles in the system. Fig.21
shows the multifragmentation probability as a function of time for
different impact parameters, b. For 0<b<6 fm there is no difference in

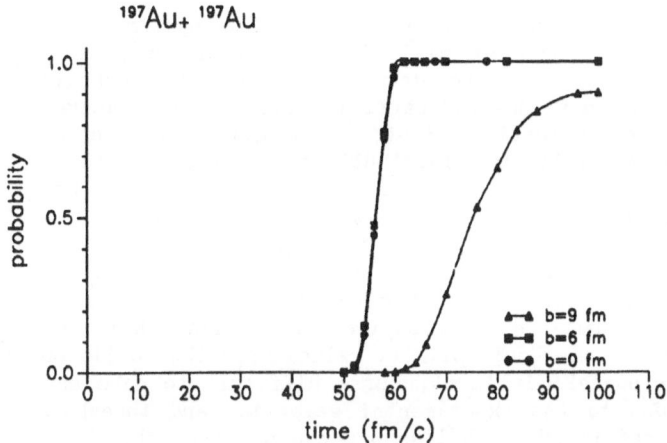

^{197}Au + ^{197}Au

Figure 21. Probability of multifragmentation as a function of time for the 200 MeV/u system at different impact parameters. From ref.[60].

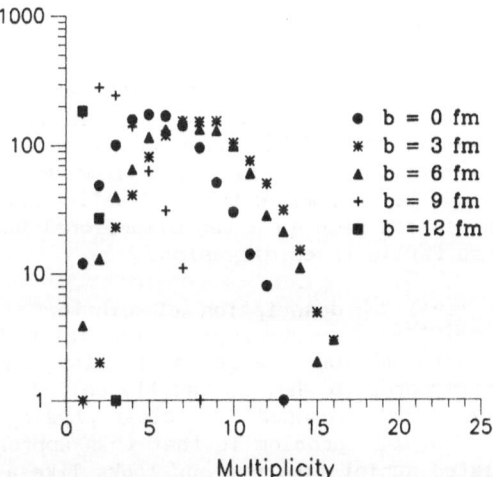

Figure 22. Calculated fragment multiplicity distribution as a function of the impact parameter for the 200 MeV/u Au+Au system at 60 fm/c. From ref.[60].

the curves, the transition occurs around 55 fm/c in less than 10 fm/c. At b=9 fm one never reaches a probability equal to one and the evolution is very slow. In fact, the mass distribution at b=9 fm shows that there is always a fragment of mass around 100 left associated to spectators in a participant-spectator picture. At b=12 fm there is no multifragmentation and the two incident nuclei reseparate without loosing there identity because the collision is too peripheral. The model indicates that multi-fragmentation occurs during the expansion phase when the density of the system has dropped down to values around 0.4 times the initial one. The same observation is made when one studies the onset of multifragmentation in a single expanding liquid drop[42].

It is interesting to come back to the system ^{197}Au+^{197}Au at 200 MeV/u measured by Doss et al[16]. In this experiment, intermediate mass fragments with $3 \leq Z \leq 10$ were detected in the forward center of mass hemisphere. The total charged particle multiplicity measured in a plastic wall gives an indication of the centrality of the collision. The experimental multiplicity distributions of intermediate mass fragments were already shown in fig.4 for different total charged particle multiplicity bins. The mass distributions, displayed in fig.22 for different impact parameters, are taken at a time of 60 fm/c when the system is unstable with respect to multifragmentation. The general trends of the data are reproduced : for central collisions (impact parameters between 0 and 6 fm) the distributions are similar while their maximum is shifted towards lower multiplicity values when the collision becomes more and more peripheral. It is of course difficult to make any quantitative comparison due to the experimental selection and thresholds present in the data of ref.16. Nevertheless one can say that the overall picture is there. The mean value of the multiplicity is also correctly reproduced by the model of Leray et al[60] if one notes that the experimental multiplicity has to be multiplied by two since the fragments were detected in half the space only.

4.4 Molecular dynamics

The formation of clusters is a phenomenon which occurs at low densities because of the short range attractive part of the nuclear force. Their formation is out of the scope of mean fields theories. Any model aiming to describe cluster formation at a microscopic level should take into account explicitly of the nucleon nucleon interaction as well as of the many body correlations. This is a tremendous quantum many body problem which cannot be solved without drastic approximations. Some quantum attempts have been made in a two dimensional world[61] but this has not been extended so far in three dimensions.

The simplest level of description of cluster formation considers that nucleons behave like classical particles which interact by a two body interaction with a long range attraction and a short range repulsion. This corresponds to what is usually called molecular dynamics. This method has been first proposed in nuclear physics by Wilets et al[62] and Bodmer et al[63]. The main problem is that this approach is not able to describe real isolated nuclei (the nucleus looks like a cristal). Several molecular dynamical models have been used since then[64,65]. They are quite useful for a gross description of a nuclear collision as far as the formation of clusters are concerned.

The most involved approach[65] takes care of some quantum effects like the Pauli principle and the Pauli blocking when two nucleons collide. Irreversibility is introduced through a stochastic scattering between the nucleons. Among other things the approach of ref.65 allows to rather well describe the multifragmentation of nuclear systems. However, the large computing time needed to solve the molecular dynamical equations prevent it to be used in a systematic way. Compared to the landau Vlasov approach, molecular dynamics allows to reproduce cluster formation. However it is certainly worse to treat the initial phase of the collision. One of the reason is that the Pauli principle is violated to some extent during the collision. Although a momentum interaction between nucleons is able to simulate to some extent the Pauli exclusion principle, some violations still exist. Several observables can be computed like the flow angle in heavy ion collisions at high bombarding energies. As far as multifragmentation is concerned, the model of ref.65 is able to calculate the mass distribution of the fragments. This is illustrated for example in fig.23 which shows the result of a simulation for the 800 MeV/u Ne+Au system. The calculated mass distribution is in good agreement with the experimental one. More details about this approach are given in the lectures of Horst Stöcker.

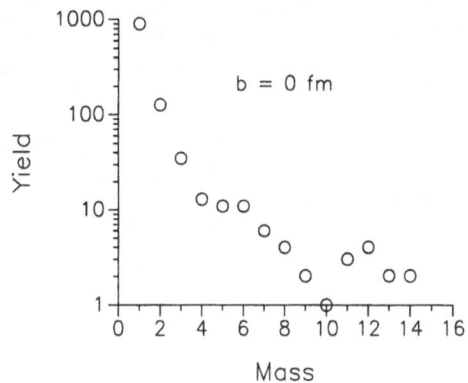

Figure 23. Calculated mass yield in a central collision for the Ne+Au system at 800 MeV/u. From Ref. [66].

5. CONCLUSION

It is interesting to study the multifragmentation of nuclei because it is a new decay mode of highly excited nuclei. This was not foreseen before the first experiments and one thought that a very excited nucleus would rather boil off. Even if the existence of multifragmentation is already quite interesting in itself, the next question is to study under which conditions it can be observed. We have seen that it can occur in proton as well as in heavy ion induced collisions provided enough excitation energy is brought into the system. The critical excitation energy above which multifragmentation takes place has been estimated to be of the order of several MeV per nucleon. Systematic studies are now needed to better precise this limit. In particular it is interesting to investigate the special influence of the compression by comparing proton and heavy ion induced multifragmentation. Since it is a process in which several fragments are obtained in the exit channel, the question is to characterize a multifragmentation event. In particular it is important to know what are the important characteristics to be considered.

As far as theoretical investigations are concerned, the situation seems to be better than for the experiments. One now understands semi quantitatively what happens and how nuclei break up. A microscopic theory is still missing. Two points can probably be improved in the near future: first, one has to better understand the growth of fluctuations. Second, one has to improve molecular dynamical approaches in order to better satisfy basic laws like the Pauli exclusion principle for example. May be one solution would be to use a mean field simulation approach, like the Landau Vlasov equation, for the first phase of the collision and then switch to a molecular dynamical approach when the density of the system becomes low. The first phase would then be better simulated and one could then describe rather well the formation of clusters in the later stages of the reaction. This is the philosophy of ref.60 in which the restructured aggregation model represents somehow the poor man molecular dynamical model.

Finally, it is worth to point out that multifragmentation could be interpreted as a critical phenomenon occuring in a microsystem. Compared to macroscopic systems, everything is smoother and nuclear physics is one of the few domains where such processes could be studied. The existence of long range forces (the Coulomb forces), together with the small number of particles contained in the system, leads to differences with macroscopic systems which are worth to be studied. Within a percolation interpretation the connection with critical processes occuring in other fields of physics, chemistry, biology is obvious. It is also interesting to note that these theoretical methods can also be applied to other fields than science. For these reasons the investigations of nuclear multifragmentation belongs to the broad area of critical phenomena studies and not only to nuclear physics.

ACKNOWLEDGEMENTS

A large part of the results presented in these lectures has been obtained in a fruitful collaboration with several physicists belonging to different institutes. I would like to thank them a lot for the pleasure I had in this collaboration which has, hopefully, not ended by multifragmentation. In particular I would like to especially acknowledge : Manuel Barranco, Rodolphe Boisgard, Christian Cerruti, Jean Desbois, Sylvie Leray, Judith Nemeth, Hélène Ngô and Maria Elena Spina for larger interactions. Finally, I thank Drs. J.Amundson, P.Freier and C.J Waddington for providing me with their emulsion data from which figure 7 could be drawn.

REFERENCES

[1] For a review see for example :
M.Lefort and C.Ngô, Ann., Phys. (Paris) 3 (1978) 5
W.U.Schröder and J.R.Huizenga, treatise on heavy-ion Science (ed. D.A. Bromley), Vol.2 (1984).
[2] B.Müller, Lectures Notes in Physics, 225 (1985)
J.Rafelski, Phys. Rep. 88 (1982) 331 and refs. therein.
[3] J.Hüfner, Phys. Rep. 125 (1985) 129; J. Phys. C4 (1986) 3.
C.Ngô Nucl. Phys. A488 (1988)233c.
C.Ngô International Summer School on Heavy Ion Physics, August 1988, Lanzhou, China and refs. therein.
[4] P.Grangé,Li Jun-Qing and H.A.Weidenmüller, Phys. Rev. C27 (1983) 2063
H.Delagrange, C.Grégoire, F.Scheuter and Y.Abe, Z. Phys. A323 (1986) 437.

[5] W.A.Friedman and W.G.Lynch, Phys. rev C28 (1983) 16.

[6] C.Ngô, R.Boisgard, J.Desbois, J.Nemeth, M.Barranco and J.F.Mathiot, Nucl. Phys. A471 (1987) 381c.

[7] C.Cerruti, J.Desbois, R.Boisgard, C.Ngô, J.Natowitz and J.Nemeth, Nucl. Phys. A476 (1988) 74.

[8] B.Jacoksson, G.Jönsson, B.Lindkvistand and A.Oskarsson, Z. Phys. A307 (1982) 293

B.Jakobsson, G.Jönsson, L.Karlsson, N.Morén, K.Söderström, E.Monnand and F.Schussler, J. Phys. C4 (1986) 369

B.Jakobsson, G.Jönsson, L.Karlsson, V.Kopljar, N.Morén, K.Söderström, E.Monnand, F.Schussler, H.Nifenecker, G.Fai, J.P.Bondorf, K.Sneppen and X.Campi, preprint (1988).

[9] C.J.Waddington and P.S.Freier, Phys. Rev. C31 (1985) 888.

[10] A.S.Hirsch, A.Bujak, J.E.Finn, L.J.Gutay, R.W.Minich, N.T.Porile, R.Scharenberg, B.C.Stringfellow and F.Turkot, Phys. Rev. C29 (1984) 508.

[11] N.T.Porile, A.T.Bujak, D.D.Carmony, Y.H.Chung, L.J.Gutay, A.S. M.Mahi, G.L.Paderewski, T.C.Sangster, R.P.Scharenberg and B.C. Stringfellow, Nucl. Phys. A471 (1987) 149c.

[12] T.J.Schlagel and V.R.Pandharipande, Phys. Rev. C36 (1987) 162.

[13] A.D.Panagiotou, M.W.Curtin, H.Toki, D.K.Scott and P.J.Siemens, Phys. Rev. Lett. 52 (1984) 496 and refs. therein.

A.D.Panagiotou, M.W.Curtin and D.K.Scott, Phys. rev. C31 (1985) 55.

[14] R.J.Charity, D.R.Bowman, Z.H.Liu, R.J.McDonald, M.A.McMahan, G.J. Wozniak, L.G.Moretto, S.Bradley, W.L.Kehoe and A.C.Mignerey, Nucl. Phys. A476 (1988) 516.

[15] G.Rudolf, J. Phys. C4 (1986) 351 and ref. therein.

[16] K.G.Doss, H.Gustafsson, H.Gutbrod, J.Harris, B.V.Jacak, K.H.Kampert, B.Kolb, A.Poskanzer, H.Ritter, H.R.Schmidt, L.Teitelbaum, M.Tincknell, S.Weiss and H.Wieman, Phys. Rev. Lett. 24 (1987) 2720

B.V.Jacak, Nucl. Phys. A488 (1988) 325.

[17] B.Berthier, R.Boisgard, J.Julien, J.M.Hisleur, R.Lucas, C.Mazur, C. Ngô and M.Ribrag, Phys. Lett. 193B (1987) 417.

[18] R.Boisgard, E.Bellini, B.Berthier, P.Bouissou, J.Julien, J.M. Hisleur, R.Lucas, C.Mazur, C.Ngô and M.Ribrag, work in progress.

[19] S.Leray, J. Phys. C4 (1986) 275 and refs. therein.[19]

[20] T.Sikkeland, phys. Lett. 27B (1968) 277.

[21] S.Leray, G.Nebbia, C.Grégoire, G.La Rana, P.L'Hénoret, C.Mazur, C.Ngô, M.Ribrag, E.Tomasi, S.Chiodelli, J.L.Charvet and C.Lebrun, Nucl. Phys. A425 (1984) 345 .

S.Leray, O.Granier, C.Ngô, E.Tomasi, C.Cerruti, P.L'Hénoret, R.Lucas, C.Mazur, M.Ribrag, J.L.Charvet, C.Humeau, J.P.Lochard, M.Morjean, Y.Patin, D.Guinet, L.Vagneron and A.Peghaire, Z. Phys. A320 (1985) 533.

[22] M.Conjeaud, S.Harar, M.Mostefai, E.C.Pollacco, C.Volant, Y. Cassagnou, R.Dayras, R.Legrain, H.Oeschler and F.Saint Laurent, Phys. Lett. 159B (1985) 244.

[23] J.Galin, H.Oeschler, S.Song, B.Borderie, M.F.Rivet, I.Forest, R. Bimbot, D.Gardès, B.Gatty, H.Guillemot, M.Lefort, B.Tamain and X.Tarrago, Phys. Rev. Lett. 48 (1982) 1787.

[24] K.Alkelett, W.Loveland, L.Sihver, Z.Xu, C.Casey, D.J.Morissey, J.O.Liljenzin, M.De Saint Simon and G.T.Seaborg, 3rd Int. Conf. on Nucleus Nucleus Collisions, Saint Malo, Contributed paper p.195

J.Dreute and W.Heinrich, 3rd Int. Conf. on Nucleus Nucleus Collisions, Saint Malo, Contributed paper p.89.

[25] X.Campi, J. Phys.A : Math. Gen. 19 (1986) 1917.

[26] E.Bedermann, J.Lühning, U.Lynen, U.Milkau, K.Müller, H.Sann, H. Stelzer, W.Trautmann, P.Kreutz, A.Kühmichel, C.Pinkenburg, J.Pochodzalla, J.Friese, H.J.Körner, W.Wagner, Z.Zeitelhack, L.Moretto, W.F.J.Müller, G.Wozniak, G.Imme, G.Raciti, I.Iori, A.Moroni, R.Boisgard, P.Bouissou, C.Cerruti, C.Ngô, S.Leray, G.C.Adloff, B.Bilwes, R.Bilwes, M.Michel, C.Massé, G.Rudolf, F.Scheibling and L.Stuttge, GSI report ISSN 0171-4546 (1988).

[27] J.Bondorf, contribution to this school.

[28] R.G.Palmer and P.W.Anderson, Phys. rev. D9 (1974) 3281
W.G.Kupper, G.Wegmann and E.R.Hilf, Ann. Phys. (N.Y.) 88 (1974) 454
G.Sauer, H.Sandra and U.Mosel, Nucl. Phys. A264 (1976) 221. M.Barranco
and J.R.Buchler, Phys. Rev. C22 (1980) 1729.

[29] G.Bertsch and P.J.Siemens, Phys. Lett. 126B (1983) 9
J.A.Lopez and P.J.Siemens, Nucl. Phys. A431 (1984) 728 and refs. therein
B.A.Li, S.Pratt and P.J.Siemens, Phys. rev. C37 (1988) 1473.

[30] H.R.Jaqaman, A.Z.Mekjian and L.Zamick, Phys. Rev. C29 (1984) 2067
A.L.Goodman, J.I.Kapusta and A.Z.Mekjian, Phys. Rev. C30 (1984) 851.

[31] D.H.E.Gross, Zhang Xiao-Ze and Xu Shu-Yan, Phys. Rev. Lett. 56
(1986) 1544.
Zhang Xiao-Ze, D.H.E Gross, Xu Shu-Yan and Zheng Yu-Ming, Nucl. Phys.
A461 (1987) 641 and Nucl. Phys. A461 (1987) 668.
D.H.E.Gross and Zhang Xiao-ze, Phys. Lett. 161B (1985) 43.

[32] Yu-Ming Zheng, H.Massmann, Shu-Yan Xu, D.H.E.Gross, Xiao-Ze Zhang,
Zhao-Qi Lu and Ben-Hao Sa, Phys. Lett. 194B (1987) 183.

[33] J.P.Bondorf, R.Donangelo, I.N.Mishustin, C.J.Pethick, H.Schülz and
K.Sneppen, Nucl. Phys. A443 (1985) 321.
J.Bondorf, R.Donangelo, I.N.Mishustin and H.Schülz, Nucl. Phys. A444
(1985) 460.

[34] S.E.Koonin and J.Randrup, Nucl. Phys. A356 (1981) 223.
G.Fai and J.Randrup, Nucl. Phys. A381 (1982) 557.

[35] J.Aichelin and J.Hüfner, Phys. Lett. 136B (1984) 15
J.Aichelin, J.Hüfner and R.Ibarra, Phys. Rev.C30 (1984) 107.

[36] D.Stauffer, Introduction to percolation theory (London, Taylor &
Francis, 1985).

[37] X.Campi and J.Desbois, Proc.23rd Int. Winter Meeting on nuclear
physics, Bormio (1985) p.497.
W.Bauer, D.R.Dean, U.Mosel and U.Post, Phys. Lett. 150B (1985) 53.
W.Bauer, U.Post, D.R.Dean and U.Mosel, Nucl. Phys. A452 (1986) 699.
O.Knospe, R.Schmidt and H.Schulz, Phys. Lett. 182B (1986) 293.

[38] J.Desbois, Nucl. Phys. A466 (1987) 724.

[39] R.Boisgard, J.Desbois, J.F.Mathiot and C.Ngô, Nucl. Phys. (in
press).

[40] J.Desbois, R.Boisgard, C.Ngô and J.Nemeth, Z. Phys. A328 (1987) 101.

[41] J.Nemeth, M.Barranco, J.Desbois and C.Ngô, Z. Phys. A325 (1986) 347.

[42] C.Ngô, H.Ngô, S.Leray and M.E.Spina, preprint.

[43] X.Campi, Phys. Lett. B208 (1988) 351.
X.Campi, 107 Course of Int. School of Physics "Enrico Fermi" on the
Chemical Physics of Atomic and Molecular Clusters. Varenna (1988).

[44] J.Nemeth, M.Barranco, C.Ngô and E.Tomasi, Z.Phys. A323 (1986) 419.

[45] C.Gregoire, contribution to this school.

[46] A.Dhar and S.Das Gupta, Phys. rev. C30 (1984) 1545
H.Sagawa and G.F.Bertsch, Phys. Lett. 155B (1985) 11.

[47]C.J.Pethick and D.G.Ravenhall, Nucl. Phys. A471 (1987) 19c.

[48] H.Schulz, B.Kämpfer, H.W.Barz, G.Röpke and J.Bondorf, Phys. Lett.
147B (1984) 17.

[49] W.A.Friedman, Phys. Rev. Lett. 60 (1988) 2125.

[50] G.Bertsch and J.Cugnon, Phys. Rev. C24 (1981) 2514.

[51] D.Hahn and H.Stöcker Nucl. Phys. A476 (1988) 718.

[52] E.Madelung, Z. Phys. 40 (1926) 322.

[53] P.Bonche, S.Koonin and J.W.Negele, Phys. rev. C13 (1976) 1226.

[54] K.C.Chung, R.Donangelo and K.Schechter, Phys. Rev. C36 (1987) 986.

[55] C.Cerruti, R.Boisgard, C.Ngô and J.Desbois, Nucl. Phys. A (in press)

[56] G.D.Harp, J.M.Miller and B.J.Berne, Phys. Rev. 165 (1968) 1166
G.D.Harp and J.M.Miller, Phys. Rev. C3 (1971) 1847.

[57] O.Granier, R.Boisgard, C.Cerruti and J.Desbois, Bormio (1989).

[58] J.J.Molitoris, D.Hahn and H.Stöcker, Prog. in Particle and Nuclear
Physics, Vol.15 (1985) 239.
J.Aichelin, J. Phys.C4 (1986) 63 and refs. therein.

J.J.Molitoris, D.Hahn and H.Stöcker, Prog. in Particle and Nuclear Physics, Vol.15 (1985) 239.

[59] C.Grégoire, B.Remaud, F.Scheuter and F.Sébille, Nucl. Phys. A436 (1985) 365.

[60] S.Leray, C.Ngô, M.E.Spina, B.Remaud, F.Sébille and C.Grégoire, 5th Int. Conf. on Clustering aspects in Nuclear and Subnuclear Systems, Kyoto (July 1988).

S.Leray, C.Ngô, M.E.Spina, B.Remaud, F.Sebille Int. Workshop on Nuclear Dynamics at Medium and High Energies, Bad Honnef (1988). To appear in Nucl. Phys. and Bormio (1989).

[61] J.Knoll and B.Strack, Phys. Lett. 149B (1984) 45.

[62] L.Wilets, E.M.Henley, M.Kraft and A.D.Mackellar, Nucl. Phys. A282 (1977) 341.

L.Wilets, Y.Yariv and R.Chestnut, Nucl. Phys. A301 (1978) 359

[63] A.R.Bodmer, C.N.Panos and A.D.MacKellar, Phys. Rev. C22 (1980) 1025.

[64] G.E.Beauvais, D.Boal and J.C.K.Wong, Phys. Rev. C35 (1987) 545

G.E.Beauvais, D.H.Boal and J.Glosli, Nucl. Phys. A471 (1987) 427c.

S.Das Gupta, Nucl. Phys. A471 (1987) 417c.

D.H.Boal and J.Glosli, Phys. Rev. C37 (1988) 91.

[65] H.Stöcker, contribution to this school.

J.Aichelin and H.Stöcker, Phys. Lett. 176B (1986) 14

G.Peilert, A.Rosenhauer, H.Stöcker, W.Greiner and J.Aichelin, Mod. Phys. Lett. 3 (1988) 459

D.Hahn and H.Stöcker, Phys. Rev. C37 (1988) 1048

Stöcker, Int. Workshop on Nuclear Dynamics at Medium and High Energies, Bad Honnef (1988). To appear in Nucl. Phys.

[66] A.Rosenhauer, J.Aichelin, H.Stöcker and W.Greiner, J. Phys. C4 (1986) 395 and refs. therein.

[67] J.Randrup, Int. Workshop on Nuclear Dynamics at Medium and High Energies, Bad Honnef (1988). To appear in Nucl. Phys.

NUCLEAR MATTER AND FRAGMENTING NUCLEI

J. P. Bondorf and K. Sneppen

The Niels Bohr Institute, University of Copenhagen
Blegdamsvej 17, DK-2100 Copenhagen Ø, Denmark

Abstract

In these lectures we approach the problem of extracting information on properties of nuclear matter through violent collisions between finite nuclei. In such processes the nuclei will undergo fragmentation into many pieces. We discuss a model of statistical multifragmentation.

1. Nuclear Matter and Finite Nuclei

Introduction

The substance in the atomic nucleus is called *nuclear matter*. There are good reasons for this. First of all the nucleus in its ground state is composed of the same constituents, *nucleons*. Secondly the matter is saturating which means that the addition of one nucleon is associated with approximately the same energy, independent of the size of the nucleus. For a long time the study of this matter was limited to the properties near the ground state. Only very few parameters could be determined, the most important ones being the average density $\rho \approx 0.16 nucleons/fm^3$ and the binding energy (apart from surface and Coulomb energy) $\approx 16 MeV/nucleon$. The compressibility could only be determined with big uncertainty from properties of density vibrations. Nuclear matter probably exists in large quantities on a completely different scale, in the neutron stars which are the end products of stellar evolution. The matter in the neutron stars is only accessible by indirect observations such as abundances and radiation of the pulsars. Also in the big bang, extended nuclear matter must have been widespread during a certain period. The study of this form of nuclear matter can be made from the composition and distribution of matter in the present state of the universe.

When trying to study nuclear matter in the laboratory one encounters immediately a lot of problems. First of all we can only work with very small systems. The biggest possible "piece" of nuclear matter can be made from colliding the two biggest nuclei, giving the maximum size of a nuclear system around 500 nucleons. There are other problems. We can only influence the state of the matter considerably by means of collisions between nuclei. Therefore there is usually a lot of excess heat, and that prevents us from reaching certain parts of the phase space. Another problem is that the lifetime of the colliding system is very short. This means that the use of conventional thermodynamic concepts of equilibrium states is often problematic.

In these lectures we shall try to pave parts of the road from the signatures of experiments to properties of nuclear matter. This road is usually very long, and it has plenty of pitfalls and blind sideroads. For some of the questions which we may ask there is maybe no way at all. In this section we give a short introduction to the equation of state for infinite nuclear matter. We then proceed to a discussion of the properties of finite systems, and discuss the effect of having a finite number of constituent particles. In section 2 we first discuss the dynamics of heavy ion collisions in which instability processes just prior to the fragmentation breakup show up and develop. Then we introduce the model for the statistical break up of the nuclei into many fragments. Features and limitations of the model are discussed, besides some calculated results.

Equation of State of Nuclear Matter

There exists a lot of theoretical work on nuclear matter, and we refer to the other series of lectures in this school. In this section we shall only remind the reader about some elementary properties of the matter and we discuss only a simple representation of an equation of state for not too hot nuclear matter. The basic thermodynamics can f. ex. be studied in the book of Landau and Lifshitz [1]. We shall often, when presenting the concepts and the equations, refer to the equivalent expressions for the ideal gas (Boltzmann gas) since this kind of matter should be known by everyone.

Nucleons obey Fermi statistics, and each nucleon with mass m and energy ϵ_k has the probability of occupying a quantum state

$$n_k = \frac{1}{1 + \exp\left(\frac{\epsilon_k - \mu}{T}\right)} \tag{1}$$

$$\epsilon_k = \frac{\hbar^2 k^2}{2m}$$

In eq. (1), and in the following the Boltzmann constant has been put equal to 1, and the temperature T and the energy have the same unit. From eq. (1) we can calculate the total number of particles N in a volume V, the total kinetic energy E_{kin}, and the average kinetic energy per nucleon e_{kin} as integrals over **k**- space

$$N = Vg \int n_k d^3 k \tag{2}$$

This equation defines the chemical potential μ. For T=0 the value of μ is called the Fermi energy ϵ_F.

$$E_{kin} = Vg \int n_k \epsilon_k d^3k \equiv N e_{kin} \tag{3}$$

The number g is the degeneracy factor for the fermions. For nucleons, $g = 2 \cdot 2 = 4$ because there are two spin states and two isospin states.

It is convenient to use the variables, temperature T and particle number density $\rho = N/V$, for the nuclear matter, and we shall do so in the following. From equations (1)-(3) one can find the chemical potential μ and the kinetic energy per nucleon e_{kin} as functions of these variables. One finds the expressions (4) for $T \ll e_F$ and (5) for bigger values of T.

$$e_{kin} = \frac{3}{5}\epsilon_F(\rho) + \frac{\pi^2}{4}\frac{T^2}{\epsilon_F(\rho)} \tag{4}$$

$$e_{kin} = \frac{3}{2}T \sum_{j=1}^{\infty} C_j \left(\frac{\rho \lambda_T^3}{g}\right)^{j-1} \tag{5}$$

where $C_1 = 1; C_2 = 0.177; C_3 = -0.0033;...;\lambda_T = \left(\frac{2\pi\hbar^2}{mT}\right)^{\frac{1}{2}}$ and $\epsilon_F = \frac{\hbar^2}{2m}\left(\frac{6\pi^2}{g}\rho\right)^{\frac{2}{3}}$. For $\rho = \rho_0 \approx 0.16 fm^{-3}$ the Fermi energy corresponds to the wave number $k_F \approx 1.36 fm^{-1}$. The average kinetic energy per particle for the ideal Boltzmann gas is $e_{kin} = \frac{3}{2}T$, which is independent of ρ.

We now go to the equation of state. For the ideal gas this equation is often expressed just as Boyle Mariottes law $PV = NT$, or, in our preferred variables, $P = \rho T$, where P is the pressure of the matter. We choose the unit for ρ as fm^{-3}). Similarly we want to express the pressure P (units MeV/fm^3) for the Fermi gas as function of ρ and T. We use the general law which is valid for a non interacting gas of either Fermi, Bose or Boltzmann statistics:

$$P = \frac{2}{3}\rho e_{kin} \tag{6}$$

By inserting (4) or (5) into (6) one can now get the desired equation of state $P = P(\rho, T)$ for the noninteracting Fermi gas. Another way of specifying the state of the matter is just by using the energy expressions (4) or (5).

Only very thin nucleonic gas obeys the expressions above, and they cannot at all be used for nuclear matter with normal density. This is because nucleons interact strongly through the N-N force. It looks much like the conventional van der Waals force for condensed matter. The force acts with repulsion at short distances below 0.5 fm, with attraction around the average N-N distance at normal density, and decreasing rapidly towards zero at larger distances. In reality the N-N force is very complicated, as it depends on relative spin and isospin states, relative momentum besides on multiparticle correlations. The theories of nuclear matter aim at finding f. ex. the energy or the pressure from the force as functions of the matter state variables. We shall not at all try to do so here but just use a parametrization of an average potential energy for a particle in the matter, as function of the variables ρ and

T. We notice, that when the density is high, the average nucleonic distances are small, and in that case the potential energy should be large and repulsive. For intermediate density around the normal nuclear density ρ_0, the potential energy should be negative (attractive), and when ρ approaches 0 the matter should behave as the noninteracting gas discussed above.

A convenient expression for the potential energy which has the properties mentioned above was introduced by Zamick [2], based on the Skyrme interaction:

$$e_{pot}(\rho) = -\frac{1}{2}d_1\frac{\rho}{\rho_0} + \frac{6}{13}d_2\left(\frac{\rho}{\rho_0}\right)^{7/6} \tag{7}$$

In (7) the constants are chosen to fit nuclear saturation properties: $d_1 = 368.3 MeV$ and $d_2 = 315.2 MeV$. The expression (7) depends only on the density. If momentum dependent forces were taken into account then a temperature dependence could not be neglected. The only temperature dependence which is left in the matter is then the one from the kinetic energy (4) or (5). We now have the resulting *average energy per nucleon* for our simple "nuclear matter",

$$e(\rho, T) = e_{kin}(\rho, T) + e_{pot}(\rho) \tag{8}$$

If we want the *pressure* in this case we cannot use (6) for the potential energy part of (8). Another definition which is more convenient here, is

$$P = \rho^2 \frac{\partial f(\rho, T)}{\partial \rho}\bigg|_T \tag{9}$$

where

$$f(\rho, T) = e(\rho, T) - s(\rho, T)T \tag{10}$$

is the free energy per particle. The quantity $s(\rho, T)$ is the entropy per particle (specific entropy). Since e_{pot} in (8) is temperature independent, the whole potential energy is a free energy. The kinetic energy part of (9) leads to (6), and therefore we have the pressure

$$P = \rho^2 \frac{de_{pot}}{d\rho} + \frac{2}{3}\rho e_{kin} \tag{11}$$

$$P = \rho\left[-\frac{1}{2}d_1\frac{\rho}{\rho_0} + \frac{7}{13}d_2\left(\frac{\rho}{\rho_0}\right)^{7/6}\right] + \frac{2}{3}\rho e_{kin} \tag{12}$$

In figs. 1a and 1b we show curves of the energy and the pressure of the nuclear matter in the above parametrization. One notices that the pressure is zero for the ground state condition $\rho = \rho_0, T = 0$. For temperatures less than a certain value, T_{crit}, the pressure slope $\frac{\partial P}{\partial \rho}|_T$ can be negative. This means that the matter tends to

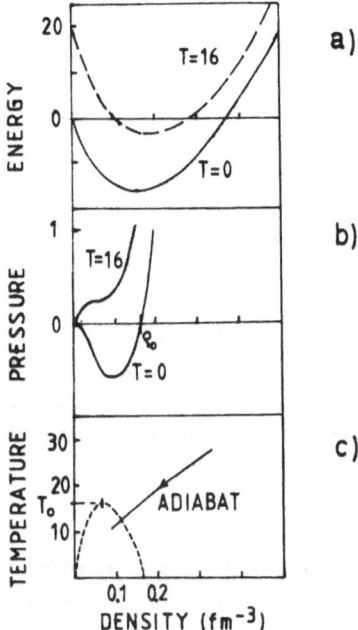

Fig.1 Parametrization of nuclear matter equation of state. Energy (8) in a) and pressure (11) in b) as functions of the density for two different temperatures. Phase diagram (T, ρ) in c). Units: energy and temperature in MeV, pressure in MeV/fm^3.

separate in a liquid and a gas phase. In fig. 1c we show the phase diagram for the matter in the (ρ, T) plane. For a discussion of the liquid gas composition of expanded nuclear matter see Siemens [3].

The matter also exhibits compressibility, since the pressure changes with density. One defines the isothermal compressibility module as

$$K = 9\frac{\partial P}{\partial \rho}\bigg|_T \tag{13}$$

For $T = 0$ one finds $K \approx 200 MeV$ for the applied parametrization of the equation of state.

In this section we have presented a very simple example of an equation of state for infinite nuclear matter. Despite the apparent accuracy of the constants in (7) the expression basically only fits one point, the conditions near the ground state. All predictions made on the basis of the example will therefore be very crude.

In the example above, and in the remaining part of this paper we only consider relatively low excitation for which it is sufficient to use the nucleon degrees of freedom. For higher excitation energy, however, the intrinsic degrees of freedom of the nucleons start to play a role. When the temperature reaches approximately the pion mass, it will no longer rise appreciably with energy. This is because of the formation of many pions, which effectively cool the matter (Hagedorn temperature). For much higher excitation energy and density, there is a chance that the nucleonic structure will be dissolved completely. The dominant degrees of freedom of the matter will then be the quarks and gluons. A lot of experimental and theoretical efforts are in these years devoted to investigations of such very hot matter through relativistic collisions between heavy nuclei.

Finite Nuclei

The infinite nuclear matter is mainly fiction. For the study of nuclear matter in the laboratory, we have to use finite nuclei as the source of information. The most versatile way to bring the matter in the nuclei away from the conditions in the ground state is through collisions between two nuclei. In these lectures we think of reactions which can change the nuclear temperature up to several tens of MeV, and the density by a few times. This means reactions with beam energy up to a couple of hundred MeV per nucleon. The lower limit of the reactions which are of interest here is around 20 MeV per nucleon. This is close to the limit below which the reacting nuclei remain approximately as they were before the collision or fuse to a conventional compound nucleus. We are, unfortunately, not able to decide in advance whether a collision event will have a big impact parameter (a grazing collision) or a small one (a near central collision). For the nuclear matter studies there are several reasons to prefer the more central collisions. Firstly, the deposited energy in the combined system is maximum, secondly there is a chance that the energy is deposited in a large amount of matter and maybe also more evenly spread over the constituents. There are in principle no big problems of selecting the central events experimentally. That can be done by demanding a combination of high multiplicity of reaction products and a momentum and fragment distribution which within certain statistical limits exhibits azimuthal symmetry. This is, however, expensive because it requires 4π detectors.

The major course of a collision event proceeds within a time interval which is of the order of $(R_1 + R_2)/v_{rel})$, the sum of the radii of the two nuclei divided by the relative velocity before the collision. This time $\approx 10^{-22} - 10^{-21}$ sec is only little longer than the single particle time in nuclei, and therefore there is usually not time enough to make proper equilibrium among the constituents.

A collision event can be characterized by various phases. First the *encounter*, when the two ions meet, and the first nucleons from either side start to penetrate the other nucleus. Some of these rapid nucleons may move relatively undisturbed through or around the other nucleus and escape. These *promptly emitted particles* (PEPs) do not share their energy with the total system, and they therefore cause a certain early energy drain in the process. After the encounter the nuclei *fuse temporarily* as they receive large mutual longitudinal acceleration, and high pressures build up. In this phase there can be *compression* of the matter, and *sidewise flow* can happen. After that the system starts to *expand*, and there continues to be a loss of fast particle evaporation from the surface. As the expansion proceeds the density falls. By inertial flow the density continues to fall, down even below normal density. During this phase there will be a tendency in the matter to development of *instabilities* which will finally result in separation of the matter into several fragments. This is what we call *multifragmentation*, see section 2. Finally, the fragments will after their mutual separation and after shorter or longer time release what they may have of excess energy by *secondary decay*.

Finite Particle Number

The first finite nuclear effect which we shall discuss is the fluctuations in the temperature due to a finite number of constituents in a system. This is a standard textbook problem which was recently discussed in [4]). Consider a finite system of A particles in thermal equilibrium with temperature T. One finds the relative temperature fluctuation

$$\frac{\Delta T}{T} = \left(\frac{\partial U}{\partial T}\right)^{-\frac{1}{2}} = (C_V)^{-\frac{1}{2}} \tag{14}$$

For the low excited Fermi gas the internal energy of the A particles is $U = A\frac{T^2}{\epsilon_0} = Ae^*$ which gives $\frac{\Delta T}{T} = \left(\frac{\epsilon_0}{4e^* A^2}\right)^{\frac{1}{4}}$. For the ideal Boltzmann gas $U = A\frac{3}{2}T$. This gives $\frac{\Delta T}{T} = \left(\frac{2}{3}\frac{1}{A}\right)^{\frac{1}{2}}$.

The calculated relative temperature fluctuations for the Fermi gas from (14) are not negligible. Even for a large system of 100 particles the fluctuation is almost 10%, and for a system with 10 particles it is rather 30% . In this connection it is very important to notice that when using (14) for a reaction, A is the particle number of a *subsystem* and practically never of the whole system because of the very different dynamical evolutions of different parts of the nuclear matter. One should also remember that the reaction products in an event represent all the phases of the collision, from the beginning to the end, and therefore one may have only a part of the measured information in the event representing each phase and location of the colliding matter.

This means that thermodynamic concepts should be applied with caution for the heavy ion reactions. Nevertheless the concepts are often used, and we shall also do so in these lectures, in the particular case of studying the development of big nuclear systems towards the end of the development in the collision event. We hope that at this stage of the reaction some degree of thermal equilibrium has been established. Another reason to use a thermodynamical approach is that it is a good way to handle the full phase space. It gives a kind of minimum bias theoretical description of the reaction.

Surface, Charge and Isospin

The smallness of the nuclei has a dramatic effect on the binding energy of the nucleus. First of all this is connected with the existence of a surface. Secondly, the protons are charged, and therefore repel each other with the long range Coulomb force. The short range nuclear force prefers isospin symmetry which means the same number of neutrons and protons. The interplay between the Coulomb and the isospin forces causes an excess of neutrons for the bigger nuclei, and the interplay between the surface force and the Coulomb force gives the upper limit of the nuclear size. We define the excess energy E_W as the total energy minus the rest mass energy of the constituent nucleons. It is given by for a nucleus with N neutrons, Z protons and $A = N + Z$ nucleons by the Weizssäcker mass formula as

$$E_W = -b_{bulk}A + b_{surf}A^{\frac{2}{3}} + b_{sym}\frac{(N-Z)^2}{A} + \frac{3}{5}\frac{e^2 Z^2}{r_0 A^{1/3}} \tag{15}$$

In the formula we neglect special structural binding effects which are of minor interest in this connection. The result of these balances is that nuclei are only bound by approximately 8 MeV per nucleon and only a very limited range of nuclei around the beta stability curve are stable. The binding energy has a strong effect on the way in which a nuclear system can break up into smaller pieces. This is the main outcome of the theory of multifragmentation which we shall discuss in section 2.

Neutron Stars and Nuclei

The biggest nucleus which we know is the neutron star. It is basically believed to be a sphere with approximately nuclear density held together by the gravitational force. The matter in the neutron star is close to what we can call infinite nuclear matter. Assuming that the matter in the star is incompressible, and that the components of the binding energy in eq. (15) are applicable also for such big systems we get an extra term to the excess energy from the gravitation:

$$E_{grav} = -\frac{3}{5} \frac{GA^2 m_N^2}{r_0 A^{\frac{1}{3}}} \tag{16}$$

giving the excess energy of the neutron star

$$E_{n.s.} = E_W + E_{grav} \tag{17}$$

As the baryon number A increases, the relative role of the surface decreases. Assuming that the charge of the neutron star is negligible i. e. $Z = 0$, we get a big symmetry energy so that E_W becomes positive (≈ 9 MeV per nucleon). When calculating $E_{n.s.}/A$ from (17) as function of $R = A^{-\frac{1}{3}}$ one finds that this energy is negative and the system bound for $R > 5$km. In fact the typical neutron star radius is around 10km corresponding to a mass of the order of the solar mass. When the mass is big one cannot longer assume incompressibility, and now the radius of the star depends on the compressibility module of the matter. Also the course of stellar collapse depends on the equation of state of nuclear matter [5]. It is important for stellar physics what one can learn about the equation of state from heavy ion collision experiments in the laboratory.

2. Statistical Multifragmentation Model (SMFM)

A certain fraction of the reaction events in heavy ion collisions with beam energy above ≈ 50 MeV per nucleon exhibit a striking feature. They are characterized by a large multiplicity, with several fragments of intermediate size with a broad momentum distribution [6]. These are the events which we want to associate with central collisions and simultaneous break up of the major part of the nuclear system in what is usually called *multifragmentation*. Before the break up, the earlier stages of the collision process between the two nuclei have taken place (see chapter 1, sub-section *Finite Nuclei*). The fragments should carry information from these earlier stages, but in order to be able to sort out the information, we must first learn about the physical signals from the break up process itself.

Instabilities before Breakup

In the heavy ion process, after the initial build up of high pressure and density, the system starts to expand freely while the density falls. Such free expansion of compressed matter is probably nearly adiabatic (has constant entropy). Expansion of infinite matter is similar. In fig. 1c we show an adiabat along which the expanding matter can develop. Just after hitting the gas liquid coexistence zone the matter is superheated and it may separate into gas and drops of liquid. This only happens after a sufficient disturbance. During the expansion along the adiabat the bulk matter may evolve to densities and temperatures where $\left.\frac{\partial P}{\partial \rho}\right|_S < 0$, and the homogeneous matter becomes unstable. How instabilities can occur and grow has become a subject which has caught considerable interest recently. One can imagine various mechanisms:

1) The fluctuations in the matter develop small amplitude density variations of some extension. The criterion for instability is that these amplitudes *grow* during the natural evolution of the system. If the amplitudes decrease by the evolution, the matter is stable [7].

2) Another and different way is a short wavelength (down to the size of the nucleon) high amplitude fluctuation in the matter. This gives rise to a spontaneous bubble formation. Depending on the condition of the matter and the size of the bubbles, they can either shrink or expand [8].

Both 1) and 2) concern infinite matter and are therefore relevant for the interior of especially big nuclei.

3) For real nuclei there are also surface and Coulomb forces. The surface needs not stay spherical but can develop deviations from the spherical shape. A well known example is the nuclear binary fission. It is conceptually analogous to the instability type 1). One can easily imagine similar processes for nuclear fragmentation into more than 2 fragments. These are sometimes called ternary, quattuornary etc... fission.

4) Usual evaporation of light particles is a decay mode which is of course very important. In a certain way it is analogous to the volume- high amplitude short wave length instability 2).

Much work is needed for investigating the instability problems. We have here only given a qualitative picture of the main questions in that field.

Breakup Scenario

Whatever the reasons are for the break up of the matter into several pieces, the colliding system will be in a quite chaotic state during the time interval in which the fragments form and separate. The statistical multifragmentation model (SMFM) has been suggested to simulate the physical state of such a nuclear system. We shall outline the physics of the model in the following. Such a model was first suggested by Koonin and Randrup [9], Bondorf et al. [10], and by Gross [11]. In our treatment we shall mostly follow [10],[14],[15].

We assume that at the stage of breakup, instabilities have already occurred in the nuclear volume and caused the matter to break into many fragments which are about to leave each other. There are as many ways to form the set of fragments out of the total system as one can imagine. The set of fragments is called a mass-charge partition or just a partition. The fragments of the partition are assumed to move around in thermodynamical equilibrium in a growing volume V_b.

The model has four input variables, the mass and the charge of the system A_0 and Z_0, the total excitation energy above the ground state E^*, and the breakup volume V_b. The rest of the input are constants such as binding energy and level density constants.

These quantities are characteristic for the system after the emission of early emitted particles such as PEP-nucleons, gamma ray Bremsstrahlung and pions.

The number of possible partitions is usually very big. A system of 10 identical constituent particles can be fragmented i 42 different partitions, and a system of 100 particles in 190562992 different partitions! To that comes that each fragment can usually exist in a large number of excited states, and can have many different positions and momenta in the breakup volume. We characterize a partition by a vector $N(A, Z)$, where N is the number of fragments of mass A and charge Z. It fulfills the conditions

$$A_0 = \sum_{A=1}^{A_0} \left(A \sum_{Z=1}^{Z_0} N(A, Z) \right) \tag{18}$$

$$Z_0 = \sum_{A=1}^{A_0} \sum_{Z=1}^{Z_0} Z \, N(A, Z) \tag{19}$$

In the present paper we shall simplify the problem to using the mass partition vector

$$N(A) = \sum_{Z=1}^{Z_0} N(A, Z) \tag{20}$$

and then assume an analytical connection for the average charge $Z = Z(A)$.

The breakup volume is a sum of the free volume V_f in which the fragments move, plus the total volume V_0 of the fragments themselves. Thus,

$$V_b = V_0 + V_f = V_0(1 + \kappa) \tag{21}$$

where the nuclear matter at normal density in the fragments occupies the volume $V_0 = \frac{4}{3}\pi R_0^3 = \frac{4}{3}\pi r_0^3 A$. We want the typical distance between the fragments at breakup to be around the distance at which interaction between nuclear surfaces vanishes. This crack width 2d replaces the breakup volume V_b as one of our four input parameters. A guide for the choice of 2d is the range of the proximity force [12]. In accordance with this the radius of the breakup volume scales as

$$R_b = R_0 + d(M^{1/3} - 1) \tag{22}$$

The quantity M in eq. (22) is the multiplicity, or the total number of fragments which is defined for a given partition as

$$M = \sum_{all A, Z} N(A, Z) \tag{23}$$

Eq's (22) and (23) ensure that the free volume in which the fragments move is approximately equal to the inner surface times the distance between fragments. With the ansatz (22) the breakup volume is dependent on the multiplicity M, and thus on the partition itself. When working with an ensemble of partitions one has therefore also a system dependent ensemble of breakup volumes.

Energetics

In eq's (18) and (19) we have already stated two conservation laws, baryon and charge conservation at breakup. Momentum conservation can be taken into account by correcting for the CM-motion for each partition (see ref. [13]). We now go to the energy conservation. In the model we assume that at this stage the fragments interact only as in a dilute gas, besides of course through the long range Coulomb force. The energy of the system for a given partition (i) characterized by the partition vector $N(A, Z)$ is then(after subtraction of the rest masses of the nucleons):

$$E_{tot} = \frac{3}{5R_b}(Z_0 e)^2 + \sum_{A=1}^{A_0} \sum_{Z=1}^{Z_0} N(A,Z)\, E(A,Z) = E_{ground} + E^* = E_{ground} + e^* A_0 \quad (24)$$

where R_b is given by (22). Note that the non-linear part of the Coulomb energy is taken into account by the first term in (24). In this simplest version of the statistical model we completely neglect flow of the matter. A possible flow energy should accompany the non linear Coulomb energy in (24). One could also introduce a gradual change of the interaction energy between the fragments during breakup, by having a crack width dependent potential energy. This has not been done here. The additional average energy $E(A, Z)$ in (24) for each fragment of mass number A and charge number Z is for temperatures not too large (cf eq. (4)):

$$E(A, Z) = E_{transl} + E_{bulk} + E_{surf} + E_{sym} + E_{Coul} \quad (25)$$

The energy terms in (25) are just the Weizsäcker energies from eq. (15) generalized to $T > 0$, and they can be found in table 1. For $A \leq 4$ we used the actual binding energies $B_{A,Z}$. Nuclear structure effects such as shell effects and pairing can be included as corrections to $B_{A,Z}$ and ε_0. The intrinsic spins of the fragments have not been included in the present version of the theory. Notice that the surface energy decreases with temperature [10],[14] in accordance with the properties of the nuclear matter. The surface energy vanishes at the critical temperature T_0 (fig 1c). The excitation energy constant ε_0 in E_{bulk} corresponds to infinite matter (cf chapter 1) and is larger than the usual experimental value for finite nuclei. This is compensated by the temperature dependence of E_{surf} so that the effective level density constant comes in agreement with the experimental one $\approx 8 - 10 MeV$. The radius of fragment A is $R = r_0 A^{1/3}$; $r_0 = 1.15$ fm. We have assumed that the additional Coulomb energy of the fragments can be calculated by the Wigner Seitz approximation for which $R/R_{cell} = R_0/R_b$, where R_b is determined in eq.(22). Notice that the additional Coulomb energies enter in expressions (24)-(25) linearly similar to all the other energies in $E_{A,Z}$. That approximation is good except for systems with only a few fragments where the special spatial arrangements could play a role.

Thermodynamical Quantities

In the thermalized mixture of fragments the thermodynamic probability of a macro state f is [1]:

$$W_f = \exp[S_f] \tag{26}$$

where S_f is the entropy of the state. By using the W_f's of all states as weights one can calculate weighted probabilities of the various physical observables. Because the phase space for the system is so overwhelmingly big it is necessary to think of simplifications. In the present model we have, instead of considering all combinations of all states in the fragments, deposited only the most probable energy averaged over many states in each fragment. This means that energy is only conserved on the average.

In order to calculate the thermodynamical probability exponent (the effective number of quantum states) for a given partition (i) one must first perform a sum of the exponent in (26) over all the states which are represented in (i) within the given constraints. This means summing over both the intrinsic states of the fragments and the states for their motion in the free volume. The thermodynamical probability for the partition is now

$$W_i = \exp S_i \tag{27}$$

where S_i is the entropy for the partition.

We shall now discuss some thermodynamical relations. All the relations in the rest of this section until eq. (35) refer to one selected partition (i), and we shall therefore omit the index i. The entropy of the partition can be determined by using the general thermodynamic relation for the energy

$$E = F + TS \tag{28}$$

where F is the free energy, T is the temperature and S the entropy. As already mentioned, all the energy components in (24) are additive except the global Coulomb energy term in the first line of (24) which depends only on the overall properties of the system. We call the component indices A, Z *chemical* in analogy with the terminology of chemical mixtures. Similarly the energy component names, *transl, bulk, surface, symmetry and residual Coulomb* are called the *physical* component names (denoted p). Now eq.s (24)-(25) can be written

$$E_{tot} - \frac{3}{5R_b}(Z_0 e)^2 \equiv E = \sum_{A,Z,p} E(A,Z,p)N(A,Z) \tag{29}$$

and similarly for the free energy and entropy

$$F = \sum_{A,Z,p} F(A,Z,p)N(A,Z) \tag{30}$$

$$S = \sum_{A,Z,p} S(A,Z,p)N(A,Z) \tag{31}$$

Besides (28) we have a similar equation for each component (analogous to eq. (10),

$$E(A, Z, p) = F(A, Z, p) + TS(A, Z, p) \qquad (32)$$

The entropy components are generally determined from the free energy by the relation

$$S(A, Z, p) = -\frac{\partial F(A, Z, p)}{\partial T}\bigg|_{V_b} \qquad (33)$$

which is valid for a system with many independent components. Eq. (33) is used to determine the entropy components from the free energy components, most of which can be guessed directly from the energy expressions (25). The translational free energy was calculated directly from the definition

Table 1. Energy, free energy and entropy components for one fragment (A, Z).

p	$E_{A,Z}$	$F_{A,Z}$	$S_{A,Z} = -\frac{\partial F_{A,Z}}{\partial T}\big|_{V_b}$
transl.	$\frac{3}{2}T$	$-T \ln\left(\frac{\varrho_{A,Z} V_f A^{3/2}}{\Lambda^3}\right) + \frac{T \ln N_{A,Z}!}{N_{A,Z}}$	$\frac{3}{2} + \ln\left(\frac{\varrho_{A,Z} V_f A^{3/2}}{\Lambda^3}\right) - \frac{\ln N_{A,Z}!}{N_{A,Z}}$
bulk	$\left(-B_{A,Z} + \frac{T^2}{\varepsilon_0}\right)A$	$\left(-B_{A,Z} - \frac{T^2}{\varepsilon_0}\right)A$	$\frac{2T}{\varepsilon_0}A$
surf.	$\left(\beta(T) - T\frac{\partial \beta(T)}{\partial T}\right)A^{\frac{2}{3}}$	$\beta(T)A^{\frac{2}{3}}$	$-\frac{\partial \beta(T)}{\partial T}A^{\frac{2}{3}}$
sym.	$\gamma\frac{(A-2Z)^2}{A}$	$\gamma\frac{(A-2Z)^2}{A}$	0
Coul.	$\frac{3}{5}\frac{(Ze)^2}{R_{A,Z}}\left(1 - \frac{R_{A,Z}}{R_{cell}^{(A,Z)}}\right)$	$\frac{3}{5}\frac{(Ze)^2}{R_{A,Z}}\left(1 - \frac{R_{A,Z}}{R_{cell}^{(A,Z)}}\right)$	0

$$\Lambda = \left(\frac{2\pi\hbar^2}{mT}\right)^{\frac{1}{2}}$$
$$\beta(T) = \beta_0 \left(\frac{T_0^2 - T^2}{T_0^2 + T^2}\right)^{\frac{5}{4}} \qquad \beta_0 = 18 MeV \qquad T_0 = 16 MeV$$
$$\gamma = 25 MeV$$
$$\varepsilon_0 = 4\epsilon_F/\pi^2 \approx 16 MeV$$

$$F = -T \ln \sum_n \exp\left(-\frac{E_n}{T}\right) \qquad (34)$$

We show all the chemical and physical components of E, F and S in table 1. The term $-\ln N(A, Z)!/N(A, Z)$ in the translational entropy is necessary because we are dealing with $N(A, Z)$ identical fragments in each chemical component.

The Thermodynamical Probability

In order to calculate the thermodynamical probability for a partition (i) we first select the partition randomly among the possible partitions. In [16],[17] the Metropolis method has been used in order to make a representative sample in the full partition-energy space, but in this treatment we shall just stick to the more simplified way of ref. [10],[14],[15]. With the chosen partition vector $N(A, Z)$ for partition (i) we calculate the temperature T from (24) and (25). The temperature will depend on (i). Now the thermodynamical probability W_i (27) can be calculated

$$W_i = \prod_{A,Z} \left[g(A, Z) V_f A^{\frac{3}{2}} \frac{1}{\Lambda(T)^3} \exp\left(\frac{3}{2} + S_{int}(A, Z)\right) \right]^{N(A,Z)} \frac{1}{N(A, Z)!} \qquad (35)$$

$$\Lambda(T) = \left(\frac{2\pi\hbar^2}{mT}\right)^{\frac{1}{2}}$$

where the "intrinsic" fragment entropy is $S_{int}(A, Z) = (2T/\epsilon_0)A - (d\beta/dT)A^{2/3}$ for $A > 4$ as seen from table 1. The statistical factor $g(A, Z)$ is specified individually for the fragments with $A \le 4$ and otherwise 1. The quantity $\Lambda(T)$ in (35) is the thermal wave length for a nucleon with mass m (cf eq.(5)). The probabilities W_i vary strongly from one partition to another.

Calculated Results

When the W_i's are found one can calculate various physical averages of observables Q from the expression, where the sums run over all the partitions (i) in the sample

$$Q_{av} = \frac{\sum_i W_i Q_i}{\sum_i W_i} \qquad (36)$$

We shall now, for some selected examples, show various characteristic calculated results from the theory. In fig. 2 we show how the temperature varies with the excitation energy per nucleon e^*. It is seen that the parabolic variation characteristic of a low excited Fermi gas changes drastically above a certain temperature. This temperature stays almost constant around 5 MeV over a long e^* interval from around 4 to 10 MeV per nucleon and then rises. We call this temperature *the crack temperature*.

In fig. 3 the multiplicity of primary fragments is shown as function of e^*. The function rises nearly linearly with the energy. Fig.4 shows how the freeze out density decreases with excitation energy. One notices that figures 2, 3 and 4 have "error" bars on the calculated points. These indicate the fluctuations which one automatically has in the thermodynamic variables of a small finite system. We refer to the discussion in chapter 1.

Fig. 5 shows some calculated primary mass distributions. It is seen how the slopes depend strongly on the excitation energy. The value of the crack width parameter $2d$ in eq. (22) was chosen to be 2.8 fm which is of the order of the range of the force

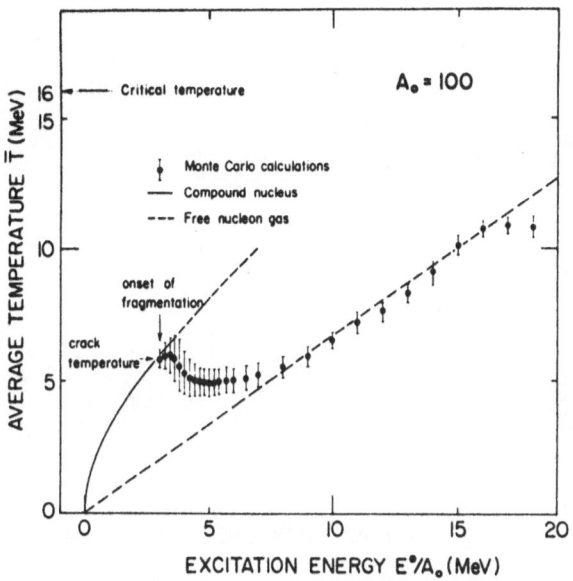

Fig.2 *Calculated average temperature as function of the excitation energy per nucleon. The fluctuations result from the use of the many different partitions, and do not represent calculational errors.*

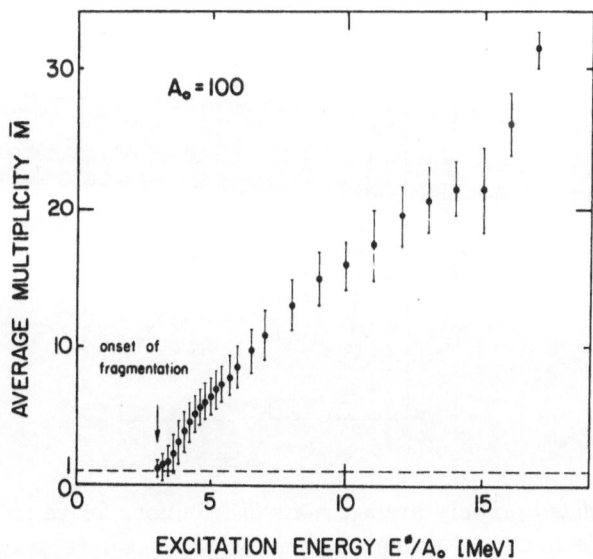

Fig.3 *Calculated average primary multiplicity as function of the excitation energy per nucleon. The fluctuations have the same meaning as in fig. 2.*

between fragments. (For $A = 1$ the value of $2d$ was chosen only half of that value). The parameter d is in principle fixed and cannot be altered very much. In order to demonstrate, however, the effects of the free volume, we have in fig. 6 shown the calculated average mass distributions for four different values of the free volume V_f. As V_f increases, the fragments become generally smaller. In this last calculation we used fixed values of the free volume, and not the multiplicity dependence (22).

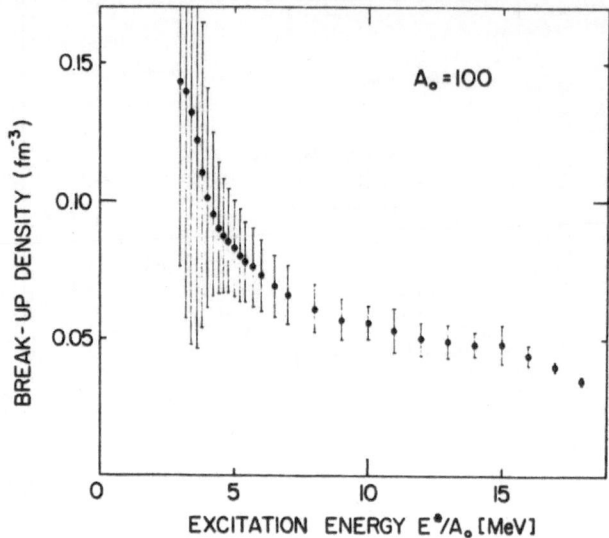

Fig.4 Freeze-out density dependence of excitation energy as defined in the SMFM.

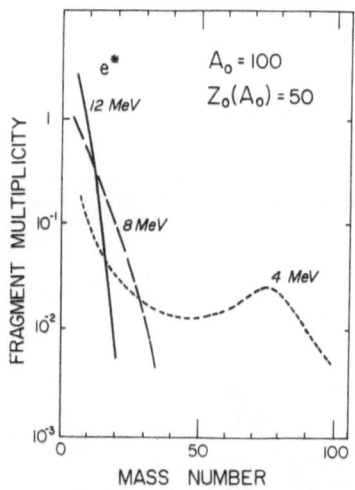

Fig.5 Calculated primary average mass distributions for various excitation energies. The freeze out volume is given by (22).

In connection with nuclear fragmentation it is often discussed to what degree the fragmentation phenomenon is a gas liquid phase transition connected to the critical point in the phase diagram, fig. 1c. In the present version of the model we use a critical temperature of the nuclear matter which is of the order 15-20 MeV. This is the temperature at which the nuclear surface energy vanishes. It is much higher than the calculated crack temperature of the order of 5 MeV where the surface energy (see table 1) is close to that of the ground state. Therefore it is not reasonable to connect the fragmentation directly with the critical temperature of infinite nuclear matter.

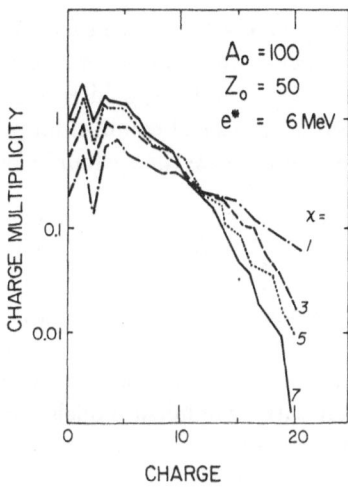

Fig.6 Calculated average charge distributions for four different values of the break up volume, defined by fixed values of the parameter κ in eq. (21).

Concluding Remarks

In the present overview we have investigated consequences of a statistical break up of an excited nuclear compound. Although there is no experimental proof for such an equilibrium break up of excited nuclei, the physics involved in the equilibrium break up probably reflects main ingredients in the understanding of the nuclear disintegration process. We emphasize that the applied parametrization of the relevant degrees of freedom are not unambiguous. Other models have been put forward, which differ in the internal fragment excitation and in the freeze out assumption [9] and [11].

What concerns the internal fragment excitation we use independent surface and bulk degrees of freedom which favours fragmentation more than when these two modes are coupled. Both [9] and [11] in this way count a smaller phase space for the fragmented system than the Copenhagen model. In the same connection it is also important to keep in mind that [11] only counts excited states of the fragments up to the threshold of charged particle emission. In [9] the population of excited levels are limited by a factor $\propto e^{-\Delta e/T_{lim}}$ where $T_{lim} \approx 10 MeV$ plays the role of a limiting temperature. Thus in [11] the intrinsic excitations play a smaller role than in [9] or in our model where no upper limit of the fragment excitation is introduced.

Another important ingredient in the present fragmentation model is the choice of a freeze out volume given by (22). Because this volume increases with excitation energy as shown in fig. 4, it lets the freeze out temperature stay around 5 MeV for a broad range of excitation energies (see fig.2). Thus the observation of a limiting temperature of about 5 MeV deduced from the populations of excited fragment levels (for a recent review, see [18]) is in agreement with an equilibrium fragmentation model where the freeze out density decreases with excitation energy. If instead the freeze out volume were kept constant, the temperature would tend to increase with excitation energy. The model [11] leaves open the possibility for a slight temperature decrease with excitation energy in a limited domain for certain systems and freeze out volumes. This "phase transition" is connected with the onset if fission.

When comparing the multifragmentation model to experiment one must be able to isolate a homogeneous sample of events with nearly identical entrance channel properties. This is a standard heavy ion collision problem which is rather model independent. Furthermore one must search for collisions in which there is hope for establishing some degree of thermodynamical equilibrium prior to the fragmentation process. This is not always possible because the system is in dynamical evolution.

Instead of using the model in this way one could maybe use it as a tool to disentangle the equilibrized degrees of freedom from the non equilibrized ones. The model so to speak describes the "noise" which is formed in the nuclear system towards the final stage of a collision, and this noise should be understood and removed when one wants to find the signals from the earlier stages. In order to do so one could apply the statistical multifragmentation model to a subset of the degrees of freedom of the colliding system, in combination with dynamical models which describe the collision at other stages than the freeze out stage.

Acknowledgements

We want to thank H. Barz, R. Donangelo, I. Mishustin, H. Schulz and members of the the CHIC collaboration for inspiring discussions. The Danish Natural Science Research Council is thanked for support.

References

1. L. D. Landau and E. M. Lifshitz, Statistical Physics part 1, Pergamon Press, Oxford, 1980.

2. L. Zamick, Phys. Lett. **B45** (1973) 313.

3. P. J. Siemens, Nature **305** (1983) 418.

4. H. Feshbach, Physics Today (Nov. 1987) 9.

5. G. E. Brown, Nuclear Physics **A488** (1988) 689c.

6. B. Jakobsson et al. Nuclear Physics **A488** (1988) 251c.

7. C. J. Pethick and D. G. Ravenhall, Nucl. Phys. **19** (1987) 33 H. Heiselberg, D. G. Ravenhall and C. J. Pethick Phys. Rev. Lett. **61** (1988) 818.

8. J. P. Bondorf, R. Donangelo and K. Sneppen Phys. Lett. **B214** (1988) 321.

9. S. E. Koonin and J. Randrup, Nucl. Phys. **A356** (1981) 223. G. Fai and J. Randrup, Nucl. Phys. **A381** (1982) 557.

10. J. P. Bondorf, Nucl. Phys. **A387** (1982) 25c. J. Bondorf, I. N. Mishustin and C. Pethick, Proc. Int. School ed. Yu. Ts. Oganessian et al. Dubna (1983) 354.

11. D. H. E. Gross, Phys. Scr. **T5** (1983) 213, also Sa Ban-Hao and D. H. E. Gross, Nucl. Phys. **A437** (1985) 643.

12. J. P. Blocki et al., Ann. of Phys. **105** (1977) 427.

13. J. P. Bondorf, R. Donangelo, H. Schulz and K. Sneppen Phys. Lett. **162B** (1985) 30.

14. J. P. Bondorf, R. Donangelo, I. N. Mishustin, C. Pethick, H. Schulz and K. Sneppen, Nucl. Phys. **A443** (1985) 321. Also Phys. Lett. **150B** (1985) 57, Nucl. Phys. **A444** (1985) 460. Also H. W. Barz et al. Nucl. Phys. **A448** (1986) 753.

15. A. S. Botvina et al. Nucl. Phys. **A475** (1987) 663.

16. D. H. E. Gross et al. Phys. Lett. **161B** (1985) 47 and Phys. Rev. Lett. **56** (1986) 1544. Also Zhang Xiao-Ze et al. Nucl. Phys. **A461** (1987) 641.

17. S. T. Koonin and J. Randrup, Nucl. Phys. **A474** (1987) 173.

18. C. K. Gelbke and D. H. Boal, Prog. Part. Nucl. Phys. **19** (1987) 33.

COLLECTIVE PHENOMENA IN RELATIVISTIC HEAVY-ION COLLISIONS: THE EXPERIMENTAL SITUATION

Karl-Heinz Kampert

University of Münster
Wilhelm-Klemm-Straße 9
D-4400 Münster, West Germany

1 Introduction

The key mechanisms for producing hot dense matter in high-energy heavy-ion colli-sions were first formulated more than a decade ago in a series of papers by Chapline, Johnson, Teller, and Weiss [Cha73], Scheid, Müller, and Greiner [Sch74], and later also by Sobel, Siemens, Bondorf, and Bethe [Sob75]. At that time only very little had been known about the properties of hadronic matter at finite temperatures and densities other than the nuclear ground state density $\rho_0 = 0.16$ fm^{-3}. Scheid et al. argued that high density nuclear shock waves should occur in violent head-on collisions of heavy nuclei leading to a preferential collective sidewards emission of the compressed hot matter. Particularly these suggestions set the stage for an investigation of the nuclear equation of state (EOS) by performing heavy-ion collisions at sufficiently high kinetic energies.

At laboratory projectile energies of about 1 GeV per nucleon the speed of inter-prenetation of the two colliding nuclei exceeds the sound velocity of ordinary nuclear matter ($v_s \cong 0.25c$), i.e. the penetration velocity is higher than the velocity of informa-tion propagation inside the medium. Hence, if nuclei stop each other, one encounters a typical shock scenario where the participant matter cannot escape rapidly enough from the interaction zone, resulting in a pile-up, i.e. in a traveling shock front of highly compressed nuclear matter. The goal as formulated in the papers noted above, was to learn about the mechanism for high compression and heating of nuclear matter, and, if possible, also about the nuclear matter EOS, i.e. the response of the nuclear medium to these extreme conditions of temperature, energy- and baryon density.

Probing the nuclear matter EOS in regions of high densities and temperatures ($\rho/\rho_0 \simeq 2$–5, $T = 50$–100 MeV) is of fundamental importance not only in nuclear physics (possible phase transitions such as liquid-vapour, pion condensation, Δ-isomers, etc.) and field theory (QCD phase transition to a Quark-Gluon-Plasma), but also in astrophysics for an understanding of the structure and dynamics of stars (in particular supernovae (SN) explosions and neutron star stability). Extreme conditions similar to those created in high-energy heavy-ion collisions probably existed at the birth of the universe during the first fractions of a second after the 'Big-Bang'. Both the mechanism of supernovae dynamics, as well as the stability of neutron stars depend strongly on the compressibility of dense nuclear matter [Baro85, Wil85]. Heavy-ion collisions are the first opportunity to study these conditions in the laboratory. Important information would be accessible from this kind of experiments, if one could apply the knowledge about the conditions achieved for very short time spans of $t \cong 10^{-22}$ seconds, to the much larger space- and time dimensions relevant in astrophysical events.

The pioneering experimental research in this new field — which has been started particularly at the Berkeley Bevalac — has served to provide orientation about the produced particle densities and the populated phase-space of heavy-ion reactions [Bau75, Gut76, Gos77, San80]. Because of the huge number of exit channels viewed by single particle detectors, nuclear collisions looked basically 'thermal'. They could be fairly well reproduced by the nuclear fireball model [Wes76], which is based on simple geometrical, kinematical, and statistical concepts, and ignores dynamical effects entirely. In parallel, more advanced approaches were further developed, such as hydrodynamics [Sch74, Ams75, Bau75, Stö79a, Buc81, Cse82], cascade [Bon76, Yar79,81, Cug80,81,82a], and classical (Newtonian) equations of motion [Cal79, Bod80]. Although the ingredients and assumptions of these models were quite different, all of them were fairly successful in reproducing the inclusive proton and light fragment spectra on a qualitative level. By concentrating therefore on central collisions with high particle multiplicities [Gos77, Nag80, Wol82], investigating global emission patterns [Sto80], and searching for correlations between different fragment groups [Mey80], it had been recognized that particle *exclusive* data were needed to gain detailed insight into the underlying reaction mechanism, and thus being able to unambiguously discriminate between different models.

In this article I shall first describe some experimental facilities which have delivered important experimental data for investigating nuclear collisions at high bombarding energies. After this brief survey on basic experimental techniques we will develop a qualitative view on the reaction dynamics. This will enable us to discuss various observables pertaining to the high density stage of the collision and their possible relation to the nuclear equation of state. Section 5 is addressed to the question of thermalization — global or local — and will lead us to the nuclear stopping power, a key observable of heavy-ion collisions. Cluster production and entropy which can be inferred from relative particle abundances will give us first evidence for the presence of compression effects (Sec. 6). Nuclear matter flow as the most direct measurement of compression will be discussed in detail in section 7. Finally, in section 8 I briefly shall summarize the present information derived from the analysis of 4π data.

2 Experimental Techniques

Full event characterization by identifying and measuring the momenta of all emitted particles with energies of up to several hundred MeV was first attempted for large scale data collection in two 4π detectors at the Bevalac; the Streamer Chamber facility [Str83] and the Plastic-Ball/Wall set-up [Bad82]. Some years ago from now, the Diogene detector [Ala87] at Saturne (Saclay, France), started taking more data with light projectiles. Low statistic data samples, furthermore, are delivered by several emulsion experiments.

In this section I will briefly discuss various solutions of 4π particle detection in high energy heavy-ion collisions.

2.1 Visual Detectors

Visual detection of particles is known in a large variety. In high multiplicity environments of high energy heavy-ion collisions two alternatives have been proven successful; nuclear emulsion stacks and streamer chambers.

Nuclear emulsions may be regarded the simplest approach for investigating nucleus-nucleus collisions. Generally, they consist of a stack of photographic plates serving simultaneously as target foils ($AgBr_2$) and tracking material. Reaction products of a collision with any absorber nucleus are — after photographic development — identified by their ionisation track in microscopic analysis. This method provides excellent spatial, i.e. angular resolution and also limited information on the charge and velocity of the fragments by their blob density along the track. The major drawback, however, is the very time consuming optical analysis of the emulsions, which in practice limits substantially the available statistics of those data.

A streamer chamber consists of a gas filled volume, typically 90% Ne, 10% He, $\simeq 30$ ppm SF_6 at atmospheric pressure, inside an intense triggerable electric field of about 10–20 kV/cm. The electric field can initiate the growth of luminous filament discharges from the sites of primary ionisation along the particle trajectory. These luminous discharged, or streamers, can be sufficiently bright to be photographed directly without image intensification. Most importantly, the electric field may be applied for some nanoseconds within a few hundred nanoseconds after an event; a time delay which is short compared with the recombination time of electron-positron pairs. Thus, streamer chambers need to be triggered by an external detector. Large scale chambers on the order of $1.2 \times 0.6 \times 0.4$ m^3 with some micron spatial resolution were engaged in large scale data production experiments [San83]. Although automized computer aided 3-dimensional event reconstruction has made appreciable progress, high statistics data samples are still not achievable. The utility of such a visual technique with rigidity measurements by a magnetic field for identifying particles with a characteristic decay such as $\Lambda \rightarrow p^+ + \pi^-$, however, is evident.

2.2 Electronic Detectors

Electronic detectors generally allow much faster data collection than visual devices. The first electronic 4π detector employed in high energy heavy-ion collisions was the Plastic-Ball spectrometer [Bad82]. It consists of 815 individual modules covering the angular region from $\theta_{Lab} = 10°$ – $160°$. Each module provides particle identification by employing the $\Delta E - E$ technique, i.e. different particles are grouped on separate hyperbolic branches in a two-dimensional $\Delta E - E$ plot as is a consequence of the empirical Bethe-Bloch formula. In the Plastic-Ball the energy loss is measured in a 4 mm thick CaF_2 crystal glued directly onto a 35 cm long plastic scintillator. Because of the big difference in the decay times of the two scintillators ($\tau_{CaF_2} \simeq 1\,\mu$s, $\tau_{CH} \simeq$ 10 ns) both signals can be read out by one phototube, whose pulse is then fed to different gated ADC's. In addition, stopped positive pions are identified by their decay positron ($\pi^+ \rightarrow \mu^+ + \bar{\nu}_\mu(\tau = 26\,\text{ns})$; $\mu^+ \rightarrow e^+ + \nu_e(\tau = 2.2\,\mu\text{s})$) in a delayed coincidence. The forward region $\theta_{Lab} \leq 10°$ of the Plastic-Ball setup is covered by a time of flight wall, measuring the velocity and charge of fast projectile fragments.

A type of detector that has been proposed more recently for use in heavy-ion collisions and which is still in its development is the so called *Time Projection Chamber* (TPC). This device is essentially a three dimensional gaseous tracking chamber capable of providing information on many points of a particle track along with information on the energy loss of the particle. A similar device, which has already been proven successful, is the *Pictorial Drift Chamber* (PDC). Contrary to a TPC the electric field is normal to the magnetic field and the charge of ionizing particles is not collected on end-caps, but rather in several radial segments over the hole length of the chamber. An example of such a 3-dimensional tracking device is the Diogene detector at Saturne [Ala87]. The volume of this 0.8 m long and 0.7 m diameter chamber located inside a 1 T solenoidal magnetic field is subdivided into ten individually read-out radial sectors. Particle identification is achieved essentially as described before, i.e. the dE/dx, which is extracted now from the charge collected on the anode wires, is plotted against the momentum of a particle, obtained from the curvature of the track in the magnetic field. Employing a magnetic field permits also to distinguish between positive and negative tracks, i.e. between π^+ and π^-.

3 Observables of the High Density Fireball

Before we attempt to infer a quantitative relation from experimental observables to the equation of state of nuclear matter, we need to discuss the basic question about the *existence* of state variables characterizing a system composed of nuclear matter. Since collision times are significantly shorter at high bombarding energies than typical relaxation times of nuclear matter, global equilibrium is unlikely to be reached during a collision. Nevertheless, statistical concepts have been successfully applied to heavy-ion reactions. Different from the simplified fireball picture, the nuclear fluid dynamical model, for example, assumes that *local* rather *global* equilibrium is approached on the short time scales we are concerned with. If this assumption can be confirmed by microscopic theories which dynamically describe the evolution of the system, and furthermore, if predictions of such models are found to be in agreement with experimental data, then we may speak of a *state* of matter characterized by macrocanonical variables, namely by the density ρ, entropy S, and the temperature T. The total center of mass (CM) energy per baryon, W, as a function of density and temperature is normally divided into a thermal and compressional part:

$$W(\rho, T) = E_T(\rho, T) + E_C(\rho) + W_0 \tag{1}$$

where $E_C(\rho)$ is defined to be the compression energy at zero temperature, E_T is the thermal excitation energy per nucleon, and $W_0 = 923$ MeV is the rest energy of a nucleon at equilibrium density. Once the functional form of W is given, any other macrocanonical variable can be calculated from this by evaluating the thermal identities. $E_C(\rho)$ is commonly referred to as the *nuclear equation of state*, as will be done here.

The key question now is how to measure the density, the compressional energy, or the temperature in the brief moments of the high density stage of the reaction. Several of the state variables noted above, and necessary to describe the system can indeed be related to physical observables, accessible to some degree in high-energy heavy-ion collisions. These are listed below:

STATE VARIABLES	OBSERVABLES
$E_T(\rho, T)$	abundance of produced particles (absolute and relative yields of strange particles, pion multiplicities)
$S(\rho, T)$	relative yields (chemical composition) of light fragments (ratios of deuteron/proton, ^{3}He/proton, etc.)
$p(\rho, T)$	phase-space distribution of fragments (directed and radial nuclear collective flow)

One sensitive probe proposed in early theoretical papers [Cha73, Sob75, Hof76b, Stö78, Stö79b, Dan79] is the production of particles, such as p, K, Λ, etc., whose rate may be regarded as a 'thermometer' of the high-energy stage [Sto86]. However, due to rather low fireball temperatures ranging from 30–120 MeV at Bevalac energies, only a small fraction of the total energy content of the system is consumed for particle production, and major carrier of the overall energy flow are rather nucleons and light clusters in this bombarding energy regime.

The possible relationship between the other proposed observables such as flow and fragment ratios to the EOS will be discussed in detail in the following.

4 Charged Particle Multiplicities

Extraction of observables from the final phase-space distribution of fragments can uniquely be achieved only on an event-by-event basis. Therefore, one first needs to find a proper measure for the impact parameter of a collision. Investigations of different models (where the impact parameter of each collision is known) capable of producing nuclear fragments have demonstrated that such a scale is fairly well represented by the observed charged particle multiplicity, or even better by the extracted multiplicity of participating nucleons. In the Plastic-Ball data this number has been approximated by the participant charged baryon multiplicity, N_p. The definition of this quantity takes into account the participating protons bound in clusters $(d, t, ^3$He, ^{4}He) and removes spectator particles in a certain $p_\perp$-window around projectile and target rapidity by application of software cuts [Dos85]. Low multiplicities are associated with peripheral, high multiplicities with violent central collisions. The average multiplicity depends on the bombarding energy and projectile target mass. In order to make meaningful comparisons between these different cases, i.e. to compare constant regions of normalized impact parameters, the multiplicity distributions were divided into five bins of constant fractions of the maximum multiplicity, N_p^{max}, which is defined to be the point where the distributions drop to one-half of the plateau height [Dos86]. Table 1 contains the values of N_p^{max} for several system investigated with the Plastic-Ball.

Table 1. Maximum participant proton multiplicity, N_p^{max}, for different projectile target systems and bombarding energies. Only fragments with an energy above a software cut of approximately 15 AMeV are considered [Kam89].

E/A (MeV)	Au + Au	Au + Fe	Nb + Nb	Ca + Ca	Ne + Au	Ne + Pb
150	66	–	38	–	–	–
200	90	68	–	–	–	–
250	92	–	52	–	–	–
400	112	–	64	32	–	–
650	128	–	70	–	–	45
800	134	–	72	–	–	–
1050	142	–	74	36	–	–
2100	–	–	–	–	68	–

5 Local Equilibrium and Thermalization

In order to be able to extract a functional relationship between state variables of the system, an understanding of the dynamic behaviour of the collision is necessary. In particular, the degree to which thermal and chemical equilibrium are reached and maintained during a high-energy nucleus-nucleus collision will determine whether thermodynamic concepts may be applied for their description. A simple argument is that in high-energy heavy-ion collisions the mean free path, λ, of nucleons

$$\lambda = \frac{1}{\sigma_{eff} \cdot \rho} \qquad (2)$$

in compressed matter (with σ_{eff} being the effective in-medium nucleon-nucleon cross section) is small compared to the typical dimensions, L, of the system

$$\lambda/L \ll 1. \qquad (3)$$

Nucleons therefore will undergo several collisions before they escape from the compressed region between projectile and target nuclei, leading closely to a local equilibrium [Bau75].

5.1 Rapidity Distributions and the question of Nuclear Stopping

Qualitative measures of nuclear stopping can be extracted from the final state phase-space distribution of the emitted fragments in different ways. A direct and illustrative way is the investigation of baryon rapidity[1] distributions, dN/dy, measured over the full 4π hemisphere and different regions of impact parameters. Rapidity distributions of baryons (which are defined here to be equal to $\{1+\frac{N}{Z}\}n_p+2n_d+3\{n_t+n_{^3He}\}+4n_{^4He})$

---

[1]The rapidity of a particle is defined in terms of its momentum and total energy by $y = \frac{1}{2}\ln\frac{E+p_\parallel}{E-p_\parallel}$. The rapidity in one frame of reference is related to the rapidity in another frame of reference by an additive constant, i.e. simply speaking, it reduces the Lorentz- to a Galilei transformation in the longitudinal direction.

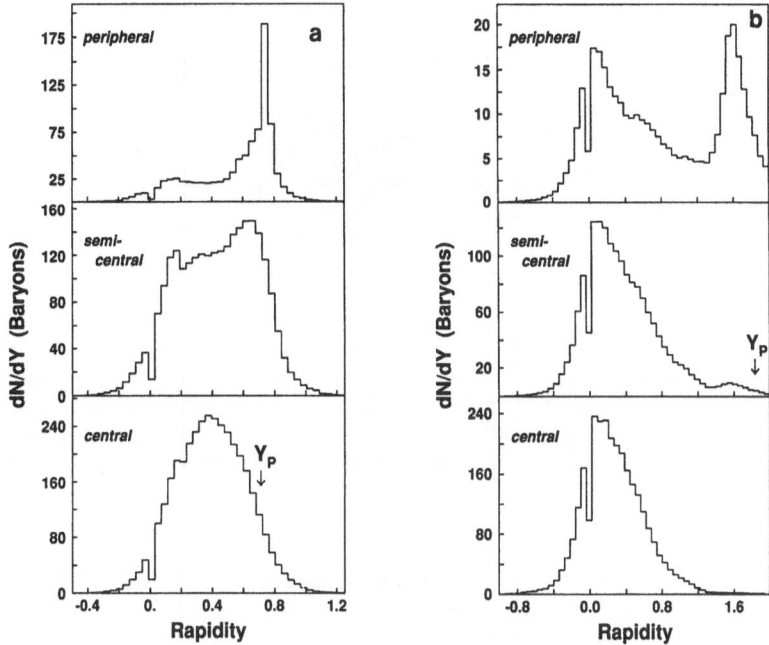

Figure 1. Baryon rapidity distributions as measured with the Plastic-Ball for Au+Au collisions at 250 AMeV (a), and Ne+Au collisions at 2100 AMeV (b). Selected are peripheral, semi-central and central collisions, respectively [Kam89].

are displayed in Fig. 1a for Au+Au collisions at 250 AMeV and different multiplicity intervals. There is a striking change in the distributions when going from peripheral (top) to semi-central (center) and further to central (bottom) collisions. In peripheral collisions (defined by $N_p/N_p^{max} < 25\%$) one observes a pronounced peak at values of the projectile rapidity, $y_{pro} = 0.72$, and only very few baryons in the fireball region around $y_{FB} = \frac{1}{2}y_{pro}$. A corresponding peak at values of the target rapidity, $y_{tar} = 0$, is absent because of the low fragment energies and the inability of the detector to measure particles at a laboratory angle of $\theta = 90°$. In semi-central collisions ($50\% \leq N_p/N_p^{max} < 75\%$) the projectile peak almost vanishes and is shifted towards lower rapidities. This reflects the stronger deceleration of the projectile by the target nucleus due to the increased number of participating nucleons in the overlap region. Central collisions ($N_p \geq N_p^{max}$), finally, lead within the experimental uncertainty to a fairly symmetric distribution around the value of the fireball rapidity, y_{FB}. The two gold nuclei therefore seem to get completely stopped within each other, forming a common excited system.

Rapidity distributions of the asymmetric system Ne+Au at a bombarding energy of 2100 AMeV are shown in Fig. 1b for the same regions of impact parameters as noted above. In resemblance to the distributions of the Au+Au system a decreasing number of projectile spectator fragments is observed with increasing centrality of the collision. For the highest multiplicities, which correspond to complete overlap of the projectile with the target nucleus, almost no fragments are observed at values of the projectile rapidity. This clearly indicates full stopping of the neon projectile by the gold nucleus up to top Bevalac energies. Recent experiments, carried out at 14.5 AGeV at the Brookhaven AGS, and at 60 and 200 AGeV at the CERN-SPS, confirm this finding and indicate a lasting of the stopping regime even up to 14.5 AGeV. Transparency,

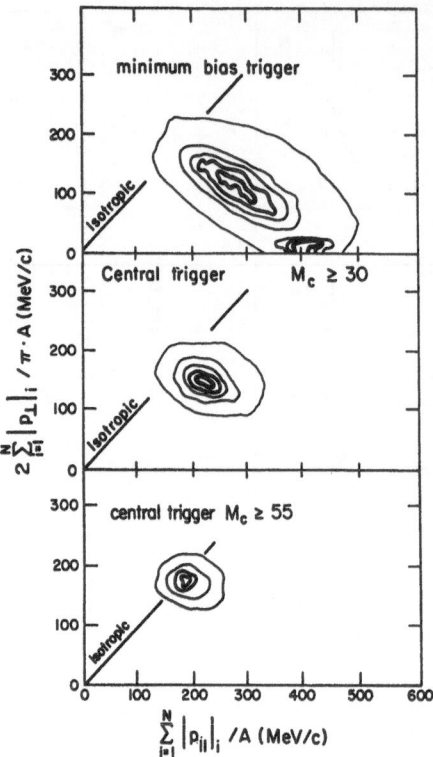

Figure 2. Contour plots of the average fragment momentum components perpendicular and parallel to the beam axis for minimum bias (top) and central (center) Ca+Ca, and central Nb+Nb collisions (bottom) at $E/A = 400\,\text{MeV}$ [Gus84a].

on the other hand, is observed at both CERN energies with a decreasing fraction of 'stopped energy' in going from 60 to 200 AGeV [Sat88].

5.2 Equipartition in Momentum Space

Thermal equilibrium may simply be viewed as equipartition of the available energy into various degrees of freedom. A minimal criterion for global thermalization thus is an isotropic distribution of the fragments momenta in the fireballs rest-frame. Experimentally, the degree of isotropy can be judged in each event from the momentum distributions of the fragments in the CM-system by calculating the ratio

$$R = \frac{2}{\pi} \sum_{i=1}^{N} |\,p_\perp(i)\,| \bigg/ \sum_{i=1}^{N} |\,p_\parallel(i)\,| \tag{4}$$

From simple phase-space considerations one finds $R = 1$ for an isotropically expanding system of N particles. In particular, global stopping of the two nuclei would result in a ratio $R \geq 1$, whilst in presence of transparency this ratio would always stay below one.

In Fig. 2 (top) contour lines of the event yield accumulated with a minimum bias trigger are shown in the $\langle p_\perp \rangle$ versus $\langle p_\parallel \rangle$ plane for the Ca+Ca system at 400 AMeV. The major part of events is far away from the $R = 1$ region represented by the diagonal line. The peak at small $\langle p_\perp \rangle$ but large $\langle p_\parallel \rangle$ corresponds to peripheral reactions and is dominated by projectile spectator fragments. This contribution vanishes as central collisions are selected (Fig. 2 center). The maximum of the yield is now shifted towards the diagonal line but is still significantly below $R = 1$, except for very few events. This behaviour is different from the heavier systems Au+Au or Nb+Nb (lower part of Fig. 2) at the same incident energy, where central collisions fulfil the condition for stopping and isotropy on the average.

5.3 Particle Spectra

Equipartition in momentum flux in central collisions of heavy nuclei is only a necessary but not a sufficient condition for global thermalization. A much stronger argument would be the observation of Maxwell-Boltzmann type energy spectra. In order to minimize possible spectator contaminations, which dominate the particle spectra at values close to the projectile and target rapidity, the following analysis has been restricted to particles observed at $\theta = 90° \pm 5°$ in the CM-system. Kinetic energy spectra of protons from Nb+Nb collisions at 400 AMeV and $\theta_{CM} = 90°$ are displayed in Fig. 3. A relativistic Boltzmann distribution was fitted to the spectra using σ_0 and T_{app} as free parameters:

$$E \frac{d^3\sigma}{dp^3} = \frac{E}{p^2} \frac{d^2\sigma}{dp\, d\Omega} = \frac{\sigma_0}{4\pi m^3} \frac{E}{\zeta(T_{app})} \exp\left(\frac{-E}{T_{app}}\right) \tag{5}$$

where E, m, and p are the total energy, rest mass, and momentum of the particle, respectively, and $\zeta(T_{app})$ is the normalization factor of a relativistic Boltzmann distribution of temperature T_{app}. The inverse slope-parameter T_{app} increases as a function of multiplicity for Ca+Ca as well as for Nb+Nb at this energy of 400 AMeV. Values ranging from 42 MeV up to 56 MeV and 66 MeV, respectively, were extracted from the data. Since the above equation describes a fireball in which *all* energy is converted into heat, the extracted T_{app}-values have to be regarded an upper limit for the temperature. The observed rise of the inverse slope parameter with increasing multiplicity might reflects again an approach towards global equilibrium, corresponding to the interpretations of figure 2. On the other hand, the increase of the apparent temperature, i.e. mean transverse energy per particle, might also be caused by an increasing formation of composite particles which reduces the total number of emitted fragments (at constant number of nucleons) and therefore the number of degrees of freedom in the system. In fact, since a strong increase of the relative abundance of composite particles is observed with increasing multiplicity (see Sec. 6), composite particle production may be responsible for at least part of the observed increase in T_{app}.

A more detailed inspection of the transverse proton energy-spectra reveals a clear deviation from the pure Boltzmann distribution being most evident at low kinetic energies. A shoulder, also observed in other experiments [Nag80], becomes more and more pronounced with increasing multiplicity. There are at least two possible reasons for this phenomenon; i) the formation of composite particles leads in terms of the coalescence model [Gut76, Gyu83] to a depletion of the primordial proton yield at low kinetic energies, and/or ii) the proton spectra are distorted by a collective radial expansion velocity of the decaying fireball. The second interpretation was formulated by Siemens and Rasmussen in the *blast wave* model [Sie79a]. In this description a part of the available energy is converted into radial flow, so that the temperatures are reduced compared to the pure fireball assumption. Fitting the proton spectra by such a superposition of an ordered radial expansion- and a uncorrelated thermal-part results in a considerable reduction of T_{app} by about a factor of two. However, the non-orthogonality of these contributions to the transverse energy spectra makes the extraction of 'temperatures' (and blast wave velocities) very problematic and those parametrizations can provide only a rough estimate of the fractions of thermal and radial flow energy. Moreover, this fraction will also not be constant during the collision time, since the ordered radial motion kinetic energy increases with expansion, while the intrinsic temperature decreases.

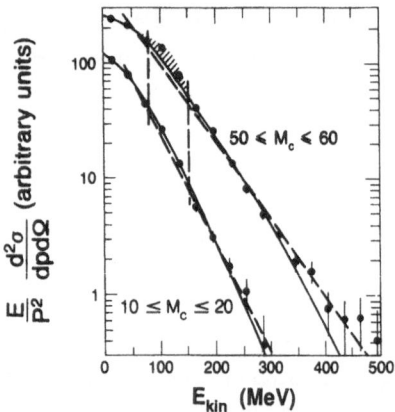

Figure 3. Proton spectra at $\theta_{CM} = 90°$ from Nb+Nb collisions at 400 AMeV. The dots are experimental data points, and error bars represent statistical errors only. The dashed lines indicate the fitted Boltzmann distribution whereas the fits from the explosion model are shown by full lines. The vertical dashed lines mark the region that is excluded from the fits because of possible deuteron contaminations indicated in the high multiplicity spectrum by the shaded area [Gus84a].

6 Composite Particle Production and Entropy

Up to now we have discussed only spectral shapes and mean transverse energies of composite particles but were not concerned about their absolute and relative yields. Already the first experimental studies of Gutbrod et al. [Gut76] have demonstrated that composite light fragments are copiously produced in high-energy heavy-ion collisions. Their cross sections were found to be two or three orders of magnitude higher than for proton-nucleus reactions at comparable incident beam energies. Interest in the mechanism of the copiously produced light fragments was driven mainly by calculations [Sie79b, Sub81, Kap84, Stö83] showing that the observed deuteron to proton ratio can be related to the entropy production in the system. If the entropy stays constant during the expansion [Ber81, Kap81, Stö86], composite particles contain information about the highly excited and compressed stage of the collision. Determination of the entropy thus might help to determine the EOS of hot dense nuclear matter. Furthermore, possible phase transitions would be perceptible by their sudden liberation of new degrees of freedom, i.e. by an extra production of entropy [Cse83a, Stö84, Bar85, Kap84, Stö86].

Over the recent years a lively and still unsettled debate took place about the significance of entropy in high-energy heavy-ion collisions and about their relationship to nuclear cluster production [Aic87a] (for a detailed review the reader is referred to Ref. [Cse86a]). Based on hydrodynamics [Kap81, Stö83] and intranuclear cascade calculations [Ber81, Gud85] it has been argued that nuclear matter expands quasi isentropically after it has thermalized at a certain baryon-density and temperature. After the collisions among the constituents of the expanding system have ceased, the phase-space density stays constant due to Liouville's theorem and the entropy determines the abundances of the produced clusters. Siemens and Kapusta [Sie79b] attempted the first estimate of entropy production. Their highly idealized model assumes that for some time during expansion nuclear matter can be treated as a dilute gas of nucleons with a small contamination of deuterons in local chemical equilibrium. Assuming equal numbers of protons and neutrons and a temperature much larger than the deuteron binding energy (i.e. $n_p \cong n_n \gg n_d$), the entropy per nucleon is given by

$$S/A \cong 3.95 - \ln(n_d/n_p) \qquad (6)$$

according to statistical mechanics of ideal gases. This simple relation, however, gives significantly larger entropy values when extracted from inclusive fragment data [Nag81] than were predicted by diverse dynamic models such as hydrodynamics and intranuclear cascade. Even the inclusion of real pions and the assumption of a super-soft EOS does not yield enough entropy. This so-called 'entropy puzzle' was resolved subsequently by Stöcker et al. [Stö83, 84], who showed that the above relation is not appropriate because of a significant change of the fragment yields after freeze-out due to the decay of particle unstable excited nuclei (see also [Bon77, Gos78, Fai82]),

$$A^* \rightarrow (A - A_X) + X \qquad (7)$$

($X = n, p, \alpha$, etc.) which becomes increasingly important at low bombarding energies ($E_{Lab} \leq 400\,\text{AMeV}$) where they dominate the chemical equilibrium contribution, and by Gutbrod and Doss et al. [Gut83, Dos85], who showed in first exclusive experiments, that the cluster to proton ratios depend strongly on the multiplicity of the event; in peripheral collisions – which dominate the inclusive particle spectra – the ratios are much smaller than in central collisions. From these measurements it became evident that the naive use of impact parameter averaged data had been one of the major reasons of the incorrect entropy determination.

Recently, the quantum statistical model (QSM) [Stö83, Stö84] has been further developed by Hahn and Stöcker [Hah88]. This model is based on the classical chemical equilibrium model of Mekjian [Mek78, Ran81, Fai82], incorporates Fermi energy, and takes into account, simultaneously, particle unstable nuclides up to mass 20 (see above) and ground state nuclei up to mass 130, as well as Bose condensation of integer spin nuclides, and excluded volume effects. Assuming that thermal and chemical equilibrium are established during expansion of the system, the model describes the abundance ratios ($x = d, t, {}^3\text{He}, \alpha, ...$) as a function of the specific entropy of the system with the breakup temperature, T_b, and the breakup density, ρ_b, as free parameters. Different from previous models, the curves $n_x/n_p(S/A)$, which are calculated for a

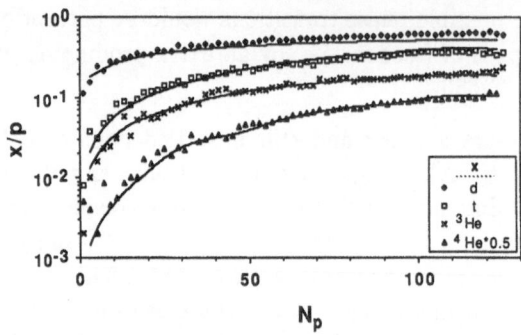

Figure 4. n_x/n_p ratios for the reactions Au+Au at 400 AMeV (where x stands for d, t, ^{3}He and ^{4}He) as a function of the participant proton multiplicity, N_p. The solid lines are fits to the data in the framework of the QSM.

grandcanonical ensemble, i.e. for infinite nuclear matter, can also be employed at finite multiplicities [Hah88]. Deviations from a classical microcanonical treatment were proven to be relevant only for very low particle numbers, e.g. for $n = 10$ they are of the order of 20% [Hah88]. This demonstration of the applicability of the QSM at finite multiplicities constitutes a decisive improvement over the previous methods to extract specific entropies from experimental data. Instead of extrapolating the experimental n_x/n_p-data to infinite multiplicities in order to extract S/A, all cluster ratios are fitted simultaneously as a function of multiplicity with T_b and ρ_b as the only free parameters.

Results of such least square fits to the Plastic-Ball data are presented in Fig. 4 for Au+Au at 400 AMeV [Dos88]. Figure 5 displays the resulting S/A values for central collisions ($N_p/N_p^{max} = 1$) together with two model calculations. The experimental values are smaller than the pure thermal fireball model, indicating that compression effects play an important role for entropy production in heavy-ion reactions. The curve, labeled as mean-field, is a hydrodynamic calculation using an EOS based on relativistic mean-field theory [Bog83]. It lies below the data, reflecting mainly the lower entropy per nucleon for infinite nuclear matter, for which the calculation was done. An extrapolation of the experimental S/A-values to infinite multiplicities, subject to the aforementioned uncertainties, would yield numbers in agreement to the hydrodynamic model. Furthermore, the similarity of the extracted entropy values for the two systems at a given bombarding energy and constant N_p shows that the specific entropy is mainly dependent on the number of particles in the reaction volume. This finding is supported also by recent preliminary results of the Diogene group [Ala88].

In a recent paper, studying the effects of momentum dependent interactions [Aic87b], it was claimed that the deuteron-to-proton ratio, at least for heavy systems, is sensitive to the nuclear EOS. The theoretical value of $(n_d/n_p)_{max} = 0.62$ assuming either a rather hard EOS neglecting momentum dependent forces or a soft EOS including momentum dependent forces (see also subsection 7.4 and other contributions to this winter school for a discussion of these effects), is in good agreement with the data at all bombarding energies. However, three-body correlations should also be considered in the formation of deuterons, as was suggested recently by Brown [Bro88].

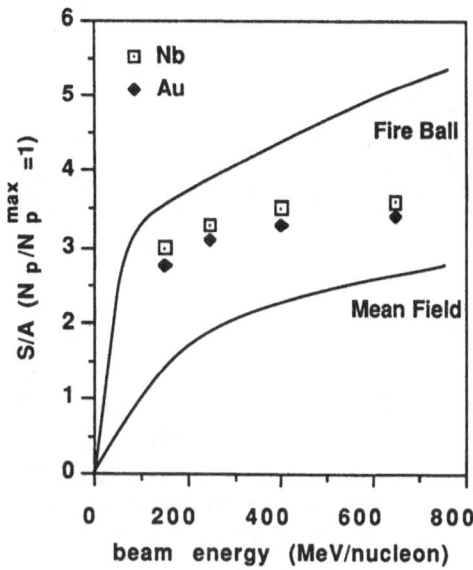

Figure 5. Comparison of the bombarding energy dependence of S/A with the fireball and hydrodynamic mean-field model. The experimental points cannot be compared directly with the models, because the data show results for finite multiplicities, $N_p/N_p^{max} = 1$.

6.1 Multifragmentation

The investigation of intermediate mass fragment (IMF) production $5 \leq A \leq 30$ is topic of much current interest and is driven, for example, by questions of how much excitation energy nuclei can support before they break apart, and how the disassembly of highly excited systems takes place. The widely differing mechanisms proposed in various models include fragment formation via purely statistical processes [Ran81, Bon85], emission from a gas of nucleons and fragments in equilibrium [Mek77, Gut76, Cse80, Stö83], breakup as a result of dynamic instabilities or partial equilibrium processes [Ber83], or the treatment of the expansion on a microscopic level in form of TDHF methods [Str84], or molecular dynamics [Kis87]. Another group of models, finally, assumes neither global nor local equilibrium but rather treats fragment formation by methods of percolation theory (see lectures given by Ngo [Ngo89]) or assumes a cold breakup similar to the shattering of glass. Furthermore, at intermediate bombarding energies of $E_{Lab} \simeq 100\,\text{AMeV}$, i.e. for moderate temperatures of the system, the expansion may lead to lower densities and possibly to a liquid-gas phase-transition [Dan79, THa85]. The study of fragment yields thus offers the possibility to investigate the EOS at higher temperatures and lower densities than the ground state.

Experimentally, a total breakup of the system into a large number of nuclear fragments, known as multifragmentation, was first observed in proton induced reactions at bombarding energies of $E_p > 1\,\text{GeV}$ [Pos71, Fin82]. In contrast to sequential evaporation processes, multifragmentation may be characterized by nearly simultaneous breakup of a nucleus into several pieces. Until recently, heavy-ion experiments identifying IMFs have only studied single particle inclusive measurements or few particle correlations. To address the questions of fragment formation as a function of impact parameter, a dedicated run has been performed with the Plastic-Ball detector

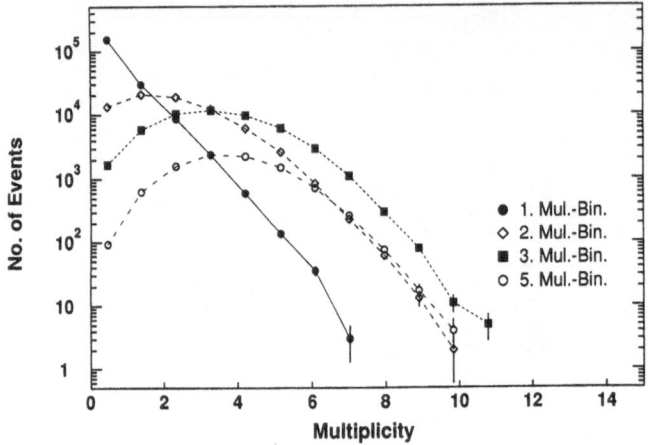

Figure 6. Fragment multiplicity ($Z > 3$) distributions for 200 AMeV Au+Au for four multiplicity bins of N_p , increasing from peripheral to central collisions. The multiplicities correspond to fragments observed in the forward part of CM-hemisphere only [Dos87].

at 200 AMeV bombarding energy. In this experiment all elements up to neon could be resolved in the forward hemisphere of the CM-system [Dos87]. In peripheral collisions fragments were observed with rapidities close to the beam rapidity, consistent with expectations for fragmentation of an excited projectile spectator. Sampling more central collisions reduces the average rapidity of the fragments to values intermediate between that of projectile and target. In the most central collisions, finally, they were found to be centered with respect to y_{CM}. The large acceptance of 2π solid angle for intermediate mass fragments and 4π solid angle for light fragments achieved in this experiment allows to investigate the properties of the fragments event-by-event. Multiplicity distributions of fragments with $3 \leq Z \leq 10$, observed in the forward part of the CM-hemisphere, are shown in Fig. 6 for different bins in participant proton multiplicity. In central collisions, where a large fraction of the projectile charge is contained in light and intermediate mass fragments, an average of 3–4 $Z \geq 3$ fragments was observed at $\theta_{CM} \leq 90°$. Extrapolation to 4π leads to about 8 IMFs, with a significant number of events yielding up to 20 fragments. These numbers are still slight underestimates because of the low energy-cutoff of the detector. Figure 6 clearly illustrates, that multi-fragmentation is a dominant process in central collisions of heavy nuclei at this bombarding energy. The onset of multi-fragment emission, i.e. the amount of excitation energy, E_X, in the composite system that is necessary to increase the production cross section of more than one IMF dramatically, was recently reported to be located near $E_X \simeq 300$ MeV [Tro88]. Furthermore, this value was found to be almost independent of the projectile-target mass system.

7 Nuclear Matter Flow

The comparison of the measured lower specific entropy compared to values of a pure thermal fireball, revealed first evidence that only a certain part of the available CM-energy is converted into thermal degrees of freedom during the collision. This argument is, as well in the case of Ref. [Sto82], based on a subtraction of the energy component one considers thermal from the total available kinetic CM-energy. A physical observable that is believed to be directly related to the compressional energy fraction, is the collective ordered motion, i.e. the flow of nuclear matter in the final stage of the reaction. This is probably the most obvious manifestation of the EOS in the experimental data.

Collective fluid like behaviour of nuclear matter in high-energy heavy-ion collisions was first predicted on the basis of hydrodynamics by Scheid, Müller, and Greiner [Sch74], and co-workers of the Frankfurt school [Bau75, Hof76a, Stö78]. Siemens and Rasmussen [Sie79a] argued that the hydrodynamical flow results in a symmetrical blastwave associated with the explosion of the system, being observable by a 'shoulder-arm' behaviour in the shape of the energy spectra. Indications for such effects, which however left the mechanism behind still open, have been reported from inclusive experiments [Nag80] and were discussed to some extend already in Sec. 5.3. In this paragraph we will now concentrate on the directed collective sidewards emission, called transverse matter flow.

7.1 Sphericity Method

Data from 4π detectors arē ideally suited to study emission patterns and event shapes in high multiplicity heavy-ion reactions. Similar problems of event topology analyses and jet-finding algorithms have been treated earlier already in high-energy particle physics. The thrust and sphericity analysis have successfully been applied to those data [Bra79]. Since the thrust vector cannot be calculated analytically, the sphericity method has generally been used and was also proposed for the analysis of flow patterns in heavy-ion reactions [Cug82b, Gyu82]. The 3-dimensional sphericity tensor

$$F_{ij} = \sum_{\nu=1}^{M} p_i(\nu)\, p_j(\nu)\, \omega(\nu) \tag{8}$$

$(p_i(\nu); i, j \equiv x, y, z$, are the cartesian components of particle ν, and ω is an appropriate weight factor) is calculated event-by-event from the CM-momenta of all measured particles. Choosing $\omega(\nu)$ in such a way that composite particles obtain the same weight per nucleon as individual nucleons at the same velocity, i.e. $\omega(\nu) = 1/2m(\nu)$, yields the kinetic energy flow tensor. The sphericity tensor approximates the event shape by an ellipsoid whose orientation in space and whose aspect ratios can be calculated by diagonalizing the tensor. In particular, the polar angle θ_F of the major principal axis to the beam axis is called the flow angle. Although being mathematically elegant, the sphericity method suffers substantially from finite particle number distortions for $M \leq 100$, which in practise restricts the only useful information to the Jacobian-free weighted flow angle distribution, $dN/d\cos\theta_F$.

7.1.1 Symmetric Systems

The sphericity analysis requires each event to be transformed into the CM-system. Therefore, it is most uncritical applicable to symmetric projectile-target systems where the effective CM-system of the reaction can to a good approximation be identified with the nucleon-nucleon CM-system. Events from collisions of Ca+Ca, Nb+Nb, and Au+Au, accumulated with the Plastic-Ball detector and analyzed in terms of the sphericity tensor [Gus84b, Rit86], are shown in Fig. 7. Different from the zero-peaked flow angle distributions of the Ca data, a net deflection angle, increasing with multiplicity, is observed for the heavier systems Nb+Nb and Au+Au. This experimental observation, a typical example of the virtue of 4π detectors, provided the first conclusive evidence of collective sidewards flow and substantiated the long standing prediction of the nuclear fluid dynamical model [Sch74, Stö80, Buc83]. Thus, the key mechanism for an investigation of the nuclear EOS has been experimentally established. Taking into account the finite particle number effects of the flow tensor and the acceptance cuts of the experimental apparatus, the model reproduces the experimental data reasonable well [Buc84] as shown in the right hand column of Fig. 7. Plotted are also results of the same analysis [Gus84b, Buc84] performed with events generated by cascade calculations from the Yariv-Fraenkel and Cugnon code [Yar79, Cug82a] filtered to the Plastic-Ball acceptance. In this case the distributions are peaked at zero degrees for all impact parameters, leading to the conclusion that collective phenomena, definitely appearing in higher mass systems, are not accounted for by the cascade model [Bra86, Mol86].

The fact that finite flow angles are seen in the data indicates that in those events a reaction plane exists that is defined by the flow-axis and the beam-axis. Thus, the event shapes can be studied in more detail by rotating all events by the azimuthal angle $-\varphi_F$, determined from the flow analysis, so that their individual reaction planes all fall into the x-z plane, with z being the beam-axis. The invariant cross section in the reaction plane, $d^2\sigma/dy\, d(p_x/m)$, [Stö82, Buc83] where p_x is the projection of the transverse momentum onto the reaction plane, is shown in Fig. 8 for 400 AMeV Ca+Ca and Nb+Nb data together with cascade calculations. The depletion near the target rapidity is due to the limited acceptance of low-energy particles in the lab-system which enhances the flow angle artificially. The difference between measured and cascade data is obvious and reflects again the results of the flow angle distributions. Figure 8 also shows the appearance of another collective effect; the highest level contour which results largely from the projectile remnants is shifted away from the beam axis into the flow direction, indicating a definite *bounce-off* effect. Because the flow angle is definitely larger than one would expect from the position of the bounce-off peak of about 50 AMeV/c, one can conclude that the observed strong sidewards flow seen in Fig. 7 is mainly due to mid-rapidity particles. The data also show that the bounce-off and side-splash effects appear to be in the same reaction plane. A schematic view of both effects is depicted in Fig. 9.

The bombarding energy dependence of the flow angle distributions has been studied in Ref. [Rit86] for Nb+Nb and Au+Au data. In both systems the peak position in the frequency distribution of the flow angle was found to increase with projectile energy up to about 400 AMeV followed by a slight fall-off towards higher energies. These findings are confirmed by high multiplicity 900 AMeV U+U events observed in the Bevalac streamer chamber [Bea85].

In order to extract the magnitude of the available CM-energy being converted into collective sidewards motion, the 400 AMeV Nb+Nb data were fitted with the statistical event simulation code of Fai and Randrup [Fai83], suitably modified to include the two observed collective phenomena [Gus84b]. From these simulations it was estimated that about 10% of the total kinetic energy available in the CM-system is contained in the *directed* collective motion. This relative small amount explains why it is so extremely difficult to observe the effect in inclusive data. It should be stressed, that this number is not to be mistaken with the higher amount of the total compressional energy in the system, as the sphericity method is not capable of detecting isotropical flow.

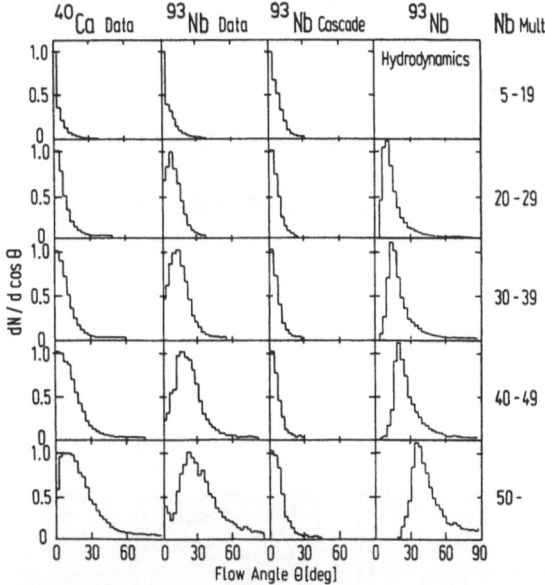

Figure 7. Frequency distribution of the flow angle θ_F for Ca+Ca and Nb+Nb collisions at 400 AMeV for five multiplicity bins [Gus84b]. The two right hand columns show results for the intranuclear cascade simulation [Gus84b, Yar79] and for the hydrodynamical calculation [Buc84], respectively. Finite particle number distortions and experimental inefficiencies are taken into account in both models.

7.1.2 Asymmetric Systems

Data of asymmetric projectile-target systems have mainly been accumulated by several Streamer Chamber groups at LBL, Riverside, and MSU, by emulsion experiments, and recently also by the Diogene detector at Saturne. A particular problem, however, arising in the sphericity analysis is that one does not immediately know the velocity of the effective CM-system, as it obviously depends on the impact parameter. Therefore, one needs to achieve almost full kinematical reconstruction in each event in order to calculate y_{CM} from the sum of the momenta of all (observed) reaction products.

Such an analysis has for the first time been performed by Renfordt et al. [Ren84, Str83]. These data — supporting the previous findings — exhibit again a definite sidewards deflection of nuclear matter away from the beam axis, and are in disagreement to simulated events of the cascade code of Ref. [Cug81], reflecting again a lack of compressional energy in this model. Similar results of triple differential cross sections have been reported from the Diogene collaboration [Val88, Hot88b]. Here the $1/e$ contours of the 2-dimensional gaussian distribution, as shown in Fig. 8, were fitted in regions of full acceptance to the momentum distribution in the y vs p_x/A plane. This analysis revealed also increasing flow angles with increasing multiplicity and indicated an approach to pancake shaped events for rather central collisions.

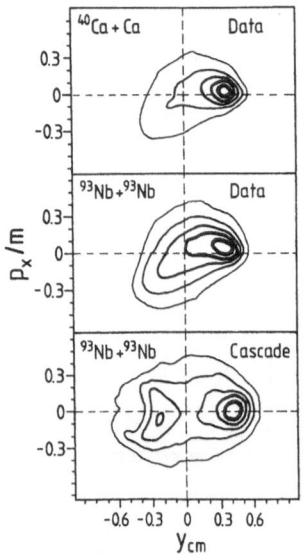

Figure 8. Linear contour plot of p_x/A as a function of the CM-rapidity for semi-central collisions of Ca+Ca, Nb+Nb and cascade simulation data, each at a bombarding energy of 400 AMeV [Gus84b].

7.2 Global Transverse Momentum Analysis

The beauty and some drawbacks of the sphericity method have briefly been mentioned in the previous section, already. Because of the strong finite particle number distortions being most disturbing for the light Ca+Ca system, no decisive conclusions about the possible presence or absence of the flow effect could be drawn so far for this system. Only a weak indication has been found in the most central collisions. Furthermore,

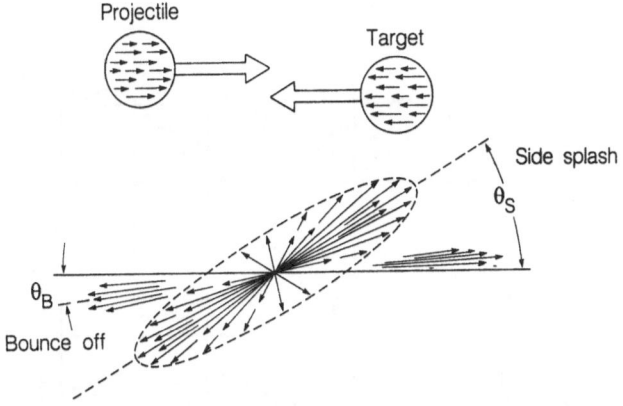

Figure 9. Schematic view of the side-splash and bounce-off effect.

since all experimental biases and inefficiencies, and in addition also the two separate collective phenomena (see Fig. 9) are folded into one single observable, it is extremely difficult to compare the experimental results of the sphericity analysis with theoretical model predictions.

To avoid most of these difficulties, a novel, more sensitive method has been introduced by Danielewicz and Odyniec [Dan85]. In this transverse momentum analysis the reaction plane is constructed individually for each single particle μ from the transverse momentum components $p_\perp$ of all remaining particles of the same event:

$$\vec{Q}(\mu) = \sum_{\nu \neq \mu} \vec{p}_\perp(\nu)\, \omega(\nu) \tag{9}$$

where $\omega(\nu)$ is a weight factor chosen to be ± 1 for fragments emitted in the forward and backward hemispheres in the CM, respectively. If $\omega(\nu)$ would be equal for all particles, then the transverse momentum vector would result (in case of an ideal detector) in $\vec{Q} = \vec{0}$ by transverse momentum conservation. In the next step of the analysis, the transverse momentum vector of particle μ is projected onto this approximated reaction plane by evaluating the scalar product

$$p_x(\mu) = \vec{p}_\perp(\mu) \cdot \frac{\vec{Q}(\mu)}{|\,\vec{Q}(\mu)\,|} \tag{10}$$

yielding in in-plane transverse momentum, p_x, which is generally plotted as a function of rapidity. The definition of $\vec{Q}$ ensures that autocorrelations are removed and that the method is sensitive only to dynamic multiparticle correlations.

This method was first applied to 1.8 AGeV Ar+KCl Streamer-Chamber data [Dan85]. Although employing the standard sphericity analysis revealed no significant effect for this reaction, the transverse momentum method evidently showed flow effects, substantially stronger than in the cascade model, but weaker than in the hydrodynamical model. Moreover, since averaging is done over particles and not over events, this method exhibits collective flow effects also in much smaller data samples than is necessary for the sphericity analysis. It is therefore much more appropriate for the analysis of emulsion and streamer chamber data [Bea86], for example.

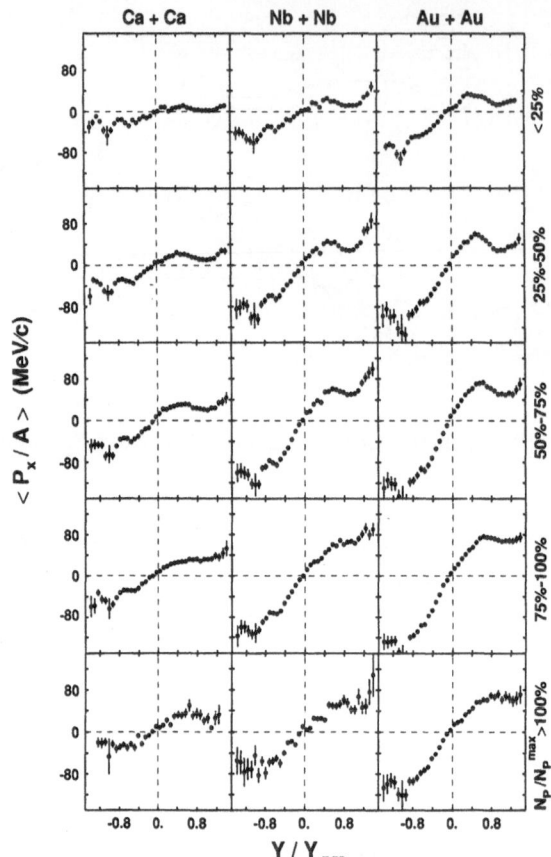

Figure 10. Mean transverse momentum per nucleon projected onto the reaction plane as a function of the CM-rapidity for different reaction systems at 400 A MeV and different impact parameter regions, respectively.

7.2.1 Symmetric Systems

Figure 10 shows as an example the mean transverse momentum per nucleon projected onto the reaction plane, $\langle p_x/A \rangle$, as a function of the normalized CM-rapidity, y/y_{proj}, for Ca+Ca, Nb+Nb, and Au+Au collisions, each at a bombarding energy of 400 A MeV [Gus88]. The error bars represent statistical errors only and data points are corrected for the deviation from the true reaction plane, as described in Ref. [Dan85]. The corresponding values of $\langle \cos \frac{1}{2}(\phi_1 - \phi_2) \rangle$, which are determined by splitting each event into two randomly chosen subevents and calculating their individual reaction planes, typically varied between 0.82 and 0.98. Particles near midrapidity ($\delta y = \pm 0.1$) which are expected to be only weakly correlated to the reaction plane because of symmetry reasons were excluded from the summation of $\vec{Q}$ to reduce fluctuations. The characteristic S-shape of the resulting curves, clearly evident for all systems and in all multiplicity bins now, demonstrates the traceability of the reaction plane from the fragments momenta and thus is a clear sign of collective nuclear matter flow. Statistically uncorrelated (i.e. pure thermal) particle emission would result in a 'null-effect' or even in a slightly reversed curve (negative values in the forward CM-hemisphere and positive values in the backward CM-hemisphere) due to transverse momentum conservation [Kam86]. The measured in-plane transverse momenta are found to increase with increasing projectile-target mass and to reach a maximum in semi-central collisions. The dip in the distributions just below the projectile rapidity in all but the most central collisions results largely from the bounce-off effect. It directly demonstrates that this effect is a small sidewards deflection of the projectile remnants into the reaction plane and should consequentially be distinguished from the side-splash of the participants. Since the corresponding low energy target fragments are mainly excluded by the detector acceptance, the same dip does not show up in the target rapidity region.

Curves of the type presented in Fig. 10 can directly be compared to event generating models (after applying the filter simulating the experimental apparatus to the calculated data) without further reduction of information. However, it is also of general interest to extract a quantitative measure of the transverse momentum flow containing as little of the detector bias as possible from those data, thus enabling us to compare peripheral and central collisions, and different projectile-target system at different beam energies with each other. The scaling properties of such a variable provides a crucial test for the origin of the flow and may help to discriminate between different models [Sch87a, 88, Bon88]. As discussed in Ref. [Dos86], the maximum of the collective momentum transfer occurs close to the target and projectile rapidities, where the relative contributions of spectator and participant particles depend strongly on the impact parameter (see for example the influence of the bounce-off effect on the $\langle p_x/A \rangle$ distributions in Fig. 10), and where experimental biases are most disturbing. However, to a good approximation all S-shaped curves are straight lines near midrapidity, so that their slope,

$$\mathcal{F} = \frac{d\langle p_x/A \rangle}{d(y/y_{proj})}\bigg|_{y=y_{CM}} \tag{11}$$

measured in MeV/c, may be considered a good measure of the directed flow [Dos86].

The dependence of $\mathcal{F}$ on the impact parameter is shown in Fig. 11 for Nb+Nb and Au+Au at different beam energies ranging from 150 AMeV up to 800 AMeV. As already seen from the S-shaped curves and the flow angle distributions, the flow increases with projectile-target mass and scales roughly with $A^{1/3}$ between Nb and Au. The multiplicity dependence shows the directed flow peaking in semi-central collisions. This behaviour is different from the flow angles, which were found to increase steadily with multiplicity. There are at least two explanations for this effect; i) the sphericity analysis approximates the *whole* event by an ellipsoid and is therefore considerably affected by spectators, i.e. particularly by the bounce-off effect at low multiplicities which tends to reduce the effective flow angle, and ii) the transverse momentum method, on the other hand, is sensitive only to φ-asymmetries in the event and is unable to distinguish between prolate and oblate event shapes. To be more precise, the directed flow $\mathcal{F}$ and the flow angle θ_F are by no means proportional to each other. Analytically they are connected to each other via the aspect ratio of the

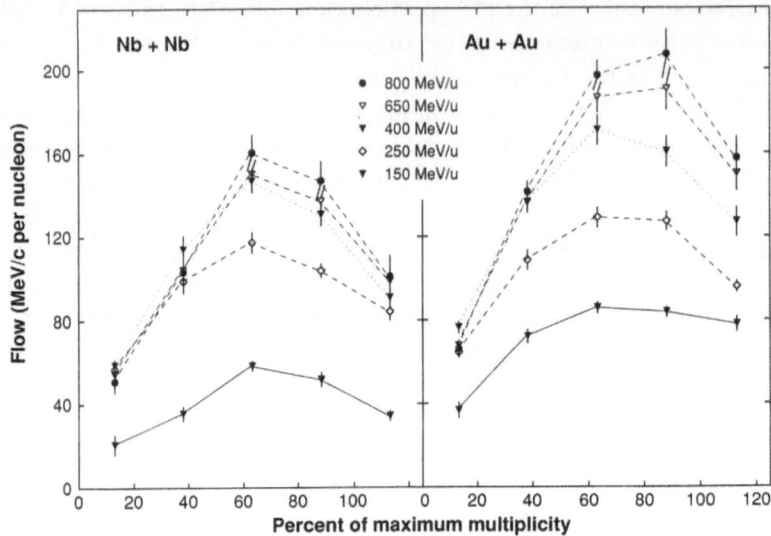

Figure 11. Directed flow as a function of the participant proton multiplicity, N_p/N_p^{max}, for Nb+Nb and Au+Au at different beam energies.

largest to smallest half axis of the associated ellipsoid, R_{13},

$$\mathcal{F} = \left(\frac{dp_x}{dp_z}\right)_{p_z=0} = \frac{(R_{13}-1)\sin\theta_F\cos\theta_F}{R_{13}\cos^2\theta_F + \sin^2\theta_F} \qquad (12)$$

which directly follows from the assumption of ellipsoid shaped events [Kam86] and as was recently obtained also empirically in the framework of an almost analytic transport model [Sch87a]. Therefore, increasing flow angles and simultaneously decreasing aspect ratios R_{13} (a behaviour that is indeed indicated when selecting central collisions) can result in decreasing flow-values $\mathcal{F}$.

The energy dependence of the maximum directed flow, i.e. the mean values of the third and fourth multiplicity bin, is displayed in Fig. 12 for all symmetric systems investigated with the Plastic-Ball. Plotted in this way, the flow increases gradually with increasing beam energy; rather rapidly up to 400 AMeV and turning into a saturation curve at the highest bombarding energies. The error bars plotted are of statistical source only including the uncertainty from the straight line fits to the data. If the impact parameter averaged minimum bias data would be plotted instead of the maximum values in Fig. 12, then one would find a slight fall-off towards higher beam energies above 650 AMeV [Dos86], which can be attributed mainly to the weaker flow effect in peripheral collisions above 650 AMeV.

7.2.2 Asymmetric Systems

The global transverse momentum analysis may directly be applied to data of asymmetric systems, because it does not necessarily require a transformation of the event into the effective CM-system. One simply needs to estimate y_{CM}, e.g. on the basis of the fireball model, and to assign the weight $\omega(\mu) = 0$ to all particles within a certain

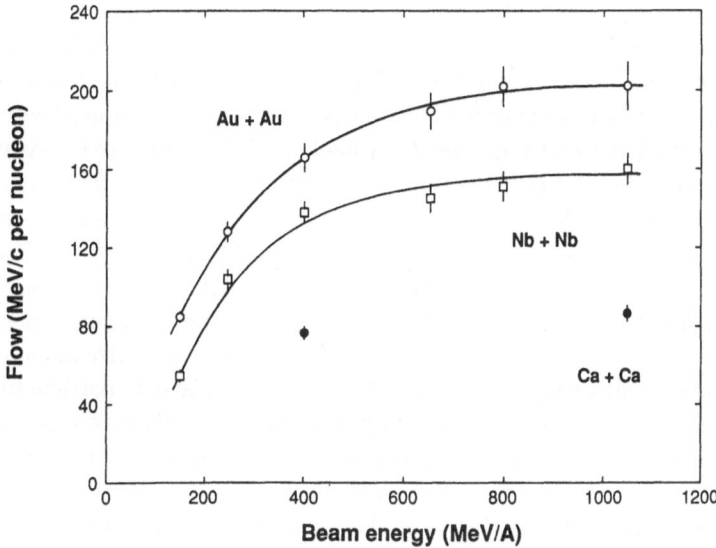

Figure 12. Flow of semi-central collisions as a function of beam energy. The error bars are of statistical nature only including the uncertainty from the straight line fits, as described in the text. The solid curves are to guide the eye.

rapidity window, $y_{CM} - \delta y \leq y_{Lab}(\mu) \leq y_{CM} + \delta y$, with δy being sufficiently large to ensure that only particles of the forward and backward hemisphere, respectively, are taken into account with their proper signs for the reaction plane calculation. Plotted is then as usual $\langle p_x/A \rangle$ as a function of y_{Lab}, where the crossing point should lie well inside the δy-window around the estimated y_{CM}, otherwise, one or both quantities need to be readjusted and the analysis repeated. If full particle information cannot be achieved in an experiment, one even might replace the 'true' rapidity of a particle by it's pseudorapidity, $\eta = -\ln(\tan(\theta/2))$, or the 'real' $p_\perp$ by the 'pseudo' $p_\perp = p_{proj} \cdot \tan\theta$, as done for example with emulsion data in Refs. [Cse86b, Jai88]. However, such an analysis can in general provide only qualitative results.

Examples of the transverse momentum analysis applied to asymmetric systems have been reported in Refs. [Dan86, Bea86, Dan88] and were in addition shown by D. Keane and D. L'Hôte on this winter school [Kea89, Hot89]. Generally, all these curves exhibit the same trends as discussed before, however, they are not symmetric around the effective CM-rapidity. Comparisons with intranuclear cascade and VUU predictions, furthermore, demonstrate again the lacking flow in the cascade model and fairly good agreement to the VUU model when assuming a stiff EOS without momentum dependent interactions. Further results of asymmetric systems at comparable and lower bombarding energies have been reported in Refs. [Gos87, Dan88], and at higher bombarding energies of 2.1 and 3.37 AGeV in Refs. [Kam86, Bia86].

7.2.3 Flow of Intermediate Mass Fragments

Up to now we have discussed collective flow effects only for the entire event by averaging over protons and composite particles with a proper weight factor. However, several calculations, capable of producing nuclear fragments with $A > 1$, predict that a stronger flow effect should be observed for nuclear fragments than light particles emitted in the reaction [Stö81, Cse83b,84]. This effect has in fact been expected rather early [Bau75] by the simple argument that heavier clusters are produced with relatively lower (undirected) thermal velocities than light particles. In order to investigate this effect experimentally, the $\langle p_x/A \rangle$ distributions have been analyzed differentially as a function of the mass of the fragments. Results of such an analysis are shown in Fig. 13 for the reaction Au+Au at 200 AMeV and the multiplicity bin containing between 50% and 75% of N_p^{max} [Kam86, Gus88]. Fragments heavier than ^{4}He were only measured in the forward CM-hemisphere. The data clearly show an increasing transverse momentum flow *per nucleon* as the fragment mass increases. This rise amounts to more than 40% when comparing $Z = 1$ to $Z > 6$ fragments, and is particularly exciting since the φ-averaged transverse momenta per nucleon, $\langle p_\perp/A \rangle$, were found to decrease with increasing fragment mass. The stronger flow of intermediate mass fragments becomes even more pronounced if the correlation is studied in position space rather than in momentum space. This is clearly visible in figure 13b. Here the same data have been analyzed, but plotted is now the fraction of the particle's transverse momentum that lies in the reaction plane, $\langle p_x/p_\perp \rangle$. If the particles were emitted exactly into the reaction plane, then this alignment function would yield plus or minus one for ideal positive or negative alignment, respectively. Similar results were obtained from the φ-distributions of different fragments measured relative to the reaction plane [Dos87].

Comparing the relative increase of the alignment in momentum and position space in Fig. 13, one might interpret the increasing transverse momentum flow with increasing fragment mass as at least partly being caused by the substantially stronger spatial correlations of the intermediate mass fragments. This would be in line with the simple idealized interpretation of the fragment mass dependence of the collective flow, where one assumes nucleons and fragments stemming from a common thermalized source, characterized by a certain temperature, being boosted by a common collective asymmetric expansion velocity caused by the inherent asymmetry in the pressure in non-zero impact parameter collisions. The thermal energy component thus defines the A-independent thermal energy per particle, whereas the flow energy, i.e. the originally built-up compressional energy, is determined by the expansion velocity and therefore should have a linear A-dependence. The flow energy therefore gains an increasingly larger fraction of the fragments energy, and the random undirected thermal motion becomes less important, as the fragment mass increases.

The dependence of the directed flow on the mass of the emitted fragment has recently been studied in the framework of the QMD model by Peilert et al. [Pei88] for Au+Au at 200 AMeV. The increase of the p_x/A-values with the mass of the fragments is fairly well reproduced by these calculations, and good agreement to the data is found when employing a stiff EOS. Furthermore, the difference between the soft and hard EOS is most significant for the intermediate mass fragments ($Z > 6$), where it amounts to a doubling of p_x/A when going from the soft to the hard EOS. Since the increased transverse momentum flow p_x/A may result both from a stronger spatial correlation of the IMFs and/or from higher average transverse momenta, $p_\perp$, in directions of the reaction plane, it would be interesting to see whether the strength of the azimuthal alignment, $p_x/p_\perp$, comes out consistently with the data when the same hard EOS is employed in the model.

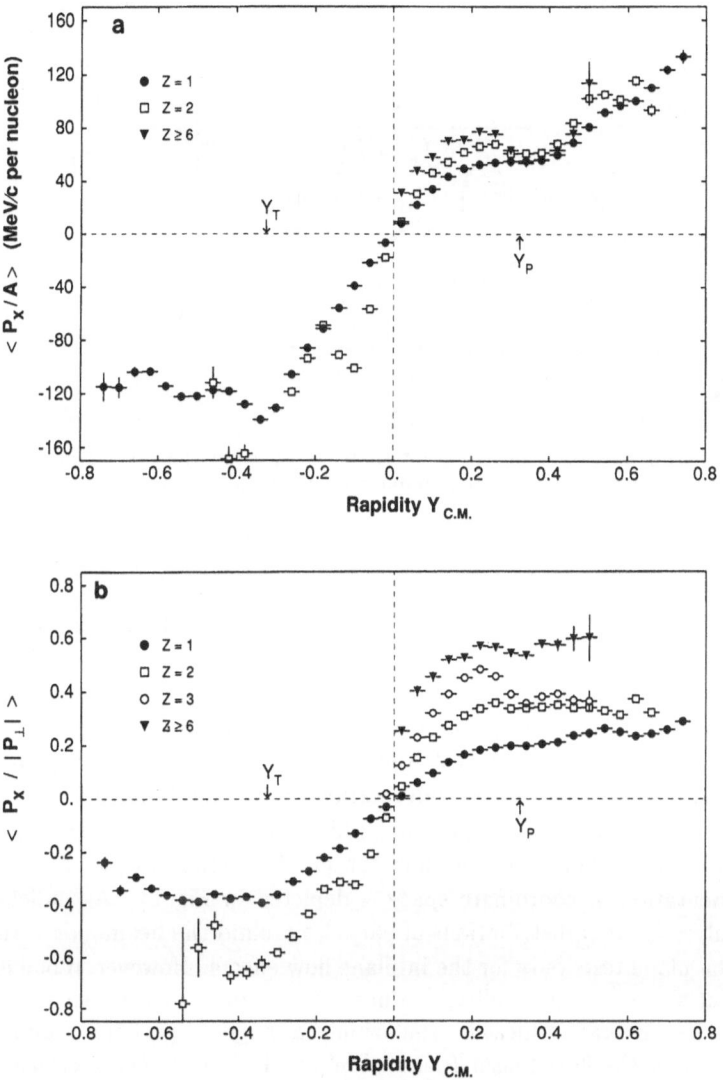

Figure 13. Mean in-plane transverse momentum per nucleon (a) and fraction of the particle's transverse momentum that is emitted into the reaction plane (b) for different fragments as a function of the CM rapidity for Au+Au collisions at 200 AMeV.

7.3 'Squeeze-out' of Particles: Another Component of Collective Flow

The methods discussed above do not yet exhaust all information that is in principle available from 4π spectrometers and allow only for an investigation of collective phenomena appearing *in* the reaction plane. In particular, the successful global transverse momentum method results in $\langle p_x \rangle = \langle p_y \rangle = 0$ at $y_{CM} = 0$ because of symmetry reasons. On the other hand, it is anticipated that strong collective phenomena take place at midrapidity, because of the large compression effects achieved these reactions. The direction perpendicular to the reaction plane is the only coordinate where nuclear matter might escape during the whole collision time without being hindered by either the target or projectile nucleus. As a consequence, this might lead to a jet-like emission pattern, also referred to as *out-of-plane squeeze-out* [Stö82, Buc83, Stö86], and allow an unique investigation of the interior of the hot dense region of the collision.

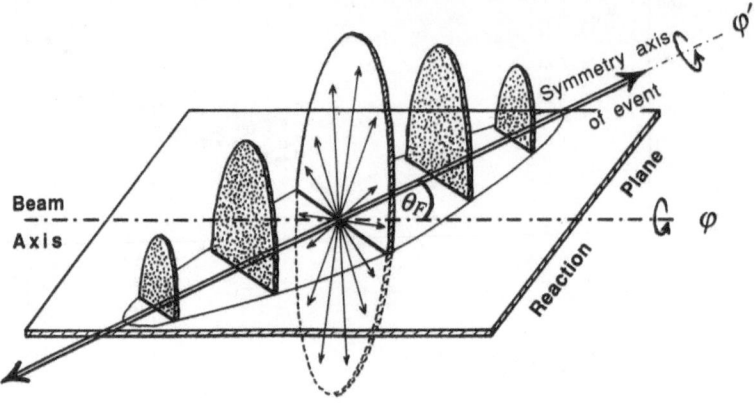

Figure 14. Schematic view of the event shape illustrating the orientation of the event in coordinate space, and indicating the difference between the azimuthal distributions φ and φ', respectively.

An indication of an out-of-plane peak in the particle distribution at midrapidity has recently been reported by the Diogene group for Ne-induced reactions [Hot88a]. In order to look for such effects in data of symmetric systems, a novel analysis, combining aspects of the sphericity- and transverse momentum analysis, has recently been proposed and applied to 400 AMeV Au+Au data by Gutbrod et al. [Gut89]. In this approach, intended to reveal new information about the 3-dimensional event shape, the φ-distributions of particles are analyzed in the coordinate system given by the principal axis of the kinetic energy flow tensor. A schematic view of the event shape and its orientation in coordinate space is depicted in Fig. 14. As indicated in that figure, analyzing the φ-distributions of particles around the beam-axis with respect to the reaction plane tests only for the in-plane flow effects. However, if one is interested in the event shape at midrapidity, one needs to account for the well known non-zero flow angle, i.e. one has to measure the azimuthal angle of particles around the major principal axis of the flow tensor (denoted φ' in Fig. 14). Because each event is now rotated not only by $-\varphi_F$ into the reaction plane, but also by $-\theta_F$ into the major symmetry axis of the event, the method allows for the first time to superimpose several events without effectively distorting and smearing out the characteristic individual event shapes. Thus, it evades the difficulty of the standard sphericity method to extract aspect ratios because of finite particle number distortions. Moreover, average ratios of the associated half-axes may now be inspected also as a function of rapidity.

Figure 15a shows as an example of this analysis the φ'-distribution as a function of the normalized momentum per nucleon along the major axis of the ellipsoid, p'_z. A projection of the φ'-distribution at $p'_z = 0$ is in addition plotted in Fig. 15b. The data clearly demonstrate that there is a preferential emission of particles into the out-of-plane direction ($\varphi' = 190°$, $270°$), and that this pattern even extends over the hole p'_z-range. Large distortions at projectile and target rapidity are also visible in Fig. 15a, showing up as a depletion at $\varphi' = 0°$ and $p_z < 0$ due to the target cuts, and as a peak structure at projectile momentum and $\varphi' = 180°$ caused by the bounce-off particles.

Figure 15c displays in addition the azimuthal dependence of the transverse momentum per nucleon $\langle p'_\perp /A \rangle$. Again, the same anisotropy in φ' at $p'_z = 0$ is observed, revealing that not only the density of particles is enhanced in the out-of-plane direction, but that these particles are also emitted with a higher average transverse momentum per nucleon!

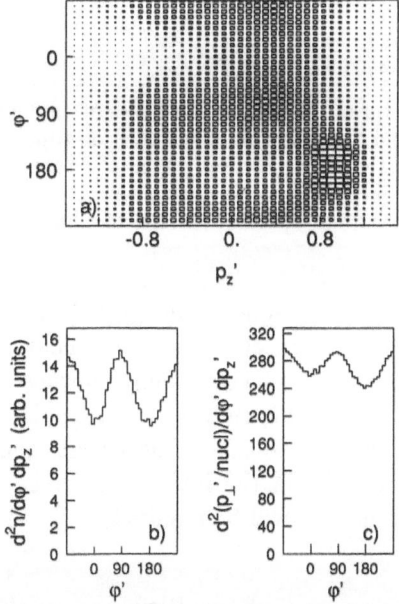

Figure 15. Plastic-Ball results for 400 AMeV Au+Au [Gut89].
a) Distribution of the number of particles in the φ' versus p'_z plane. The size of the squares is proportional to the number of particles.
b) φ' distribution of the number of particles for $-0.1 < p'_z < 0.1$
c) φ' distribution of the average transverse momentum per nucleon for $-0.1 < p'_z < 0.1$.

The anisotropy in φ' corresponds to an elliptical distribution of the particles both in position and momentum space. Thus, fitting the distributions of Figs. 15b,c with $f(\varphi') \sim 1 + \alpha \cos(2\varphi')$, and α being the free parameter, yields the aspect ratio of the two shorter half-axes of the associated ellipsoid as extrema of f. Figure 16 (left) shows the multiplicity dependence of this aspect ratio at midrapidity for the position density (filled circles) and the average momentum per nucleon (open circles). Both ratios exhibit the same tendency; they are found to reach a maximum in semi-central collisions, resembling the impact parameter dependence of the directed in-plane flow. In the limit of impact parameter $b = 0$ (which cannot be reached in practice) symmetry requires that the ratios become equal to one. The right hand part of Fig. 16 shows

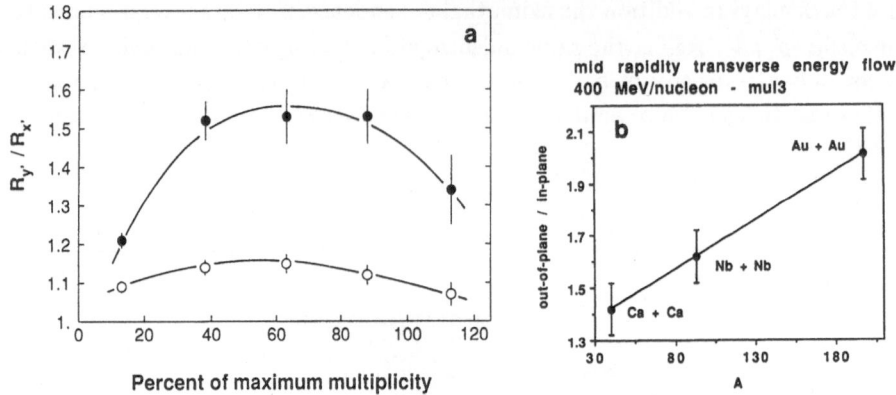

Figure 16. a) Aspect ratio of the associated half axes of an ellipsoid describing the density of particles (filled circles) and the momentum per nucleon (open circles) in a slice normal to the major symmetry axis of the event (see Fig. 14) at $p'_z = 0$ as a function of multiplicity for 400 AMeV Au+Au, and b) projectile-target mass dependence of the energy squeeze-out (no of particles × transverse Energy) at 400 AMeV. The solid lines are to guide the eye.

the mass dependence of the *energy squeeze-out* $(\sum_i E'_\perp(i))$. The aspect ratio increases here fairly proportional to A and reaches a value of about 2 for Au+Au. This means that one observes twice as much transverse energy flowing out-of-plane as opposed to the in-plane flow! However, the A-dependence also rises the question on what fraction of this anisotropy is due to possible shadowing effects, and which due to dynamics caused by the pressure built up in the interaction zone. To answer this question detailed comparisons with models will be required.

7.4 Comparison with Theory and Implications on the EOS

The unambiguous experimental confirmation of the existence of dynamic collective flow phenomena, extracted in several ways from the data, establishes the formation of highly compressed nuclear matter. However, although observables for investigating the properties of high density matter have been identified, the most important problem remains, namely how to derive quantitative information about the nuclear EOS from these data. This task can in principle be accomplished via detailed comparisons with model predictions assuming different equations of state. Because of substantial non-equilibrium effects in the data, this aspect may best be investigated in the frame of microscopic approaches, which require neither global nor local equilibrium (for detailed reviews on microscopic theories the reader is referred to Refs. [Stö86, Ber88] and several lectures of this school). Such approaches are in particular the time dependent Dirac equation [Cus85], the Newtonian force model (i.e. classical molecular dynamics) [Mol84], the Vlasov-Uehling-Uhlenbeck (VUU) approach [Kru85, Mol85,88a, Ber84], and the Quantum Molecular Dynamics approach (QMD) [Aic86,88, Pei88]. In fact, non equilibrium nuclear transport theories including mean field aspects have become fundamental to nuclear physics in the Bevalac energy regime. Different from the fluid dynamical model, the nuclear EOS serves not as immediate input to these models, but is simulated by different parametrizations for the density dependent potential field, $U(\rho)$. Common to all these approaches is that the observed flow is caused by the nuclear compression energy resulting from the short-range repulsion of the nuclear force. A net sidewards flow can in principle also result in the absence of high compression as has been discussed by Schürmann et al. [Sch87b]. However, those values are considerably too small as compared to the experimental data.

The VUU-model e.g. predicts an almost linear dependence of the in-plane transverse momentum spectrum $p_x(y)$ on the 'stiffness' of the EOS [Mol85,87]. From a quantitative comparison with the observed flow-angles and p_x-slopes at $y_{CM} = 0$ it has been concluded evidence for a stiff EOS [Stö86, Mol88a]. Instead of running immediately into the yet on-going discussion about the 'best-fit' EOS, we rather shall first try to differentiate between the predictive power and performances of various models by comparing their predicted scaling properties of collective flow variables. Such an analysis, recently pursued by Schürmann and Bonasera et al. [Sch87a, Bon88], reveals that at present only transport models (VUU, BUU, QMD, etc.) on the one hand and viscous hydrodynamics on the other, remain to be consistent with the observed bombarding energy and projectile-target mass dependence. It should be emphasized that the increasing flow values measured at constant bombarding energy with increasing projectile-target mass do not a priori imply an increase of the maximum density reached in the system. The observed rise is rather an effect of the increasing number of collisions fragments suffer before they escape from the reaction zone (thus explaining the observed $A^{1/3}$ dependence between Nb and Au), while the maximum density is determined mainly by the EOS [Mol87].

An interesting observation demonstrating the influence of the bombarding energy dependent interplay between the nuclear mean-field and nucleus-nucleus collisions to the in-plane transverse momentum distribution, p_x, has been reported in Refs. [Mol85, 88b]; at very low bombarding energies of $E_{Lab} \simeq 50\,\mathrm{AMeV}$ the p_x-distributions were predicted to be inverted compared to the well known shape of positive slopes found at higher bombarding energies. This transition from the nucleon-nucleon dominated dynamics of high-energy reactions to mean-field dominated dynamics at low-energy reactions was expected to occur at incident projectile velocities comparable to the Fermi velocity when Pauli-blocking becomes very effective. This negative angle scattering (well known also from deep-inelastic reactions at energies of a few MeV above the Coulomb-barrier) was in fact confirmed experimentally by measuring the circular polarization of γ-rays, emitted from the residual nucleus, in coincidence with light particles from $^{14}N + ^{154}Sm$ reactions at 20 and 35 AMeV [Tsa86]. The point of highest bombarding energy where observations of collective flow effects have been reported from is at present set at 3.37 AGeV [Bia86]. However, as indicated already when going to the highest Bevalac energies, the amount of collective flow seems to decrease at these high projectile velocities. Therefore, it would be of particular importance and interest to extend these studies to Brookhaven (14.5 AGeV) and CERN-SPS energies (60–200 AGeV) and to find out whether the occurrence of flow is a phenomenon limited to a certain window in bombarding energy.

Recently, it has been recognized [Aic87b, THa85,86, Ain87, Ber87, Koc88] that in-medium effects, until then largely been ignored, lead to drastic consequences in the extraction of the EOS. The largest correction results from the inclusion of momentum dependent interactions (MDI). This causes an additional repulsion because of the large momentum difference between projectile and target nucleons in the initial stage of the reaction, and hence simulates a stiffer EOS. Another important in-medium effect is the reduction of the effective nucleon-nucleon (NN) cross section because of the Pauli principle [THa87]. At present, still not all effects might be taken into account, as was pointed out in a recent article [Ber89] where, similar to earlier suggestions of Ruck et al. [Ruc76], it is argued that the inelastic NN cross-section may be enhanced again by pion collectivity in the nuclear medium.

The sensitivity of global observables thought to be relevant for the EOS to in-medium effects has recently been studied by the Frankfurt-Heidelberg collaboration [Aic87b, Ros88, Pei88]. These investigations demonstrated that in-medium effects influence most strongly observables connected to particle production, e.g. p, K and Λ-yields, n_d/n_p-ratios, etc., and are less substantial but nevertheless significant to flow observables. Inclusion of momentum dependent interactions leads to an increase of flow variables at bombarding energies $E_{Lab} > 400\,\text{AMeV}$ and is negligible at lower energies. The reduction of the effective NN scattering cross-section, on the other hand, tends to decrease its values at all energies. Simultaneous inclusion of both in-medium effects results in decreasing flow-values at all bombarding energies. This apparent dilemma in the determination of the EOS, however, might partly being solved by measuring particle rapidity distributions, dN/dy. This is hoped to serve as an unambiguous tool for determining the effective NN in-medium scattering cross-section experimentally. Since these distributions do — different from the flow variables — not depend significantly on the EOS, one might proceed in that way to adjust first the NN cross-section by means of rapidity distributions, plug in this result, and finally adjust the EOS to fit the flow variables. This should be done simultaneously for all systems at different bombarding energies and preferably also by selecting IMFs rather than nucleons, because fragments have been proven to exhibit the predicted enhanced sensitivity to the nuclear collective flow. Their measured p_x/A-values exceed the predicted model values by a factor of two, if the soft EOS is employed, but are well reproduced with standard NN cross sections, reduced by Pauli blocking of the final states, and a hard EOS [Pei88]. The new calculations with in-medium corrections included demonstrate that the stiffness of the EOS or an appreciable increase in the NN scattering cross-section would be needed to explain the data on the directed transverse momentum flow.

Another very important piece of information became just available from the observation of the out-of-plane squeeze-out of particles. New viscous fluid results exhibit a dramatic dependence of this effect on the nuclear viscosity: for large values the directed emission is damped out, and a maximum is not visible in $dN/d\varphi$ [Schm88]. Thus, such data will be essential in determining this unknown property of hot nuclear matter. The same sensitivity is also found in microscopic calculations (VUU, QMD), where $dN/d\varphi$ depends strongly on the EOS and σ_{eff}, i.e. parameters which enter directly into the coefficients of the viscosity [Har88]. It will be most interesting in the next future to see whether these microscopic models can consistently, i.e. using the same EOS and σ_{eff}, describe the p_x/A, $p_x/p_\perp$, dN/dy, and $dN/d\varphi'$ distributions. This would be a large step forward in the ultimate goals of relativistic heavy-ion collisions, namely the determination of the bulk properties of nuclear matter.

7.4.1 Comparison to Astrophysical Evidence on the EOS

Heavy-ion reactions probe the EOS at high temperatures $(W(\rho, T \gg 0))$, astrophysical objects, on the other hand, are composed of rather cold nuclear matter. An EOS necessary to describe the latter objects, can therefore not immediately be compared with the EOS from relativistic nuclear collisions. In fact, it was claimed that the EOS necessary to blow up stars and achieve an explosion energy as high as that seen in SN 1987A, the supernova event of early 1987 [Baro87], is considerably softer than the EOS necessary to explain the experimental data, discussed in this article.

Nuclear in-medium corrections were originally suggested to account for these different nuclear environments, and thus to explain the apparent discrepancy. Although inclusion of such effects indeed brought results obtained via the two different paths to the EOS closer together, their discrepancy still remains. As emphasized recently by Glendenning [Gle88], the problem could as well lie on the astrophysical side, because there are many uncertainties in the physics of prompt-bounce mechanism in supernovae explosions either. For example, as recently pointed out by Hillebrandt and co-workers [Hil88] the early calculations of Baron et al. [Baro87] have to be modified substantially in various respects so that the early conclusions seem to be premature. Thus, in total it seems clear that much more elaborate work is still needed particularly on the astrophysical side and that one is much further from being able to draw conclusions about the EOS than in high-energy heavy-ion collisions.

8 Summary and Conclusions

In this lecture some exciting recent progress made in the understanding of the fundamental physics of relativistic heavy-ion reactions in the bombarding energy range from about 100–2000 MeV per nucleon has been reviewed. It was shown that statistical concepts and the assumption of a (global) thermal equilibrium achieved during the collision may be used as a guide-line in describing the bulk properties of nuclear matter. However, a large body of experimental evidence has now been accumulated which validates the picture that a large amount ($\approx 40\%$) of the available center of mass energy is being converted into compressional degrees of freedom in the moment of highest density, demonstrating the presence of strong compression effects. These conclusions have been drawn independently from different observables, such as (strange) particle production (π, K, Λ) yields, composite particle to proton ratios and their related specific entropy, transverse particle energies as compared to the intrinsic temperature of the system, and from the collective flow pattern of fragments. The nuclear EOS determines the fractions of available energy going into thermal and compressional degrees of freedom. Since the first group of observables listed above is influenced by the produced thermal energy, while the flow observables are connected to the compressional energy, it is hoped that their measurements provide some information about the underlying EOS. The quantitative discussion, however, is heavily aggravated by a number of not yet fully understood in-medium effects of hot dense nuclear matter as well as by final-state effects influencing mainly the particle yields. Although all obstacles have not yet been completely surmounted, enormous progress has been made during recent few years, both, theoretically in describing the dynamical evolution of the highly excited strongly interacting system, as well as experimentally by delivering the necessary exclusive 4π observables. The most outstanding discovery of nuclear collective flow effects, which confirmed theoretical expectations of shock-wave phenomena from nuclear fluid dynamics, must be considered the manifestation of the key mechanism for heating and compression of nuclear matter. Thus, it provided the grounds for extracting first information on the equation of state of hot and dense hadronic matter from data of high-energy heavy-ion collisions; the result as it appears at this stage are large viscous effects and a surprisingly stiff EOS.

Acknowledgement

It is pleasure to warmly thank the conference organizers M. Soyeur, H. Flocard, and B. Tamain for inviting me to Les Houches and giving me the opportunity to participate on this extremely stimulating and fruitful school taking place in a pleasant and relaxed atmosphere.

References

Aic86: J. Aichelin and H. Stöcker, Phys. Lett. 176B (1986) 14.
Aic87a: J. Aichelin and E.A. Remler, Phys. Rev. C35 (1987) 1291.
Aic87b: J. Aichelin, et al., Phys. Rev. Lett. 58 (1987) 1926.
Ain87: T.L. Ainsworth, et al., Nucl. Phys. A464 (1987) 740.
Ala87: J.P. Alard, et al., Nucl. Instr. Meth. A261 (1987) 379.
Ala88: J.P. Alard, et al., contributed paper to Saint-Malo Conference, France, (1988) p. 156.
Ams75: A.A. Amsden, et al.,Phys. Lett. 35 (1975) 905.
Bad82: A. Baden, et al., Nucl. Instr. Meth. 203 (1982) 189.
Bar85: H.W. Barz, et al., Phys. Rev. C31 (1985), 268.
Baro85: E. Baron, J. Cooperstein, and S. Kahana, Phys. Rev. Lett. 55 (1985) 126.
Baro87: E. Baron, et al., Phys. Rev. Lett. 59 (1987) 736.
Bau75: H.G. Baumgardt, et al., Z. Phys. A273 (1975) 359.
Bea85: D. Beavis, et al., Phys. Rev. Lett. 54 (1985) 1652.
Bea86: D. Beavis, et al., Phys. Rev. C33 (1986) 1113.
Ber81: G. Bertsch and J. Cugnon, Phys. Rev. C24 (1981) 2514.
Ber83: G. Bertsch and P.J. Siemens, Phys. Lett. 126B (1983) 9.
Ber84: G. Bertsch, H. Kruse, S. Das Gupta, and Phys. Rev. C29 (1984) 673.
Ber87: G. Bertsch, W.G. Lynch, and M.B. Tsang, Phys. Lett. 189B (1987) 384.
Ber88: G.F. Bertsch and S. DasGupta, Phys. Rep. (1988) 189.
Ber89: G.F. Bertsch, et al., Nucl. Phys. A490 (1989) 745.
Bia86: H. Bialowska, et al., Phys. Lett. 173B (1986) 349.
Bod80: A.R. Bodmer, C.N. Panos, and A.D. MacKellar, Phys. Rev. C22 (1980) 1025.
Bog83: J. Boguta and H. Stöcker, Phys. Lett. 120B (1983) 289.
Bon76: J.P. Bondorf, et al., Z. Phys. A279 (1976) 385; and Phys. Lett. B65 (1976) 217.
Bon77: R. Bond, et al., Phys. Lett. 71B (1977) 43.
Bon85: J.P. Bondorf, et al., Nucl. Phys. A443 (1985) 321, and Phys. Lett. 150B (1985) 57.
Bon88: A. Bonasera, L.P. Csernai, and B. Schürmann, Nucl. Phys. A476 (1988) 159.
Bra79: S. Brandt and H.D. Dahmen, Z. Phys. C1 (1979) 61.
Bra86: E. Braun and Z. Fraenkel, Phys. Rev. C34 (1986) 120.
Bro88: G.E. Brown, Proceedings to Saint Malo Conference 1988.
Buc81: G. Buchwald, et al., Phys. Rev. C24 (1981) 135.
Buc83: G. Buchwald, et al., Phys. Rev. C28 (1983) 2349.
Buc84: G. Buchwald, et al., Phys. Rev. Lett. 52 (1984) 1594.
Cal79: D.J.E. Callaway, L. Wilets, and Y. Yariv, Nucl. Phys. A327 (1979) 250.
Cha73: G.F. Chapline, et al., Phys. Rev. D8 (1973) 4302.

Cse80: L.P. Csernai, B. Lukacs, and J. Zimanyi, Nou. Cimento Lett. 27. (1980) 111.

Cse82: L.P. Csernai, et al., Phys. Rev. C26 (1982) 149.

Cse83a: L.P. Csernai and B. Lukas, Phys. Lett. 132B (1983) 295.

Cse83b: L.P. Csernai, et al., Phys. Rev. C28 (1983) 2001.

Cse84: L.P. Csernai, G. Fai, and J. Randrup, Phys. Lett. 140B (1984) 149.

Cse86a: L.P. Csernai and J.I. Kapusta, Phys. Rep. 131 (1986) 223, and references therein.

Cse86b: L.P. Csernai, et al., Phys. Rev. C34 (1986) 1270.

Cug80: J. Cugnon, Phys. Rev. C22 (1980) 1885.

Cug81: J. Cugnon, T. Mizutani, and J. Vandermeulen, Nucl. Phys. A352 (1981) 505.

Cug82a: J. Cugnon, D. Kinet, and J. Vandermeulen, Nucl. Phys. A379 (1982) 553.

Cug82b: J. Cugnon, et al., Phys. Lett. 109B (1982) 167.

Cus85: R.Y. Cusson, et al., Phys. Rev. Lett. 55 (1985) 2786.

Dan79: P. Danielewicz, Nucl. Phys. A314 (1979) 465.

Dan83: P. Danielewicz and M. Gyulassy, Phys. Lett. 129B (1983) 283.

Dan85: P. Danielewicz and G. Odyniec, Phys. Lett. 157B (1985) 146.

Dan88: P. Danielewicz, et al., Phys. Rev. C38 (1988) 120.

Dos85: K.G.R. Doss, et al., Phys. Rev. C32 (1985) 116.

Dos86: K.G.R. Doss, et al., Phys. Rev. Lett. 57 (1986) 302.

Dos87: K.G.R. Doss, et al., Phys. Rev. Lett. 59 (1987) 2720.

Dos88: K.G.R. Doss, et al., Phys. Rev. C37 (1988) 163.

Fai82: G. Fai and J. Randrup, Nucl. Phys. A381 (1982) 557.

Fai83: G. Fai and J. Randrup, Nucl. Phys. A404 (1983) 551.

Fin82: J.E. Finn, et al., Phys. Rev. Lett. 49 (1982) 1321.

Gle88: N.K. Glendenning, Preprint, LBL-25375 (1988).

Gos77: J. Gosset, et al., Phys. Rev. C16 (1977) 629.

Gos78: J. Gosset, J.I. Kapusta, and G.D. Westfall, Phys. Rev. C18 (1978) 844.

Gos87: J. Gosset, et al., Saclay preprint, DPh-N/Sacaly 2469B (1987).

Gud85: K.K. Gudima, et al., Phys. Rev. C32 (1985) 1605.

Gus84a: H.A. Gustafsson, et al., Phys. Lett. 142B (1984) 141.

Gus84b: H.A. Gustafsson, et al., Phys. Rev. Lett. 52 (1984) 1590.

Gus88: H.A. Gustafsson, et al., Mod. Phys. Lett. A3 (1988) 1323.

Gut76: H.H. Gutbrod, et al., Phys. Rev. Lett. 37 (1976) 667.

Gut83: H.H. Gutbrod, et al., Phys. Lett. 127B (1983) 317.

Gut89: H.H. Gutbrod, et al., Phys. Lett. 216B (1989) 267.

Gyu82: M. Gyulassy, K.A. Fraenkel, and H. Stöcker, Phys. Lett. 110B (1982) 185.

Gyu83: M. Gyulassy, K. Fraenkel, and E.A. Remler, Nucl. Phys. A402 (1983) 596.

Hah88: D. Hahn and H. Stöcker, Nucl. Phys. A476 (1988) 718.

Har88: Hartnach, et al., to be published, and H. Stöcker, private communication.

Hil88: W. Hillebrandt, et al., Preprint MPI-Munich, MPA 386, July 1988; R. Mönchmeyer and E. Müller, Preprint MPI-Munich, MPA 374, June 1988.

Hof76a: J. Hofman, et al., Phys. Rev. Lett. 36 (1976) 88.

Hof76b: J. Hofman, W. Scheid, and W. Greiner, Nuovo. Cimento. 33A (1976) 343.

Hot88a: D. L'Hôte, private communication and talk at Fifth Gull Lake Nuclear Physics Conference, Gull Lake, MI (1988).

Hot88b: D. L'Hôte, Nucl. Phys. A488 (1988) 457c

Hot89: D. L'Hôte, proceedings to this winter school.

Jai88: P.L. Jain, K. Sengupta, and S. Singh, Phys. Rev. C37 (1988) 637.

Kam86: K.H. Kampert, Doctoral Thesis, University of Münster, (1986).

Kam89: K.H. Kampert, submitted to J. Phys. G (1989).

Kap81: J.I. Kapusta, Phys. Rev. C24 (1981) 2545.

Kap84: J.I. Kapusta, Phys. Rev. C29 (1984) 1735.

Kea88: D. Keane et al., to be published

Kea89: D. Keane et al., proceedings to this winter school.

Kis87: S.M. Kiselev, Phys. Lett. 198B (1987) 324.

Koc88: V. Koch, et al., Phys. Lett. 206B (1988) 395.

Kru85: H. Kruse, et al., Phys. Rev. C31 (1985) 1770; H. Kruse, B.V. Jacak, and H. Stöcker, Phys. Rev. Lett. 54 (1985) 289.

Mek77: A.Z. Mekjian, Phys. Rev. Lett. 38 (1977) 640.

Mek78: A.Z. Mekjian, Nucl. Phys. A312 (1978) 491.

Mey80: W.G. Meyer, et al., Phys. Rev. C22 (1980) 179.

Mol84: J.J. Molitoris, et al., Phys. Rev. Lett. 53 (1984) 899.

Mol85: J.J. Molitoris and H. Stöcker, Phys. Rev. C32 (1985) 346; J.J. Molitoris, D. Hahn and H. Stöcker, Nucl. Phys. A447 (1985) 13c.

Mol86: J.J. Molitoris, et al., Phys. Rev. C33 (1986) 867.

Mol87: J.J. Molitoris, H. Stöcker, and B.L. Winer, Phys. Rev. C36 (1987) 220.

Mol88a: J.J. Molitoris, et al., Phys. Rev. C37 (1988) 1014.

Mol88b: J.J. Molitoris, et al., Phys. Rev. C37 (1988) 1020.

Nag80: S. Nagamiya, et al., Phys. Rev. Lett. 45 (1980) 602.

Nag81: S. Nagamiya, et al., Phys. Rev. C24 (1981) 971.

Ngo89: C. Ngo, proceedings to this winter school.

Pei88: G. Peilert, et al., Mod. Phys. Lett. A3 (1988) 459; UFTP-Preprint, University of Frankfurt (FRG), 214/1988; G. Peilert, Diploma Thesis, University of Frankfurt, 1988 (unpublished).

Pos71: A.M. Poskanzer, G.W. Butler, and E.K. Hyde, Phys. Rev. C3 (1971) 882.

Ran81: J. Randrup, and S.E. Koonin, Nucl. Phys. A356 (1981) 223.

Ren84: R.E. Renfordt, et al., Phys. Rev. Lett. 53 (1984) 763.

Rit86: H.G. Ritter, et al., Nucl. Phys. A 447 (1986) 3c.

Ros88: A. Rosenhauer, Doctoral Thesis, University of Frankfurt; GSI report GSI-88-09; UFTP Preprint, University of Frankfurt (FRG), 203/1987.

Ruc76: V. Ruck, M. Gyulassy, and W. Greiner, Z. Phys. A277 (1976) 391.

San80: A. Sandoval, et al., Phys. Rev. C16 (1977) 629.

San83: A. Sandoval, et al., Nucl. Phys. A400 (1983) 1365c.

Sat88: H. Satz, H.J. Specht, and R. Stock, eds., Z. Phys. C38 (1988), Quark-Matter Proceedings 1987.

Sch74: W. Scheid, H. Müller, and W. Greiner, Phys. Rev. Lett. 32 (1974) 741.

Sch87a: B. Schürmann and W. Zwermann, Phys. Rev. Lett. 59 (1987) 2848.

Sch87b: B. Schürmann, W. Zwermann, and R. Malfliet, Phys. Rep. 147 (1987) 1.

Sch88: B. Schürmann, Mod. Phys. Lett. A3 (1988) 1137.

Schm88: W. Schmidt, et al., to be published, and H. Stöcker, private communication.

Sie79a: P.J. Siemens and J.O. Rasmussen, Phys. Rev. Lett. 42 (1979) 880.

Sie79b: P.J. Siemens and J.I. Kapusta, Phys. Rev. Lett. 43 (1979) 1486.

Sob75: M.I. Sobel, et al., Nucl. Phys. A251 (1975) 502.

Sto80: R. Stock, et al., Phys. Rev. Lett. 44 (1980) 1243.

Sto82: R. Stock, et al., Phys. Rev. Lett. 49 (1982) 1236.

Sto86: R. Stock, Phys. Rep. 135 (1986) 259.

Stö78: H. Stöcker, W. Greiner, and W. Scheid, Z. Phys. A286 (1978) 121.

Stö79a: H. Stöcker, J.A. Maruhn, and W. Greiner, Z. Phys. A293 (1979) 173.

Stö79b: H. Stöcker, J.A. Maruhn, and W. Greiner, Phys. Lett. 81B (1979) 303.

Stö80: H. Stöcker, J.A. Maruhn, and W. Greiner, Phys. Rev. Lett. 44 (1980) 725.

Stö81: H. Stöcker, A.A. Ogloblin, and W. Greiner, Z. Phys. A 303 (1981) 259.

Stö82: H. Stöcker, et al., Phys. Rev. C25 (1982) 1873.

Stö83: H. Stöcker, et al., Nucl. Phys. A400 (1983) 63c.

Stö84: H. Stöcker, J. Phys. G 10 (1984) L111; Nucl. Phys. A418 (1984) 587c.

Stö86: H. Stöcker and W. Greiner, Phys. Rep. 137 (1986) 277.

Str83: H. Ströbele, et al., Phys. Rev. C27 (1983) 1349.

Str84: B.J. Strack and J. Knoll, Z. Phys. A315 (1984) 249.

Sub81: P.R. Subramanian, et al., J. Phys. G7 (1981) L241.

THa85: B. ter Haar and R. Malfliet, Phys. Rev. Lett. 56 (1985) 1237.

THa86: B. ter Haar and R. Malfliet, Phys. Lett. 172B (1986) 10.

THa87: B. ter Har and R. Malfliet, Phys. Rev. C36 (1987) 1611.

Tro88: R. Trockel, et al., Phys. Rev. C39 (1989) 729.

Tsa86: M.B. Tsang, et al., Phys. Rev. Lett. 57 (1986) 559.

Val88: O. Valette, et al., contributed paper to Saint-Malo Conference, France, (1988) p. 141.

Wes76: G.D. Westfall, et al., Phys. Rev. Lett. 37 (1976) 1202.

Wil85: J.R. Wilson and H.A. Bethe, Astrophys. J. 295 (1985) 14.

Wol82: K.L. Wolf, et al., Phys. Rev. C26 (1982) 2572.

Yar79: Y. Yariv and Z. Fraenkel, Phys. Rev. C20 (1979) 2227.

Yar81: Y. Yariv and Z. Fraenkel, Phys. Rev. C24 (1981) 488.

INCLUSIVE EXPERIMENTS, CORRELATIONS AND PION PRODUCTION DATA

Denis L'Hôte

Service de Physique Nucléaire - Moyenne Energie
CEN Saclay
91191 Gif-sur-Yvette Cedex, France

1. INTRODUCTION

In this lecture, I would like to present part of the experimental data measured in Relativistic Heavy Ion experiments. The main goal of RHIC experiments is to extract information on the properties of the hot and dense nuclear matter drop that has been created during non-peripheral collisions of two nuclei. As this task in itself raises several important questions, we shall discuss them in the introductory part. In the second part, we shall describe the first type of experimental results that were obtained in the field: inclusive and correlation data. Finally, in the third part, we shall review some of the pion production data (above threshold).

<u>Making dense and hot nuclear matter in the laboratory</u>

From very general considerations,[1,2] it can be shown that, during the central collision of two nuclei (at a lab. incident energy larger than $\sim$ 100-200 MeV per nucleon), "hot and dense nuclear matter may be formed in a transient state"[1]. Furthermore, dynamical models like hydrodynamics,[3-6] intranuclear cascade,[7-14] VUU,[15-18] BUU,[19-21] QMD,[22,23] RVU,[24] etc.... predict that the c.m. density at the geometrical center of a symmetric system reaches values ranging from $\sim$ 1.5 to $\sim$ 4 times normal nuclear density[25,26]. Therefore, accelerating heavy nuclei offers the fascinating possibility to make small samples of highly compressed and heated nuclear matter, and to infer from the experimental data what are its properties[3,4,19,27-33] (Equation of state (EOS): heat capacity and compressibility; transport properties: viscosity and heat conductivity; stopping power; in-medium behaviour of nucleons, deltas, and pions, etc...).

But this task is not easy because we are dealing with finite systems, for which surface effects, geometry, and finite number effects are unavoidable ingredients of the reaction process: small spheres collide, not slabs! In addition, non-equilibrium effects are undoubtly important, at least during the first stage of the collision, and also for those of the nucleons that are close to the surface. Finally, the high density and temperature stage is only part of the whole collision history. The dynamical models quoted above predict the collision to proceed via a three step scenario: firstly interpenetration and compression of the nuclei, secondly high density and temperature stage, and thirdly expansion of the system. Such division simplifies the whole phase space time evolution, but it is a good starting point for depicting the collision. The main point now is that experimentally, we get information only on the final distribution function in **p** space. Of course, one could argue that the system will begin to emit particles even before the end of the collision, but, if those particles interact stongly with the system, we expect them to be emitted mainly by its surface; and this raises the problem of extracting information on volume properties from surface radiation (this is why measuring particles that interact only weakly with the system, like for instance positive kaons, is of special interest).

In order to infer a dynamical information on the high density stage from what we measured after the final stage, we need "good" observables. The best ones would be "primordial observables" whose values remain constant during the expansion stage, and that are determined by the properties of the high density part (not the peripheral one) of the system. For instance, it has been suggested in the past that the pion multiplicity could be "primordial"[30,34]; I will come to this point in part 3. Other observables have been also proposed, like "flow", or entropy (related to the composite to proton yield ratios)[35]. In fact, we can also accept observables that evolve during the expansion provided that they keep the memory of stage 2: different high density stage conditions must give different final values of the observable. This leads us to the conclusion that we have to understand also the expansion stage in order to ascertain its influence on the observable evolution. In addition, the first stage (interpenetration and compression) is also of interest since it determines which point (density, etc...) will be reached during the second one. Finally, we must note also that the spatial distribution of the system plays an important role: we can expect the high density zone to be surrounded by lower density nuclear matter (the transition between the two being continuous), and by spectator pieces that can influence the final observable values. To summarize this discussion, the whole reaction process has to be studied and understood in order to be able to gather information on the high density and temperature nuclear matter from experimental results.

Experimental systematics

To study the reaction process, systematics in incident energy, target mass, and projectile mass from 1 to ~ 200 are important. This is because several unknown parameters need many measured observables to be constrained; but we can illustrate this general statement by more precise examples. Firstly, the surface versus volume contribution to the observables can be investigated using mass or multiplicity systematics. Secondly, how far from equilibrium the system is can also be studied with such systematics. Thirdly, varying the ratio between target and projectile masses can help understanding the effects of spectator pieces on the observables; etc....

More fundamental considerations show the need for systematics. Firstly, the equation of state involves a two variable dependence, the energy per nucleon as a function of the density ρ and temperature T. Hence, it is clear that we need an incident energy systematics in order to vary the densities and temperatures reached during the collision. But in doing so, the two variables ρ and T are varied at the same time. Probably, looking also at asymmetric systems (projectile mass < target mass) can help to disentangle the ρ and T dependence because such systems can be expected to go on different trajectories in the ρ,T plane during the collision (the target nucleons may cool down and confine more efficiently the participant nuclear matter). The incident energy systematics can be also useful in order to study the relative contributions of momentum dependent forces and compression effects on observables such as the flow[20,23,27]. Secondly, the transport properties of nuclear matter (mainly viscosity) can be investigated from a systematics in mass and energy, following the philosophy of similarity properties of fluid dynamics[11,36,37]. Thirdly, the stopping power of nuclear matter is of interest, not only because it determines to what extent the system equilibrates, but also because as recently stressed, it can be related to the in-(dense)-medium propagation and interactions of nucleons, deltas, and mesons[38,67]. In such calculations, an enhancement of the nuclear matter stopping is expected for rather large incident energies. Hence, a systematics of the stopping power versus incident energy is of great interest. Finally, in what concerns the observable systematics, I would like to quote the importance of measuring different particles (p,n,d,t,He,...,$\pi^{\pm}$,π^{o},K,Λ,γ, e+e-...) because their different interaction cross sections make them sensitive to different aspects of the dynamics. In most cases, the results of the experimental systematics will be used in order to constrain free parameters of the models (for instance compressibility and in-medium cross sections in VUU-like models, or density dependent Landau-Migdal parameters in relativistic transport theory of fluctuating fields,[41] etc...).

From the adjusted values of the models parameters, fundamental nuclear matter properties can then be derived. In the last 15 years, many experimental results were obtained, mainly at the Berkeley Bevalac, and also at Saturne in Saclay. In the near future, experiments at SIS (Darmstadt) will start. Let us quote also the Dubna Synchrophasotron that works at higher energies (3-4 GeV per nucleon instead of 0.1-2). This experimental effort has permitted important progresses in our understanding of the collision process: models describing the collision have been tested and improved, different observables expected to be "good" have been measured, and their properties have been investigated. The final goals -determining the dense and hot nuclear matter properties- still need theoretical and experimental work, but the first attempts have started recently, encouraging us to go forward[30,34,35,42-45].

Impact parameter selection

The impact parameter **b** is the projection onto the transverse plane of the vector that goes from the center of the target nucleus to the center of the projectile nucleus in the initial state. Selecting its magnitude is of great interest because as it decreases, the amount of compressed nuclear matter produced and the degree of equilibration reached increase. However, b = |**b**| cannot be selected directly, and is usually inferred from participants or spectators multiplicity selections[46] (see section 2). Note however that we cannot expect a one to one correspondence between b and the multiplicity, and that the correlation between the two quantities is known only from the models.

It is also very interesting to determine event by event the direction **n** = **b**/b of the impact parameter (**n** defines the azimuth Φ_b of the reaction plane) because this information allows us to measure the one-body distribution functions (of p,d,π,etc...) in the reference frame of the reaction plane, while the valuable information contained in this function is partially lost when averaging over Φ_b. If, in the initial state, the system is symmetric with respect to the reaction plane and if this symmetry is conserved during the collision, then linear combinations of the particles transverse momenta must lie in the reaction plane. The method commonly used to reconstruct **n** is to write[35,47] :

$$\mathbf{n} \simeq \mathbf{Q}/|\mathbf{Q}|$$

with:

$$\mathbf{Q} = \Sigma\ \omega_i \cdot \mathbf{P}_{ti}$$

where the sum runs over the measured particles (baryons); $\mathbf{P}_{ti}$ and ω_i being respectively the transverse momentum of the ith. particle and a weight chosen to be positive for c.m. forward particles and negative for backward ones.

From inclusive to exclusive experiments

The first experiments were inclusive (only one particle is measured per collision). The observables they provided (p,d,π,... inclusive double differential cross section) proved to be only weakly sensitive to the dynamics of the compressed nuclear matter. For instance, hydrodynamics calculations using three different equations of state gave similar proton inclusive cross section values for the Ne+U reaction at 400 MeV per nucleon[32]. Recently, comparisons of several nuclear transport models (cascade,VUU,BUU,QMD,RVU) with inclusive proton-like experimental data[48,49] for the heavy system La+La at 800 MeV per nucleon has shown that in spite of the different basic assumptions made, all the predictions concentrated on the same lines. However, it was also shown that varying (strongly) the elementary N-N cross sections in the cascade model influences the final La+La inclusive cross sections.

A systematic, detailed and precise (not only the data have error bars, but also the predictions of the models that use Monte-Carlo techniques) comparison between the data and the models would tell us whether any partial information on dense and hot nuclear matter can be extracted from inclusive cross sections. But it is probably more interesting to understand what is the problem with inclusive data. The first point is that they result from an averaging over b and Φ_b. As the weight $2\pi.b.db$ favours large impact parameters, the latter contribution washes out the interesting information from central collisions. Note however that the relative contributions of peripheral and central collisions depends strongly on the region of $\mathbf{p}$ space we are looking at. The second point is that inclusive data lack a major part of the experimental available information contained in the 2,3,...,n-particles correlations: the experimentally measurable n-body distribution function in $\mathbf{p}$ space has to be integrated over the n-1 particles momenta in order to give the inclusive cross section. A striking illustration of this point is the fact that the $\mathbf{Q}/|\mathbf{Q}|$ vector (see previous paragraph) cannot be determined from inclusive experiments, but requires multiparticle ones.

In order to overcome the inclusive experiment limitations, large solid angle (4π) detectors have been built[35]. At the Berkeley Bevalac, the LBL Streamer chamber[50] was first used. Let us quote also the SKM-200 streamer chamber and the 2-meter bubble chamber at Dubna[51,52]. Electronic detectors, have been subsequently developped: the GSI/LBL Plastic Ball/Wall detector system[53] was used at the Bevalac, and the Diogene Pictorial Drift Chamber and Wall[54] was used at Saturne. In the future, a 4π multidetector[55] is planned at SIS, and a TPC has been proposed at the Bevalac[56]. The development of such devices represents a decisive progress in the field. However, the analysis of multiparticle observables is more complicated than for inclusive experiments: in order to compare the data to the predictions of the models, a program simulating the so-called "experimental filter" (the biases, cuts, inefficiencies and resolutions of the detector) has to be used in many cases. In practice, such program must also take into account some shortcomings of the models (for instance the non prediction of composite yields and spectra). In the past few years, only exclusive 4π experiments have allowed the first attempts to extract information on the compressed nuclear matter from the data[30,34,35,42-45].

Between inclusive and exclusive, we find what we can call semi-inclusive or correlation experiments. A first type consists in measuring or detecting 2 or 3 particles per collision. Such experiments are briefly presented in section 2. Recently, The "Phi-ometer" has been used at the Bevalac in order to measure neutron flow,[57] using the fact that forward fragments lie in the reaction plane[35,58]. A second type consists in measuring one particle per collision in coincidence with a rough estimate of the participant multiplicity, obtained from arrays of scintillator or telescopes[49,59-61]. The correlation experiments provide less information than the 4π ones, but their "experimental filter" can be simpler. As a conclusion, let us note that a 4π detector using a non restrictive trigger provides also the inclusive and semi-inclusive information.

2. ABOUT INCLUSIVE AND CORRELATION EXPERIMENTS

Introduction

Inclusive and correlation experiments provide less information than the 4π ones, but they represent a good introduction to the field, as several relevant concepts have been developed in order to understand their results, and also because they represent still an important set of data that must constrain the models. We shall not attempt to review them,[32] but rather we shall concentrate only on four concepts that were put forward in order to analyze the experimental results: participants and spectators separation, isotropy and equilibrium, coalescence, and quasi-elastic scattering. We shall also discuss the relationship between correlations and impact parameter vector.

Participants and spectators

Fig. 1 shows contours of double differential cross section in the rapidity-transverse momentum plane for protons measured in the Ar+Pb reactions at 800 MeV per nucleon. If we compare the left part (inclusive) to the right one (only events where at least five telescopes among nine placed at 40° are hit), we see that the multiplicity selection suppresses strongly the protons with velocity close to the projectile one. Many experimental results[32] show that proton inclusive spectra exhibit significant enhancements of the cross section around the target and projectile velocities. The nucleons that contribute to those enhancements are called spectators, while the other ones are the participants. The "clean cut-off" picture[62] qualitatively explains what is expected to cause this separation: the incident energy is large enough to neglect the effects of the binding energy, and to get a nucleon mean free path smaller than the nuclei radii. As a consequence, the system will geometrically split in three parts: the projectile and target spectators and the participants. According to such a picture, when the impact parameter b decreases, the participant multiplicity increases. For the system Ar+Pb (Fig. 1), M_{TAG} is a rough estimate of this multiplicity, and selecting values larger than 5 would correspond to a complete overlap of the two nuclei, hence to no projectile spectators. The participant-spectator geometrical picture is more or less confirmed by cascade calculations,[46] but its limits of validity (towards low incident energies, low nuclei masses, and low impact parameters) have not been precisely investigated. It is a rather crude picture because experimentally, the kinetic boundary between participants and spectators is not so clean. In addition, note that the "bounce-off" effect[35] shows that the spectator-like nucleons are kicked away in the reaction plane. Also, at not too high incident energies, it is very likely that a small projectile is stopped in the target[63]. Nevertheless, the geometrical picture of participants-spectators separation proved to be a nice and not unrealistic tool for understanding the collision process.

Isotropy and equilibrium

On the right part of Fig. 1, the broken contour line corresponds to an isotropic emission from a source whose rapidity is intermediate between the target and the projectile ones. We can see that large transverse momentum protons exhibit an approximately isotropic pattern. This can be qualitatively accounted for in a cascade picture: to get a large transverse momentum, a nucleon must make several binary collisions, and hence it will be "equilibrated" and isotropically emitted in the participant reference frame. The study of the deviation from isotropy of the proton and pion emission patterns in the participant center of mass has been carried on by several experimental groups[49,59,60,64,65] because in a simple picture of the participant fluid, complete equilibration implies isotropy and also because studying the deviation from isotropy can be considered as a way to evaluate the stopping power of nuclear matter. However, let us note that the shape of the emission pattern is also inflenced by the non isotropic component of the collective motion. The results show that the emission is less and less forward and backward peaked when the masses of the nuclei or the participant multiplicity increase. The continuation of such analyzes using 4π detectors consists in studying the event by event shape of the emission pattern, and also its multiplicity dependence[30,35,65-69]. The isotropy studies have been often carried on at the same time as 90° c.m. spectra analysis because an other proof for equilibration would be Boltzmann shapes of such spectra. However they are inflenced by the radial expansion of the system,[70] and also by finite number effects, Coulomb distortion, and composite production effects. Finally note that not only the thermal equilibrium must be considered, but also the chemical equilibrium, which involves pion, delta and composite abundancies[32].

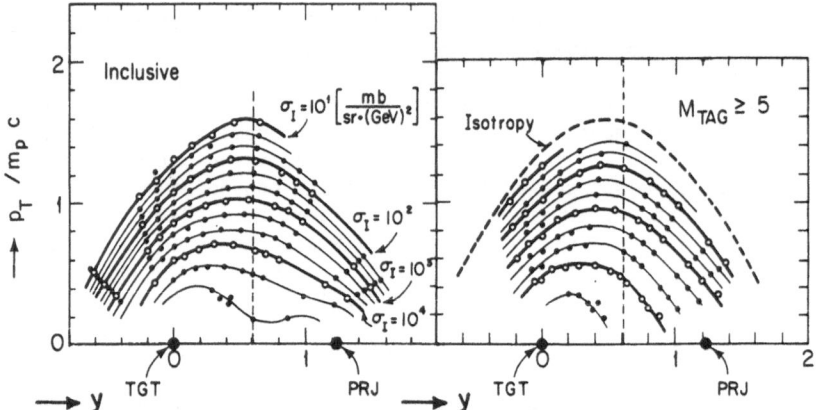

Fig. 1. Contour plots of proton invariant cross sections as a function of the rapidity and of the transverse momentum divided by the proton mass for Ar+Pb at 800 MeV per nucleon[59]. The lines are labelled with the cross section values on the left part. The latter corresponds to inclusive data and the right part to "high" multiplicity events.

Coalescence

The particles emitted by colliding nuclei are mainly nucleons, pions, and light nuclei (composites: d,t,He,...). Heavier fragments (A > 4) have been also measured, but in the participant zone, their cross sections are very low. A first task was to understand the composite yields and spectra compared to nucleon emission. The coalescence model assumes that an A mass composite is produced if A nucleons come close to each other in phase space[29,32,35,64,71,72]. As a consequence, the A mass composite differential cross section can be related to the proton one through the relation:

$$E_A(d^3\sigma_A/d\mathbf{p}_A) = C_A \cdot [E_p(d^3\sigma_p/d\mathbf{p}_p)]^A , \qquad (2.1)$$

where E and $\mathbf{p}$ are the total energy and momentum of the particles, and where the cross sections are considered for the same emission angle and velocity. Formula (2.1) is rather well verified for many inclusive results in the sense that the empirical C_A (A=2,3 or 4) is weakly sensitive to the emission angle, to the velocity of the particles and to the projectile incident energy. However, the consistency of the coalescence picture implies that the cross sections in formula (2.1) are not those of the measured protons and composites, but those of the "primordial" particles. The latter include not only the free particles, but also the particles that are bound in heavier composites. Using primordial particles, formula (2.1) is not so well verified. The quantity C_A can be related to the size of the emitting system. This is also true for equilibrium models, and for the "sudden approximation" model[29,32,72]. As a consequence, inclusive experiments where varying sizes are mixed up for varying b are not well suited for checking formula (2.1). The study of composite spectra in relation with the proton ones, and in coincidence with impact parameter (multiplicity) selection remains to be done. Keeping the coalescence model, but integrating the yields over p-space leads to the fact that the ratio of the (primordial) "deuteron-like" to "proton-like" yields increases when the participant number increases,[32,73,74] in accordance with the picture of an increasing source size. Finally, let us mention the important connection that can be established between the composite to proton ratios and the entropy[29,32].

Particle correlations

Two-particle correlations provide a very rich set of significant informations on the reaction. I would like to discuss two of them (quasi elastic scattering and impact parameter averaging), but there are several others. Two (or three) pion interferometry or two proton final state interaction generate momenta correlations whose magnitude and shape give information on the lifetime, degree of coherence, size, and spatial structure of the emitting source[75-83]. However, it seems that those measurements are mainly sensitive to the final state. Let us quote also the correlations due the decay of unstable light nuclei[84]. May be, new dynamical information could be gathered by looking at other correlations. For instance, it has been suggested that three body forces could be investigated by looking at deuteron-proton correlations[38].

The experimental study of quasi-elastic scattering consists in looking at pairs of protons whose total four-momentum is close to the four-momentum of two incoming nucleons belonging to the projectile and the target respectively. Such measurements have been carried on for C + C, C + Pb, Ne + NaF, Ar + KCl, and Ar + Pb at 800 MeV per nucleon[85,86]. They consisted in determining the ratio of in plane pairs to out of plane ones. Except for Ar + Pb and C + Pb, a quasi-elastic peak was found, from which the fraction of nucleons that have made only one collision can be extracted. Thus, such an analysis can provide a valuable information on the degree of equilibration or stopping that has been reached during the collision. In fact, it showed that the first version of the Yariv and Fraenkel cascade in which no reinteraction of the nucleons was allowed, was in disagreement with experiment, while the second one was not[11]. May be, the quasi-elastic scattering studies could be continued by looking at 3, 4 ... proton correlations in order to study 2, 3 ... collision processes, or clusters of interacting nucleons. An other extension would be to look at quasi-free pion production by studying correlations between two nucleons and one pion.

As we have seen, systems with heavy targets do not exhibit the quasi-elastic peak, but the structures observed could be qualitatively explained in terms of target "shadowing" or of collective flow. A similar explanation was suggested in order to interpret measured correlations between fast and slow particles in Ne + Au reactions[87]. Recent results from the Berkeley Streamer Chamber for Ar + KCl, La + La, and Ar + Pb semicentral collisions provided the observables characterizing the collective flow of nuclear matter ($< p_x > = f(y)$, and sphericity tensor[35]) from two and three particles (deuterons) correlations and give a method to correct for the correlations due to momentum-energy conservation[88]. The basic idea of such calculations is that the measured two or three-particle distribution function contains correlations due to the fact that the original one-particle distribution function has been averaged over Φ_b. Note that other two or three-particle correlations could also exist. Recent results from two-particle correlation measurements at Ganil[89] have also been analysed in terms of the one body function averaging, but in this case, the averaging is made not only on Φ_b, but also on b.

3. PION PRODUCTION

Introduction

Experimentally, the mean number of pions ($\pi^+ + \pi^- + \pi^0$) measured at a typical incident energy of 800 MeV per nucleon is roughly 10% of the participating nucleon multiplicity. Measuring pions is of great interest for studying dense and hot nuclear matter properties, since pions and deltas play a fundamental role in the nucleon-nucleon interaction at those energies. Pion production represents roughly half of the N-N total cross section at 800 MeV per nucleon. An interesting picture is to view the pions emitted by the compressed nuclear matter as the light radiated by an explosion: its intensity and spectrum can be related to the properties of the exploding system[30]. However, the questions adressed in part 1 (surface versus volume emission, final stage versus dense stage probe) still hold for pion production. If, in a model, their yield or spectra depend on the high density and temperature zone and stage, it is important that the connections between the center and the surface of the fireball, and between the second and third stage be accurately described. In a 4π experiment, an event contains a certain number of baryons and pions. Hence, the first observable to study is the pion multiplicity distribution and its mean value, $< n_\pi >$, which can be related to the pion production cross section σ_π through the relation: $< n_\pi > = \sigma_\pi /\sigma_t$, where σ_t is the total cross section for the reactions we consider. In what follows, the first two paragraphs are devoted to those topics, while the third deals with pion spectra. In the fourth, we present recent measurements of pion triple differential cross section. Finally, in the last paragraph, the search for delta resonances is presented.

Multiplicity distributions

A purpose of the pion multiplicity distribution studies is to look for possible deviations from a Poisson law. It can be shown[90] that at a given impact parameter, such deviation would be a signal of unusual coherent pion processes. Experimentally, this was investigated through the study of the relation between the pion multiplicity variance and mean value (they are equal for a Poisson law), at a given participant proton multiplicity (b selection). Until now, no significant deviation from Poisson has been found[91-100]. Recent Diogene results for He + Cu, He + Pb, Ne + Pb reactions still give a variance equal to the mean[101,102].

Pion yields

The study of the mean pion multiplicity $< n_\pi >$ as a function both of the projectile-target masses and participant proton multiplicity is of fundamental interest since it can tell us whether the pions are emitted by the surface or by the volume of the system (resp. $A^{2/3}$ or A^1 dependence). A volume emission would attest the validity of pion yield as a probe of compressed nuclear matter. Fig. 2 shows the experimental π^- multiplicity dependence for La + La[44]. Such linear shapes favour strongly a volume emission. In addition, for the lighter system Ar + KCl a similar linear dependence has been found,[43] and the authors mention that, at a given incident energy, the slopes are the same for the two systems La + La and Ar + KCl. Recently, the ratio of the π^- to the participant proton-like production cross sections has been also measured for La + La at 800 MeV per nucleon in an inclusive and semi-inclusive experiment[49]. Within the 20% error bars, this ratio is the same for inclusive and high multiplicity cases.

In addition, it is roughly the same than the C + C, Ne + NaF and Ar + KCl ones. Such results still favour an A^1 (A = participant number) dependence of $< n_\pi >$, but notice that, because of the Lanthanum isospin asymmetry, a volume dependence would imply larger $< n - >$ values for La + La than for lighter systems[49]. In this problematics, we must not forget the emission duration. Final pions are emitted during a certain time[103] that we can assume to scale like A^α, α being an exponent still unknown. An empirical A^β dependence can thus be understood as follows: instantaneous pion yield scales like $A^{\beta-\alpha}$, and emission duration scales like A^α. If $\beta = 1$ and $\alpha = 1/3$ (the latter assumption is not unreasonable as cascade calculations support the idea that the whole collision time scales roughly like the nuclear radii[11]), the surface emission hypothesis would be favored. The question remains open because we need a phenomenological study of the space-time evolution of pion production during the collision and because the systematics has to be completed. An obvious connection could be made with interferometry experiments that shoud provide the pion source lifetime, hence permitting an experimental determination of the α exponent from mass and multiplicity systematics. Finally, let us quote also empirical target and pro-jectile mass dependence studies[64,104].

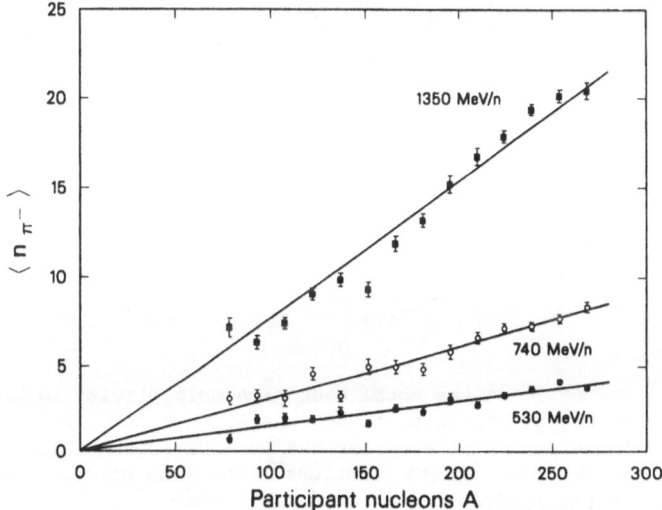

Fig. 2. Mean π^- multiplicity as a function of the number of participant nucleons A, for the La + La collisions at three incident energies, from the Berkeley streamer chamber[44]. The lines are straight line fits to the points.

The ratios of the $\pi^\pm$ multiplicities to the total charged multiplicity M have been measured at Diogene for He, Ne, and Ar projectiles on several targets. For asymmetric systems, this ratio decreases when M increases. For instance, Ne + Pb at 0.8 GeV per nucleon gives a 40-50% decrease when M goes from 2 to 40. This result can be explained by the fact that when M increases, the parti-cipant nuclear matter will contain more and more nucleons from the target, thus leading to a decrease of the mean energy available for pion production. An other effect can also come from π absorption in the spectator nuclear matter (shadowing)[105,106]. Such results favour the statement that mass scaling studies must not mix results from both symmetric and asymmetric systems[30].

Now, let us go to the mean pion multiplicity incident energy dependence. It has been empirically studied for many systems and energies, but we shall rather concentrate on results from central Ar + KCl and La + La collisions[43,44,92]. It had been conjectured that production of abnormal nuclear matter states could imply a jump in the $< n_\pi > = f(E_{cm})$ dependence. Fig. 3 shows that on the contrary a smooth linear dependence is observed. On this figure, we can see that the pion yields are overpredicted by an equilibrium model. Many models predict too many pions. This is the case for three intranuclear cascade codes[8,9,12]. At Dubna energies, the overestimate of π multiplicities by intranuclear cascade predictions has also been examined[51,52].

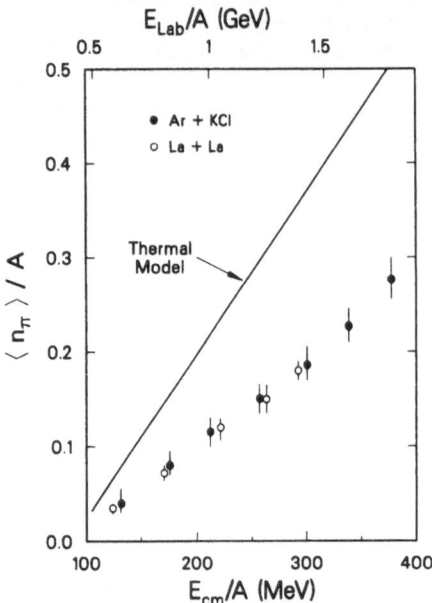

Fig. 3. The ratio of the total mean pion multiplicity to the number of participant nucleons A, as a function of incident c.m. energy for the La + La (open circles) and Ar + KCl (dots) reactions[44]. The line corresponds to the predictions of a thermal model.

The pion streamer chamber data have been analyzed in terms of compressional vs. thermal energy. The basic idea[1,34,107] is that the number of pions is just determined by the amount of thermal energy present in the system (pion "thermometer"). In such a picture, the discrepancy mentionned above is due to the fact that the models neglect the compressional (potential) energy, and assumes too large a thermal energy. The discrepancy permits an evaluation of both the thermal and compressional energies, providing a fundamental information on the nuclear equation of state[30,34,43,44,108]. Clearly it is also necessary to assume that the mean value of the total pion multiplicity (including the pions bound in deltas) is a "primordial" observable (see section 1) in the sense that this quantity must remain constant after the high density was reached. Cascade calculations support this hypothesis,[30,31] but it rose much controversy because the cascade simplifies the description of in medium π's and Δ's. As a matter of fact, the pions are expected to interact strongly with the nuclear matter, which implies that their number during the high density stage would be much larger than their final number, as most of them should be absorbed during the expansion[38].

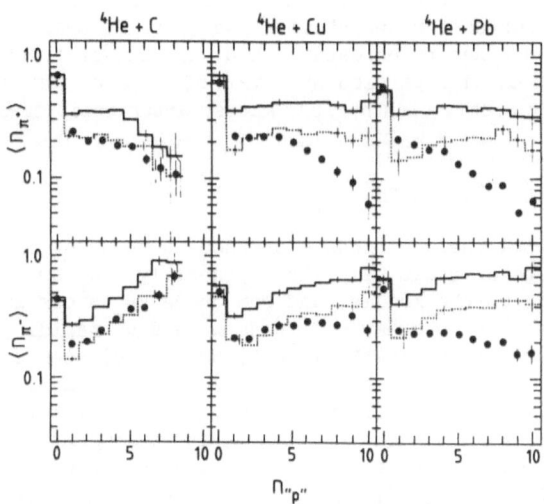

Fig. 4. Mean π^+ (upper plots) and π^- (lower plots) multiplicities as a function of the proton-like multiplicity in Diogene,[102] for the ⁴He + (C, Cu, or Pb) reactions at 800 MeV per nucleon. The experimental data (dots) are compared to the standard Cugnon cascade[9] (solid lines) and to an improved version including a binding energy prescription[10] (dotted lines).

This beautiful investigation, which was the first attempt to extract the EOS from the experimental data, stimulated many studies and addressed many questions. Firstly, it was noticed that, in the cascade codes (where π's are described mainly through the reactions NN↔NΔ and Δ↔Nπ) several prescriptions (binding energy, "variable" Δ width, type of elementary processes simulated, etc...) could influence the pion multiplicities[9,10,13,14,109,110]. Secondly, several models (VUU,[15] BUU,[19,111] QMD,[22,23] hydrodynamics[5]) predicted a weak sensitivity of the π yield on the equation of state (compressional energy). Thirdly, other quantities (N, Δ effective masses, NN in-medium cross sections, parameters of the in-medium Δ simulation, etc...) than the EOS could influence the π yield[21,112-114]. Note that a positive counterbalance of the latter results is that the π yield is sensitive to nuclear matter dynamical assumptions. Finally, more fundamental considerations indicate that, as the in-medium properties of N, π, and Δ should differ from the free ones, and as multiparticle interactions can take place, the basic assumptions of cascade, equilibrium, BUU, VUU, or QMD models (i.e. pion production described through free NN↔NΔ and Δ↔Nπ reactions) are questionnable[38,40,41,67,115]. As a consequence, pion absorption could be underestimated in such models.

The pion absorption underestimation by the Cugnon cascade has been discussed in a recent study of the code predictions for π + A, p + A and A + A reactions[109]. An other indication is given in Fig. 4, which shows that in α + A reactions where the compressional energy should be small, the discrepancy between the cascade predictions and the data increases when the target mass or the multiplicity increases. As a conclusion, it seems that the EOS extraction from pion yields deserves further studies. Whether the mean pion multiplicity is "primordial" or not has to be established. May be, restricting ourselves to the measured pion yield in a particular region of **p** space (corresponding to pions that interacted only weakly with nuclear matter after their creation) could help us to progress.

Pion double differential spectra

In order to solve the questions addressed in the previous paragraph, the pion spectra could be of great help. More generally, the spectra represent an important constraint on the models, and can guide us in our understanding of the reaction process. As an example Diogene π^+ and π^- rapidity distributions for the non-peripheral reactions Ne + (NaF, Nb, or Pb) at 800 MeV/nucleon show that, as the target mass increases, the curves shift towards lower rapidities[68]. This shift proves that the pion production cannot be described just by free nucleon-nucleon first collision, and that pion absorption by target spectators plays probably a role. Many inclusive double differential charged pion spectra have been measured, mainly for projectile masses lower than 40,[32,34,60,64,116] but also for La + La[49]. In their presentation, I will skip the comparisons between π^+ and π^-, due to the lack of space, but note that those comparisons can provide informations on the emitting source first for asymmetric systems (Z/A differs for target and projectile), and second through Coulomb effects (due to the positively charged source and to the spectators)[30,32,116].

If the transverse momenta are not too low, the spectra exhibit a structureless shape, and an exponential-type decrease as a function of the pion energy, for a given laboratory angle. At low transverse momenta, some smooth maxima can appear: for instance, for symmetric systems, a small bump at intermediate rapidities and ∼ 60 MeV/c transverse momenta has been investigated both experimentally and theoretically[30,60,116,117]. The π spectra have been calculated within many models. Most of them (equilibrium, cascade, VUU, BUU, QMD) assume their production and absorption to proceed via free two nucleon processes, with delta dominance. Note that VUU, BUU, and QMD predictions have been only poorly compared to the spectra. Intranuclear cascade models are in reasonnable agreement with the inclusive spectra, provided that not only the NN ↔ NΔ reactions are incorporated, but also Δ↔Nπ[13,14,118]. The 90° c.m. Ar + KCl (1.8 GeV per nucleon) central collision π^- spectrum is rather well reproduced by the Cugnon cascade calculations[65]. The question whether the discrepancy observed between the measured and the calculated pion yield is due to the lack of compressional energy or to an underestimation of the pion absorption (see previous paragraph) could be investigated through the spectra analysis, as suggested by a calculation involving the two possibilities[119].

Pion spectra exhibit, like the proton ones (see section 2), a tendency towards isotropy (in c.m.) when the participant proton multiplicity increases: Streamer Chamber results[30,65,66] for Ar + KCl at 1.8 GeV per nucleon show that a parameter describing the π c.m. angular distribution goes from ∼ 1.6 to 0.38 when the participant proton multiplicity goes from 10 to 36. The pion emission c.m. isotropy has also been studied as a function of the pion energy in several inclusive[49,64] and exclusive[65] experiments. In the latter case, for central

Ar + KCl collisions at 1.8 GeV per nucleon, the experimental data exhibit much more isotropy than cascade predictions, specially for high pion energies. An other study of interest is to look at 90° c.m. spectra: except for their low energy part, they can be fitted by exponential shapes. At 800 MeV per nucleon, and for inclusive data, the π^- slopes do not depend strongly on the masses of the nuclei for symmetric systems. The decrease is steeper for pions than for protons, even for high multiplicity events, and the pion spectra do not deviate strongly from the exponential shape at low energy, while on the contrary, the proton ones do (see Fig. 5)[32,49,59,64].

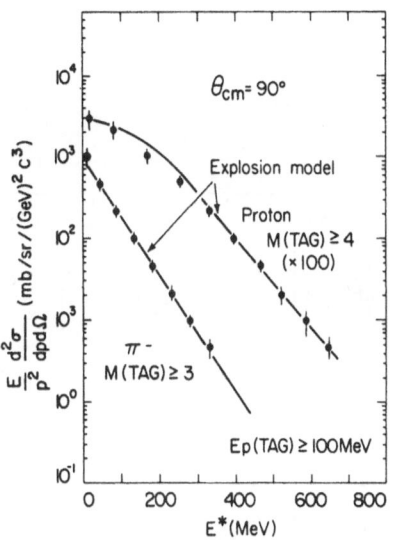

Fig. 5. Proton and π^- 90° c.m. spectra (dots) for the Ar + KCl
"central" reactions at 800 MeV/nucleon, as a function
of the particles c.m. energy[59]. The centrality of
the reactions results from a selection on the
multiplicity M(TAG). The lines are the "blast-wave"
model predictions,[70] whith an absolute normalization
fit.

In order to understand the differences between the shapes of pion and proton spectra, "trivial" effects must be taken into account (Coulomb distortions, the different shapes for elementary NN process, composite production, finite number effects), but it seems that this is not enough. A first possible explanation was that, as pions interact more strongly than protons with nuclear matter, they are emitted later, hence exhibiting lower "temperatures" (inverse slopes)[120]. An effect that can explain the proton spectra deviation from exponential shapes is that experimental spectra result from superpositions of thermal and radial collective motions ("blast-wave", see Fig. 5)[70]. The contribution of delta decay kinematics has also been stressed: the 90° c.m. π spectra for high energy Ar + KCl and La + La central reactions were fitted using the sum of two Boltzmann distributions. The distribution accounting for most of the cross section can be interpreted as resulting from delta decays, while the second one, corresponding to higher apparent "temperatures" would be associated to "direct pions escaping from the early stages of hot, dense nuclear matter"[65,121]. The question has been re-investigated recently[122].

Pion triple differential spectra

New observables were measured recently: the $\mathbf{n} \simeq \mathbf{Q}/|\mathbf{Q}|$ vector (see section 1) can be reconstructed using the proton-like particles in order to

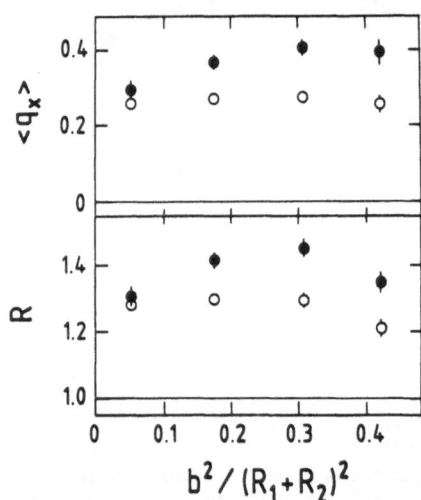

Fig. 6. Squared impact parameter dependence of $< q_x >$ and R (ratio between the numbers of pions emitted with $q_x > 0$ and $q_x < 0$) for π^+ (full symbols) and π^- (open symbols) produced in Ne + Pb collisions at 800 MeV per nucleon, from the Diogene experiment[106]. The impact parameter b is estimated from the proton-like multiplicity and it is divided by the sum of the two nuclear radii.

determine approximately the reaction plane azimuth. The pion triple differential cross section can thus be measured.

The triple differential cross section is defined as follows: the first two differential variables are the usual ones that we can choose among energy, momentum, emission angle, rapidity, transverse momentum, etc..., while the third is the azimuth of the pion momentum in the reaction plane reference frame. From this triple differential cross section, the non radial pion "flow" can be extracted,[88,106] as shown on Fig. 6 for the Ne + Pb reaction at 800 MeV per nucleon. To study a possible effect of the collective flow of nuclear matter on pion emission is of course of great interest. In particular, this question should be related to the problematics presented in the previous paragraph ("blast-wave" contributing to the differences between pion and proton spectra). The results obtained at Diogene for the Ne + (NaF, Nb, or Pb) reactions at 800 MeV/nucleon show the following trends[106]. Firstly, the standard flow plots[35,47] ($< q_x > = f(y)$ where q_x is the projection of the pion transverse momentum on $\mathbf{n}$ and y its rapidity) exhibit a positive slope and an always positive $< q_x >$ value for several asymmetric systems. This corresponds to a preferential emission in the $\mathbf{n}$ direction. Secondly, the effects are larger for positive pions than for negative ones. Thirdly, for the Ne + NaF system, no (or weak) effects are observed. The Cugnon intranuclear cascade does not predict any effect ($< p_x >$ is always compatible with zero for the reaction Ne + Pb at 800 MeV per nucleon, using the most recent version of the code), but on the contrary recent QMD calculations do[123].

Those data are qualitatively in agreement with a shadowing effect by the target spectator nuclear matter: the pions going in the impact parameter vector direction are less stopped and absorbed by the target than those going in the opposite direction, because they have less matter to go through. This explanation is qualitatively consistent with the observed differences between π^+ and π^- results because of the target isospin asymmetry. If it is valid, it means that pion absorption in the cascade simulation is underestimated. More generally, those results represent important constraints on the models, and can help to understand the influence of spectators shadowing on the pion yields and spectra and the space time evolution of the pion source. However, the question of "true" pion flow has still to be investigated.

Delta measurements

Before concluding this section on pion production, I would like to quote recent preliminary results from Diogene that show experimentally for the first time the presence of deltas in non-peripheral relativistic heavy ion collisions. An experimental measurement of the Δ's is a very significant information. We have seen in the paragraph devoted to pion double differential cross sections that the contribution of the delta kinematics to the pion spectra is of great interest.

Experimentally, the delta signature arises from proton-pion pairs invariant mass distribution analysis. However, as the number of available pairs in each event is rather large, the combinatorics weakens the signal. To recover it, it is worth studying the difference between the number of pairs p-π from the same event and the number of pairs from two different events, both numbers being determined in the same invariant mass bin. Fig. 7 displays such analysis in the case of the Ne + NaF reaction. For larger target masses or higher multiplicities, the signal is much weaker. This is also the case for p-π$^-$ pairs, which can be qualitatively understood by simple isospin considerations. The study of the mass distribution (mean value and width) and of the delta production cross section (note that even if the signal is compatible with zero, an upper limit can be determined) represents an important piece of information for the models, especially for pion production.

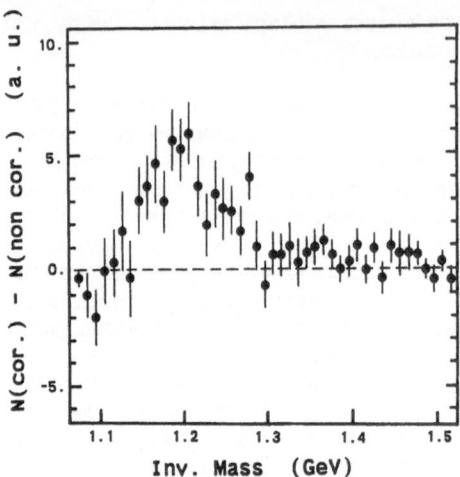

Fig. 7. Proton-π^+ invariant mass distribution from Diogene (preliminary results) for Ne + NaF at 800 MeV per nucleon, with a proton-like multiplicity larger than 3 and smaller than 7. A background subtraction procedure has been used which consists in plotting the difference between the number of pairs in a same event N(cor.) and the number of pairs in different events N(non cor.).

4. CONCLUSION

Relativistic Heavy Ion experiments offer the unique opportunity to make and study strongly compressed and heated nuclear matter in the laboratory. But we are dealing with small finite sytems and we measure only the final p state. As a consequence, we must understand the reaction process in order to extract information on dense and hot nuclear matter. We have also seen the importance of impact parameter selection. A definite experimental progress has been to go from inclusive to exclusive experiments. We have overviewed a few inclusive and semi-inclusive experimental data, selecting several chief points: participant-spectator distinction, isotropy and equilibrium, coalescence, quasi-elastic scattering. Those data have permitted progress in the understanding of the reaction process and in the development of models describing the collision. Finally, we have reviewed several pion production data (above threshold). The pion yield mass scaling and energy dependence have been presented, and we have discussed the possible interpretations of pion spectra. The experimental pion data represent an important constraint on the models because of the importance of π's and Δ's at those energies, but the connection between the final and the in-medium pions if not straightforward. Finally, pion triple differential cross sections and preliminary results on delta measurements have been presented.

REFERENCES

1. G. F. Chapline, M. H. Johnson, E. Teller, and M. S. Weiss, Phys. Rev. D8:4302 (1973).
2. W. Scheid, H. Muller, and W. Greiner, Phys. Rev. Lett. 32:741 (1974).
3. H. Stoecker and W. Greiner, Phys. Rep. 137:277 (1986).
4. R. B. Clare and D. Strottman, Phys. Rep. 141:177 (1986).
5. R. B. Clare, D. Strottman, and J. Kapusta, Phys. Rev. C33:1288 (1986).
6. J. A. Zingman, T. L. McAbee, J. R. Wilson, and C. T. Alonso, Phys. Rev. C38:760 (1988).

7.   J. Cugnon, Nucl. Phys. A387:191c (1982).
8.   K. K. Gudima and V. D. Toneev, Nucl. Phys. A400:173c (1983).
9.   J. Cugnon, D. Kinet, and J. Vandermeulen, Nucl. Phys. A379:553 (1982).
10. M. Cahay, J. Cugnon, and J. Vandermeulen, Nucl. Phys. A411:524 (1983).
11. J. Cugnon and D. L'Hôte, Nucl. Phys. A452:738 (1986).
12. Y. Yariv and Z. Fraenkel, Phys. Rev. C24:488 (1981).
13. Y. Kitazoe, M. Sano, H. Toki, and S. Nagamiya, Phys. Lett. 166B:35 (1986).
14. Y. Kitazoe, M. Sano, H. Toki, and S. Nagamiya, Phys. Rev. Lett. 58:1508 (1987).
15. H. Kruse, B. V. Jacak, and H. Stoecker, Phys. Rev. Lett. 54:289 (1985).
16. J. J. Molitoris, H. Stoecker, and B. L. Winer, Phys. Rev. C36:220 (1987).
17. J. J. Molitoris et al., Phys. Rev. C37:1014 (1988).
18. J. J. Molitoris, A. Bonasera, B. L. Winer, and H. Stoecker, Phys. Rev. C37:1020 (1988).
19. G. F. Bertsch and S. Das Gupta, Phys. Rep. 160:189 (1988).
20. C. Gale, G. Bertsch, and S. Das Gupta, Phys. Rev. C35:1666 (1987).
21. C. Gale, Phys. Rev. C36:2152 (1987).
22. J. Aichelin, A. Rosenhauer, G. Peilert, H. Stoecker, and W. Greiner, Phys. Rev. Lett. 58:1926 (1987).
23. J. Aichelin et al., Phys. Rev. C37:2451 (1988).
24. C. M. Ko, Q. Li, and R. Wang, Phys. Rev. Lett. 59:1084 (1987).
25. H. Stoecker, Lecture given at this School.
26. G. Welke, Lecture given at this School.
27. G. E. Brown, Phys. Rep. 163:167 (1988).
28. B. Ter Haar and R. Malfliet, Phys. Rep. 149:207 (1987).
29. L. P. Csernai and J. I. Kapusta, Phys. Rep. 131:223 (1986).
30. R. Stock, Phys. Rep. 135:259 (1986).
31. B. Schurmann, W. Zwermann, and R. Malfliet, Phys. Rep. 147:1 (1987).
32. S. Nagamiya and M. Gyulassy, in: "Advances in Nuclear Physics", Volume 13, J. W. Negele and E. Vogt, ed., Plenum Press, New-York (1982).
33. D. L'Hôte, Nucl. Phys. A488:457c (1988).
34. R. Stock et al., Phys. Rev. Lett. 49:1236 (1982).
35. K. H. Kampert, Lecture given at this school.
36. J. Cugnon and D. L'Hôte, Nucl. Phys. A447:27c (1985).
37. A. Bonasera, L. P. Csernai, and B. Schurmann, Nucl. Phys. A475:159 (1988).
38. G. E. Brown, Nucl. Phys. A488:689c (1988).
39. C. Gregoire, Lecture given at this School.
40. R. Malfliet, Lecture given at this School.
41. P. Siemens, Lecture given at this School.
42. D. Keane et al., Phys. Rev. C37:1447 (1988).
43. J. W. Harris et al., Phys. Lett. 153B:377 (1985).
44. J. W. Harris et al., Phys. Rev. Lett. 58:463 (1987).
45. M. Sano, M. Gyulassy, M. Wakai, and Y. Kitazoe, Phys. Lett. 156B:27 (1985).
46. J. Cugnon and D. L'Hôte, Nucl. Phys. A397:519 (1983).
47. P. Danielewicz and G. Odyniec, Phys. Lett. 157B:146 (1985).
48. J. Aichelin et al., Lawrence Berkeley Laboratory preprint, LBL-26383, Berkeley (1989).
49. S. Hayashi et al., Phys. Rev. C38:1229 (1988).
50. K. Van Bibber and A. Sandoval, Streamer Chamber for Heavy Ion, in: Treatise on Heavy-Ion Science, Vol. 7, D. A. Bromley, ed., Plenum Press (1983).
51. M. Kh. Anikina et al., Phys. Rev. C33:895 (1986).
52. V. Boldea et al., Sov. J. Nucl. Phys. 44:94 (1986).
53. A. Baden et al., Nucl. Inst. and Meth. 203:189 (1982).
54. J. P. Alard et al., Nucl. Inst. and Meth. A261:379 (1987).
55. P. Kienle, Lecture given at this school.
56. H. Wieman et al., in: Proc. 8th. High Energy Heavy Ion Study, 16-20/11/87, Lawrence Berkeley Laboratory, Report LBL-24580 (1988).
57. D. Keane, communication at this school.
58. G. Fai, Wei-ming Zhang, and M. Gyulassy, Phys. Rev. C36:597 (1987).

59. S. Nagamiya, M.-C. Lemaire, S. Schnetzer, H. Steiner, and I. Tanihata, Phys. Rev. Lett. 45:602 (1980).
60. K. L. Wolf et al., Phys. Rev. C26:2572 (1982).
61. W. G. Meyer, H. H. Gutbrod, Ch. Lukner, and A. Sandoval, Phys. Rev. C22:179 (1980).
62. G. D. Westfall et al., Phys. Rev. Lett. 37:1202 (1976).
63. H. H. Gutbrod, A. I. Warwick, and H. Wieman, Nucl. Phys. A387:177c (1982).
64. S. Nagamiya et al., Phys. Rev. C24:971 (1981).
65. R. Brockman et al., Phys. Rev. Lett. 53:2012 (1984).
66. H. Stroebele et al., Thermalization and stopping in nuclear collisions, in: Proc. 7th. High Energy Heavy Ion Study, Darmstadt, Oct. 1984, Report GSI-85-10 (1985).
67. G. F. Bertsch, G. E. Brown, V. Koch, and Bao-An Li, Nucl. Phys. A490:745 (1988).
68. J. Poitou et al., Current analyses of particle emission in Ne-nucleus collisions, in: Proc. 17th. Int. workshop on Gross Properties of Nuclei and nuclear excitations, Hirschegg, Jan. 1988, H. Feldmeier, ed., GSI Darmstadt (1988).
69. O. Valette et al., Nuclear collective flow in Ne-nucleus collisions, contribution in: Proc. 3rd. Int. Conf. Nucleus-Nucleus collisions, Saint-Malo, June 1988, Ganil, Caen (1988).
70. P. J. Siemens and J. O. Rasmussen, Phys. Rev. Lett. 42:880 (1979).
71. H. H. Gutbrod et al., Phys. Rev. Lett. 37:667 (1976).
72. H. Sato and K. Yazaki, Phys. Lett. 98B:153 (1981).
73. G. Montarou, These à l'Université Blaise Pascal de Clermont-Ferrand, April 1988, unpublished.
74. J. Marroncle, private communication.
75. W. A. Zajc et al., Phys. Rev. C29:2173 (1984).
76. J. Bartke, Phys. Lett. 174B:32 (1986).
77. Y. M. Liu et al., Phys. Rev. C34:1667 (1986).
78. A. D. Chacon et al., Phys. Rev. Lett. 60:780 (1988).
79. S. E. Koonin, Phys. Lett. 70B:43 (1977).
80. H. A. Gustafsson et al., Phys. Rev. Lett. 53:544 (1984).
81. T. C. Awes et al., Phys. Rev. Lett. 61:2665 (1988).
82. P. Dupieux et al., Phys. Lett. 200B:17 (1988).
83. D. Ardouin, communication at this school.
84. D. Guerreau, lecture given at this school.
85. I. Tanihata, M.-C. Lemaire, S. Nagamiya, and S. Schnetzer, Phys. Lett. 97B:363 (1980).
86. S. Nagamiya, Nucl. Phys. A400:399c (1983).
87. W. G. Meyer, H. H. Gutbrod, Ch. Lukner, and A. Sandoval, Phys. Rev. C22:179 (1980).
88. P. Danielewicz et al., Phys. Rev. C38:120 (1988).
89. D. Ardouin et al., Z. Phys. A 329:505 (1988).
90. M. Gyulassy and S. K. Kauffmann, Phys. Rev. Lett. 40:298 (1978).
91. R. Szwed, in: Proc. 5th. conf. on Relativistic Heavy Ions, Berkeley 1981, Report LBL-12652 (1981).
92. A. Sandoval et al., Phys. Rev. Lett. 45:874 (1980).
93. S. Y. Fung et al., Phys. Rev. Lett. 40:292 (1978).
94. D. Beavis et al., Phys. Rev. C27:910 (1983).
95. D. Beavis et al., Phys. Rev. C28:2561 (1983).
97. D. Beavis et al., in: Proceedings of the 7th. high energy heavy ion study, Darmstadt, Oct. 1984, Report GSI-85-10, (1985).
98. Collab. SKM-200, Nucl. Phys. A348:518 (1980).
99. N. Angelov et al., Sov. J. Nucl. Phys. 30:518 (1980).
100. G. R. Gulkanyan, T. Kanarek, E. N. Kladniskaya, S. A. Korchagin, and A. P. Cheplakov, Sov. J. Nucl. Phys. 40:479 (1984).
101. C. Cavata, private communication.
102. D. L'Hôte et al., Phys. Lett. 198B:139 (1987).
103. T. J. Humanic, Phys. Rev. C34:191 (1986).

104. J. Bartke, Nucl. Phys. A335:481 (1980).

105. E. Grosse, Lecture given at this school.

106. J. Gosset, O. Valette et al., Phys. Rev. Lett. 62:1251 (1989).

107. H. Stoecker, W. Greiner, and W. Scheid, Z. Phys. A286:121 (1978).

108. J. Harris et al.,Pion production and the equation of state, in: Proceedings of the 7th. high energy HI study, GSI Darmstadt, Oct. 1984, Report GSI-85-10, (1985).

109. J. Cugnon and M. C. Lemaire, Nucl. Phys. A489:781 (1988).

110. E. L. Medeiros, S. J. B. Duarte, and T. Kodama, Phys. Lett. 203B:205 (1988).

111. G. F. Bertsch, H. Kruse, and S. Das Gupta, Phys. Rev. C29:673 (1984).

112. B. M. Waldhauser, J. A. Maruhn, H. Stoecker, and W. Greiner, Z. Phys. A328:19 (1987).

113. M. Cubero, M. Schonhofen, H. Feldmeier, and W. Norenberg, Phys. Lett. 201B:11 (1988).

114. N. K. Glendenning, Phys. Rev. C37:1442 (1988).

115. M. Thies, Lecture given at this School.

116. K. A. Frankel et al., Phys. Rev. 32:975 (1985).

117. J. O. Rasmussen, Nucl. Phys. A400:383c (1983).

118. J. Cugnon, T. Mizutani, and J. Vandermeulen, Nucl. Phys. A352:505 (1981).

119. R. Malfliet and B. Schurmann, Phys. Rev. C28:1136 (1983).

120. S. Nagamiya, Phys. Rev. Lett. 49:1383 (1982).

121. G. Odyniec et al., Pion spectra in central La+La collisions at 530,740, and 1350 MeV/A, in: Proc. 8th. High Energy Heavy Ion Study, Lawrence Berkeley Laboratory, Berkeley, Nov. 1987, Report LBL-24580 (1987).

122. D. Hahn and N. K. Glendenning, Phys. Rev. C37:1053 (1988).

123. C. Hartnack, H. Stoecker, and W. Greiner, Analysis of Diogene's 4π data on Δ, π^+/π^- using QMD, in: Proc. Int. workshop on gross properties of nuclei and nuclear excitations XVI, Hirschegg, Jan. 1988, Darmstadt (1988).

CONFRONTATION OF THEORETICAL APPROACHES AND EXPERIMENTAL DATA ON HIGH ENERGY HEAVY ION COLLISIONS

M.Berenguer, C.Hartnack, G.Peilert, A.Rosenhauer*,
W.Schmidt, J.Aichelin**, J.A.Maruhn, W.Greiner, and
H.Stöcker [1]

Institut für Theoretische Physik, Johann Wolfgang Goethe-Universität
Postfach 111932, D-6000 Frankfurt am Main, West Germany
*present address: Institut for Theoretical Physics, University of Tel-Aviv,Israel
**Institut für Theoretische Physik, Universität Heidelberg
D-6900 Heidelberg, West Germany

Abstract

We give an overview of high energy heavy ion collisions. The merits and drawbacks of macroscopic and microscopic theoretical approaches (Fluid Dynamics, TDHF, Cascade, Vlasov-Uehling-Uhlenbeck, Classical and Quantum Molecular Dynamics) are discussed. The importance of nonequilibrium transport properties (viscosity, mean free path, effective in-medium cross sections) and of the nuclear potential (equation of state) is pointed out. The liquid-vapour phase transition and multifragmentation have been studied. The possibility of meassuring Machshock fragments in inverse kinematics experiments is also pointed out. It is demonstrated that the projectile and target are stopped at Y_{CM} if central collisions are studied. The stopping is only sensitive to σ^{eff}. The predicted bounce-off of the rather cold fragments in the reaction plane and the predicted accompanying squeeze-out of the hot participant baryons perpendicular to the reaction plane are experimentally discovered. These effects are sensitive both to the viscosity $(\sigma^{eff}(\rho, E, \Omega))$ and to the generalized equation of state (optical potential $U(\rho, E)$). The data clearly ask for a repulsive potential interaction. We conclude that nuclear matter produced in relativistic collisions is a hot, dense, viscous and rather incompresible fluid, with important quantum properties.

THE NUCLEAR EQUATION OF STATE

The knowledge of the bulk properties of nuclear matter, the nuclear equation of state and the coefficients of viscosity and heat conductivity is a basic prerequisite for the understanding of many astrophysical problems. The dynamics of the early universe

---

[1] invited speaker

shortly after the big bang as well as the dynamics of supernova explosions or the structure of neutron stars depend strongly on the nuclear equation of state over wide regions of densities, temperatures, and Z/A ratios. The key mechanism to produce and to study the properties of nuclear matter under such extreme conditions in the laboratory is the formation of shock waves which are predicted to occur in heavy ion collisions [1]. A directed collective sideward flow of the shocked matter is predicted as a direct signature for the shock wave formation. First indications for this collective sideward flow were obtained already in 1975 by Schopper and coworkers [2]. This idea was heavily critized over many years on the basis of the argument that the mean free path of nucleons is too long (7 fm at 100 MeV/n) so that heavy ion collisions will more resemble gases passing though each other rather than droplets splashing off each other [6].

Only after electronic 4π detectors have been developed a decade later, the collective sideward flow [7,8] has been experimentally unambigously discovered. Up to now also the more detailed predictions of the hydrodynamical model like the bounce off of the projectile spectators due to transverse communication with the reaction zone [3] and the squeeze out perpendicular to the reaction plane have been experimentally observed [13,9](see below). A pictorial view of the squeeze out and of the bounce off is given in fig.1.

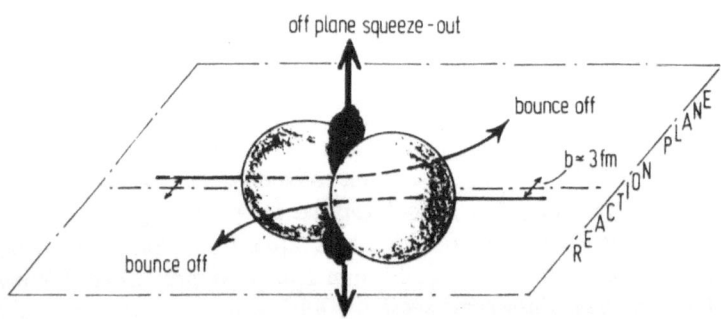

Figure 1. Perspective view of bounce off and off plane squeeze out of nuclear matter

The basic assumptions of the hydrodynamical model (short mean free path, rapid local thermalization) seem to be surprisingly well fulfilled: for very central collisions (high multiplicities) the rapidity distributions are indeed centered around mid rapidity (nearly complete stopping is observed) and the particle spectra exhibit a roughly thermal behaviour [9,12] (see below).

The concept of a nuclear equation of state (EOS) relies on global or at least local equilibrium, characterized by macroscopic variables like density and temperature. However, in the initial stage of heavy ion collisions local equilibrium is not yet reached and even in the later stages there might be a residual anisotropy in the momentum distribution of the not completely thermalized hot participant matter. In these cases the equation of state has to be generalized to non-isotropic momentum distributions. This generalization is especially important if the nucleon-nucleon interaction is momentum dependent. Besides density and temperature then additional macroscopic variables characterizing the deviations from the single spherical Fermi distribution are necessary like deformations or the relative collective flow velocity. In the presence of momentum dependent interactions (MDI) the flow contribution can become much larger than the trivial kinetic flow energy at least in the early stage of the collision due to a substantial repulsive potential for large relative momenta. In this way a stiff equation of state could be mocked up by the MDI. The inclusion of such interactions with anisotropic momentum distributions in the equation of state can have drastic effects on the observables like reduced pion and kaon yields (less thermal energy) and increased transverse momentum transfer [31].

A generalization of the equation of state, which is applicable also in the non-equilibrium case of two interpenetrating slabs of nuclear matter [15], has been proposed, where the relative collective flow velocity (momentum anisotropy) is introduced as an additional thermodynamic variable. However, in order to follow the temporal evolution of heavy ion collisions from the early nonequilibrium state through the high density phase to the subsequent expansion state dynamical models must be involved.

1. THEORETICAL APPROACHES

Several models have been developed so far, which describe heavy ion collisions in different energy regimes. At low energies (just above the Coulomb barrier), where mean-field dynamics is most important, TDHF describes phenomena like compound nucleus formation, fusion, or deep inelastic reactions. But this model gives too much transparency at higher energies ($\approx 100 AMeV$): here two body collisions due to the short range hard core repulsion come into the game, which are not included in TDHF. At these energies hydrodynamics predicts the bounce off of the spectators, the squeeze out of the participants from the reaction plane and especially the large collective flow [1,2,5]. This is in contrast to the intranuclear cascade [30] (ideal gas with two body collisions, no mean-field) which, although being a microscopic approach, gives too small a collective flow. However, it is difficult to include quantum effects, particle production, fragment formation, etc. in the hydrodynamical model, where the nuclei are represented as drops of compressible, viscous, thermoconductive, charged and bound fluid. Also the restriction to small deviations from local equilibrium and to very heavy nuclei (100-500 nucleons) are shortcomings of this model. A comparison of this models with the experimental GSI-LBL Plastic Ball is given in fig.2.

A more general microscopic nonequilibrium quantum statistical model is based on Vlasov-Uehling-Uhlenbeck equation which governs the time development of the single particle distribution function $f(\vec{x}, \vec{p}, t)$ in phase space [21,43]. It is a modification of the Boltzmann equation, which is regained if the $(1 - f)$ factors in the collision term on the right hand side are omitted. These factors ensure that two body collisions are suppressed according to the occupation probability of the final states (Pauli principle).

$$\dot{f}_1(\vec{r}, \vec{p}, t) = \frac{\partial f_1}{\partial t} + \vec{v} \bullet \frac{\partial f_1}{\partial \vec{r}} - \nabla U \bullet \frac{\partial f_1}{\partial \vec{p}} = -\frac{1}{(2\pi)^6} \int d^3 p_1' d^3 p_2' d^3 p_2$$

$$\delta(\vec{p}_1 + \vec{p}_2 - \vec{p}_1{}' - \vec{p}_2{}')\sigma v_{1,2}[f_1 f_2(1 - f_1')(1 - f_2') - f_1' f_2'(1 - f_1)(1 - f_2)]. \quad (1)$$

The VUU equation contains single particle dynamics (mean-field) as well as two body scattering but no many body correlations, because f is the single particle distribution function. From the Boltzmann equation the hydrodynamical equations of motion can be obtained if f can be approximated locally by the equilibrium distribution and by integration over momentum space. Therefore one can consider VUU as a micro-

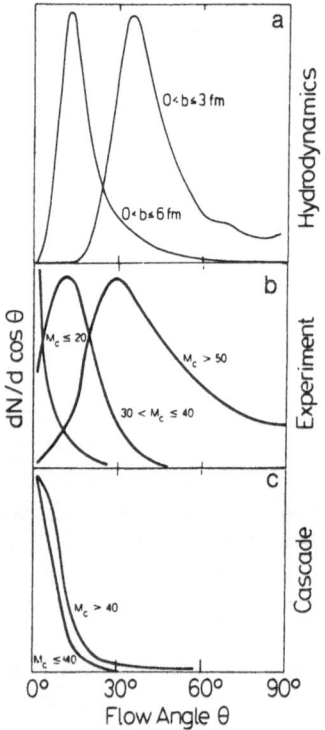

Figure 2. Distributions of flow angles $dN/\cos\Theta_f$ for the reaction Nb(400 A MeV) + Nb. (a) Result for the hydrodynamical calculation. The finite-multiplicity distortions are taken into account. The given impact parameter ranges correspond to the multiplicity cuts indicated on the experimental curves. (b) Plastic Ball data for various multiplicity cuts given on the curves. (c) Result of the cascade simulation after multiplicity selection.

scopic equivalent of nonequilibrium fluid dynamics. The time evolution of the reaction ^{197}Au(200 MeV) $+^{197}$Au in momentum space is shown in fig.3. The VUU model is quite successful in describing single particle observables (angular distributions and energy spectra, flow angles and pion yields) [16,17,20,21].

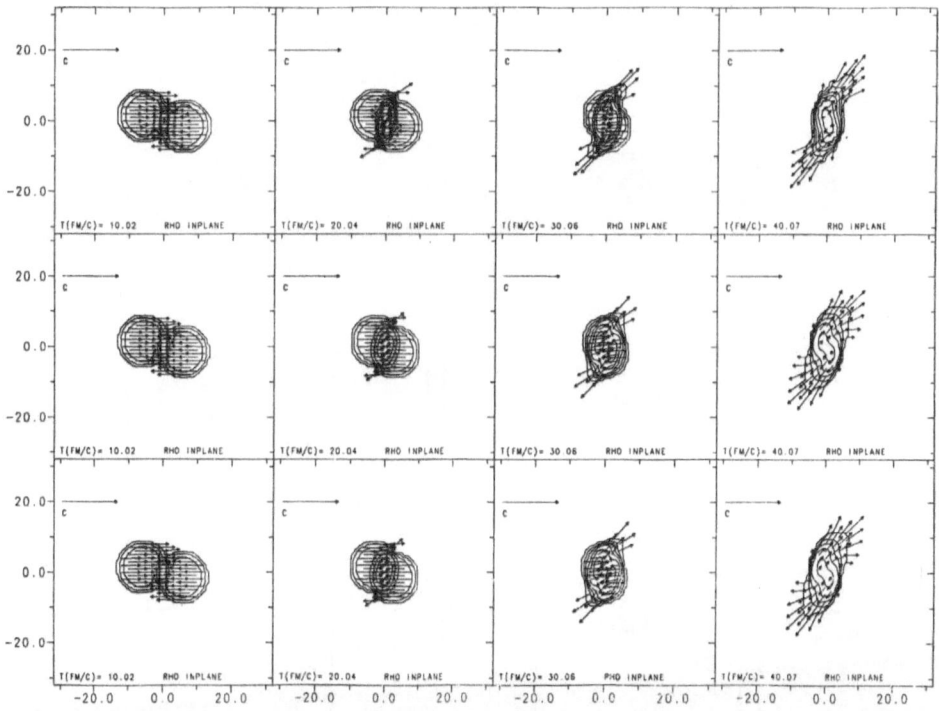

Figure 3. The different rows show the time evolution of the density profiles at b=3, 5 and 7 fm for the reaction Au(200 A MeV) + Au with the soft EOS for the viscosities $\eta=0$ MeV/fm^{2}c (upper row), $\eta=30$ MeV/fm^{2}c (medium row) and $\eta=60$ MeV/fm^{2}c (lower row). The velocity arrows refer to the lab system.

Phenomena like fragment formation and multifragmentation cannot be calculated in the single particle theories (VUU or hydrodynamics) since they are intimately connected to many body correlations. These N-body correlations can, in practice, only be investigated by solving the classical equations of motion with a reasonable interaction between the nucleons. This idea is used by the classical Molecular Dynamics [19,17,28,18] where two body interactions with a long range attraction and short range repulsion are introduced, and all quantum features like Fermi motion, Pauli principle and uncertainty principle or s-wave-scattering are neglected. The non anti-symmetrized cross sections lead to a very large flow and yield too many n-n collisions, which in turn produce too much stopping of the incoming matter. However, it is difficult to include the Pauli principle and particle production in the purely classical equations of motion.

Both the classical many body correlations and the quantum statistic UU collision are the ingredients of the quantum molecular dynamics model (QMD) [23,24,25,26]. Here gaussian wave packets are propagated according to their classical equations of motion instead of the point particles of VUU. With these wave packets smooth phase space distributions are obtained in single events (in the VUU model several test particles per nucleon are used to obtain smooth functions in phase space). It can be shown [22] that the pure Vlasov equation (omitting the collision term in (1), mean-field only)

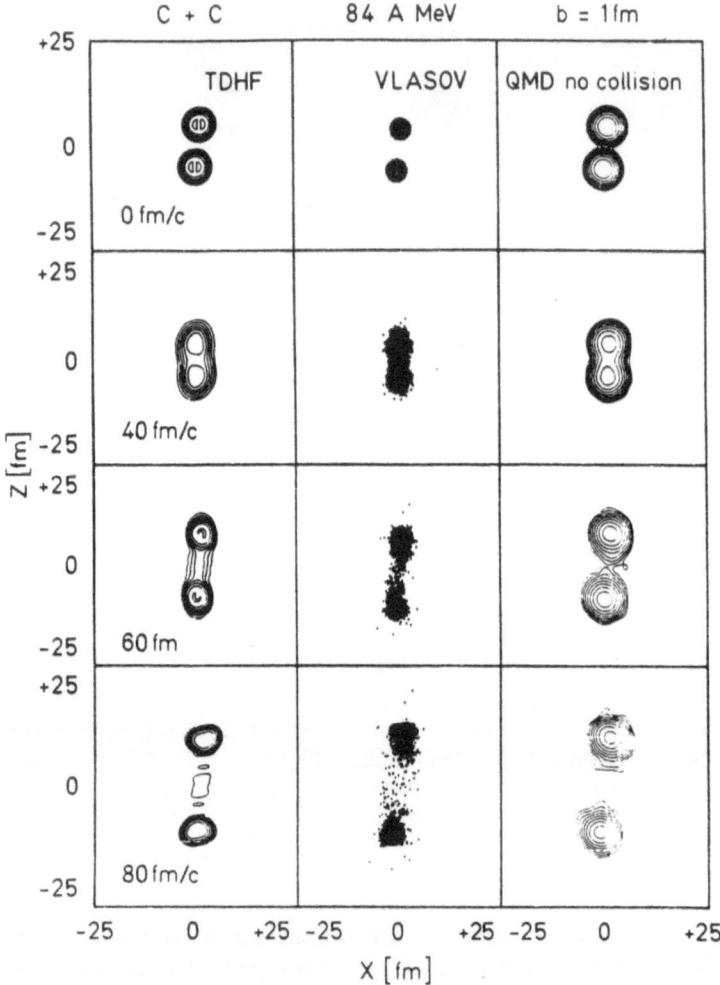

Figure 4. Time evolution in configuration space of C(85 MeV/n.) + C at b=1 fm for TDHF, the Vlasov equation, and QMD without collision term. In all cases transparency is observed.

exhibits essentially the same behaviour in the time development as its quantum mechanical analog, the TDHF equation, for $E_{lab} \simeq 10$ MeV per nucleon [36] (see fig.4). This indicates that details of the wave packets and also interference effects are not too important at high energies.

348

The basic ingredients of the QMD model can be summarized as follows:

1. The microscopic nucleon-nucleon interaction is given by a local Skyrme interaction, a Yukawa and a Coulomb part, the momentum dependent interaction, and the Pauli potential.

The Yukawa term has been added to improve the surface properties of the interaction, which are very important in the process of fragment formation.

The other parameters are determined by the ground state properties of nuclear matter (see below). With the Pauli potential we put special emphasis on the momentum space since the repulsive short range interaction in coordinate space is supposed to be included in the scattering cross section. The Pauli and Yukawa potentials are used in the following only for calculations in the low energy regime ($E < 20AMeV$).

2. Each nucleon is represented by its Wigner density in phase space corresponding to a boosted Gaussian wave packet

$$f_i(\vec{r}, \vec{p}, t) = \frac{1}{\pi \hbar^3} \exp\{-\frac{(\vec{r} - \vec{r}_i(t))^2}{2L^2} - (\vec{p} - \vec{p}_i(t))^2 \frac{2L^2}{\hbar^2}\} \qquad (2)$$

3. The center of momentum of each Gaussian $f_i(\vec{r}_i(t), \vec{p}_i(t))$ is propagated according to the classical relativistic equations of motion .

4. A 4π random scattering is performed at the point of closest approach if $r < \sqrt{\sigma/\pi}$ according to the measured angular distributions and the measured energy dependence of the cross section for the different processes. Pion production is calculated from the formation of deltas in inelastic scattering.

5. The Pauli blocking of final states in the stochastic two body collisions is taken care of in the same way as in the VUU model.

In infinite nuclear matter this interaction leads to the following potential:

$$U(\rho) = \alpha(\rho/\rho_0) + \beta(\rho/\rho_0)^\gamma + \delta(\ln^2\{1 + \epsilon(\rho/\rho_0)^{2/3}\})\rho/\rho_0 \qquad (3)$$

where $\gamma = 2$ and the relative momenta have been determined in the Thomas-Fermi approximation (the Coulomb, Yukawa and Pauli terms are omitted). The parameters of the interaction have to be fixed carefully since fusion thresholds and the scattering behaviour depend sensitively on the two body interaction as TDHF calculations have shown. We fit the parameters to the correct binding energy and mean square root values of radius and momentum of the nuclei. The Pauli potential is essential to reach the correct root mean square value of momentum and to achieve that the phase space distribution does not exceed unity. A schematic view of the different models with their most characteristic features is given in table 1.

2. THE THREE MAIN INGREDIENTS OF THE MICROSCOPIC THEORETICAL APPROACHES

The three most important ingredients of microscopic models are: 1)the long wave length (soft) interaction i.e. the potential, 2)the hard n-n-collisions i.e. the scattering cross sections, 3)the Pauli principle, which inhibits propagation and scattering of nucleons into regions of phase space which are already occupied.

Let us now discuss the influence of these terms on the dynamics in detail.

Table 1. Different models with main features

Model	Microscopic	Scattering	Potential (EOS)	Quantum effects	clusters
Hydrodynamics	no	(∞)viscous	yes	(EOS) Fermi	no
Cascade	yes	yes	no	$\sigma(E,\Omega)$ also (s-wave-scattering)	no
VUU	yes	yes	yes → Vlasov	yes Fermi,Pauli $\sigma(E,\Omega)$	no
TDHF	yes	no	yes	yes important at $E \gg E_{coul}$	no
Classical Molecular Dynamics	yes	(yes)	yes	no	(yes) classical correlations + fluctuations
QMD	yes	yes	yes	yes quantum statistics Pauli,Fermi,$\sigma(E,\Omega)$	(yes) classical correlations + fluctuations

1.THE COLLISION TERM

The collision term gains in significance with increasing bombarding energy and it becomes indispensable at certain energies to reproduce the experimental data, as studies made with the C(85 MeV/nucl.) + C system at b= 1 fm [22] have shown. A very similar behaviour in both the quantum mechanical and classical pure mean field theories is observed: both TDHF and Vlasov calculations exhibit nearly identically small longitudinal and transverse momentum transfers. The lack of two body collisions results in strongly forward peaked angular distributions, in sharp contrast to the data in this energy regime . Both theories predict that for central collisions of C + C the nuclei slip through each other and survive the reaction rather intact. About 85% of the initial longitudinal momentum is conserved in the projectile- and target-like nuclear fragments. Hence the low angular momentum fusion window previously observed in TDHF calculations persists up to energies of 85 MeV/nucl.

This is drastically changed by the inclusion of the Uehling-Uhlenbeck collision integral into the Vlasov equation. Each individual reaction can now be separated into two clearly distinct components. First we observe again the slipped-through projectile- and target-like fragments which, however, now retain only about 40% of the nucleons and less than 20% of the initial longitudinal c.m. momentum. As we see in fig.5, these slipped-through residues contain mostly particles which have not scattered at all.

Also a comparison of measured and calculated proton spectra for 42 and 92 MeV/nucl. Ar + Ca shows that the calculated absolute cross sections and the slopes of the spectra are in reasonable agreement with the data [43]. In contrast, a simple cascade simulation, though appropriate for high energies, cannot reproduce the medium energy data.

The QMD model has been applied to the description of heavy ion collisions around the Coulomb barrier [36]. The scattering angles and the kinetic energy dissipation of collision of the system $^{86}Kr + ^{139}La$ at E/A=7 MeV compares favorably with data and TDHF calculations.

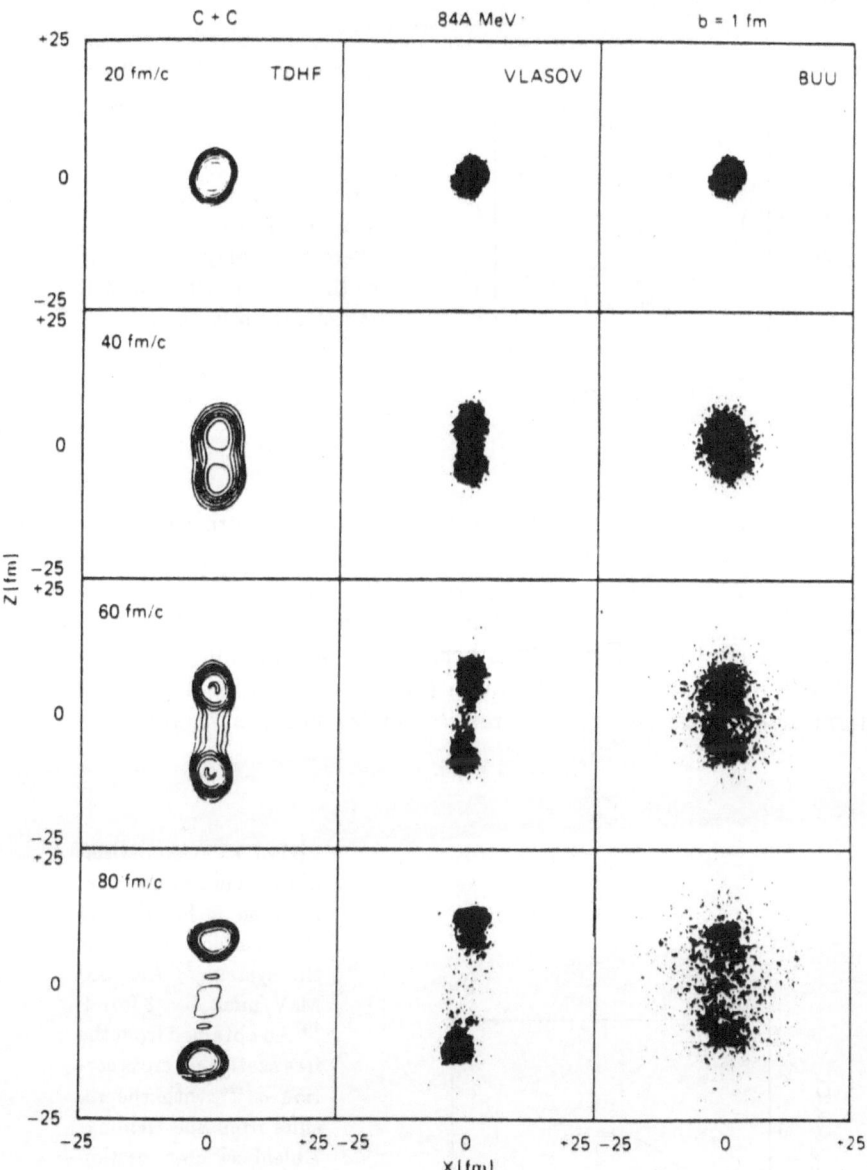

Figure 5. Time evolution in configuration space of C(85 MeV/n.) + C at b=1 fm for TDHF, the Vlasov equation, and the Vlasov-Uehling-Uhlenbeck theory. Transparency occurs in both cases with a mean field only.

2. PAULI BLOCKING

A very important quantum feature of a heavy ion collision is the Pauli blocking of the final states. The implementation of Pauli principle is quite essential for quantitative predictions as can be seen from fig.6.

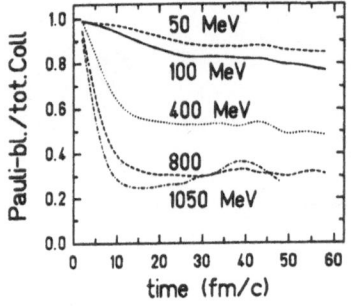

Figure 6. Time dependence of the Pauli blocking ratio at different bombarding energies for the reaction Nb+Nb at 3 fm using the soft EOS.

VUU and also QMD take the Pauli principle into account with the Pauli blocking factor $(1-f)$, where $f(\vec{x}, \vec{p}, t)$ represents the density in phase space. At intermediate bombarding energies, the Pauli blocking factor $(1-f)$ causes a substantial reduction of the number of n-n collisions: a doubling of the number of n-n collisions is observed if this quantum effect is neglected. This behaviour is displayed in fig.7, where results obtained with a 30% reduced cross section (due to the additional Pauli blocking of the intermediate scattering states as predicted by [27,30,57]) σ_{eff}, are shown.

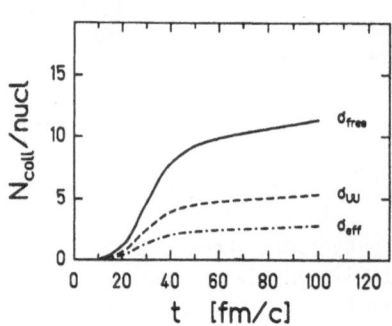

Figure 7. Comparison of the collision numbers N_{coll} as a function of the reaction time for the system ^{197}Au (200 MeV/nucl., b = 3 fm) + ^{197}Au obtained from the free scattering cross section σ^{free} with the results from the Uehling-Uhlenbeck cross section σ^{UU} taking into account the Pauli blocking of the final states in a n-n scattering, and also for the reduced Uehling-Uhlenbeck cross section σ_{eff} which additionally takes into account the Pauli blocking of the intermediate scattering states.

3. THE STOPPING POWER AS A MEASURE OF THE SCATTERING CROSS SECTION $\sigma(\Omega, E)$ AND THE MEAN FREE PATH

The influence of the cross section on the dynamics of the reaction can most clearly be seen from the mean free path of the nucleons. Classical calculations give the mean free path as $\lambda \sim 1/\sigma\rho \approx 2$fm at $(\rho_0, \sigma_{NN}^{eff})$. However, the quantum features and the nonequilibrium effects lead to different results in nuclear collisions: λ increases at low energies (due to the Pauli principle) and decreases at higher energies, because the density exceeds $2\rho_0$. This is of drastic importance for the thermalisation and the stopping power. Fig.8 shows the energy dependence of the mean free path obtained from the QMD model [59] for the most central collisions of Au+Au. The results are shown for the Uehling-Uhlenbeck cross section σ^{UU} and also for a 30% reduced cross section σ^{eff}. One clearly observes that the reduction of the cross section causes at all energies a increase of the mean free path of the nucleons.

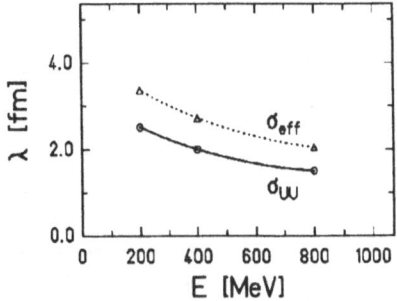

Figure 8. The mean free path of the nucleons is shown for the very central Au + Au collisions using different n–n cross sections as indicated.

While in the early stages of the reaction, the rapidity spectra indicate an incomplete stopping of the incoming matter, nearly all matter has been stopped and thermalized at the end of the compression stage ($t \approx 30$ fm/c)[26]. Consequently, as can be seen from fig.9, almost no remnants of the projectile and the target survive the reaction at b= 1 fm.

Reduced in-medium cross sections increase the transparency of the system (there are fewer collisions and therefore less thermalization). Fig. 10 shows the total rapidity distribution of baryons for the collision Au(200AMeV)+Au at b=3fm impact parameter using the different interactions S (soft EOS), SM (soft + MDI), and SIM (soft + MDI + reduced σ). We find that the local potential (EOS) has little influence on the rapidity spectra: the distributions obtained with the interactions H are almost identical to those obtained with the interactions S [24,26]. Also the MDI has only a small influence on dN/dY (compare curves for S and SM). On the other hand, the distributions obtained with and without in-medium cross sections exhibit a clear increase of transparency when in-medium cross sections are switched on. This is most pronounced for the intermediate mass fragments and could be used to determine the effective scattering cross sections experimentally. Recent dN/dY data [12] indicate no substantial modification of the Uehling-Uhlenbeck cross section, hence $\sigma^{eff} \approx \sigma^{UU}$. This is of great importance because the magnitude of the transverse flow depends sensitively on the effective cross sections. Thus one can use the dN/dY-distributions to determine experimentally the in medium cross section σ^{eff} and use this information to study the nuclear EOS in the flow analysis.

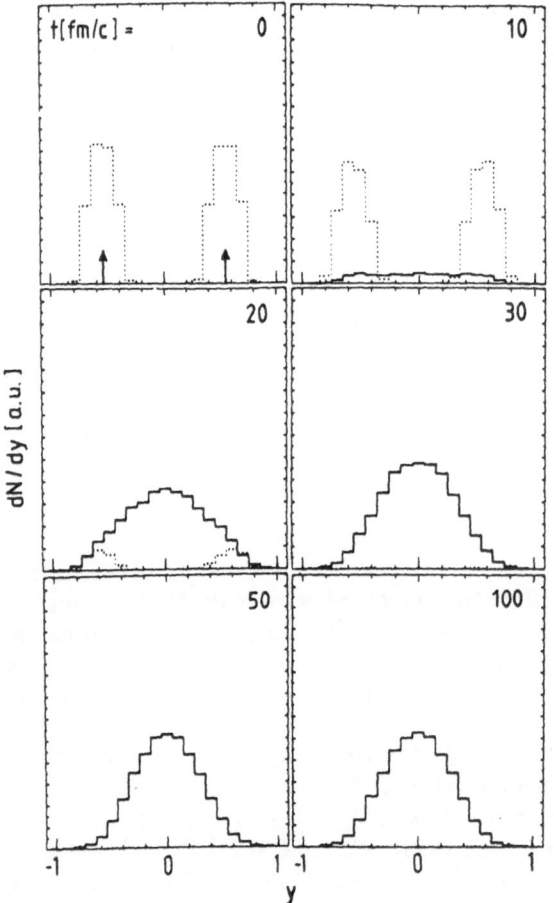

Figure 9. Rapidity distribution dN/dy of the fireball participants (full lines) and of the projectile and target spectators (dashed lines) at different reaction times in near central collisions ^{197}Au (200 MeV/nucl., b = 3 fm) + ^{197}Au. The initial projectile and target rapidities are indicated by arrows.

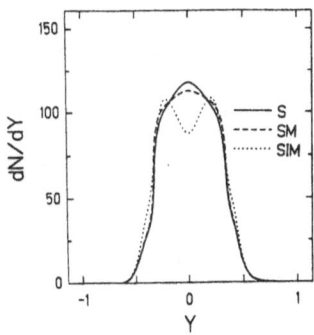

Figure 10. Total rapidity distribution dN/dy in the reaction ^{197}Au (200 MeV/nucl., b = 3 fm) + ^{197}Au.

The mass number dependence of the stopping power can be seen by inspection of fig.11. In agreement with experimental data [9], the rapidity spectra for the Ca+Ca , Nb+Nb and Au+Au reactions at the correspondingly scaled impact parameters $\tilde{b} = b/(R_P + R_T)$ for 400 MeV/nucl. bombarding energy become much flatter with increasing mass number; for A=40 a plateau is observed (similar results are obtained above 1 GeV/nucl. [12]). This effect can be understood within a hydrodynamic picture in terms of the quantity λ/R, where λ is the mean free path and R the nuclear radius. The mean free path depends strongly on the n-n cross section (cf.fig.8) , the value λ/R, however, decreases with increasing mass of the system due to the increase of the reaction volume from $\sim$ 0.5 for Ca+Ca to 0.25 for Au+Au. For heavy systems $(\lambda/R < 1)$ one should be close to a local equilibrium situation where viscous hydrodynamics , which yields large flow effects, is applicable. When considering lighter systems (A $\approx$ 20, $\lambda/R \approx$ 1) non-equilibrium effects could be taken into account via the three fluid model [26] .

In microscopic models like VUU and QMD, the quantity λ/R is directly related to the number of n-n collisions which in turn determine the amount of nuclear stopping and thermalization.

3. HOT NUCLEAR LIQUID AND MULTIFRAGMENTATION

The extension of the QMD model to low energies is quite important since it allows to study fusion, fast fission and deep inelastic collisions, where TDHF does not contain the classical fluctuations present in QMD (TDHF provides only information on mean values, not on the width of distributions).

At higher energies the Pauli and Yukawa potentials are less important. The two body final state gives way to a many cluster final state (i.e. multifragmentation).

Let us first consider the reaction ^{197}Au+^{197}Au at an energy of 200 A MeV . We follow the evolution of the reaction for 200 fm/c. This time is long enough to study the final fragment distributior, since as we will see later, the mass yields reach their asymptotic values after approximately 150 fm/c reaction time [60].

Two nucleons are considered to be bound in a fragment if the centroids of their wave packets have a spatial distance $d_0 \leq 3 fm$. We checked that the results do not depend sensitively on d_0; nucleons which belong to different clusters are widely spread in configuration space after 200 fm/c. Doubling of d_0 leads to a $\approx$ 20% shift of the yields of the complex fragments. Because of the strong correlation between position

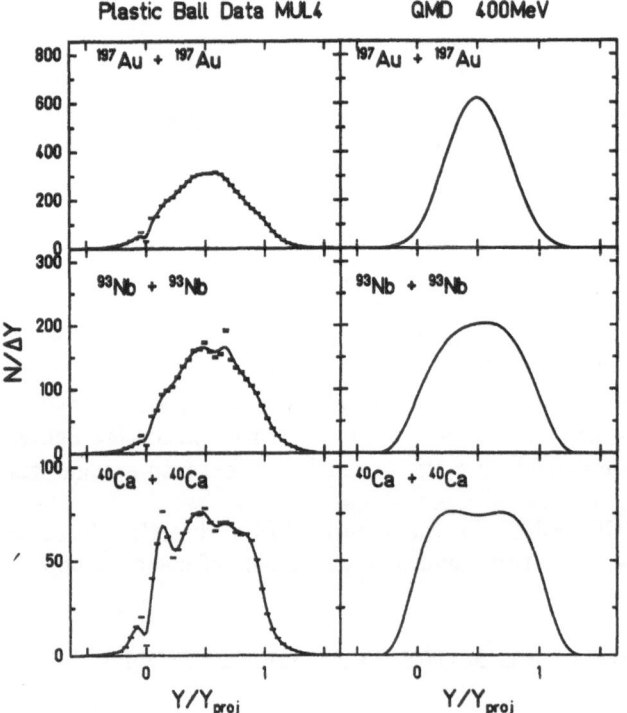

Figure 11. The experimental rapidity spectra of the reactions ^{197}Au (400 MeV/nucl.) $+^{197}$Au, ^{93}Nb (400 MeV/nucl.) $+^{93}$Nb and ^{40}Ca (400 MeV/nucl.) $+^{40}$Ca for the multiplicity interval MUL4 corresponding to half overlap are compared to the final rapidity distributions dN/dy for the system ^{197}Au (400 MeV/nucl., hard EOS, b=3 fm) $+^{197}$Au, ^{93}Nb (400 MeV/nucl., hard EOS, b=2.3 fm) $+^{93}$Nb and ^{40}Ca (400 MeV/nucl., hard EOS, b=1.75 fm) $+^{40}$Ca , obtained from the QMD calculation.

and momentum of the nucleons forming a fragment, it is not necessary (although it is trivial to do [43]) to consider the relative momenta of the nucleons forming a cluster at late times.

We have grouped the fragments into seven mass intervals. The first interval contains all free nucleons ($A = 1$), the second light fragments with $A = 2 - 4$, then follow the mass ranges with $A = 5 - 15, 16 - 30, 31 - 50, 51 - 70$ and $A > 70$.

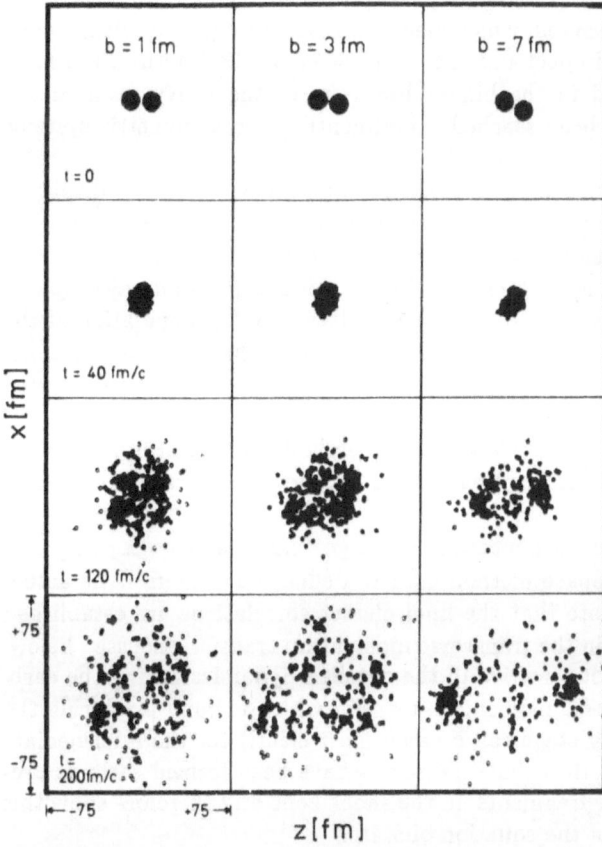

Figure 12. Time evolution of the particle distribution in configuration space for the reaction Au (200 A MeV) + Au for impact parameters of b = 1, 3 and 7 fm. The projection of all particles in the reaction plane (x–z) is displayed for four different times as indicated.

The spatial evolution of an Au+Au collision at 200 A MeV energy can be followed in fig.12 for impact parameters 1, 3 and 7 fm. The beam axis coincides with the z-direction. Observe the formation of one blob of matter for all impact parameters ($t \leq 40 fm/c$), which then disintegrates and yields the fragments. For half overlap collisions ($b = 7 fm$), two massive projectile and target remnants survive the reaction. In this case, one observes almost no intermediate mass fragments. When going to more central collisions the projectile and target–like fragments become smaller ($b = 3 fm$) and are absent in the most central collisions, where we observe a complete disintegration of the incident nuclei. On the other hand, the intermediate mass fragments are abundantly produced in central collisions and disappear for peripheral ones.

The typical time structure of a single half overlap collision for Au$(200A\ MeV, b = 7fm)+$ Au is shown in more detail in fig.13 . Only fragments with $A \geq 10$ are considered from $t = 0$ (bottom) to $t = 200fm/c$ (top) in steps of 10 fm/c. Up to 50 fm/c one observes one blob of matter in configuration space which – for this large impact parameter – is still separated in momentum space into a projectile and a target–like residue (for central collisions ($b \leq 3fm$), this is not the case; we then observe complete stopping of projectile and target in the cm frame). After 50 fm/c, the system breaks up into these two residues. Between 30 and 80 fm/c most of the single nucleons and light fragments ($A < 10$) are emitted. Then a rather stable fragment with $A \approx 105$ remains in the projectile rapidity regime. In the target regime a second break–up is observed, which yields two stable fragments with $A \approx 80$ and 15, respectively. At large impact parameters ($b = 7fm$) the intermediate mass fragments are mostly produced in the binary break–up of the heavy residues. A similar conclusion has recently been reached experimentally for asymmetric systems [42].

The average number of fragments as a function of time for b = 3 fm is displayed in fig.14 . We observe that the mass yield distribution for all fragments ($A \geq 2$) stabilizes at $\sim$ 100 fm/c . Let us first concentrate on the heavy clusters with A>70. Here one recognizes one cluster up to 50 fm/c which is not stable and decays rapidly. The decay chain can be seen by the subsequent population and depopulation of the different mass bins for the smaller clusters. Along the decay chain the cluster emits also single nucleons and therefore these numbers increase but have almost saturated at t = 200 fm/c.

The clusters in the bins $2 \leq A \leq 30$ have a completely different history. They are formed at a very early stage of the reaction . After 100 fm/c practically all of them have been formed.

Keep in mind that the transient appearance of large "clusters" ($A \sim A_T + A_P$) reflects the simple configuration space method used to define the clusters. The actual phase space distributions indicate that the final cluster correlations are established much earlier. This can be seen in the transverse momentum transfer discussed below. The complex fragments are most sensitive to the detailed dynamics during the early compression stage. The transverse momentum transfer which is build up during the expansion from the high density stage can be seen most clearly for the intermediate mass fragments. This indicates that these fragments have been formed early as pre-fragments, i.e. particle unstable fragments, in the shock zone and therefore show this strongly enhanced sensitivity on the equation of state.

Also shown in fig.14 is the time evolution of the local density in the central region of the reaction. Observe the compression shock at $t \sim 20fm/c$. At that time also the highest temperatures are found. Subsequently the system rapidly expands out of the central region of the shocked matter. Note that the highest compression occurs at a time when fragments are not yet separated in configuration space. After the central density has decreased they are visible as individual entities only.

We have checked the dependence of the calculated fragment yields on the width of the Gaussians, i.e. on the effective range of our potential. While the light fragment yield (A = 1 –4) changes only by 10%, the number of massive fragments decreases by 30% when $2L^2$ is doubled.

The inclusive, i.e. impact parameter averaged, mass yields for the hard and the soft equation of state for Au (200 A MeV) + Au are shown in fig.15. Both curves exhibit a clear power law behaviour $Y(A) \sim A^{-\tau}$. The fragment yields, however, are not sensitive to the underlying EOS. For the constant τ we find the value $\tau \approx 2.3$. The same behaviour is found in the asymmetric system Ne + Au at 1 A GeV bombarding

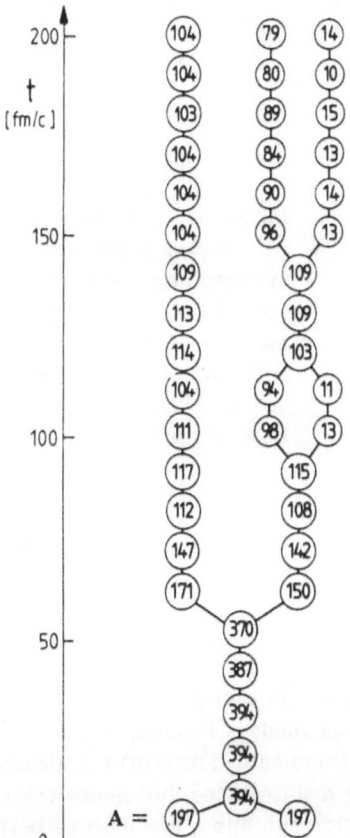

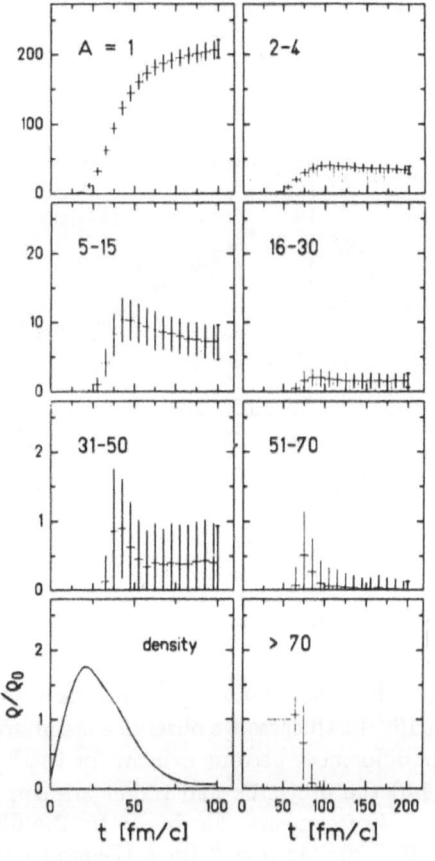

Figure 13. Time evolution of the fragmentation process in a single event for the reaction Au (200 A MeV, b=7 fm) + Au. Shown are all massive fragments (A ≥ 10) from $t = 0\ fm/c$ (bottom) to $t = 200\,fm/c$ (top). Two nucleons are considered to be members of a fragment if their spatial distance is less than $d_0 = 3\,fm$.

Figure 14. Time dependence of the fragment yields for the reaction Au (200 A MeV, b = 3 fm) + Au, based on 200 collisions. The average number of fragments for different mass classes as indicated are shown as a function of reaction time. The width of the fragment distributions are indicated by the error bars. Also shown is the time evolution of the local density in the central region of the reaction. The local density has been determined in a sphere with the radius r = 2 fm in the center of the CM–frame.

energy [41,25]. We want to emphasize that a change of the correlation lenght d_0 in the minimum spanning tree procedure does not change this power law dependence of Y(A). Such a power law dependence with an exponent $2 \leq \tau \leq 3$ has been interpreted as an evidence for a liquid–vapour phase–transition [40]–[39].

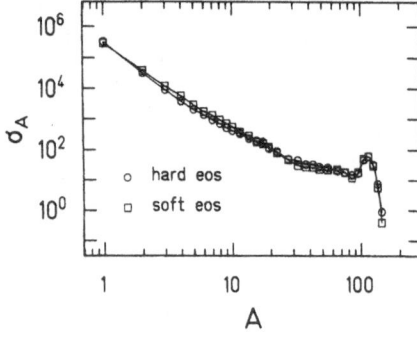

Figure 15. Inclusive, i.e. impact parameter averaged mass yield for the reaction Au (200 A MeV) + Au obtained with the soft (squares) and the hard (circles) EOS, respectively.

The dependence of the final fragment yields on the impact parameter is displayed in fig.16 . In all cases we observe a steep drop of the yields of fragments with $A \leq 10$. Large differences become evident for the heavier fragments. For central collisions (b = 1 fm) the projectile and target are completely disintegrated and hence there are no $A > 40$ fragments. For b = 5 fm the distribution exhibits a flat plateau between A = 40 - 70. At b = 7 fm a U–shaped curve with a peak at $A \approx 120$ (projectile and target residue) and almost no fragment in the $A = 20 - 80$ region result. This systematics has already been discussed in fig.12.

Hence we conclude that impact parameter averaging (rather than a liquid–vapour phase–transition) leads to an accidental power law dependence of the inclusive mass yield.

Furthermore, with respect to the liquid-vapour phase transition, we find that for noncentral collisions the fragments of different masses reside in different rapidity bins (see also fig.25) in each event. However, for very central ($b < 2$ fm) collisions of equal, massive nuclei, where we do observe a stopping and total disintegration of the projectile and the target, all fragments do reside in the same rapidity interval. Therefore we propose to study the occurence of the liquid–vapour phase transition by measuring the fragment yield excitation function for intermediate energies (E = 30–200 A MeV). One should trigger on central collisions of very heavy projectiles, which provide sufficient stopping power.

Independent of the form of the local potential there is a rapid decrease of the fragment yields for the light fragments ($A < 10$) at all energies. The slope of Y(A) becomes steeper with increasing bombarding energy in agreement with simple thermostatic predictions, which assume that the entropy of the system increases with bombarding energy [38,37]. At $E \geq 400\ A\ MeV$ and $b = 3 fm$ there are almost no fragments with $A > 20$ left. In this case only the light fragment yield exceeds one fragment per event.

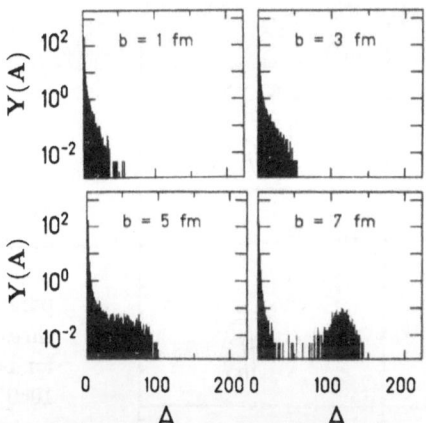

Figure 16. Impact parameter dependence of the average mass yield per event Y(A) for the Au (200 A MeV) + Au reaction.

4. POSSIBLE OBSERVATION OF DYNAMICAL EMISSION OF MACH SHOCK FRAGMENTS IN INVERSE KINEMATIC EXPERIMENTS

Light fragments have a highly non isotropic emission pattern even in central collisions if asymmetric systems are studied in inverse kinematics as we can see in fig.17, which displays the laboratory double differential cross section $d^2\sigma/dydp_t$. There are many fast particles in the forward direction which are not counterbalanced at backward angles. This fact is a great opportunity for inverse kinematics experiments (e.g. ALADIN [45]), where cold fragments formed in the shock front can be searched . For the heaviest clusters (A $\sim$ A$_P$), on the contrary, exhibit an isotropic emission pattern in their rest sytem which moves with close to beam velocity in the laboratory system.

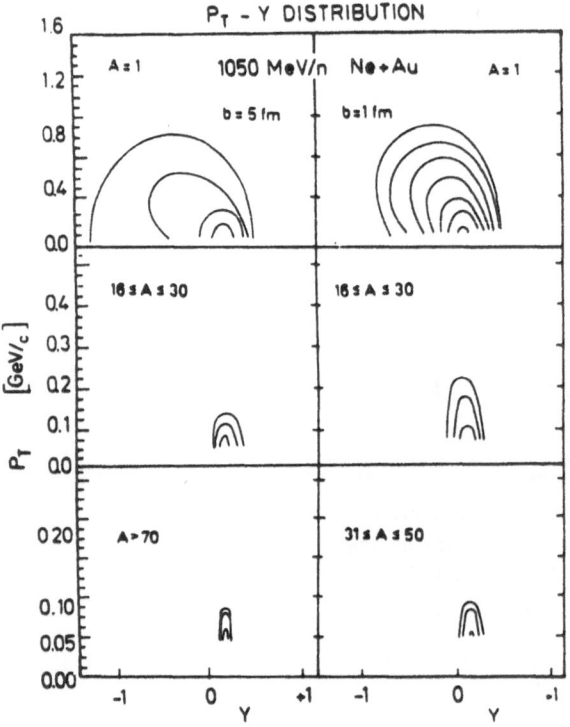

Figure 17. Contour plot of the double differential cross section $d^2\sigma/dp_t dy$ for two impact parameters (b=1, 5 fm) and three different mass intervals for b=3, 7 fm of the reaction 1050 MeV/nucl. Ne+Au.

The rapidity distributions of the different classes of fragments at two impact parameters are shown in fig.18. It is quite obvious that the rapidity distributions are not thermal.This is in contrast to the results for central collisions in the symmetric Au(200 MeV/nucl.) + Au collision [24] . There the particle distribution in very central collisions (b=1 fm) can be approximated by one single stopped midrapidity source. For the large impact parameter (b=7fm) we see clearly a projectile and a target region. For low impact parameters the projectile dived into the target and is strongly slowed down. Once more we can see that the medium mass clusters come from a single well defined rest system.

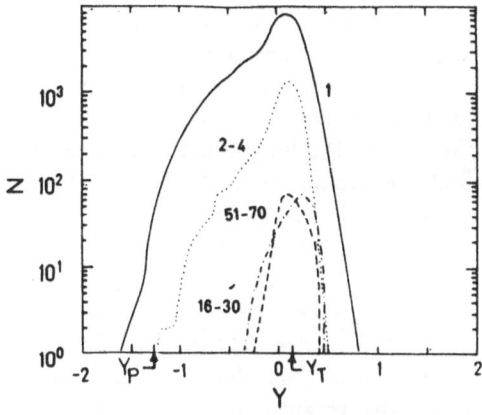

Figure 18. Rapidity distribution of different mass intervals for b=3, 7 fm of the reaction 1050 MeV/nucl. Ne+Au.

The cause of multifragmentation is an important point which we want study. As we have seen the clusters are not formed in a globally equilibrated environment, so that processes other than statistical decay have to be taken into account. The medium mass clusters are emitted from the system already quite early, long before the target residue evaporation chain ceases. This process has to be studied on a event-by-event basis, since clusters are produced by fluctuations in the system.

When a collision occurs the projectile nucleons experience strong transverse forces due to the strong density gradient . They pick up transverse velocity and are deflected to finite angles. A shock profile develops which moves inwards into the projectile nucleus - the other nucleons have already been carried away by the sidewards travelling compression wave [2,5]. The projectile causes the emission of clusters with a velocity above the sound velocity while the source itself decelerates gradually but has still supersonic velocity. In an infinite system this is the situation where a Mach cone would be formed and indeed the form of the velocity distribution at 200 fm/c resembles very much such a velocity profile. The fate of a prefragment in such a situation can be studied in fig.19, where the time evolution of mean values of different quantities for a large prefragment (A=24 left side) and for a small fragment (A=6) are shown. The upper picture displays the number of collisions per fragment nucleon. Between the arrival of the projectile and the separation of the fragment from the remnant we observe a quite high collision rate. When the fragment is formed there is still excitation energy which allows further collisions among the fragment nucleons. The next row shows the density in units of ρ_0 and the average radial force (with respect to the c.m. of the target). As already expected from the previous figure we see a strong increase in the density when the projectile matter hits the nucleons which will form the fragment. Initially the collective momentum of those nucleons is close to zero . The density, however, causes a strong radial repulsive force which reverses the direction of the average momentum.

So far it seems that collisions are not required at all to cause fragments to break off. This, however, is not true. If the n-n collisions are suppressed we do not find fragmentation. Collisions have a twofold influence on the fragmentation processes. Firstly, they decelerate the projectile nucleons and hence increase the density in the projectile region. Secondly, they may provide an additional momentum transfer to those nucleons which are going to form a fragment. To check whether the second mechanism is important we compare the actual average radial momentum with that caused by the average radial field only. The difference between both reveals the importance of the momentum transfer to the fragments due to collisions. The results are displayed in the same graph. In the case of the large cluster the final momenta are almost identical, whereas the small cluster would not be broken off at all. So the role of collisions for the actual break up process is ambiguous.

The lower graph displays the time evolution of the internal excitation of the fragment, where the temperature $\Delta = (1/A_{free}) \sum_{j \epsilon A_{free}} (m/3) < v_j - v_{av} >^2$ is not a true temperature since it also includes the Fermi momentum. We see only a small increase in the course of the reaction. So the prefragments are only moderately excited and there seems to be no equilibration between the internal degrees of freedom and the translational motion. This is in agreement with recent experiments which show that the excitation of prefragments corresponds to a temperature of 5 MeV independent of the beam energy [44]. Hence the picture which emerges is like follows: the momentum transfer to the prefragment occurs early on $t \sim 20$fm/c into the reaction , when a compression (Mach shock) wave hits the prefragment - n-n-collisions plus the large potential gradients yield a large pressure on the fragment. This determines the observable c.m. momentum of the emitted fragment. This fragment must travel, however, for quite a while through the surrounding nuclear medium, and therefore experience additional interactions with the medium. During this time span (up to ~ 60 fm/c) the internal (rather than the c.m.) momentum distribution is continuously altered. Hence the fragment spectra exibit high temperature values while their chemical composition displays the rather low 5 MeV "temperature".

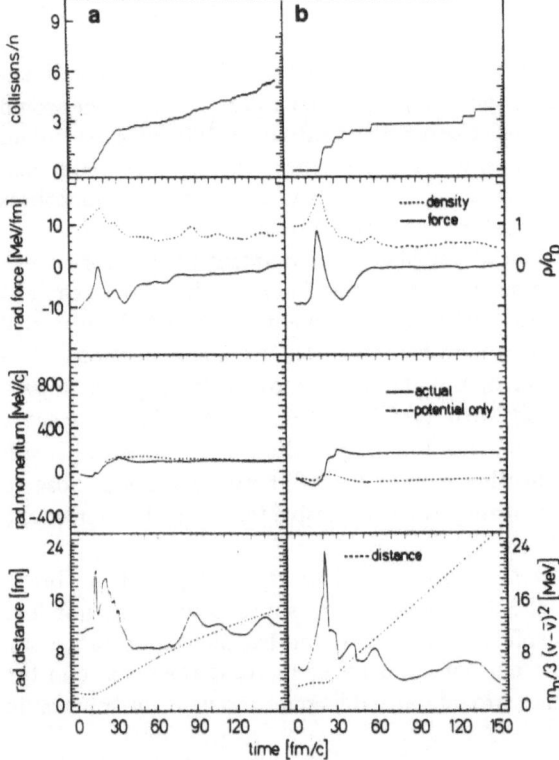

Figure 19. Time evolution of a single prefragment produced in the reaction 1050 MeV/nucl. Ne+Au b=0 fm. The properties of heavy (A=24) prefragments are displayed in (a) those of light prefragments (A=6) in (b). We show the average number of collisions, the mean radial force, the density, the mean radial momentum, the mean radial distance from the center of the target and the temperature $\Delta = \frac{m}{3} < v - v_{av} >^2$ as a function of time. The values displayed are averaged over the fragments constituents.

5. MULITPLICITY DISTRIBUTIONS OF INTERMEDIATE–MASS FRAGMENTS

Let us now come back to the Au + Au system and study the behaviour of the intermediate mass fragments in more detail. Here we want to compare our results directly with the Plastic Ball data [9,10]. Therefore we applied the experimental efficiency filters, i.e. the same low energy cut–off at 35 MeV/nucleon to all particles. The intermediate mass fragments are only counted if they were emitted with an angle of less than 30 degrees in the laboratory system, corresponding to the experimental Plastic Ball setup, which has detected such intermediate mass fragments due to higher gains than the rest of the 4π spectrometers. Fragments with masses greater then 20 (Z $\geq$ 10) are ignored [9,10].

The multiplicity distribution $(M_c(A > 4))$ of the intermediate mass fragments, i.e. fragments with $A > 4$, obtained is shown in fig.20 for four different impact parameters. In the upper part we show the unfiltered results and in the lower part those obtained with the Plastic Ball filter. In the case of the unfiltered distributions we find for peripheral collisions typically 2–4 heavy projectile and target–like remnants. The number of intermediate mass fragments increases to a mean multiplicity of about 10 fragments per event for near central collisions (b = 3 and 1 fm).

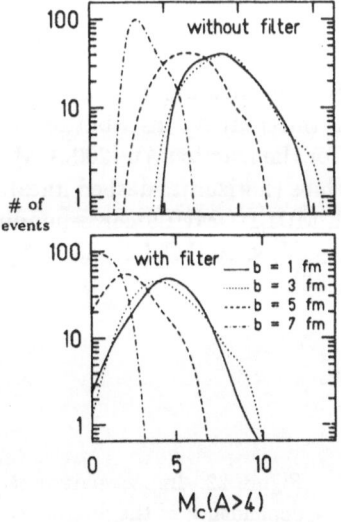

Figure 20. Calculated multiplicity distributions of fragments with $A > 4$ for the reaction Au (200 A MeV) + Au at impact parameters of b = 1, 3, 5, 7 fm. The upper part shows the unfiltered yields, while the lower figure has been obtained using a Plastic Ball filter. The Plastic Ball filter rejects all particles and fragments with $E_{Lab} < 25\ A\ MeV$. The massive fragments (up to mass 20) are only registered if the are emitted with an angle less than 30° in the laboratory frame. Fragments with masses greater than 20 (Z $\geq$ 10) are ignored.

The multiplicity distributions $M_c(5 \leq A \leq 20)$ of the remaining detectable fragments (lower part of fig.20) also show that the near central collisions (b= 1,3 fm) lead to almost identical distributions. Both curves are now peaked at much smaller mean values of $< M_c(5 \leq A \leq 20) > \approx 5$. Observe that in the half overlap reaction ($b = 7\ fm$) the maximum of the distribution $M_c(5 \leq A \leq 20)$ is at zero, i.e. the 2–4 fragments at $b = 7\ fm$ shown in the upper part have escaped the detector window of the Plastic Ball, i.e. the heavy fragments from the target and projectile hemisphere are strongly suppressed. We also find that $M_c(A > 4)$ is nearly independent of the EOS [24].

In fig.21 the filtered multiplicity distributions of central and peripheral collisions (b = 3,10 fm) are compared directly with the corresponding experimental curves from [9,10]

The data are shown for those events with maximal (MUL4) and minimal (MUL1) participant proton multiplicities N_p. The data and theory are shown for fragments with $A > 5$. Note the dependence of M_c on the width L used in the calculation.

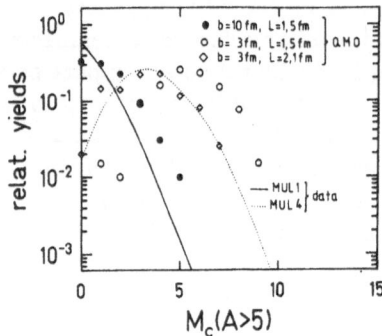

Figure 21. Calculated versus measured fragment multiplicity distributions for peripheral (MUL1, b = 10 fm) and central (MUL5, b = 1 fm) collisions of the system Au (200 A MeV) + Au. The QMD results (circles) have been obtained with the filter, while the lines represent the experimental data.

In order to relate the experimental multiplicity bins with our impact parameter we want to show the dependence of the calculated (filtered) N_p distributions on the impact parameter. This relation is shown in fig.22 for the reaction Au (200 A MeV) + Au. Also shown in this figure are the experimental bins (horizontal dashed lines). The multiplicity bins given in ref. [10] are labeled with MUL1···MUL5 corresponding to the multiplicity intervals $0 \leq N_p < 23$ (MUL1), $23 \leq N_p < 46$ (MUL2), $46 \leq N_p < 69$ (MUL3), $69 \leq N_p < 92$ (MUL4) and $N_p > 92$ (MUL5).

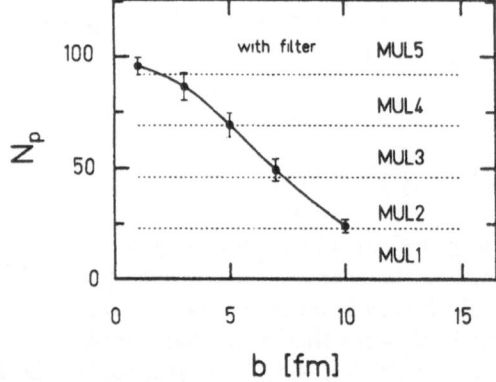

Figure 22. Impact parameter dependence of the filtered average participant proton multiplicities N_p for the reaction Au (200 A MeV) + Au. The horizontal dashed lines depict the experimental multiplicity cuts (MUL1 – MUL5). The bars indicate the statistical width of the distributions.

The calculations show that impact parameters from 1 to 10 fm cover the multiplicity distributions in the region $20 \leq N_p \leq 100$. The distributions are very narrow, there is little overlap in the multiplicity distributions . The mean values of these distributions are: $< N_p > \approx 25$ (b = 10 fm) $< N_p > \approx 50$ (b = 7 fm) , $< N_p > \approx 70$ (b = 5 fm), $< N_p > \approx 85$ (b = 3 fm) and $< N_p > \approx 95$ (b = 1 fm). With these results at hand we can finally connect the experimental multiplicity bins with the impact parameter: we find that apparently the bin MUL1 results from collisions with impact parameters $b \geq 10 fm$. The MUL2 bin corresponds approximately to collisions with $7 fm \leq b \leq 10 fm$, while the impact parameters b= 5 and 3 fm roughly represent the MUL3 and MUL4 bins. The highest experimental multiplicities (MUL5) are related to our most central collisions with b = 1 fm albeit the 3 fm collisions also populate this multiplicity region.

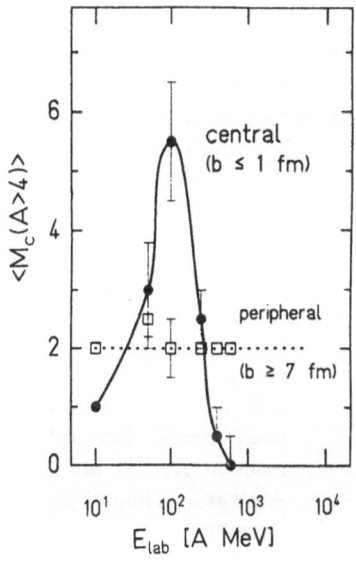

Figure 23. Excitation function of the average multiplicity of intermediate mass fragments $M_c(A > 4)$ for the system Nb + Nb obtained without filters for central and peripheral collisions as indicated. The curves are to guide the eye only.

The intermediate mass fragment multiplicities depend strongly on the bombarding energy. The average multiplicities of intermediate mass fragments (A > 4, without filter) is shown in fig.23 for the Nb + Nb system as a function of bombarding energy, for the most central and the peripheral (b = 7 fm) collisions. Observe that for central collisions the average multiplicity $< M_c(A > 4) >$ is one at low energies (fusion) and has a maximum of about 5 to 6 intermediate mass fragments at intermediate energies ($E \approx 100A\ MeV$). It is at these highest multiplicities of intermediate mass fragments that we expect the largest sensitivity to the liquid-vapour phase transition discussed above. This clearly demonstrates that the system completely breaks up into the lightest (A < 4) fragments at higher energies. This is in accordance with the quantum statistical model predictions [38,37] which observe the same dependence of $M_c(A > 4)$ on entropy and bombarding energy. This behaviour should be experimentally investigated. For peripheral collisions we find at all energies a breakup into two projectile and targetlike residues.

6. THE BOUNCE OFF – FRAGMENTS IN THE REACTION PLANE AS A BAROMETER FOR NUCLEAR COLLISIONS

Let us now turn to the average transverse momentum transfer $p_x(y)$ [32] in the scattering plane. It has been shown theoretically that $p_x(y)$ exhibits an enhanced sensitivity to the nuclear EOS as compared to other observables [17,20,16,21,26]. By its definition the transverse momentum analysis fails to detect flow effects for very central collisions; because of symmetry it is $p_x(b = 0) \equiv 0$. A clear maximum is experimentally observed in the multiplicity dependence $p_x(N_p)$ [9] for medium multiplicities (bin MUL3 and MUL4). The analogous maximum is observed in the calculated impact parameter dependence $p_x(b)$ at b = 3 fm [17,26]. Data and theory can thus be directly confronted by comparing the p_x/A values at the corresponding maxima.

The longitudinal momentum distribution, to be more precise, the rapidity distribution dN_{Frag}/dY of various fragment mass bins in the reaction Au (200 A MeV) + Au at 3 fm impact parameter is depicted in fig.24. The distinct fragment species can be related to a single source in momentum space only for the most central events $(b = 1 fm)$, where all baryons have been stopped and true multifragmentation is observed (see our discussion of the liquid–vapour phase transition in the previous sections). Already for $b = 3\ fm$ a clear separation of two sources, namely the projectile and target remnants, is observed for the heavy fragments $(A > 15)$. The two maxima of these curves directly reflect the finite impact parameters. The dN_{Frag}/dY distributions do not depend on the EOS employed.

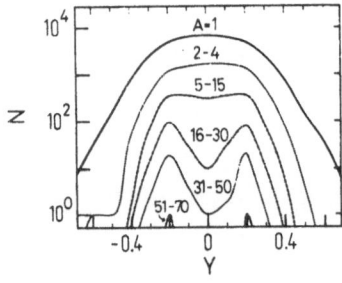

Figure 24. Fragment rapidity distribution dN_{Frag}/dY for different mass intervals from the reaction Au(200 MeV) + Au at 3 fm impact parameter.

If we look at the nucleons which come from different density regions we can observe that nucleons with a maximum density $\rho < 1.5\rho_0$ have target or projectile rapidity. These nucleons predominantely end up in complex fragments. On the contrary, nucleons which were in a high density zone are almost always stopped, i.e. they have c. m. rapidity. They are mostly emitted as free protons. In fig.25 the rapidity distributions for different maximum densities are shown.

Now we take a look at the $p_x(y)$ values of these particles; free nucleons which come from the high density zone show a small transverse momentum. On the opposite nucleons which experienced a small or medium compression have much more p_x, which can be seen in fig.26. We conclude that fragments which represent cold matter are stronger sidewards deflected due to the strong repulsion from the large potential gradient from the high density region. Hence we can get information about the compression energy of the high density region which gives these fragments (with low entropy) their transverse momentum. The latter is given by [61]

$$\Delta p_x \sim \int_t \int_A dA P(\rho, s) dt$$

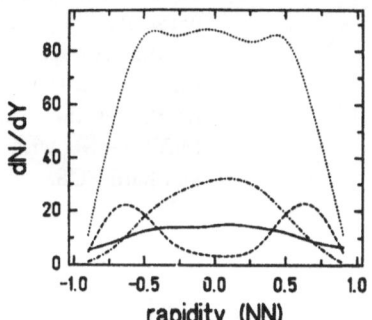

Figure 25. Rapidity distributions dN/dY for different maximal densities $(\rho/\rho_0)_{max}$ for the reaction Nb(1050 MeV) + Nb at b=3 fm with the soft EOS.

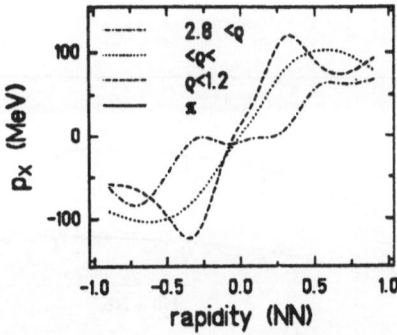

Figure 26. p_x versus rapidity for different maximal densities $(\rho/\rho_0)_{max}$ for the reaction Nb(1050 MeV) + Nb at b=3 fm with the soft EOS.

so we can consider the fragments as a barometer for the pressure i.e. the nuclear equation of state in the dense region. The p_x values are lower for the protons due to the thermal smearing, therefore the p_x observable does not give appropiate information about the thermodynamical properties of the system.

The impact parameter dependence of the transverse momentum shown in fig.27 for the soft and the hard EOS exibits a maximum at b=3 fm. The two equations of state yield quite different absolute p_x values , which is due to the different repulsive interactions employed.

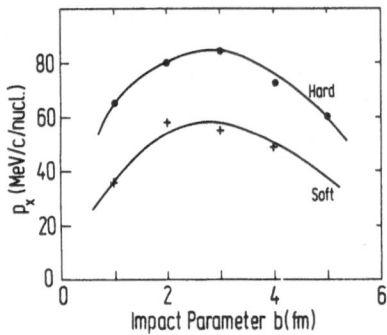

Figure 27. Impact parameter dependence of the average transverse momentum $p_x(y)$ for the reaction Nb(400 MeV) + Nb with soft and hard EOS.

A scaling analysis of the energy and mass dependence of the p_x-values has been performed in [46]. It yields a dependence $p_x \sim \sqrt{E}$ and $p_x \sim A^{1/3}$. Only VUU, QMD and viscous hydrodynamics give such a dependence, as shown in fig.28.

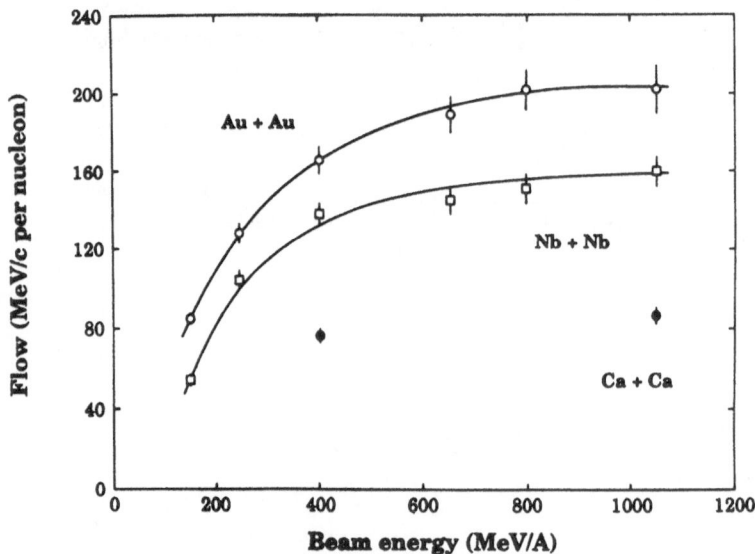

Figure 28. Energy and mass dependence of the average transverse momentum $p_x(y)$.

370

7. INFLUENCE OF IN–MEDIUM EFFECTS ON THE COLLECTIVE FLOW

It has recently been emphasized [48,47,50] that non–equilibrium effects can play an important role in a realistic treatment of heavy ion collisions. First there is the momentum dependence of the nuclear interactions, which leads to an additional repulsion between the nucleons in heavy ion collisions. It has been shown [31] that the inclusion of such momentum dependent interactions (MDI) reduces the yield of pions and kaons substantially, while the transverse momentum transfer [31,49] increases. The compressional energy is shown in fig.29 for the local soft (S) and the hard (H) equations of state and for the soft and hard equations of state with MDI (SM,HM). Note that all four equations of state give the same ground state binding ($E/A = -16MeV$ at $\varrho = \varrho_0$), but they differ drastically for higher densities. Here the hard EOS leads to much more compression energy than the soft EOS at the same density. The inclusion of the momentum dependent interactions leads for infinite nuclear matter at rest to almost no difference between the cases S, SM and H, HM respectively. This changes drastically if one considers heavy ion collisions: the additional repulsion due to the separation of projectile and target in momentum space shifts the curve for the SM (HM) interactions to higher energies. This can also be seen in the density, where we observe a decrease of the local density in the central region of the reaction of half a unit when the MDI are added to the local potential [51].

The real part of the optical potential has been parametrized in the following way [31]

$$U^{MDI} = t_4 ln^2(t_5(\vec{p_1} - \vec{p_2})^2 + 1)\delta(\vec{r_1} - \vec{r_2}) \tag{4}$$

with the parameters ($t_4 = 1.57\ MeV, t_5 = 5 \cdot 10^{-4} MeV^{-2}$). The present expression for the MDI reproduces the experimental data up to energies $E \approx 1$ A GeV. The parametrization of the real part of the optical potential together with the data is shown in fig.30. It substitutes the term proportional $\Delta\vec{p}^{\,2}$ in the Skyrme interaction, which is at striking variance with the data at $E_{Lab} \geq 150$ A MeV. Relativistic mean meson field theories [62,64,65] , however, do produce a similarly strong linear energy dependence of the optical potential and must therefore be taken seriously when the influence of the MDI on the reaction is investigated.

Another in-medium effect with a large influence on the observables is the reduction of the nucleon–nucleon scattering cross section in the nuclear medium, due to the Pauli blocking of intermediate scattering states [27,52,30].

It is well known that the n–n cross section in infinite nuclear matter is lowered as compared to the free case. One part of this reduction results from the Pauli blocking of the final states. This reduction leads to a reduced cross section σ_{NN}^{UU}, which is included in the collision term taken from the Uehling-Uhlenbeck transport equation (which includes the Pauli blocking on the final states). In a Dirac–Brueckner theory the n–n scattering also includes the Pauli blocking of intermediate states. This leads to an additional reduction of the n–n cross section which can be approximated [27,30] by a simple reduction factor

$$\sigma_{NN}^{eff} \approx 0.7\ \sigma_{NN}^{UU}. \tag{5}$$

The EOS itself is, of course, not influenced by the reduced cross sections. QMD calculations, which employ both the in–medium corrected cross sections and the MDI are denoted here by the abbreviations SIM (HIM) depending on whether a soft or a hard local potential has been used.

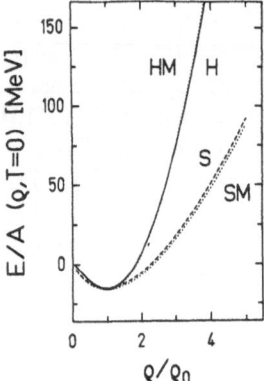

Figure 29. The density dependence of the compression energy per particle in cold infinite nuclear matter is shown for four distinct interactions: full lines correspond to a hard (H) and a soft (S) EOS, respectively. The dotted lines result from the corresponding local potential with additional momentum dependent interactions (HM), (SM).

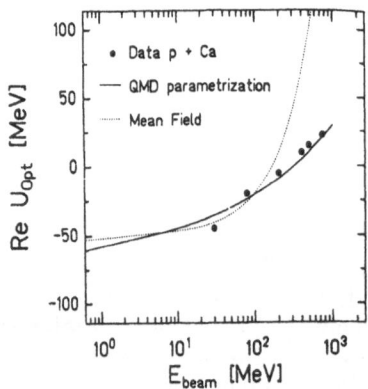

Figure 30. Energy dependence of the real part of the optical potential determined from the reaction p + Ca at different beam energies (dots). The full line depicts the parametrization of the momentum dependent potential as a fit to the data. The functional form of the interaction is indicated. Also shown is the linear energy dependence of the real part of the optical potential taken from the $\sigma - \Omega$–model

The excitation function of the flow of all nucleons for the reaction Au$(E = 200 - 800A \ MeV, b = 3fm)+$ Au for the hard (H), the soft (S), for the soft EOS plus the momentum dependent interactions (SM) and for the reduced cross sections plus the mdi (SIM) is shown in fig.31 . One observes at all energies a difference between the hard and the soft EOS. The transverse momentum of the free nucleons obtained with the hard EOS exceeds those obtained with the soft EOS by approximately 40%.

This difference between the hard and the soft local potential is not changed, no matter whether the purely local interactions H and S, or the in medium corrected interactions HIM and SIM are used [24]. The abolute value of the flow drops by a factor of almost 2 when the reduced [27,30] cross sections are employed. The third curve in the upper part of fig.31 shows the excitation function of the flow obtained with the soft EOS plus the additional momentum dependent interactions (SM). In the low energy regime ($E = 200A \ MeV$) one observes little difference between the cases S and SM, while for the high energy regime ($E \geq 400A \ MEV$) the additional repulsion of the MDI increases the transverse flow. There is still a difference of about one standard deviation between the soft EOS with MDI (SM) and the hard EOS without MDI (H).

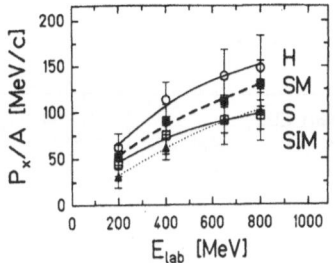

Figure 31. Excitation function of the flow (absolute value of the p_x/A–Y distribution at $Y = Y_{Pr}$) for the reaction Au (200–800 A MeV, b = 3 fm) + Au obtained from the hard and the soft EOS, respectively.

We now proceed to compare directly the data for the Au + Au reaction at 200 A MeV bombarding energy [9,10] to the theoretical results obtained with the soft (S),the hard (H) EOS,the soft local potential and also the soft local potential plus the different in medium corrections. Fig.32 shows the calculated (full circles) and the measured (full triangles) in-plane transverse momenta per nucleon p_x/A vs. rapidity for the reaction Au$(200A\ MeV, b = 3fm)$+Au, evaluated with the Plastic Ball efficiency cuts for different fragment masses and for the soft and the hard EOS, respectively. For b = 3 fm, where the maximal p_x/A values occur, the heavier fragment yields ($A > 10$) are peaked at projectile and target rapidity ($Y \approx \pm 0.3$). The light fragment distributions ($A \leq 4$) exhibit a broad maximum at $Y_{CM} = 0$.

Note the increase of the p_x/A–values with the mass of the fragments both in theory and experiment and the dependence of the p_x/A values on the equation of state. This difference is most significant for the intermediate mass fragments ($Z \geq 6$). These fragments are formed at early times in the reaction as a result of the shock wave traveling through the interpenetrating projectile and target. The heaviest fragments ($A > 30$), on the other hand, are formed in a decay chain from the excited projectile and target residue at late times. Hence they do not carry the strong signatures of the compression stage.

The measured transverse momenta p_x/A were obtained for the MUL3 bin [9,10], where again $p_x(N_p)$ is maximal, for fragments with Z=1, Z=2 and Z=6–10.

In order to draw conclusions from the comparison of the present calculations with the data the filter is applied, which has been used above to evaluate the multiplicity distributions. The left (right) column shows the p_x/A–distributions for fragments with Z=1 (A=1,2), Z=2 (A=3,4) and Z$\geq$6 (A=12–20) obtained with the soft (hard) EOS (full circles) as compared to the data (triangles). Observe the increasing transverse momentum with increasing fragment mass in both cases. We would like to point out that the soft EOS underpredicts the data. The right column shows that the agreement with the data is improved for all fragment masses if the hard EOS is employed in the QMD calculation. The inclusion of the momentum dependent interactions leads only to a small increase of p_x/A. The reduced cross sections lead to a strong reduction of the p_x/A values. We would like to emphasize that the effective cross sections should be determined consistently in experiments which simultanously measure the collective flow and baryon rapidity distribution dN/dY, i.e., different projections of the triple differential cross section [5].

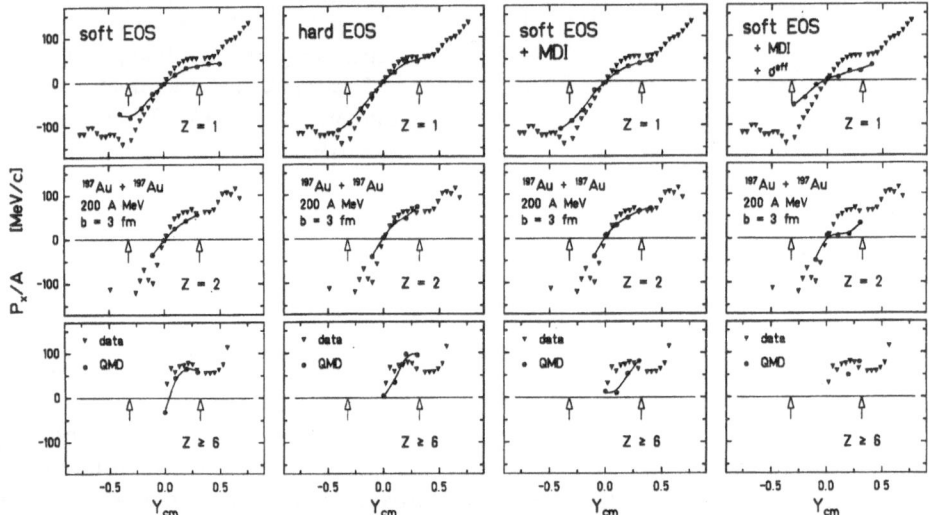

Figure 32. The transverse momentum distributions p_x/A–Y of light and intermediate mass fragments obtained from the QMD calculations for the reaction Au (200 A MeV, b = 3 fm) + Au are compared to the data. The data (triangles) have been obtained for the different fragment classes with Z = 1 (A = 1–2), Z = 2 (A = 3–4) and $6 \leq Z \leq 10$ ($12 \leq A \leq 20$). The QMD results (circles, full lines) have been obtained with the Plastic Ball filter. The columns compare the results obtained from the soft (hard EOS, with momentum dependent forces and in-medium cross sections)to the data . Both the QMD results and the data have been taken at the impact parameter (multiplicity) were the transverse momentum as a function of impact parameter (multiplicity) has a maximum (this maximum occurs at b = 3 fm in the calculation and in the experimental MUL3 bin).

Reduced in–medium cross sections could only be balanced by an even harder EOS. Large p_x values could also be created, if the effective cross sections were increased as compared to the UU cross section, which would be opposite to current theoretical understanding [27,30]. Increased in–medium cross sections[33,34], e.g. due to precritical scattering in the vicinity of a phase transition, would be a most fascinating resolution of the puzzle about the stiffness of the EOS from high energy heavy ion collisions.

Hence a systematic measurement of the flow and the dN_{Frag}/dY–distribution of complex fragments is urgently needed to open up the possibility to pin down the nuclear matter properties at the higher bombarding energies.

8. 90° SQUEEZE IN CENTRAL COLLISIONS

The measurement of the collective flow via the p_x anf the Θ_F- observable is strongly influenced by the behaviour of particles near projectile- and target rapidities, which are neither stopped nor (globally) thermalized. They correspond to those nucleons which have suffered the least compression. The most interesting region around the CM-rapidity, where the thermalized particles, which felt the highest compression, are found, is characterized by a lack of $< p_X > \approx 0$ by construction. Our aim is to look into other observables which characterize this part of the reaction.

Theoretical predictions for a squeeze out of particles at central collisions are based on the idea of shock wave formation in nuclear collisions. In the hydrodynamical model high energy particles are pressed out perpendicular to the beam and (for semicentral collisions) perpendicular to the reaction plane [1,3,5]. Further calculations have been done using both the Vlasov-Uehling-Uhlenbeck- model (VUU) as well as QMD [58]. We found a similar behaviour of protons and neutrons.

First we want to look at the overall θ distribution of the particles. Here θ is the lateral c.m. angle for each particle - it should not be misinterpreted as the flow angle Θ_f of the whole system, which is obtained in the event by event analysis.

First we consider the reaction La(246MeV/n) +La for very central collisions ($b = 1$fm). Using a hard equation of state we show on the l.h.s. of fig.33 the angular distribution $d^2\sigma d\Omega dE$ versus the c.m. angle θ . For the nucleons different cuts in the kinetic energy of the particles in the CM-frame are regarded. Particles with low energy ($E_k < 50$MeV, squares) and of intermediate energy ($50 < E_k < 150$MeV, triangles) do not show any significant peak. High energy particles ($E_k > 150$MeV, circles) show a clear maximum at 90 degrees to the beam axis. On the r.h.s. we see the behaviour of the high energy particles on a linear scale compared to the data of [55]. We see an enhancement of about 70 percent from 40^0 to 90^0 . The high energy particles are clearly squeezed out perpendicular from the stopped matter. Their momentum distributions exhibit an oblate shape. In the energy integrated angular distributions this effect can be masked by the behaviour of low energy particles. With increasing impact parameter this squeeze out is no longer visible. These predictions have recently been confirmed experimentaly by recent multiplicity selected data of Claesson et al. [55].

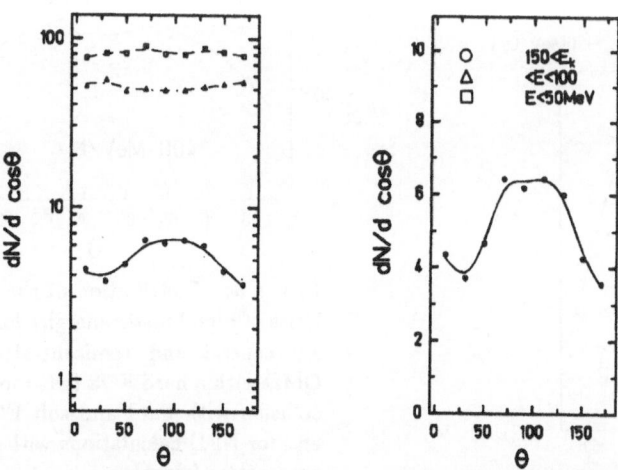

Figure 33. Distribution of the angle θ for particles of different kinetic energies for La(246MeV)+ La at a central collision. The reference system is the CM of the reaction.

Now let us describe the reaction of Au(400MeV)+Au, where experimental data could be analyzed by the Plastic Ball group. Again we find a maximum in the θ-distribution of high energy particles ($E_k > 150$ MeV) for very central collisions. The l.h.s. of fig.34 shows the distribution of a very central collision ($b = 1$fm, full line) and a slightly less-central collision ($b = 3$fm, dashed line) using a hard equation of state. We see a completely different behaviour in the two cases. While the 90 degree bin has a clear maximum for $b = 1$ it turns to a clear minimum for higher impact parameters. On the m.h.s. of fig. 34 we see the differences of a hard (full line) and a soft (dashed line) equation of state for the 90 degree peak at central collisions ($b = 1$fm). The hard equation of state has obviously a sharper peak than the soft EOS, corresponding to the higher repulsion of the hard EOS. Calculations with Nuclear Fluid Dynamics [63] are shown in the right side of fig.34. They yield similar results. However, a direct look at the squeeze out of the stopped hot participant matter can still be obtained by an analysis of the energy flow out of plane.

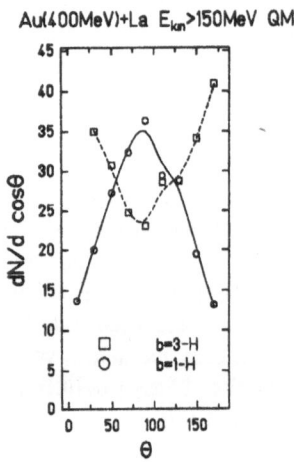

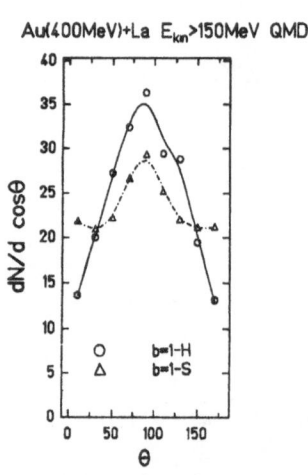

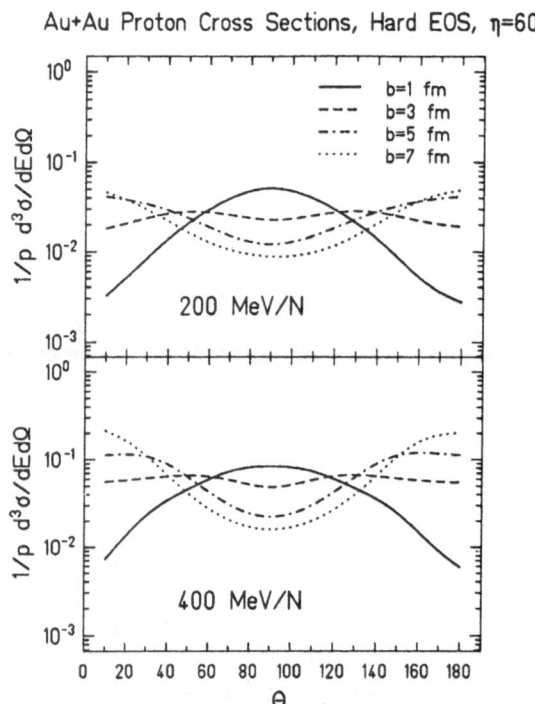

Figure 34. Distribution of the angle θ for particles of high kinetic energies for Au(400MeV)+Au central and semicentral collisions using QMD with a hard EOS (left top) and for central collision with hard and soft EOS (left botton) and for NFD calculations with different impact parameters (right).

9. ENERGY FLOW PERPENDICULAR TO THE REACTION PLANE – A DIRECT LOOK AT THE HOT PARTICIPANT SHOCK FRONT

Now let us turn to less central collisions ($b = 3fm$) of Au+Au and look at the azimuthal (φ) distribution of the particles where φ is the angle between $\vec{p}_T$ and the x-axis. Thus, $\varphi = 0$ degree means the x-axis and $\varphi = 180$ degrees means the negative axis $-x$. Here we will only regard the behaviour at CM rapidity.

Fig. 35 shows the NFD calculation for Au(400MeV)+Au at $b = 3fm$ for the azimuthal distribution of the protons. The upper part shows calculations with a hard EOS, the lower one uses a soft one. The experimental result is shown by the dotted line. The other line corresponds to different viscosities (full line $\eta = 30$, dashed $\eta = 0$ (i.e. $\sigma \rightarrow \infty$) and dash-dotted means $\eta = 60$). A sensible value is between $\eta = 30-60$. In this case the experimental data could be well reproduced . Both experiment and hydrodynamics get a clear peak at $\varphi = 90^0$ [63].

This peak becomes even clearer if we rotate the system by the flow angle Θ_{flow} in the reaction plane, that means, if we match the eigensystem of the flow ellipsoid in momentum space. There are more particles coming "out of plane" than in plane. Also mean kinetic energy and mean momentum are higher for out-of-plane particles. Much more transverse momentum and thus much more transverse energy is squeezed out perpendicular to the reaction plane than in the reaction plane.

For further comparison of the azimuthal distributions we use the ratio of the particle multiplicity out of plane and in plane

$$R_{out/in} = \left. \frac{\frac{dN}{d\varphi}(\varphi = 90^0, 270^0)}{\frac{dN}{d\varphi}(\varphi = 0^0, 180^0)} \right|_{Y=Y(CM)}$$

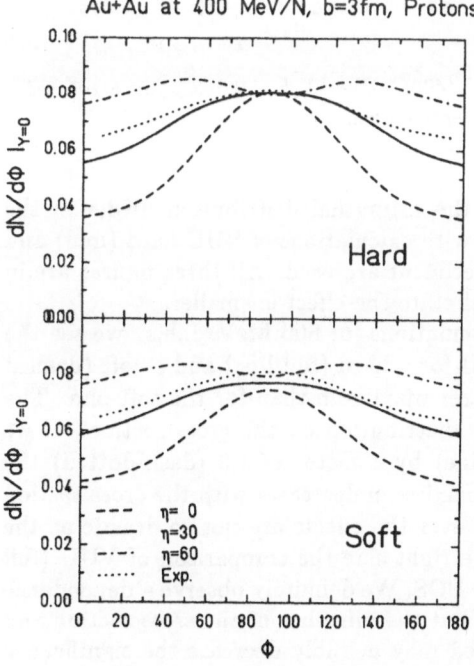

Figure 35. Distribution of the azimuthal angle φ in the unrotated system calculated with hydrodynamics.

Of course this value can again be taken in the rotated as well as in the unrotated system. In Fig.36 we see the mass dependence of $R_{out/in}(E_T)$ for Au+Au ($b = 3fm$), Nb+Nb ($b = 2.5fm$), and Ca+Ca ($b = 2fm$) ($b = 1.5fm$) all for 400 MeV incident energy in the rotated system calculated with QMD (left). The full line describes the ratio of the multiplicity distribution, the dashed line the ratio of the total transverse momentum in and out of plane and the dash-dotted line the respective ratios of the transverse energy. We see that the ratios of total transverse energy and momentum are higher than the multiplicity ratios. Note that the squeeze out increases linearly with the total size of the system. The r.h.s figure shows experimental data [56] for the ratio of the transverse energy. They are in good agreement with the predictions if a hard EOS is employed. A soft EOS yields too little squeeze-out, which demonstrates that this is not a simple shadowing effect but actually a dynamic pressing out of the hot dense participant matter [58].

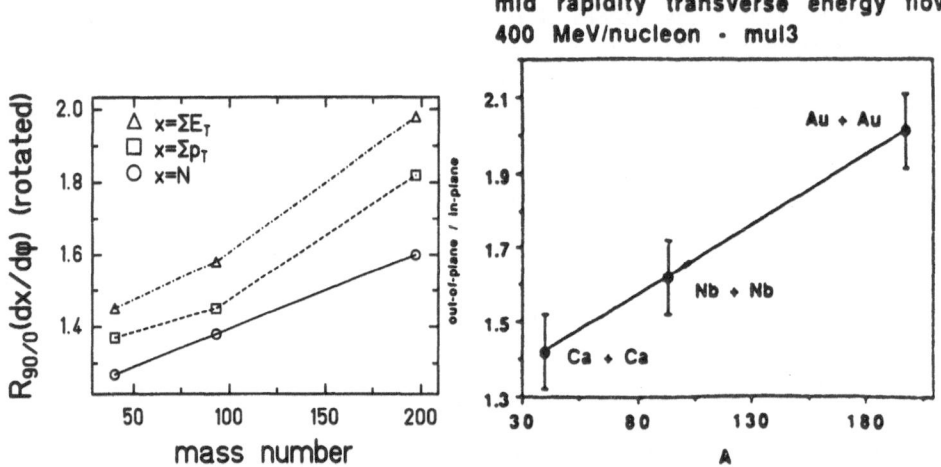

Figure 36. Mass dependence of $R_{out/in}$ for ^AX(400MeV)+^AX, $b \approx 0.25b_{max}$ in a rotated system for the multiplicity, momentum and energy ratios.

In Fig. 37 we see the comparison of the azimuthal distribution dNdφ in the rotated system for experimental data (left) with calculations of VUU hard (mid) and soft (right). The cut definitions of the experiment are used. All three figures are in fairly nice agreement. For a soft equation of state the effect is smaller.

In fig.38 we give a prediction of our calculations for 800 MeV. L.h.s. we see the azimuthal distribution calculated with VUU for a hard (full line) and a soft (dashed line) EOS. We see for the hard EOS a clearer maximum than for the soft one. The middle figure shows the dependence of the distribution on the cross section. If we increase the free cross section used (full line) by a factor of 1.3 (dash-dotted) the maximum increases as well, same way the maximum decreases with the cross section decreased by a factor of 0.7 (dashed). However the effects are not so drastic as the influence of the EOS. Finally we see on the right side the comparison of VUU (full line) and IQMD (dashed) both using a hard EOS. We definitely observe a dependence of the squeeze out on the equation of state but the influence of the cross section and the differences between the two models used may possibly decrease the significance of these differences. Fig.39 presents the energy dependence of $R_{out/in}$ for Au+Au at 400 MeV calculated with VUU-hard EOS (full line), VUU-soft (dashed), IQMD-hard (dash-dotted) and IQMD-soft (dotted) in the rotated system. Despite differences be-

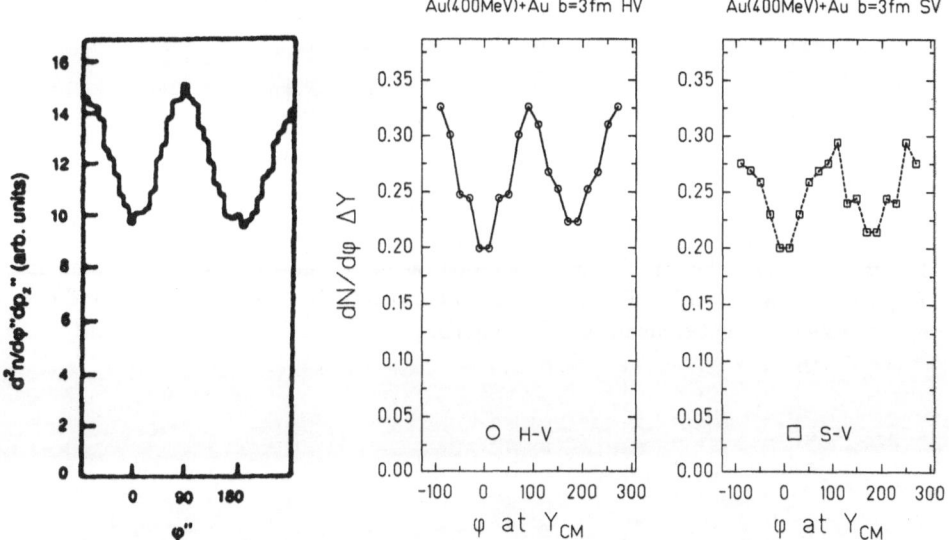

Figure 37. Distribution of the azimuthal angle φ in the rotated system for the experimental data (left), calculations with VUU hard (mid) and soft (right).

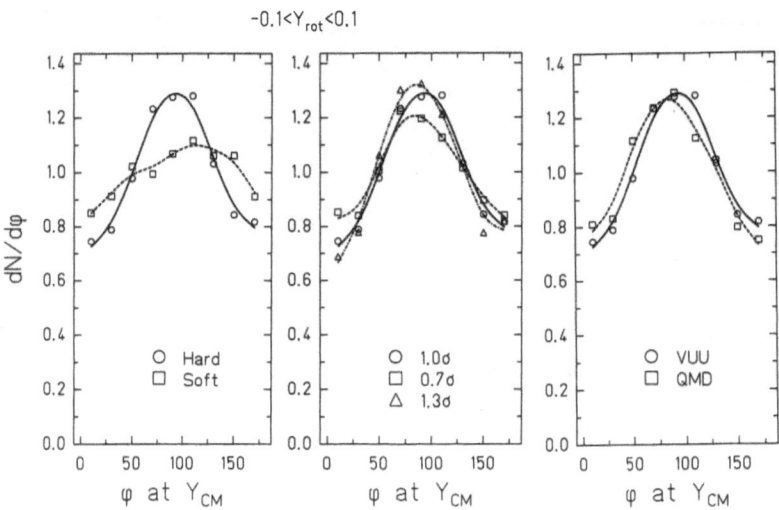

$-0.1 < Y_{rot} < 0.1$

Figure 38. Distribution of the azimuthal angle φ in the rotated system for Au(800MeV)+Au $b = 3fm$. L.h.s. comparison hard/soft EOS, mid comparison for different cross section, right comparison VUU/IQMD.

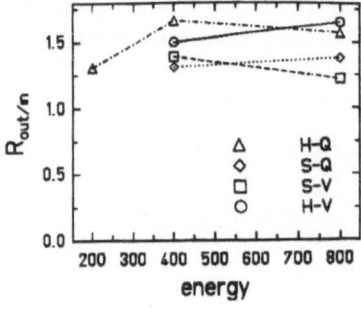

Figure 39. Dependence of $R_{out/in}$ on the incident energy for Au+Au at $b = 3fm$ in the rotated system for VUU and IQMD, hard and soft EOS.

tween both simulations, the soft EOS leads to smaller $R_{out/in}$-values than the hard EOS. Because of the smaller repulsion of the soft EOS the strength of the squeeze out peak is stronger for a hard EOS.

Fig.39 presents the energy dependence of $R_{out/in}$ for Au+Au at 400 MeV calculated with VUU-hard EOS (full line), VUU-soft (dashed), IQMD-hard (dash-dotted) and IQMD-soft (dotted) in the rotated system. Despite differences between both simulations, the soft EOS leads to smaller $R_{out/in}$-values than the hard EOS. Because of the smaller repulsion of the soft EOS the strength of the squeeze out peak is stronger for a hard EOS.

Finally we come back to the behaviour of particles with high kinetic energy. Fig.40 shows that $R_{out/in}$ increases with increasing kinetic energy. These high energy particles seem to be messengers of the compression region.

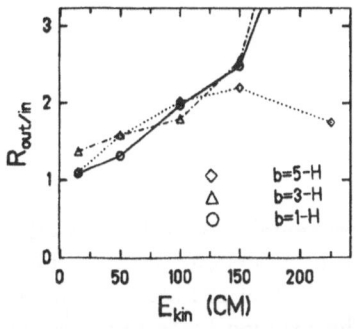

Figure 40. Dependence of $R_{out/in}$ on the kinetic energy of the particles for Au(400MeV)+Au, $b = 3fm$.

10. PION PRODUCTION IN THE MEDIUM AND THE DELTA MASS IN HOT NUCLEAR MATTER FROM PION-PROTON CORRELATIONS

Why do we regard pions? A view on the time evolution of a heavy ion collision simulated with VUU illustrates the answer fairly well. Fig.41 shows the behaviour of density and pion number for Nb (1050 MeV) + Nb at b=3fm. On the left hand side we see the evolution of the density in centre of the reaction (full line), of the mean density per particle (dashed line) and of the mean densities at which the delta production reaction occurs. The central density increases and decreases very rapidly, so that after about 20 fm/c most of the reaction is over. The mean density is also increasing but it reaches smaller values, since different particles enter the high density stage at different times. The deltas are produced at higher densities but not only at the maximum. We will soon come back to this point. On the r.h.s. of fig.41 we see that during the stage of high compression and high temperature the (total) pion number (full line) increases rapidly and stays roughly constant afterwards. Therefore, we regard the pions to be a signal from the hot high density state. Note that we regard the total pion number, counting as well pions as deltas. During the early time of the collision nearly no free pion exists and the total pion number is due to deltas (dashed line). Free pions (dash-dotted) are coming out rather late.

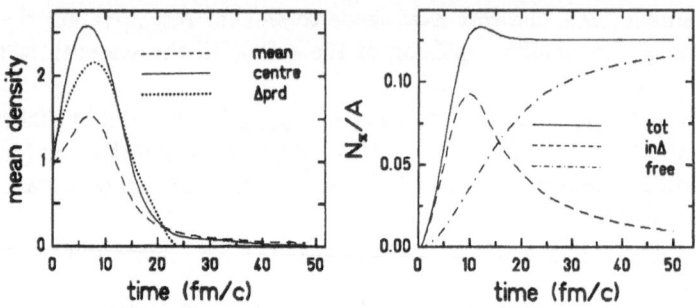

Figure 41. Time evolution of the density (l.h.s.) and of the pion number (r.h.s.) for Nb (1050 MeV) + Nb b=3fm hard EOS.

Let us first consider the way the pion production is calculated. For incident kinetic energies lower than about 1200 MeV only four reactions are relevant:

- $NN \rightarrow N\Delta$ (delta production)

- $N\Delta \rightarrow NN$ (delta absorbtion)

- $\Delta \rightarrow N\pi$ (delta decay - pion production)

- $N\pi \rightarrow \Delta$ (pion absorbtion)

For the first and the last reactions there are experimental cross sections available. The cross sections of the delta absorption are obtained via detailed balance calculations. For the pion production via delta decay a decay width of 120 MeV is assumed.

Now take a look at the time evolution of the reaction types to understand why free pions are coming out so late. On the l.h.s. of fig. 42 we see that the delta production via n-n collisions (full line) starts early while the reverse reaction (dashed) starts later. The pion production via delta decay (dash-dotted) and the pion reabsorption nearly cancel during the initial time. The free pions thus come out very late. The number of free pions is given by the time integrated difference of pion production and absorption, while the total pion number is obtained by integrating the difference of delta production and absorption via collisions. At the beginning of the collision the delta production is much larger while later the absorption is slightly dominant. This yields the time evolution of the total pion number of fig.41. The number of free pions increases slowly. The large absolute values of the multiplicities of pion production and absorption indicate a frequent change $\Delta \rightarrow N\pi \rightarrow \Delta \rightarrow N\pi \rightarrow \cdots$. The r.h.s of fig. 42 answers our question how often these changes occur for the final outcoming pions. We see that there are pions that have been reabsorbed in deltas for more than ten times.

Where do these reactions occur? Let us consider an asymmetric system like Ne (800 MeV) + Pb at $b = 3$ fm. Fig. 43 shows us the spatial distribution projected on the x-axis for two reaction types. The centre of the target is at $x = 0$, the centre of the incoming projectile is at $x = b = 3$ fm. We see a clear reaction channel around the impact parameter. For delta production (full line) via inelastic nn-collisions the size of that channel is about the size of the incoming projectile ($R_{Ne} \approx 3$ fm). The absorption of pions (dotted line) takes place in the total target nucleus ($R_{Pb} \approx 6.6$ fm) .

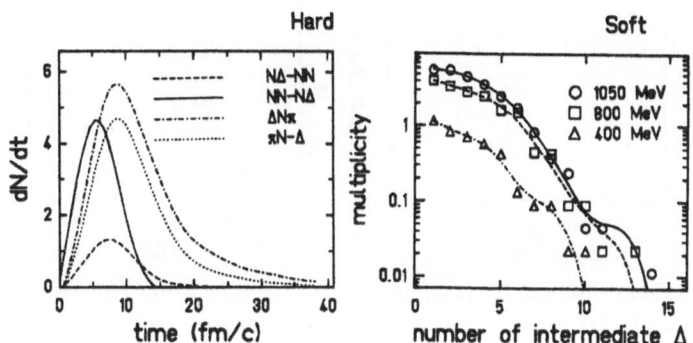

Figure 42. Time evolution of different reaction types for Nb (1050 MeV) + Nb at b= 3 fm (l.h.s). and the distribution of the number of times the surviving pions have been in a delta for different energies.

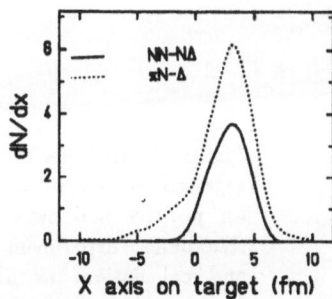

Figure 43. Projection of the spatial distribution of different reaction channels onto the x-axis (= axis of the impact parameter) for Ne (800 MeV) + Pb at b= 3 fm. $x = 0$ is laid into the centre of the target.

This yields a shadowing of the pions by the target and a positive pion flow for the pions, which has been found for this system as well by experiment [53] as by our calculations [54]. Let us briefly compare the behaviour of pions and nucleons in an asymmetric system. In fig. 44 the rapidity distribution (left) and the mean transverse momentum $p_X(Y)$ (right) show a completely different behaviour for nucleons and pions. For pions the rapidity distribution is much broader and $< p_X >$ is positive everywhere. Also there are differences between positive and negative pions.

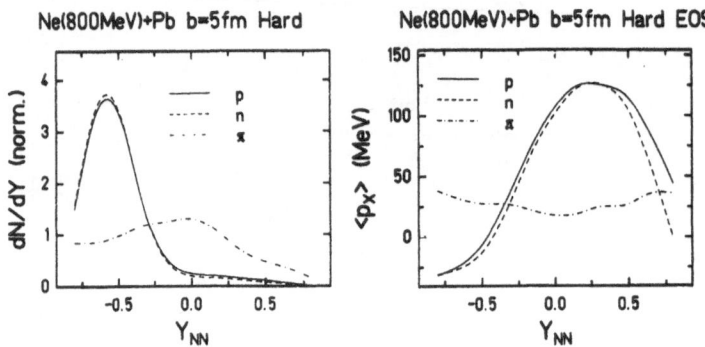

Figure 44. Rapidity distribution (left) and transverse momentum (right) for pions and nucleons for Ne(800MeV)+Pb.

Fig.45 shows the rapidity integrated averaged transverse momentum $< p_X/m >$ for different centralities of the reaction. 0 means central collisions, 1 means peripheral. The results of the calculation are filtered with the Diogéne's detector acceptances and sorted according to their multiplicity. We see that both experiment and data show a higher positive p_X for positive pions than for negative ones.

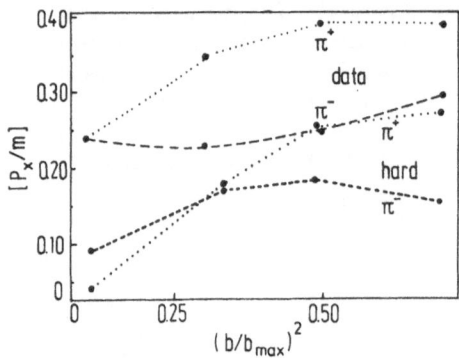

Figure 45. Rapidity integrated $< p_X/m >$ for positive and negative pions, experiment and calculation, as a function of the centrality of the reaction.

A further measurement of the Diogéne group has been the invariant mass distribution of the deltas in the reaction recalculated from the correlated outcoming nucleon-pion pairs in very peripheral collisions. The distribution (l.h.s. the experiment) shown in fig.46 has a maximum at the free delta mass of 1236 MeV minus 90 MeV. Calculations with IQMD (r.h.s.) using free delta masses in their reaction channels are getting analogous results. Therefore we conclude this lowering of the delta mass to be an effect of the phase space.

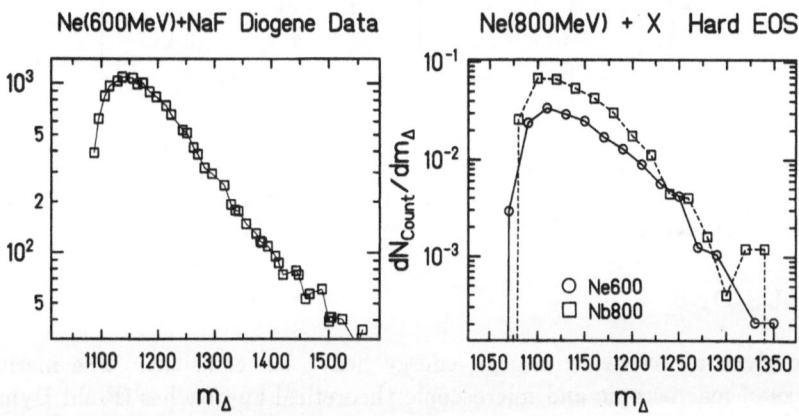

Figure 46. Distribution of the delta masses from experiment (left) and IQMD calculation (right).

We know that deltas (and thus pions) are produced at the time when the highest compression is reached. But are they really produced at that maximal density? Fig. 47 shows on the l.h.s. that the production of deltas via nn collisions occurs in a wide density range and is not peaked at maximal density. The pion absorption still takes place at very low densities. We found that the last deltas decaying to the final outcoming pions (full line) are produced at very low densities [58].

The r.h.s. of fig. 47 shows the maximal densities the outcoming pions experienced through all the production and reabsorption reactions. We see again a distribution on a wide density range as well for a hard as for a soft equation of state. Models which assume that all pions are produced at maximum (shock-) density have to be taken very cautiously.

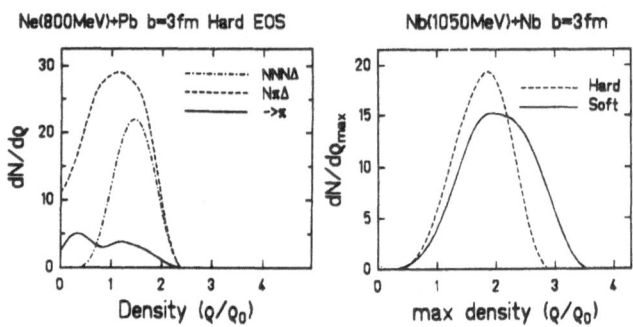

Figure 47. Density distribution of different reaction channels for a soft EOS (left) and right the distribution of the maximal density the outcoming pions have reached during the collision for hard and soft EOS.

CONCLUSIONS

We have given an overwiew of high energy heavy ion collisions. The merits and drawbacks of macroscopic and microscopic theoretical approaches (Fluid Dynamics, TDHF, Cascade, Vlasov-Uehling-Uhlenbeck, Classical and Quantum Molecular Dynamics) have been discussed. The importance of nonequilibrium transport properties (viscosity, mean free path, effective in-medium cross sections) and of the nuclear potential (equation of state) have been pointed out.

The liquid-vapour phase transition and multifragmentation have been studied. The possibility of measuring Machshock fragments in inverse kinematics experiments has been pointed out.

It has been demonstrated that the projectile and target are stopped at Y_{CM} if central collisions are studied. The stopping is only sensitive to σ^{eff}.

The predicted bounce off of the rather cold fragments in the reaction plane and the predicted accompanying squeeze out of the hot participant baryons out off the reaction plane have been experimentally discovered. These effects are sensitive both to the viscosity ($\sigma^{eff}(\rho, E, \Omega)$) and to the generalized equation of state (optical potential $U(\rho,E)$). The data clearly ask for a repulsive potential interaction.

We conclude that nuclear matter produced in relativistic collisions is a hot, dense, viscous, rather incompresible fluid, with important quantum properties.

References

[1] W. Scheid, H. Müller, and W. Greiner, Phys. Rev. Lett. 21 (1974) 741

[2] H.G. Baumgardt, J.U. Schott, Y. Sakamoto, E. Schopper, H. Stöcker, J. Hofmann, W. Scheid and W. Greiner, Z. Phys. A273 (1975) 359

[3] H.Stöcker,J.Maruhn and W. Greiner, Z. Phys. A290(1979) 297-300

[4] R.B.Clare and D.Strottman, Phys.Rep.141(1986)179

[5] H. Stöcker, J.A. Maruhn and W. Greiner, Phys. Rev. Lett 44 (1980) 725;
Phys. Rev. Lett 47 (1981) 1807;
Phys. Rev. C25 (1982) 1873;
G. Buchwald et al., Phys. Rev. Lett 52 (1984) 1594

[6] G.F. Bertsch, Phys. Rev. Lett. 34(1975) P.J. Siemens, Nucl. Phys. A335 (1980) 491

[7] H.A. Gustafsson, H.H. Gutbrod, B. Kolb, H. Löhner, B. Ludewigt, A.M. Poskanzer, T. Renner, H. Riedesel, H.G. Ritter, A. Warwich, F. Weik and H. Wieman, Phys. Rev. Lett. 52 (1984) 1590

[8] H.G. Ritter, K.G.R. Doss, H.A. Gustafsson, H.H. Gutbrod, K.H. Kampert, B. Kolb, H. Löhner, B. Ludewigt, A.M. Poskanzer, A. Warwick and H. Wieman, Nucl. Phys. A447 (1985) 3c

[9] K.H. Kampert, Ph. D. Thesis, University of Münster (1986) and J. Physic G, Topical Review, 1989 in print

[10] K.G.R. Doss, H.-A. Gustafsson, H. Gutbrod, J.W. Harris , B.V. Jacak, K.-H. Kampert, B. Kolb, A.M. Poskanzer, H.-G. Ritter, H.R. Schmidt, L. Teitelbaum, M. Tincknell, S. Weiss and H. Wiemann, Phys. Rev. Lett. 59 (1987) 2720

[11] K.G.R. Doss H.A. Gustafsson, H.H. Gutbrod, K.H. Kampert, B. Kolb, H. Löhner, B. Ludewigt, A.M. Poskanzer, H.G. Ritter, H.R. Schmidt and H. Wieman, Phys. Rev. Lett. 57 (1986) 302

[12] D. Keane, S.Y. Chu, S.Y. Fung, Y.M. Liu, L.J. Qiao, G. VanDalen, M. Vient, S. Wang, J.J. Molitoris and H. Stöcker, Phys. Rev. C. 1987 and Proc. of the 8th High Energy Heavy Ion Study, LBL Berkeley (1987) 165

[13] H.H. Gutbrod, K.H. Kampert, B.W. Kolb, A.M. Poskanzer, H.G. Ritter, H.R. Schmidt, Phys. Lett. B216 (1989) 267

[14] I. Lovas Nucl. Phys. A367 (1981) 509

[15] L. Neise, H. Stöcker, and W. Greiner, J. Phys. G13 (1987) L181

[16] H. Stöcker and W. Greiner, Phys. Rep. 137 (1986) 277-392

[17] J.J. Molitoris, J.B. Hoffer, H. Kruse and H. Stöcker, Phys. Rev. Lett 53 (1984) 899

[18] D.H. Boal and J.N. Glosli, Phys. Rev. C38 (1988) 2621

[19] T.Schlagel and V.R.Pandharipande , Phys. Rev. C36(1987) 162

[20] J.J. Molitoris and H. Stöcker, Phys. Rev. C32 (1985) 346 and Phys. Lett 162B (1985) 47

[21] J.J. Molitoris, H. Stöcker and B.L. Winer, Phys. Rev. C36 (1987) 220

[22] J. Aichelin and H. Stöcker, Phys. Lett. 163B (1985) 59 Phys. Rev. Lett. 59 (1987) 2720

[23] J. Aichelin and H. Stöcker, Phys. Lett. B176(1986) 14

[24] G. Peilert, Thesis, University Frankfurt (1988)

[25] J.Aichelin, G. Peilert, A. Bohnet, A. Rosenhauer, H. Stöcker and W. Greiner, Phys. Rev. C37 (1988) 2451

[26] A.Rosenhauer, PhD Thesis, University Frankfurt (1988)

[27] B.ter Haar, R. Malfliet and W. Botermans, Phys. Rep. 149 (1987) 207 and Phys. Lett. 172B (1986) 10

[28] L.Wilets, E.M.Henley, M.Kreft and A.D.Mackeller, Nucl.Phys. A282, 341(1977)

[29] C.Dorso, S.Duarte and J.Randrup, Phys.Lett. B188, 287(1987)

[30] J. Cugnon, A.Lejeunne and P. Grange, Phys. Rev. C35 (1987) 861

[31] J.Aichelin, A.Rosenhauer, G.Peilert, H.Stöcker and W.Greiner, Phys. Rev. Lett. 58 (1987) 1926

[32] P. Danielewicz and G. Odyniec, Phys. Lett. 157B (1985) 146

[33] M. Gyulassy and W. Greiner, Ann. Phys. 109 (1977) 485

[34] M. Gyulassy, K.A. Frankel, and H. Stöcker, Phys. Lett. 110B (1982) 185

[35] H.R. Schmidt et al., Z.Phys. C38(1988) 109

[36] Z.Li,H.Stöcker and W.Greiner , preprint

[37] D. Hahn and H. Stöcker, Nucl. Phys. A476 (1988) 718

[38] H. Stöcker, G. Buchwald, G. Graebner, P. Subramanian, J.A. Maruhn, W. Greiner, B.V. Jacak and G.D. Westfall, Nucl. Phys. A400 (1983) 63

[39] L. Csernai and J. Kapusta, Phys. Rep. 131 (1986) 223

[40] R.W. Minich, S. Agarwal, A. Bujak, J. Chuang, J.E. Finn, L.J. Gutay, A.S. Hirsch, N.T. Porile, R.P. Scharenberg, B.C. Stringfellow, F. Turkot, Phys. Lett. 118B (1982) 458

[41] A.I. Warwick, H. Wieman, H.H. Gutbrod, M.R. Maier, J. Peter, H.G. Ritter, H. Stelzer, F. Weik, M. Freedman, D.J. Henderson, S.B. Kaufman, E.P. Steinberg and B.D. Wilkins, Phys. Rev. C27 (1983) 1083

[42] D.R. Bowman, R.J. Charity, R.J. McDonald, M.A. McMahan, G.J. Wozniak, L.G. Moretto, W.L. Kehoe, S. Bradley, A.C. Mignerey, A. Moroni, A. Bracco, I. Iori and M.N. Namboodiri, Phys. Lett. B189 (1987) 282

[43] H. Kruse, B.V. Jacak and H. Stöcker, Phys. Rev. Lett. 54 (1985) 289

[44] J. Pochodzalla, C.K. Gelbke, W.G. Lynch, M. Maier, D. Ardouin,
H. Delagrange, H. Poubre, C. Gregoire, A. Kyanowski, W. Mittig, A. Peghaire,
J. Peter, F. Saint-Laurent, B. Zwieglinski, G. Bizard, F. Lefebvres, B. Tamain,
J. Quebert, Y.P. Viyogi, W.A. Friedman, D.H. Boal , Phys.Rev. C35 (1987)
1695

[45] E. Berdermann et al., Proposal for a Forward Spectrometer at the 4π Detector
GSI 88-08 Report ISSN 0171-4546

[46] B. Schürmann, Mod. Phys. Lett. A12 Vol.3 (1988) 1137-1143 and
B. Schürmann and W. Zwermann, Phys. Rev. Lett. 59 (1987) 2848

[47] B. Schürmann and W. Zwermann, Phys. Lett. 158B (1985) 366 and Phys. Rep.
147 (1987)

[48] A.R. Bodmer and C.N. Panos, Phys. Rev. C15 (1977) 1342, and Phys. Rev C22
(1980) 1023

[49] C. Gale, G.F. Bertsch and S.Das Gupta, Phys. Rev. C35 (1987) 1666

[50] T.L. Ainsworth, E. Baron, G.E. Brown, J. Cooperstein and M. Prakash, Nucl.
Phys. A464 (1987) 740

[51] A.Rosenhauer,M.Berenguer,G.Peilert,J.Aichelin,H.Stöcker, and W.Greiner,
Phys. Rev. C. in print

[52] G.F.Bertsch,W.G.Lynch and M.B.Tsang,Phys.Lett.B189(1987)384

[53] J. Poitou et al., Proc. Eighth Balaton Conference on Nuclear Physics, 1987
and Proc. Gross Properties of Nuclear Matter XVI, Hirschegg 1988

[54] C. Hartnack et. al., Proc. Gross Properties of Nuclear Matter XVI, Hirschegg
1988

[55] G. Claesson, G. Krebs, J. Miller, G. Roche, L. S. Schoeder, W. Benenson, J. van
der Plicht, J.S. Winfield, G. Landaud, H. Stöcker, submitted to Phys.Rev.Lett.

[56] H.H. Gutbrod, K.H. Kampert, B.W. Kolb, A.M. Poskanzer, H.G. Ritter,
H.R. Schmidt, Phys. Lett. 216B (1989) 267.
H.R. Schmidt H.H. Gutbrod, K.H. Kampert, B.W. Kolb, A.M. Poskanzer,
H.G. Ritter, R. Schicker, Preprint

[57] A.Fässler et al.,preprint

[58] C.Hartnack,Thesis,University Frankfurt (1989), unpublished

[59] M.Berenguer,Thesis,University Frankfurt (1989), unpublished

[60] G.Peilert et al. ,Mod,Phys.Lett. A5 Vol.3 (1988) 459 and Phys. Rev. C39
(1989) 1402

[61] H.Stöcker and B. Müller,preprint

[62] R.Y. Cusson, P.G. Reinhard, J.J. Molitoris, H. Stöcker, M.R. Strayer,
W. Greiner, Phys.Rev.Lett. 55 (1985) 2786

[63] W.Schmidt,PhD Thesis,University Frankfurt (1989), unpublished

[64] B. Blättel, V. Koch, W. Cassing, U. Mosel, Phys.Rev. C38 (1988) 1767

[65] C.M. Ko, Linhua Xia, Phys. Rev. C38 (1988) 179

MOMENTUM DEPENDENT MEAN FIELDS IN THE

BUU MODEL OF HEAVY ION COLLISIONS

Gerd M. Welke

Physics Department, State University of New York at
Stony Brook, Stony Brook, New York 11794-3800

1. INTRODUCTION

In attempts to extract the high density nuclear equation of state (EOS) from intermediate energy heavy ion collisions much attention has focused on simulating these using the Boltzmann–Uehling–Uhlenbeck [1,2] (BUU) transport equation, which describes the time evolution of the one–body nucleon phase space distribution function $f(\mathbf{r}, \mathbf{p}, t)$:

$$\frac{\partial f}{\partial t} + \mathbf{v} \cdot \frac{\partial f}{\partial \mathbf{r}} - \frac{\partial U}{\partial \mathbf{r}} \cdot \frac{\partial f}{\partial \mathbf{p}} = -\frac{1}{(2\pi)^6} \int d^3 p_2 \, d^3 p_{2'} \, d\Omega \frac{d\sigma_{NN}}{d\Omega} \, g \tag{1.1}$$

$$\left\{ \left[f f_2 (1 - f_{1'})(1 - f_{2'}) - f_{1'} f_{2'} (1 - f)(1 - f_2) \right] (2\pi)^3 \, \delta^{(3)}(\mathbf{p} + \mathbf{p}_2 - \mathbf{p}_{1'} - \mathbf{p}_{2'}) \right\}.$$

The collision (RHS) term in Eq.(1.1) describes stochastic binary collisions between nucleons of relative velocity g, with the possibility of particle production via the two–body cross–section σ_{NN}. The factors $(1 - f)(1 - f_2)$ and $(1 - f_{1'})(1 - f_{2'})$ incorporate the effects of final state Pauli blocking in the gain and loss terms respectively [1]. The nuclear EOS enters the simulation via the mean field, or single particle potential, $U(\rho, \mathbf{p})$, since it is the functional derivative of the potential energy density of the system.

Eq.(1.1) thus requires two inputs, $U(\rho, \mathbf{p})$ and σ_{NN}, which must be incorporated self–consistently under conditions more extreme than those for which they are presently best known; in a collision, densities are thought to reach values of $(2 - 3)\rho_0$ ($\rho_0 \sim 0.16$ fm^{-3}), and initially the matter is highly non–equilibrated. Heavy ion collision experiments therefore offer the possibility of studying nuclear matter and in–medium transport properties under such conditions. This task is arduous, for it requires an extraction of the collision history from many final state products. In particular, one is faced with the difficulty of disentangling the effects of the two inputs on the observables, since different combinations of these may produce the same effects for a given collision [3]. It is therefore important to construct (global, time–saturating) observables that are selectively sensitive to the nuclear EOS and collisional effects. Promising candidates include nuclear matter flow [4] from near central collisions by way of sphericity [5] and transverse momentum [6] analysis, pion [7,8] and kaon production [9], etc. Said selective sensitivity may be achieved from a variation in the size of the colliding system and the beam energy [10,11]. For example, the effect of σ_{NN} decreases with increasing mass number [5,12].

A commonly used single particle potential is a momentum independent Skyrme-like parameterization [13] :

$$U(\rho) = A\frac{\rho}{\rho_0} + B\left(\frac{\rho}{\rho_0}\right)^{\sigma} . \qquad (1.2)$$

With this form of the single particle potential, collective flow [14] and pion production [7] data is well reproduced within the BUU model if a cold nuclear matter incompressibility of $K_0 \sim 380$ MeV is assumed [15,16]. This is considerably higher than values obtained from giant dipole resonance data [17,18] and calculations that fit nucleon–nucleon scattering data [19]. A more realistic parameterization might take into account the momentum dependence of the force a nucleon feels in–medium. With such potentials and $K_0 \sim 210$ MeV several authors [20,21,22] have reproduced flow characteristics of the stiff $K_0 = 380$ MeV Skyrme–like mean field.

In Section 3 an example [21] of such a p-dependent mean field is given and compared to microscopic calculations [23] of the single particle potential in the density range $(1-3)\rho_0$. Section 4 compares flow observables from the soft momentum dependent mean field with those from the stiff Skyrme–like parameterization [24] and some recent Streamer Chamber data at 800 MeV/nucleon [16]. In particular, we shall also consider the inclusive nucleon cross-section as a function of the azimuthal angle [9,21]. An attempt is made to address the question of how the transverse momentum production mechanism differs in two type of forces. But first, we turn to the question of how the BUU equation (1.1) is simulated in practice.

2. NUMERICAL SIMULATIONS AND THE KROOK–WU MODEL

In numerical simulations of the Boltzmann equation the so called test particle method is often employed: the continuous Wigner function $f(\mathbf{p},\mathbf{r},t)$ is replaced by N discrete "test particles":

$$f(\mathbf{r},\mathbf{p}) = \frac{1}{\tilde{N}} \sum_{i=1}^{N} \delta^{(3)}(\mathbf{r}-\mathbf{r}_i)\, \delta^{(3)}(\mathbf{p}-\mathbf{p}_i) , \qquad (2.1)$$

where $\tilde{N} = N/A$ is the number of test particles per nucleon. The coordinates $\mathbf{r}_i$ and $\mathbf{p}_i$ are chosen from the specified distribution at time t and are then evolved according to Hamilton's equations of motion. At each timestep test particles are also checked for scattering with cross-section $\sigma_{NN}/\tilde{N}$ [25]. This "full ensemble" (FE) procedure enforces the local nature of the Boltzmann equation, and angular momentum conservation, to a high degree of accuracy [26].

In practice, since the FE method is time consuming, it is replaced by a "parallel ensemble" (PE) technique [25], in which the test particle propagation is still performed via a mean field calculated from the entire ensemble, but only intra-ensemble collisions are permitted; that is, each of the $\tilde{N}$ sets is separated as far as hard interactions are concerned and the scattering cross-section is now σ_{NN}.

The above two numerical methods of simulating the collisional integral may be tested in a variety of scenarios for which exact solutions exist (see also Ref.[27,28]). Here we briefly describe one such check [26], provided by the Krook–Wu model [29,30]: for a non–relativistic classical system with an isotropic and homogeneous distribution function $f(|\mathbf{v}| \equiv v, t)$, and an isotropic elastic differential scattering cross–section $\sigma(g) = \kappa/g$, where κ is a constant independent of g, it is shown in Ref.[29] that the Boltzmann equation (1.1) (with $U \equiv 0$ and no Pauli blocking factors) has a solution

$$f(v,\tau) = \rho \, \frac{e^{-v^2/(2K\beta^2)}}{2(2\pi K\beta^2)^{3/2}} \left\{ \frac{5K-3}{K} + \frac{1-K}{K^2} \frac{v^2}{\beta^2} \right\} , \quad \beta^2 \equiv \frac{T}{m} , \qquad (2.2)$$

with $K = 1 - e^{-\tau/6}$ and $\tau \geq \tau_0 = 6\ln(5/2)$. Above, $\tau = 4\pi\rho\kappa t$ is a dimensionless time variable; ρ, T, and m are the constant number density, temperature and mass, respectively. For large τ, f tends to the equilibrium Maxwell–Boltzmann distribution.

The initial test particle velocity distribution is chosen according to (2.2) with some initial τ_i, while the spatial configuration is uniform in a fixed volume of linear dimension L. Periodic boundary conditions ensure the constancy of ρ. One may compare moments (normalized to unity for the equilibrium Maxwell–Boltzmann distribution)

$$M_n(\tau) = \frac{\sqrt{\pi}}{2(2\beta^2)^{n/2}\Gamma(\frac{n+3}{2})} \frac{1}{\rho} \int d^3v \, v^n \, f(v,\tau) , \quad n = -2, -1, 0, 1, ..., \qquad (2.3)$$

derived from the exact solution (2.2) with those obtained from a numerical cascade evolution of the test particle fluid. Mass and energy conservation imply that M_0 and M_2, respectively, are unity for all τ. Clearly, since L is finite (or, equivalently, since the cascade simulation is not perfectly localized), we must ensure that the radii $r_1 = \sqrt{\langle\sigma_{NN}\rangle/(\pi\tilde{N})}$ for the FE, and $r_2 = \sqrt{\langle\sigma_{NN}\rangle/\pi}$ for the PE, are small compared to L. This implies a small cross–section κ and a high temperature. On the other hand, one might require large κ and low T so that the mean free path $\lambda \ll L$ [26].

In Fig. 1 we show comparisons of a numerical evolution to the analytical solution (solid line) for various indicated moments. The circles in 1(a) correspond to a FE, and those in (b) to a PE simulation. The system considered is a cube of $L = 6$ fm, with parameters $\rho = 4\rho_0, T = 40$ MeV, $m = 939$ MeV/c^2 and $\kappa = 0.024$ fm^{2}c. The total number of ensembles for the FE case is $\tilde{N} = 30$, while $\tilde{N} = 120$ for the PE calculation; thus

$$r_{1,\tau\to\infty} = 0.09 \text{ fm} , \quad r_{2,\tau\to\infty} = 0.49 \text{ fm} \quad \text{and} \quad \lambda_{\tau\to\infty} = 1.51 \text{ fm} . \qquad (2.4)$$

The larger $\tilde{N}$ for the PE case serves only to increase the statistics. The dashed lines are least squares fits to the numerical "data", with K in Eq.(2.2) now given by

$$K' = 1 - e^{-\tau_i/6} \, e^{-\alpha(\tau-\tau_i)/6} , \qquad (2.5)$$

where $\tau_i = 5.5$ is the initial time. The variational parameter $\alpha = \kappa'/\kappa$ measures the deviation of the observed "numerical" cross–section κ' from the "input" value κ. This α is displayed in Fig. 1 for the various cases. Clearly, for systems that are large enough in the sense of the previous paragraph, the PE technique provides a good and efficient numerical method for solving the Boltzmann equation.

Fig. 1(c) gives results for a PE simulation with parameters given as above, but now $\kappa = 0.22$ fm²c. The values of $r_{2,\tau\to\infty}$ and $\lambda_{\tau\to\infty}$ change to 1.5 fm and 0.16 fm, respectively, and we see that surface effects become important (α drops by $\sim 35\%$).

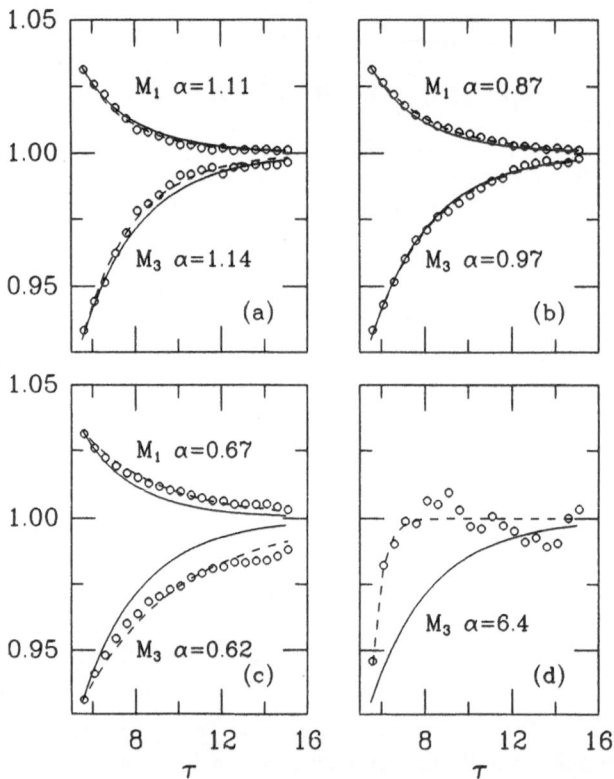

Fig. 1. Comparison of the numerical evolution of various moments (open circles) to analytic solutions (solid lines). Results in (a) are from a FE calculation, while those in (b) and (c) are for a PE simulation. In (a) and (b), $\kappa = 0.024$ fm²c, while $\kappa = 0.22$ fm²c in (c). The value of α is given in each case together with the corresponding fit (dashed lines; see text for details). For (d), a FE calculation with $\kappa = 0.024$ fm²c is presented, but using the full (physical) cross-section (see text). The numerical evolution in (a) and (b) is consistent with the analytic solution, while that in (c) and (d) is not.

In Fig. 1(d), we consider again a cross-section of $\kappa = 0.024$ fm²c. Here, though, a method is used in which all test particles collide with each other with the physical cross-section σ_{NN}, but multiple collisions are suppressed if they occur within one timestep. This technique would therefore lead to a timestep dependent relaxation rate. Since the size of this timestep is constrained by the requirement that the propagation should be sufficiently accurate, one may conclude that this method overestimates the relaxation rate.

The above "surface effect" translates into a sizable change in observables in collision simulations of light mass systems. Consider a cascade calculation (Eq.(1.1) with $U \equiv 0$) of C + C at a beam energy of 200 MeV and at an impact parameter $b = 1.2$ fm. For free–space cross-sections, both the FE and PE simulation yield an average in–plane transverse momentum $\langle p_x \rangle_{y>0}$ of 11 MeV/c. If, however, the two–body cross–section is doubled, the FE method leads to a value of $\langle p_x \rangle_{y>0}$ of 26 MeV/c, while the PE simulation gives 22 MeV. In using the PE method for heavy ion collisions, one should therefore be aware of the fact that the BUU equation is simulated only approximately if the nuclear radii are comparable to $\sqrt{\sigma_{NN}/\pi}$.

3. MEAN FIELD PARAMETERIZATIONS

The constants A, B and σ in the Skyrme–like parameterization Eq.(1.2) of U may be fixed to yield the cold nuclear matter binding energy, saturation density and incompressibility. Two commonly used parameter sets [13] take $K_0 = 200$ MeV and $K_0 = 380$ MeV. We shall refer to these as the soft (SBKD) and hard (HBKD) mean fields respectively.

The exchange term of a Yukawa interaction in nuclear matter leads to a potential energy density

$$V^{ex} = \frac{C'}{\rho_0} \int d^3p \, d^3p' \, \frac{f(\mathbf{r},\mathbf{p}) f(\mathbf{r},\mathbf{p}')}{1 + \left(\frac{\mathbf{p}-\mathbf{p}'}{\Lambda}\right)^2} \, , \tag{3.1}$$

so that one might consider instead of (1.2) a mean field of the form used in Ref.[21]:

$$U(\rho,\mathbf{p}) = A\frac{\rho}{\rho_0} + B\left(\frac{\rho}{\rho_0}\right)^\sigma + 2\frac{C'}{\rho_0} \int d^3p' \, \frac{f(\mathbf{r},\mathbf{p}')}{1 + \left(\frac{\mathbf{p}-\mathbf{p}'}{\Lambda}\right)^2} \, . \tag{3.2}$$

The five parameters are chosen [21] to give $E/A = -16$ MeV, $\rho_0 = 0.16$ fm^{-3}, $K_0 = 215$ MeV, $U(\rho_0, p = 0) = -75$ MeV and $U(\rho_0, p^2/2m = 300\text{MeV}) = 0$. This means that $U(\rho_0, p \to \infty) \to 30.5$ MeV and implies an effective nucleon mass of $m^*(p_F) = 0.67m$. We shall refer to (3.2) as the soft momentum dependent interaction (SMDI). Its form has also been used in simulations by Gale, Bertsch and Das Gupta (GBD) [20], where an averaging is done over one of the integrals in (3.1): $\mathbf{p}' \to \langle \mathbf{p}' \rangle$ and $\int f(\mathbf{r},\mathbf{p}')d^3p' \to \rho$. The two forms differ under conditions of extreme non–equilibrium, as was shown in Ref.[21].

Friedman and Pandharipande [19] have calculated the single particle potential from the Urbana v_{14} plus three nucleon interaction (UV14+TNI) [31,32] at densities below ρ_0 and find good agreement with optical potential nucleus–nucleus scattering data. In a recent paper, Wiringa [23] has extended this calculation to densities up to $3\rho_0$, and for the Argonne v_{14} [33] and UV14 potentials with the Urbana model VII three nucleon interaction (UVII) [34]. The comparison between a fit to the mean field derived from UV14+UVII and the SMDI is presented in Fig. 2, and indicates good agreement at zero temperature in the relevant range of density and momentum. The other Hamiltonians lead to similarly resonable agreement [24], and one therefore has some measure of confidence in the usefulness of the SMDI parameterization.

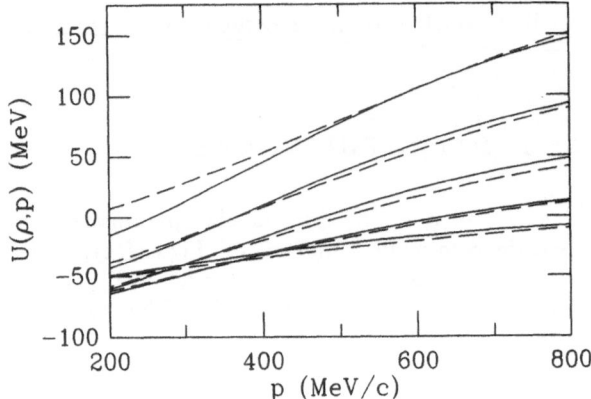

Fig. 2. Comparison of the SMDI mean field (solid lines) to one derived from UV14+UVII (dashed lines) (Ref.[23]) as a function of momentum. From top to bottom on the right–hand side the densities are: $0.5, 0.4, 0.3, 0.2$ and 0.1 fm^{-3}.

4. COLLECTIVE FLOW AND THE EOS

While extensive calculations [15,35] have shown the HBKD to reproduce collective flow data [14,35] well, it is also known that a SMDI will produce similar flow characteristics [20,21,24,36]. This shall be demonstrated below.

Measures of collective flow include sphericity analysis [5,14] transverse momentum [6,37] or velocity [38] analysis, azimuthal anisotropy [9,21,39,40], and, more recently, the so called squeeze–out effect [40]. The azimuthal anisotropy was shown to be sensitive to the nuclear EOS using near analytical models in Ref.[9] and within the BUU model in Ref.[21], where it was defined as the ratio of the maximum to minimum value of the azimuthal distribution: $\mathcal{R} = (d\sigma/d\phi)_{max}/(d\sigma/d\phi)_{min}$, where ϕ is measured from the reaction plane. The case studied in Ref.[21] was motivated by a recent experiment [41] designed to measure neutron flow, and considered the reaction Nb + Nb at 650 MeV/nucleon in the impact parameter range $0.4R \leq b \leq 0.8R$, where R is the Nb radius. It is found that $\mathcal{R}$ is most sensitive to the nuclear EOS if only particles moving in a forward cone are considered in its construction. Fig. 3 therefore shows $d\sigma/d\phi$ for the SBKD, HBKD and SMDI for participants emitted with $(y/y_p)_{lab} > 0.75$. Values of $\mathcal{R}$, the average in–plane transverse momentum $\langle \omega p_x \rangle$ (where ω is $+1$ (-1) for forward (backward) moving particles in the c.m.), the normalized mid–rapidity slope $\tilde{F}$ of the $\langle p_x(y) \rangle$ curve, and flow angles θ_{max} are summarized in Table 1. The results from SMDI and SGBD are hardly distinguishable from HBKD, while the SBKD and HMDI (a version of Eq.(3.2) with $K_0 = 380$ MeV) produce respectively less and more flow.

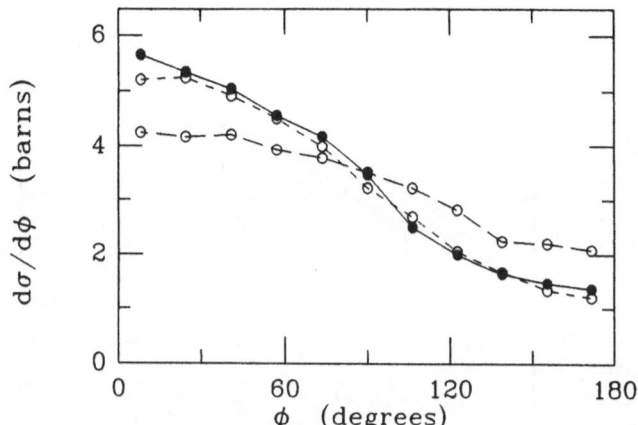

Fig. 3. Azimuthal cross–section for participants emitted within a forward direction given by $(y/y_{proj})_{lab} > 0.75$ for the SMDI (solid curve), HBKD (short–dash line) and SBKD (long–dash line). In all cases, the two–body cross–section used is the free space value.

Table 1. Collective flow parameters for various mean fields (see text). The reaction considered is Nb + Nb at $E_{lab}/A = 650$ MeV, averaged over an impact parameter range $0.4R \leq b \leq 0.8R$. Free space two–body cross–sections are used.

Mean Field	K_0 (MeV)	$\mathcal{R}_{(y/y_p)_{lab} > 0.75}$	$\tilde{F}$ (MeV/c)	$\langle \omega p_x \rangle$ (MeV/c)	θ_{max}
HBKD	380	4.3	180	68	14°
SBKD	200	2.0	140	47	7°
SGBD	215	5.2	210	80	16°
SMDI	215	4.3	200	73	15°
HMDI	380	7.4	240	92	18°

The equivalence of transverse momentum in SMDI and HBKD persists over a wide range of beam energies. This is shown in Fig. 4 for 400 and 800 MeV/nucleon. Even at energies as low as 150 MeV/nucleon, $\tilde{F}$ is found to be similar [42] in simulations using the HBKD and a soft momentum dependent mean field ($K_0 = 228$ MeV) derived from the Gogny force [43].

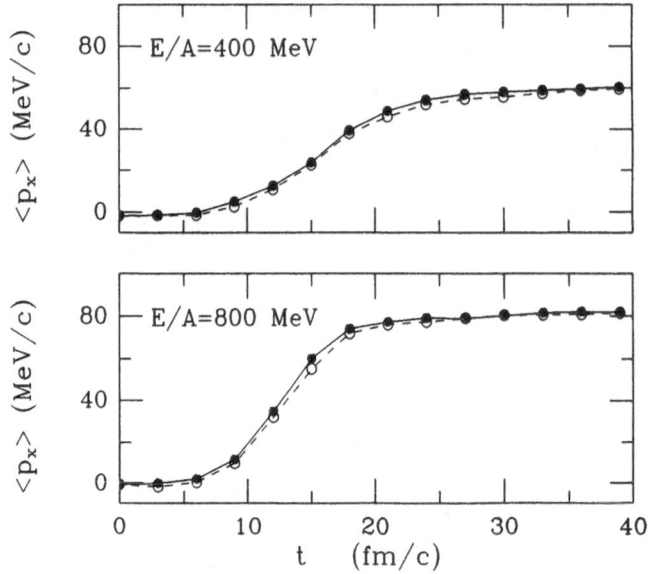

Fig. 4. Average in–plane transverse momentum as a function of time for two values of the beam energy (400 and 800 MeV/nucleon). The solid line is a SMDI, the dashed line a HBKD calculation. In both cases the reaction is La + La with $b = 2.7$ fm.

We next turn to a comparison of SMDI and HBKD simulations with Streamer Chamber deuteron flow data from near central reactions of Ar + KCl, La + La and Ar + Pb at 800 MeV/nucleon [16]. BUU is a one–body theory and does not incorporate fragment formation explicitly, but data and detector efficiency simulations done in Ref.[16] show that in–plane deuteron momenta are twice as high as as those for protons at a given rapidity. In both experiment and simulation the average in–plane momenta are calculated from particles with rapidity in the c.m. greater than 0.15, while in the simulation we further exclude those particles which have suffered no collisions. The $\langle p_x \rangle$ and especially average flow angles (here calculated from an average sphericity tensor from all events, with weight $1/p_i$) are rather sensitive to the latter cut. We consider it because spectators will produce spurious flow effects in the BUU [5], but a more detailed future analysis is warranted (see Ref.[12,35]).

The $\langle p_x \rangle$ vs rapidity plots are are shown in Fig. 5. Reasonable agreement is obtained for both the SMDI and HBKD mean fields in all three systems. The very high experimental values of $\langle p_x \rangle$ near the target rapidity region are due to difficulties in particle identification. Note that the Ar + Pb reaction is shown in the laboratory system because of uncertainty in the location of the center of mass. For this reason we show in Table 2 flow parameter comparisons for only the Ar + KCl and La + La systems. Once again, the agreement is resonable for the $\langle p_x \rangle$, but less so for the flow angles. This latter observable is dependent strongly on the final filter applied.

Table 2. A comparison of experimental collective flow data from Ref.[16] with the SMDI and HBKD BUU results (see text) at 800 MeV/nucleon. For the BUU, particles that have suffered no collisions are removed from the analysis. The second column gives the impact parameter range used for each calculation, and is obtained from the experimentally analysed fraction of the total inelastic cross–section via a geometrical picture. For the Ar + KCl case two event sets with different triggers are incorporated into the experimental data. The quoted b is therefore constructed from an event number weighted average of the cross–sections for each individual set.

System	b (fm)	θ_F^{EXP} (deg)	θ_F^{SMDI} (deg)	θ_F^{HBKD} (deg)	$\langle p_x/a \rangle_{y>0.15}^{EXP}$ (MeV/c)	$\langle p_x \rangle_{y>0.15}^{SMDI}$ (MeV/c)	$\langle p_x \rangle_{y>0.15}^{HBKD}$ (MeV/c)
Ar + KCl	≤ 3.8	9.6 ± 0.8	12	12	50 ± 4	61	57
La + La	≤ 8.5	16.5 ± 1.7	11	12	72 ± 6	76	81

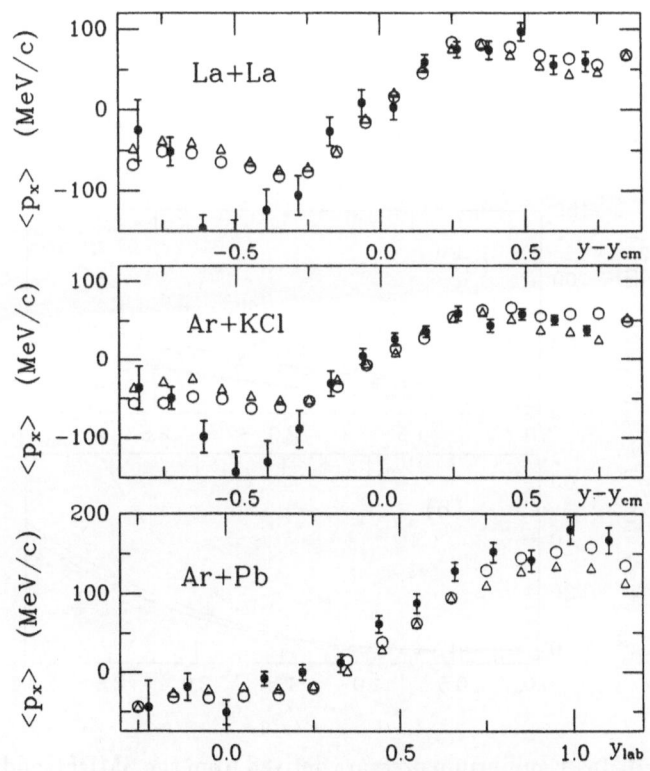

Fig. 5. A comparison of experimental data [16] (solid circles) to SMDI (open circles) and HBKD (open triangles). See text for details.

Given the fact that the SMDI and HBKD produce similar transverse momenta, one might reasonably expect the EOS's derived from them to become similar at high temperatures T. Ref.[24] calculates the equilibrium pressure $P(\rho)$ at constant temperature T; for momentum dependent forces one is required to compute the chemical potential μ self–consistently from

$$\rho = 4\pi \int dp\, p^2 f(\mathbf{r},\mathbf{p}) \quad , \epsilon(p) = \frac{p^2}{2m} + U(\rho,p) \quad , \tag{4.1}$$

where $\beta = T^{-1}$ and $f(\mathbf{r},\mathbf{p}) = 4/h^3 (e^{\beta(\epsilon(p)-\mu)} + 1)^{-1}$. Surprisingly, the HBKD differs substantially from the SMDI and SBKD (see Fig. 6(a)), even at very high temperatures. This fact is corroborated in Fig. 4: the transverse momentum is generated early in the collision, before thermal equilibrium has been achieved. We therefore consider the non–equilibrium pressure tensor [25,24] :

$$P_{xx} = \int d^3p\, p_x \left(\frac{p_x}{m} + \frac{\partial U}{\partial p_x} \right) + \delta_{ij} \left(\int d^3p\, U f - V \right) \quad , \tag{4.2}$$

calculated for the opposite extreme of two heavy nuclei overlapping completely in position space, but having two disjoint Fermi spheres in momentum space ($E/A = 400$ and 800 MeV). This P_{xx} is shown in Fig. 6(b), but once again the HBKD and SMDI are qualitatively different, given that the central densities reached in simulations with these two mean fields do not differ substantially ($\rho \sim 2.5\rho_0$).

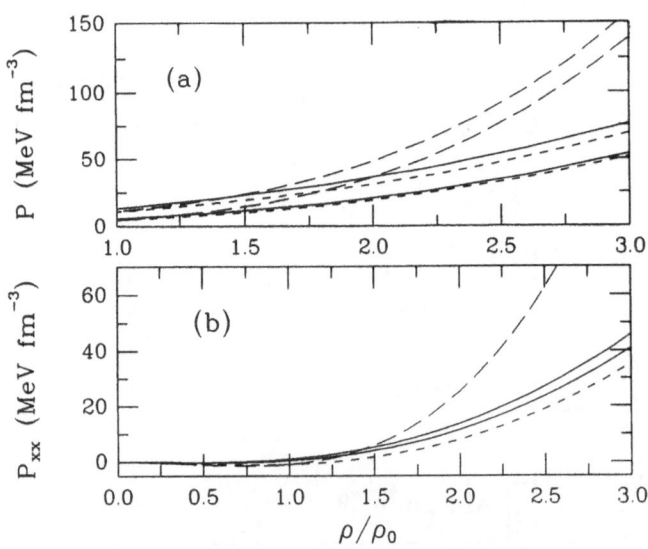

Fig. 6. (a) Equilibrium pressure derived from the SMDI (solid lines), HBKD (long–dashed lines) and SBKD (short–dash curves) mean fields, as a function of density. For each EOS, the upper curve is for $T = 80$ MeV while the lower one is for 40 MeV. (b) P_{xx} vs ρ, line–type designation for mean fields as in (a). The upper solid line is for SMDI at $E/A = 800$ MeV; the lower solid line is for SMDI with $E/A = 400$ MeV.

Hard and soft collisions cooperate in producing the transverse momentum observed in BUU; cascade ($U \equiv 0$) and Vlasov ($\sigma_{NN} \equiv 0$) models lead to less flow. A difference between the SMDI and HBKD emerges, however, if we decrease the two–body cross-section. This is shown in Table 3 for La + La with $b = 2.7$ fm: as σ_{NN} is lowered, $\langle \omega p_x \rangle$ drops more dramatically for HBKD than SMDI. Finally, when $\sigma_{NN} \equiv 0$ (Vlasov case), HBKD has negligible transverse momentum while that for SMDI remains substantial.

Table 3. The effect decreasing the two–body cross–section σ_{NN} at two beam energies and for the SMDI and HBKD cases. The reaction is La + La at an impact parameter of $b = 2.7$ fm. The variables are explained in the text.

E/A (MeV)				HBKD $\langle \omega p_x \rangle$ (MeV/c)	SMDI $\langle \omega p_x \rangle$ (MeV/c)
400	σ_{NN}	$=$	σ_{free}	60	61
	$\sigma_{NN}^{elastic}$	$=$	$0.7\sigma_{free}^{elastic}$	46	57
	σ_{NN}	$\equiv$	0	1	34
800	σ_{NN}	$=$	σ_{free}	79	82
	σ_{NN}	$\equiv$	0	6	38

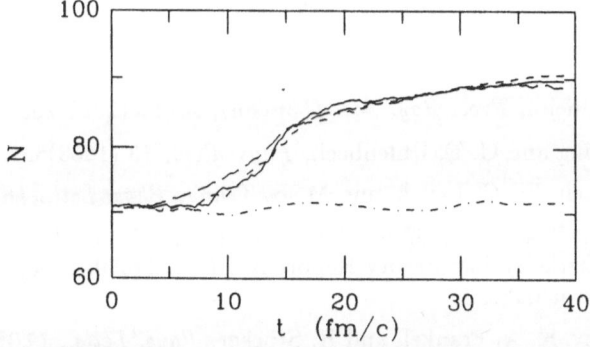

Fig. 7. Number of particles with $p_x > 0$ and $p_z > 0$ as a function of time. The system is La + La with $b = 2.7$ fm and $E/A = 800$ MeV, simulated via SMDI BUU (solid line), HBKD BUU (short dash line), SMDI Vlasov (long dash line) and HBKD Vlasov (dot–short dash line). Both BUU calculations use the free space two–body cross-sections. The situation for particles with $p_x < 0, p_z > 0$ is a reflection of the above diagram about the line $N = 70$.

Qualitatively, this effect may be understood from Fig. 7, which shows that in a SMDI Vlasov calculation more particles have $p_x > 0$ and $p_z > 0$ than in the HBKD Vlasov. The situation is reversed for the "wrong" direction ($p_x < 0$, $p_z > 0$) so that the in-plane transverse momentum is very nearly canceled in the HBKD Vlasov case. The momentum dependent term in (3.2) implies a relative attraction between nucleons that have similar momenta (an effect absent in HBKD), inducing a momentum space correlation that compensates for stronger pressure effects in HBKD.

5. CONCLUSION

A momentum dependent mean field parameterization with a soft compression modulus (SMDI) was presented and shown to match reasonably a single particle potential that fits optical potential scattering data and whose underlying Hamiltonian reproduces nuclear matter properties. The SMDI parameterization produces the same transverse flow as a stiff momentum independent single particle potential (HBKD) over a wide range of energies and impact parameters. This feature results from the correlations induced by the relative attraction between particles close in momentum–space in the case of SMDI and compensates for the higher compressional effects in the HBKD mean field. As a consequence, the SMDI transverse flow is less sensitive to changes in the two–body cross–section than in the HBKD case. While agreement with the data shown here is resonable, questions regarding impact parameter ranges and fragment formation make a detailed multiplicity bin comparison difficult.

ACKNOWLEDGEMENTS

It is my pleasure to acknowledge that much of the work presented here was performed in collaboration with S. Das Gupta, C. Gale, C. Grégoire, R. Malfliet and M. Prakash. I also wish to thank G.E. Brown for many discussions. This work was supported in part by the Department of Energy under Grant No. DE-FG02-88ER40388.

References

[1] L. W. Nordheim, *Proc. Roy. Soc. (London)*, **A119** (1928)689.

[2] E. A. Uehling and G. E. Uhlenbeck, *Phys. Rev.*, **43** (1933)552.

[3] G. F. Bertsch, W. G. Lynch, and M. B. Tsang, *Phys. Lett.*, **189B** (1987)384.

[4] G. Fai, In *Proceedings of the 8th High Energy Heavy Ion Study*, page 103, Lawrence Berkeley Laboratory Report No. LBL-24580, 1988; ed. J. W. Harris and G. J. Wozniak.

[5] M. Gyulassy, K. A. Frankel, and H. Stöcker, *Phys. Letts.*, **110B** (1982)185.

[6] P. Danielewicz and G. Odyniec, *Phys. Lett.*, **157B** (1985)146.

[7] R. Stock et al, *Phys. Rev. Lett.*, **49** (1982)1236.

[8] D. Hahn and H. Stöcker, *Nucl. Phys.*, **A452** (1986)723.

[9] B. Schürmann, W. Zwermann, and R. Malfliet, *Phys. Rep.*, **147** (1987)1.

[10] P. Danielewicz, In *Workshop on Physics of Intermediate and High Energy Heavy-Ion Reactions, Krakow*, 1987; ed. M. Kutschera, World Scientific, 1988, Singapore.

[11] A. Bonasera and L. P. Csernai, *Phys. Rev. Lett.*, **59** (1987)630; B. Schürmann and W. Zwermann, *Phys. Rev. Lett.*, **59**, (1987)2848.

[12] J. Cugnon and D. l'Hôte, *Nucl. Phys.*, **A452** (1986)738.

[13] G. F. Bertsch, H. Kruse, and S. Das Gupta, *Phys. Rev.*, **C29** (1984)673.

[14] H. A. Gustafsson et al, *Phys. Rev. Lett.*, **52** (1984)1590.

[15] J. J. Molitoris and H. Stöcker, *Phys. Rev.*, **C31** (1985)346.

[16] P. Danielewicz et al., *Phys. Rev.*, **C38** (1988)120.

[17] J. M. Cavedon et al., *Phys. Rev. Lett.*, **58** (1987)195.

[18] J. P. Blaizot, D. Gogny, and B. Grammaticos, *Nucl. Phys.*, **A265** (1976)315.

[19] B. Friedman and V. R. Pandharipande, *Nucl. Phys.*, **A361** (1981)502.

[20] C. Gale, G. F. Bertsch, and S. Das Gupta, *Phys. Rev.*, **C35** (1987)1666.

[21] G. M. Welke, M. Prakash, T. T. S. Kuo, S. Das Gupta, and C. Gale, *Phys. Rev.*, **C38** (1988)2101.

[22] F. Sebille, G. Royer, C. Gregoire, B. Remaud, and P. Schuck, to be published.

[23] R. B. Wiringa, *Phys. Rev.*, **C38** (1988)2967.

[24] C. Gale, G. M. Welke, M. Prakash, S. J. Lee, and S. Das Gupta, *Stony Brook Preprint*, (1989).

[25] G. F. Bertsch and S. Das Gupta, *Phys. Rep.*, **160** (1988)189.

[26] G. M. Welke, R. Malfliet, C. Gregoire, M. Prakash, and E. Suraud, *Stony Brook Preprint*, (1989).

[27] A. Bonasera, G. F. Burgio, and M. Di Toro, *Preprint*, (1988).

[28] G. F. Bertsch, In *Nuclear Physics with Heavy Ions and Mesons, Vol.1*, page 175, Les Houches 1977; ed. R. Balian, M. Rho and G. Ripka (Noth Holland, Amsterdam 1987).

[29] M. Krook and T. Wu, *Phys. Rev. Lett.*, **36** (1976)1107.

[30] M. Ernst, *Phys. Repts.*, **78** (1981)1.

[31] I. E. Lagaris and V. R. Pandharipande, *Nucl. Phys.*, **A359** (1981)331.

[32] I. E. Lagaris and V. R. Pandharipande, *Nucl. Phys.*, **A359** (1981)349.

[33] R. B. Wiringa, R. A. Smith, and T. L. Ainsworth, *Phys. Rev.*, **C29** (1984)1207.

[34] R. Schiavilla, V. R. Pandharipande, and R. B. Wiringa, *Nucl. Phys.*, **A449** (1986)219.

[35] D. Keane et al., *Phys. Rev.*, **C37** (1988)1447.

[36] J. Aichelin, A. Rosenhauer, G. Peilert, H. Stöcker, and W. Greiner, *Phys. Rev. Lett.*, **58** (1987)1926.

[37] K. G. R. Doss et al, *Phys. Rev. Lett.*, **57** (1986)302.

[38] G. Fai, Wei-ming Zhang, and M. Gyulassy, *Phys. Rev.*, **C36** (1987)597.

[39] K. G. R. Doss et al, *Phys. Rev. Lett.*, **59** (1987)2720.

[40] H. H. Gutbrod et al., *Phys. Lett.*, **216B** (1989)267.

[41] R. Madey et al., Bevalac Experiment 848 H (1987).

[42] C. Gregoire, private communication.

[43] D. Gogny, In *Nuclear Self Consistent Fields*, page 333, (North-Holland, Amsterdam, 1975); edited by G. Ripka and M. Porneuf.

HARD PHOTONS AND SUBTHRESHOLD MESONS
FROM NUCLEUS-NUCLEUS COLLISIONS

Eckart Grosse

GSI Darmstadt
D-6100 Darmstadt, FRG

INTRODUCTION

The study of nucleus-nucleus collisions at an energy high enough to allow for an appreciable nuclear overlap is the only experimental means to obtain information about the properties of nuclear matter at a density clearly above the saturation density of heavy nuclei $\rho_0 = 0.16$ fm^{-3}. Considering the lack of success of all attempts to reproduce ρ_0 and the nuclear matter binding energy per nucleon $\varepsilon = -16$ MeV in (non relativistic) self consistent calculations on the basis of realistic nucleon-nucleon potentials[1], experimental information about the ρ-dependence of ε, i.e. the equation of state (eos), may help to solve this long-standing problem of nuclear physics. Additionally, such information is extremely valuable for astrophysics, as the stability of neutron stars as well as the dynamics of a supernova of type II are strongly depending[2] on the nuclear eos. The density variations in nuclear ground states are very small and do not yield relevant information; but the E0-giant resonance (breathing mode) energies can be related to the compressibility K_0 of nuclear matter which determines the eos near the minimum ρ_0. A recent analysis[3] of a series of new E0-energy determinations results in $K_0 = 300$ MeV, a value which is larger than the 140 MeV used predominantly[2] in astrophysical calculations.

Heavy ion reactions allow to obtain in the laboratory nuclear densities considerably higher than that of ground state matter. Recent calculations[4] on the basis of transport equations give values of $\rho = 3\rho_0$ for the collision of medium heavy nuclei at about 1 GeV/u. Obviously, a large high density volume is assured only by a large spatial overlap of the two colliding nuclei, making the use of very heavy ions desirable.

Of course, this high density is reached only during the collision time of a few times 10^{-23} s and it is the key problem of the experiments to find the right observables for this high density phase of the collision. Newly created particles and photons of sufficiently high energy are very likely to be emitted from the early collision zone; at a later stage the projectile energy is dissipated over many nucleons ("thermalized") and only soft photons (or pions) can be produced. In these lectures it will be outlined how such hard photons from the early phase of nucleus-nucleus collisions - as well as high energy pions and heavier mesons - can be detected and identified as such, which means they have to be distinguished from those emitted thermally from the remnants of the collision at a later time. At the beginning, experimental techniques for the detection of photons (as well as π^0 and η, both decay into two photons) will be described. Then the properties of photon emission will be shown to be a good model case for the application and test of recent theoretical concepts derived for nucleus-nucleus collisions in this energy range. At the end the special features of pion and kaon production will be discussed and their possible use as a probe for high density nuclear matter.

THE PRINCIPLES OF HARD PHOTON DETECTION

A detector for high energy photons has to have one component of high Z material to convert the γ's into e^+e^- pairs; these leptons - respectively the electromagnetic shower resulting from successive bremsstrahlung and pair production processes - are to be detected via their Cherenkov radiation or the ionization and/or scintillation they induce. The detector dimensions needed to assure sufficient conversion probability and shower containment can be calculated from the well known cross sections of the different electro magnetic processes using a Monte-Carlo technique. For different materials these dimensions scale approximately with the radiation length L_R, a quantity dependent on density and Z. Because of the strong increase of the pair production cross section with photon energy the shower size only increases slowly (about logarithmicly) when going to high energy photons. For E_γ between 20 and 200 MeV 98 to 99 % of the shower are contained in a detector volume of a width of 4 L_R and a depth of 13 L_R; by the use of an active converter only showers commencing inside of it are selected and the total detector lepth can be reduced to 10 L_R. A widely used detector material is lead glass: The total Cherenkov light emitted in it by the fast leptons is nearly proportional to the incident photon energy and results in about one photoelectron registered in the photomultiplier per MeV incident energy. In (much more expensive) scintillator materials such as BaF_2 or NaJ this number is higher by about two orders of magnitude leading to appreciably better energy resolution.

HARD PHOTONS FROM NUCLEUS-NUCLEUS COLLISIONS - A MODEL CASE

Gamma rays in the range of 20 to 200 MeV ("hard photons") have become a rather widely investigated feature in the study of nucleus-nucleus collisions at beam energies of 10 - 100 MeV/u. Because of the well understood electromagnetic interaction a theoretical treatment of such radiative collision processes should be straightforward. Hard photons were first observed[5] incidentally in experiments on nucleus-nucleus collisions: In one case they showed up as uncorrelated background to the two coincident photons from π^0 decay; in other experiments[6][7] they formed a high energy tail in the γ-spectrum from the decay of giant resonances produced in heavy ion collisions (see figure 1).

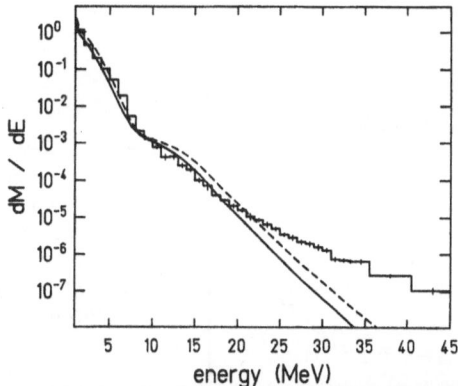

Fig. 1. Inclusive photon spectrum from 14.6 MeV/u $^{16}O + ^{184}W$. The dashed line describes the statistical decay of the compound nucleus assuming complete fusion; precompound particle emission was accounted for to obtain the drawn line (cf. ref. 6).

The decay of nuclei highly excited in giant resonances or in the underlying quasi-continuum of closely spaced levels is described in the statistical model of compound nucleus decay; the partial γ-decay width can be calculated[7] from the γ-absorption cross section by applying the detailed balance principle

$$\Gamma_\gamma(E_\gamma, A) = (\pi\hbar c)^{-2}\sigma_\gamma(E_\gamma, A)E_\gamma^2\exp(-E_\gamma/_T)dE_\gamma$$

The exponential factor accounts for the fact, that the absorption is measured on the nucleus A in its ground state, whereas in the photon emission experiments the compound nucleus - if it is formed - has a temperature

$$T = \frac{2}{3} \cdot \frac{E_p}{4A_p} \quad .$$

The products of incomplete fusion or deep inelastic scattering processes have very similar temperatures, such that the procedure (indicated in figure 1) of extrapolating the contribution of the statistical γ-decay to higher photon energies should be rather reliable not only in the case of complete fusion. This is especially so when the steeply falling low energy section as well as the giant resonance region are well reproduced by statistical model calculations.

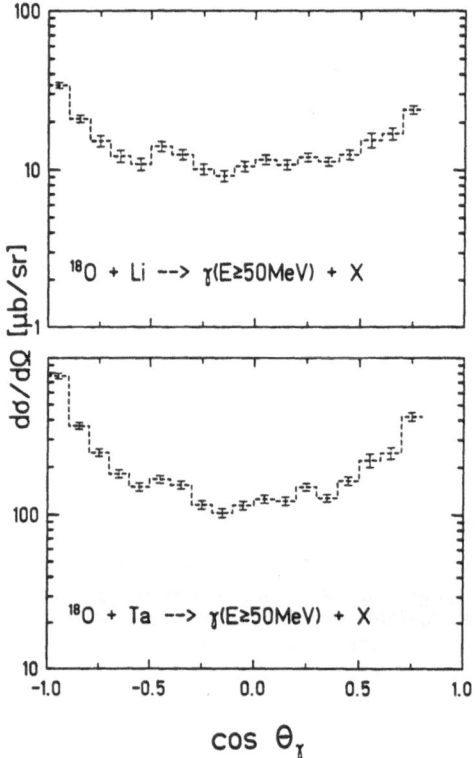

Fig. 2. Angular distributions of photons produced inclusively in $^{18}O + Li$ and $^{18}O + Ta$ at 84 MeV/u. The data have been Lorentz-transformed to a system with half the projectile rapidity; there they are symmetric around 90°.

The surplus cross section at higher photon energies - the "hard photon" yield - shows[6] the interesting feature of an angular distribution which is strongly forward peaked in the laboratory system. A similar observation was already made for hard photons produced[8] in collisions at higher energy (see fig. 2). A Lorentz transformation aiming for symmetry around 90° leads to a source rapidity which is equal to half the beam rapidity (at these low energies the rapidity $y = \text{artanh } \beta$ is not very different from the velocity β). This "half-rapidity" source of the hard photons is obviously formed from equally many nucleons out of projectile and target and can thus be identified as the initial collision zone formed in the first encounters of projectile and target nucleons.

The idea of assigning the hard photons to originate from these first collisions is supported by their systematic dependence on projectile and target mass: The "normalized" photon production cross section is proportional[9] to an exponential

$$\frac{d\sigma}{dE_\gamma d\Omega} \propto \frac{N_c \sigma_R}{E_o} \exp(-E_\gamma/E_o)$$

with a slope constant E_o. All data published so far fall on the same line[9] when σ_R is the (geometrical) total reaction cross section and N_c is the number of (first chance) collisions between a photon from the projectile and a neutron from the target or vice versa. Collisions between two protons play a minor role, since the respective elementary bremsstrahlung cross section[10] is smaller by about one order of magnitude; in a multipole expansion picture dipole radiation can only be emitted from a charge asymmetric system.

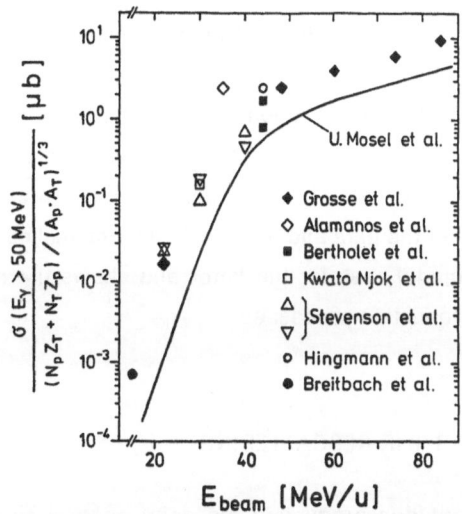

E_{beam} [MeV/u]

Fig. 3. Beam energy dependence of the photon production yield normalized to the number of p-n collisions. The drawn line shows a transport-equation calculation[11)12)].

The main features of the hard photon emission from colliding nuclei, i.e. the absolute yield, the scaling with projectile and target and the spectral shapes are reasonably well reproduced by calculations[10-13] based on transport equations for the nucleons within the mean field of the colliding nuclei. The nuclear mean fields not only lead to nucleon binding and Fermi motion before the collision, but also to a collisional acceleration due to the partial overlap of the two mean fields; they also induce a blocking of occupied phase space to those nucleons, which have lost part of their initial momentum in a collision, eventually accompanied by bremsstrahlung radiation. Of course, the elementary bremsstrahlung cross section sensitively enters the calculations; unfortunately there is some minor inconsistency[9] between different experiments on photon production from p + nucleus collisions. This inconsistency is transferred to the theoretical description of the nucleus-nucleus data, making the comparison of absolute cross sections still somewhat marginal. But the systematics of the data is well reproduced, including the exponential fall off of the photon yield with photon energy. This exponential slope resembles a (thermal) spectrum from statistical decay, but the slope parameter E_0 is significantly larger than the temperature T expected for products of - complete or incomplete - fusion, fusion-fission or deep inelastic processes. Slopes corresponding to the temperatures T of these processes are observed in the photon spectra below 15 MeV; for the hard photon part ($E_\gamma \geq$ 20 MeV) slope parameters E_0 were reported[5] to be larger than T by about 4 MeV, another clear indication of the pre-equilibrium non-thermal nature of this radiation. It should be mentioned here, that collective nuclear bremsstrahlung[14] to be emitted from the two nuclei (as a whole) during their collisional deceleration would also show up at $E_\gamma <$ 15 MeV, where the photon wavelength is comparable to the dimensions of the emitting system.

The photon angular distributions are rather flat in the frame comoving with the source at half the beam rapidity; data[15] at variance to this are in contradiction to several independent recent experiments[9]. The more pronounced structure in the elementary pn → pnγ angular distribution[10] gets lost in the nucleus-nucleus collision due to Fermi motion and relativistic effects - as indicated by the transport equation calculations[12]. The overall consistency between the data and these calculations makes the photon production a sensitive test and model case for the latter and shows the reliability for their extension to the meson production processes.

PIONS FROM THE COLLISION ZONE AND THEIR RESCATTERING

In the preceding discussion about photon production the good general accord between data and calculations on the basis of transport theory was emphasized. It becomes transparent from figure 3 which displays the beam energy dependence of the photon production cross section properly normalized to the geometrical situation of a nucleus-nucleus collision. When comparing this cross section to the one for the elementary process pn → pnγ a strong enhancement at low bombarding energies is observed; this can be depicted as the result of medium effects present in the nucleus-nucleus-collisions: binding and Fermi-motion, collisional acceleration and Pauli-blocking. Medium effects in the exit channel can be neglected as the mean free path of photons in nuclear matter is huge. For pion production the situation is largely different from the photon case: there not only the coupling to the three isospin channels has to be considered but there also is the strong pion rescattering by the Coulomb and nuclear fields of the projectile and target nucleons.

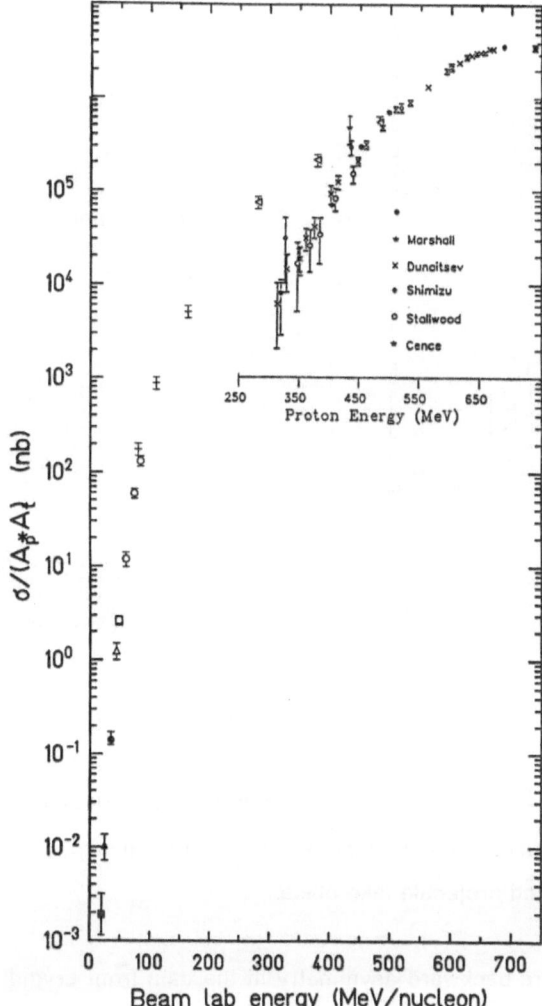

Fig. 4. Excitation function for π-production from nucleus-nucleus collisions normalized to the number of nucleon-nucleon collisions. The data[18] for the "elementary" process $pp \rightarrow pp\pi^0$ are shown for comparison.

A general overview[16][17] of the integrated pion production yields (fig. 4) has to cover many orders of magnitude to show the multitude of data taken so far. On that scale differences between charged and neutral pion data or data from collision systems of different mass do not become obvious. In comparison to the elementary cross section[18] there is a clear enhancement in the nucleus-nucleus collision data taken at lower beam energy; it is especially spectacular due to the much higher quality and sensitivity of these as compared to the p-p data.

From a comparison[17] of π^+ and π^- data to those obtained for π^0's from $^{12}C + ^{12}C$ collisions at 84 MeV/u isospin effects can clearly be seen. There are shifts in the charged pion spectra due to the finite state Coulomb interaction at high pion momenta and differences at small momenta which are possibly due to the different reabsorption of π^+ and π^-.

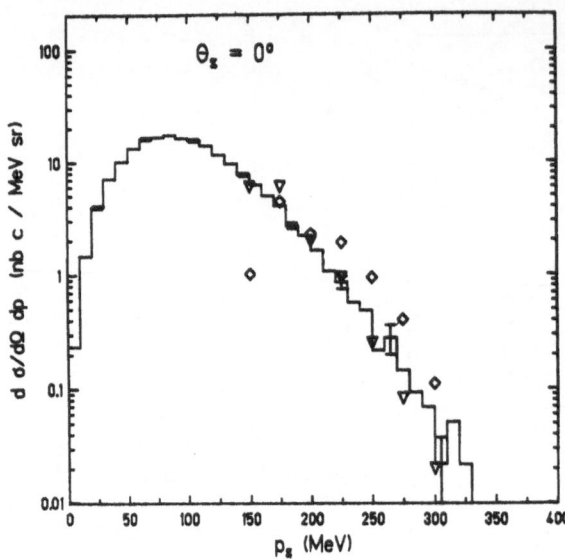

Fig. 5. Comparison of spectra of π^+, π^- and π^0 observed for $^{12}C + ^{12}C$ at 84 MeV/u.

An inspection of the complete π^0 angular distribution for the asymmetric collision systems $^{18}O + Li$ and $^{18}O + Ta$ (cf. Fig. 6) shows such reabsorption effects very clearly. For the light Li target the cross section - Lorentz transformed to the half rapidity system - is forward peaked whereas the Ta target causes a strong decrease of the forward cross section. This can be explained straightforward by the reabsorption of the pions in the heavy nucleus after being produced on its surface, where the first chance collisions between nucleons from target and projectile take place.

In principle, the different forward-backward asymmetry in the data from Li and Ta targets could also be interpreted as following from different velocities of the pion source formed in the collisions with the different target nuclei. But an inspection of the spectral shape shows[17] that an angle independent slope parameter is only obtained after transformation into the half-rapidity frame. Additionally the similarity of the π^0 spectra to those obtained for γ's (cf. fig. 7) suggests similar production mechanisms. Similarly as for photons it has been shown[17] that in the energy range below 100 MeV/u also the pion production yields are proportional to the number of first chance nucleon-nucleon collisions, after correcttion for reabsorption.

As can be seen from the comparison of the spectra, pions are more abundant than photons at a given total energy; taking into account, that both photon spin directions are compared to only one pion isospin and that pions are strongly reabsorbed, the primordial pion enhancement is in the order of 20-30. Regarding the coupling constants of electromagnetic and strong interaction, a larger π/γ-ratio is expected.

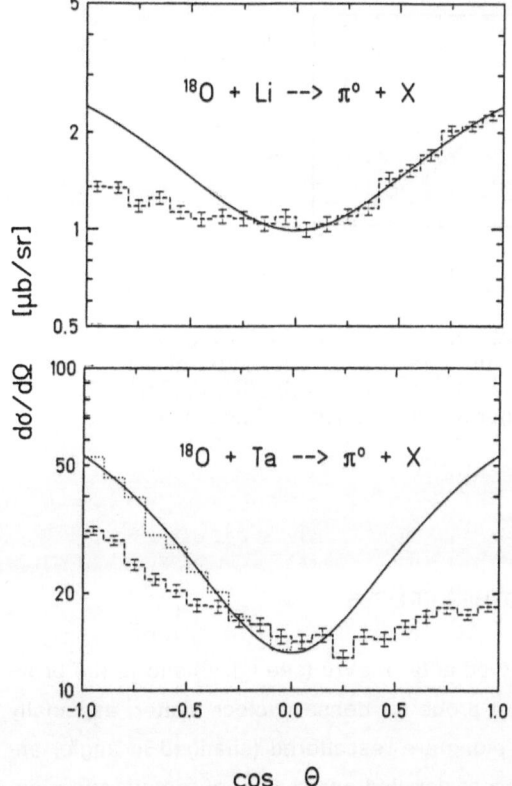

Fig. 6. π^0-angular distributions for 84 MeV/u ^{18}O projectile on the light Li (top) and the heavy Ta-target (bottom). The data are transformed into the half-rapidity system.

Calculations[13)19)] have not yet been developed as far as in the photon case; the "elementary" cross sections are not well known for all isospin channels and it is not clear if s-wave production is to be added to the p-wave going through the Δ-resonance. Since also rescattering and charge exchange in the outgoing channel are not well determined the existance of more exotic phenomena like the pion-collectivity proposed recently[20)] can not yet be identified in the data.

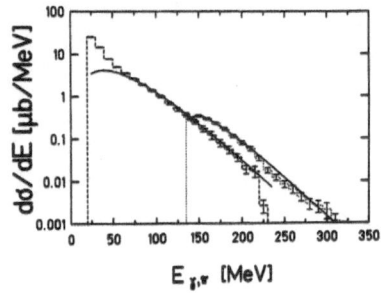

Fig. 7. Photon and π^0-spectra from $^{18}O + Li$ at 84 MeV/u. The cross sections are plotted vs. the total energy allowing a direct comparison at a given energy taken out of the relative motion. The drawn lines depict the energy dependence of phase space.

OUTLOOK: EMISSION OF KAONS AND OTHER PROBES

The strong pion reabsorption observed at 84 MeV/u (see fig. 6) shows the problems arising when envisaging pions as a probe for dense nuclear matter, especially since it is not obvious, if the "missing" pions are rescattered (changed in angle, energy, or isospin) or truly obsorbed. A recent detailed analysis[21)] of proton- and pion-nucleus collisions has demonstrated the difficulties of a consistent treatment of medium effects in Δ-propagation and pion production in nuclei. In view of the small effects expected[22)] to arise in pion emission yields from compression effects in heavy ion collisions a detailed knowledge about the nuclear eos can probably not be extracted from them alone.

Due to strangeness conservation kaons (with the exception of low momentum K^-) have a low absorption probability in nuclei, which makes them a good probe for the nuclear interior. The excitation function for K^+-production in nuclear collisions has been predicted[23)] to rise steeply up to a projectile energy of 1.5 GeV/u. This makes

K$^+$-production especially sensitivity of this process to effects reducing the energy effective in the moment of a nucleon-nucleon encounter. The compression of the nuclear matter in the collision zone has been described[22] as using up part of the projectile energy very early in the collision. K$^+$-emission could thus become a powerful probe for the nuclear eos as manifest in nucleus-nucleus collisions; up to now only very scarce data exist[24], most of which have been taken above the nucleon-nucleon threshold at 1.58 GeV/u. The production of K$^+$ at low bombarding energy not only proceeds via the correlated hyperon production channel:

$$NN \rightarrow N\Lambda K \text{ or } N\Sigma K$$

but also via two step processes like

$$NN \rightarrow \pi NN$$
$$N\pi \rightarrow \Lambda K;$$

these are depending quadratically on the density in the collision zone and on its volume.

To make experimental investigations unambiguous in their interpretation, the collision volume has to be well defined experimentally; this can be done by varying projectile and target mass or by the coincident observation of all the participant nucleons. The simultaneous study of the emission of different probe particles seems very desirable as well; pions will be present abundantly at energies around 1 GeV/u, where compression effects are strong. Very probably this will make the observation of directly produced photons rather difficult because of the combinatorical background from π^0's. Large arrays of scintillators with good time and energy resolution might allow to subtract this background; they also can be used to observe the production of η's, which are similar in mass to kaons but do not carry strangeness. At the new heavy ion synchrotron SIS a strong effort in this direction will be made and there also will be magnetic detection systems especially suited for charged pion, kaon, lepton or even antiproton experiments.

REFERENCES

1. C. Mahaux, this volume
2. H.A. Bethe in: Unified Concepts of Many-Body Problems, p. 3,
    T.T.S. Kuo and J. Speth ed., Amsterdam 1987
3. M.M. Sharma et al., Phys. Rev. C 38 (1988) 2562 and to be published
4. A.L. DePaoli et al., proceedings Hirschegg (1988)
5. E. Grosse et al., Europhys. Lett. 2 (1986) 9
6. G. Breitbach et al., submitted to Phys. Rev. Lett.
7. N. Hermann et al., Phys. Rev. Lett. 60 (1988) 1630
8. P. Grimm and E. Grosse, Progr. in Part. and Nucl. Phys. 15 (1985) 339
9. H. Nifenecker and J.A. Pinston, to be publ. in Reports on Progr. in Physics
10. K. Nakayama and G.F. Bertsch, Phys. Rev. C34 (1986) 2190;
    K. Nakayama, to be published in Phys. Rev. C
11. V. Metag, proceedings St. Malo 1988, Nucl. Phys. A488 (1988) 483
12. T.S. Biro et al., Nucl. Phys. A475 (1987) 579

13. W. Bauer, MSUCL-672, to be published
14. R. Heuer et al., Z. Phys. A330 (1988) 315
15. N. Alamanos et al., Phys. Lett. 173B (1986) 392
16. H. Noll et al., Phys. Rev. Lett. 52 (1984) 1284
    H. Heckwolf et al., Z. Phys. A315 (1984) 243
17. E. Grosse, Nucl. Phys. A447 (1985) 611; id., Varenna lectures 1987, to be publ.
18. T. Reposeur, thesis Paris 1989; T.D.S. Stanislaus, thesis Vancouver 1987
19. M. Tohyama et al., Nucl. Phys. A437 (1985) 739
20. G.F. Bertsch et al., MSUCL-644, to be published
21. J. Cugnon and M.C. Lemaire, Nucl. Phys. A489 (1988) 781
22. R. Stock, Phys. Reports 135 (1986) 259
23. B. Schürmann and W. Zwermann, Europhys. Lett. (1988)
24. J.B. Carroll, this volume

MEASUREMENTS OF e+e- PAIR PRODUCTION AT THE BEVALAC

Jim Carroll

Physics Department, University of California at Los Angeles
Los Angeles, CA 90024 *

INTRODUCTION

It has come to be generally recognized that a relativistic nuclear collision (RNC) can only be described as a complex dynamic evolution of the not-so-many-body system, whether the components of the system are taken at the hadronic level or at the quark level. The evolution proceeds from an initial state of two separated nuclei, through stages of: first nucleon-nucleon collisions; increasing rates of collisions and particle production (heating and compression); gradually decreasing rates of collision and particle production rates (cooling and expansion); and, finally, decoupling of various particle types from the overall system (freeze out). (Parentheses enclose the adiabatic thermodynamic terms that may in some circumstances be relevant descriptions for parts of this process.) It is, in particular, the operation of the dynamics during the stages of high excitation energy and high particle density that are of interest in studying such collisions. Unfortunately we cannot build a detector that will produce information about a particular part of this process. Our detectors integrate over the entire space-time history of a collision - only the dynamics itself determines how the final-state phase space will be populated with particles.

If the detected particles are hadrons, then the dynamics of the strong interaction makes it probable that the last interaction occurred late in the history of the collision and in the outer parts of the system. To infer the state that the system had earlier in its history from information that is mostly about its latest stages requires that one understand the dynamics. (But then one would not need the experiments!) While a complete theory accounting for particle production, scattering, absorption and escape must be consistent with the final-state hadronic observables, the necessity to understand all stages of the evolution in order to study the most interesting parts make this a daunting task.

A desirable probe should not be subject to multiple interactions within the nuclear medium (therefore not a hadron) but must have a detectable yield, which eliminates particles with only weak interactions. Thus only the electromagnetic interaction remains, and one must consider real photons and charged leptons. Detection of single real photons is difficult because of the large yields from $\pi^\circ \to \gamma\gamma$, and it is not possible to arrange that an external lepton scatter from the system at a relevant space-time point during the collision. The only remaining candidate is the virtual photon, which decays into a lepton pair. The masses and momenta of the virtual photons will be of the same magnitude as the excitation energies and the momenta associated with bulk motion, which are thought to be in the range of a few hundred MeV. For fixed target experiments with beam energies $T_{beam} < 15$ A GeV, it

* Supported by USDOE under contract DE-AT03-81ER40027, PA DE AM03-765SF00034

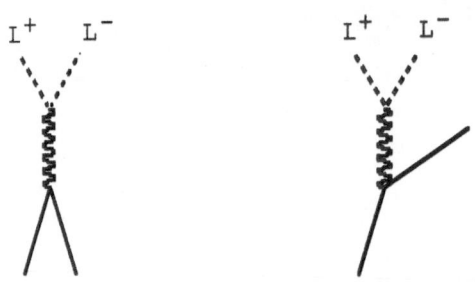

Figure 1. Diagrams for annihilation and bremsstrahlung.

is easier to measure (with adequate kinematic resolution) electron pairs than muon pairs. Thus there are strong arguments that fix the choice of the relevant probe for nuclear collisions in this energy range. The yield of lepton pairs is, however, even lower than that of real photons because of the extra electromagnetic vertex; the experimental requirements that result from this will be discussed below.

There are only two diagrams which describe the mechanisms that produce lepton pairs - annihilation and bremsstrahlung (Fig. 1). If the parent particles are point-like (quarks or leptons), then QED gives a complete description of the cross sections. In a Fermi gas of such particles, the yield ofelectron pairs increases rapidly with temperature and density (ie, energy and frequency of the interactions). (The increased yield that occurs during the interesting stages of the collision will be offset somewhat by the larger space-time extent of the later stages, but the kinematic information conveyed from these two regions will be quite different.) If the parents of the virtual photon are hadrons, then the vertex coupling the hadrons to the virtual photon must be described by a form-factor that accounts for the complexity of the underlying QCD structure. Thus, to evaluate electron pair production in nuclear collisions, one first needs to understand (or be able to parametrize) the 'elementary' production by free hadrons.

A complete kinematic description of a lepton pair (virtual photon) requires 6 independent variables, (eg, the 3-momenta of the two particles, the masses being fixed) in contrast to the case of a real photon, where there are only 3 independent variables. The mass and momentum of the virtual photon may be specified independently, with the remaining variables being the two angles specifying its production direction, and the two angles of the decay in its rest frame. It has become customary (at least for data taken at high incident energies) to use as kinematic variables: the mass, M; the transverse momentum, P_t; and a longitudinal variable, either the rapidity - Y, or the scaling variable - X. Data are almost always presented as a function of only one of these variables. This may be due to the difficulty inherent in presenting multi-dimensional data, to poor statistics, to limited experimental acceptance, or to some combination of these reasons. It should be noted, however, that these one-dimensional projections always represent integrations over the remaining variables - integrations within which the shape and limits of the acceptance of the experimental apparatus must be taken into account. The effects of the acceptance must be understood in comparing experimental data either to theory or to results from experiments with different coverage of the total phase space. (Comparisons of differing regions of phase space can be made *only* with the aid of a theory or model.)

Although we have a strong interest in studying nucleus-nucleus collisions using this tool (some first steps in this direction will be discussed later), most of the discussion presented here will be about our measurements in p+Be collisions. It was necessary for us to make these measurements because, in the relevant range of incident energies, the yield of electron pairs from nucleon-nucleon collisions had not been measured, nor was the hadronic production mechanism understood. A brief discussion of lepton production in hadronic collisions will help put our results in a proper framework.

I begin by listing the leptons which have *not* been of interest in discussions of previous

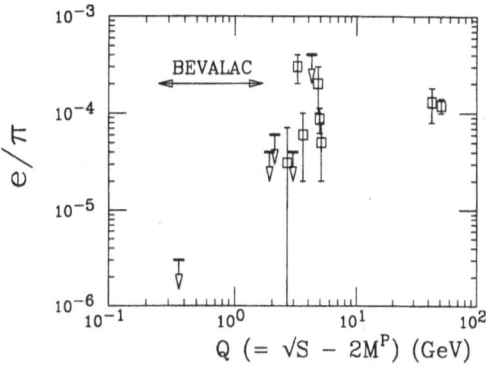

Figure 2. Energy dependence of (e/π) ratio, showing possible threshold behavior.

data - ie, those which arise from the decay of known hadrons. Leptons produced in Dalitz decays of π°'s (η's, etc), in two-body decays of vector mesons, or in semi-leptonic decays, etc, are not the object of discussion here. (Nor are the leptons arising from photon conversion in material of the target or detectors.) The leptons which remain after these sources have been excluded are referred to as 'direct' (ie non-decay) leptons even though they may also come from decays of yet-undiscovered resonances. Direct production has been observed in measurements of inclusive production of 'single' leptons at energies of [2.3 < Q (= $\sqrt{s}$ - 2M_p) < 51] GeV.[1] ('Single' lepton - one that is not a member of a 'low-mass' pair - the mass limit varies between experiments; leptons from 'high-mass' pairs are not excluded from these data.) In this energy range, and for P_t > 1 GeV/c, the ratio of the cross section $(d^2\sigma/dP_t dY)_{Y^*=0}$ for direct electron production to that for pion production (frequently called simply the 'e/π' ratio) is found to be approximately 10^{-4}. A measurement of this ratio at Q = 0.36, on the other hand, found an upper limit of 3×10^{-6}, raising the question of a threshold in the production process.[2] This data is summarized in Figure 2, which also indicates the range over which the LBL Bevalac can provide data.

Measurements of lepton-pair production offer information supplementary to that obtained from the yield of 'singles'. Figure 3a illustrates the typical features observed in high-statistics, high-energy experiments. Peaks from the two-body decays of the vector mesons stand above a continuum, which at large masses is well described by the Drell-Yan model.[3] (In this model the virtual photon is produced in the annihilation of a quark from one hadron with an anti-quark from the other. The approximation of asymptotic freedom used to permit the calculation of cross sections is expected to limit the validity to pair masses above 3-4 GeV. The extrapolation to lower masses thus reflects only what might be expected if the quarks within the hadrons always behaved as if they were unconfined.) At lower masses, the part of the continuum that is not produced by the kinds of decays mentioned above (ie the 'direct' part), has been referred to as "anomalous low-mass pairs", indicating that it has not been possible to account for this yield on the basis of known sources. For example, calculations of hadronic bremsstrahlung fail to account for these pairs.[4] Calculations of the yield from annihilation of partons[5] or virtual mesons[6] that are produced in the collisions reproduce some features of the existing data, but because of poor statistics, no strong tests of these theories has yet been possible.

Figure 3b shows an example of the data available for the production of electron pairs, in this instance from p+Be collisions at 12 GeV.[7] Several typical features are apparent: the poorer statistics of the electron sample; the calculated subtraction necessary to remove the contribution of the η and ω Dalitz decays (dot-dash curve); and the dashed curve due to the phenomenological model of Kinoshita, Satz and Schildknecht (KSS).[8] The KSS model, which pieces together ad hoc assumptions about the E_t and mass dependences with an overall phase space factor to account for energy conservation, gives a good representation of the existing data. Reviews of lepton production in hadron collisions may be found in Mikamo[9] and Stroynowski[10].

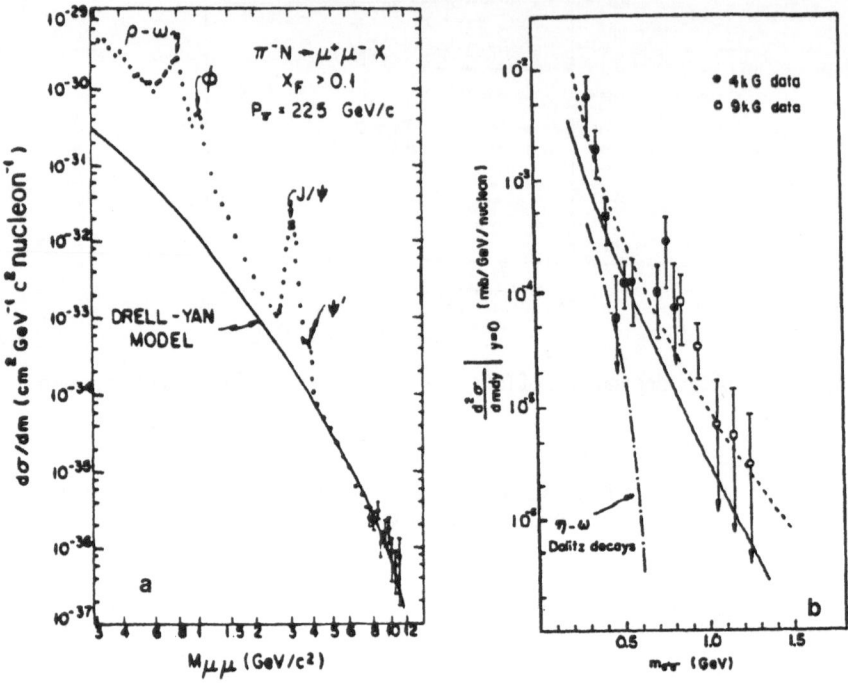

Figure 3. Mass spectrum of a) muon pairs from the Chicago-Princeton collaboration, and b) electron pairs produced in 12 GeV p+Be collisions (Ref. 7).

DLS DATA

The DLS (Di-Lepton Spectrometer)[13] was designed to carry out a program of measuring electron-pair production in both p+Be and nucleus-nucleus collisions. It consists of two arms (each mirror-symmetric around its central axis) positioned symmetrically about the beam axis. Within each arm are a large aperture dipole, two scintillator hodoscopes, two one-atmosphere Cherenkov arrays, and three drift chamber stacks. The solid angle of each arm is about 200 msr. The granularity of each detector element was chosen to be adequate for the multiplicities expected in 2.1 A GeV Ca+Ca collisions. The data shown here were obtained during three running periods: December, 1986; May 1987; and January, 1988. These and other data are included in a number of papers from the DLS collaboration[14], and the reader is referred to these for more detail.

The pair statistics from p+Be collision accumulated during this time are shown in Table 1. The DLS data sample was accumulated with significantly less running time than any of the three preceding experiments; the increased sample size is due entirely to the large

Table 1. Existing low mass electron-pair data.

Source	Reaction	# Unlike-sign	# Like-sign	# True
Adams[11]	17 GeV π+p			165
Blockus[12]	16 GeV π+p			107
Mikamo[7]	12 GeV p+Be			144
DLS	4.9 GeV p+Be	732	201	531±31
	2.1 GeV p+Be	567	148	419±27
	1.05 GeV p+Be	263	111	152±19

acceptance of the spectrometer. The '# Unlike-sign', '# Like-sign' columns of this table refer to the necessity of removing the predominant background to these measurements - events in which an e+ and an e- are detected, as members of anelectron pair, but which came from different parents. Since essentially all electrons in the interesting energy range arise as pairs (ie, the weak decay contribution is negligible), this happens when one detects the electron (but not the positron) from pair A, and the positron (but not the electron) from pair B. (Events containing more than two detected electrons, about 1% of the final data, are discarded.) The inherent charge symmetry of this background and a charge-symmetrized operation of the spectrometer (equal amount of beam on target for each of the 4 possible magnet polarities) permits us to calculate the number of 'true pairs' (those which came from a single virtual photon) by subtracting the like-sign ('false') pairs from the opposite-sign pairs. Some fraction of the remaining 'true pairs' will be from sources that are of no particular interest in themselves (eg, high-mass Dalitz decays of known hadrons). These contributions must be calculated from known cross sections, branching ratios, form factors, etc. The contributions from the two-body decays of the vector mesons, which in themselves may be of considerable interest in nucleus-nucleus collisions, are readily identifiable as peaks in the mass spectrum.

Before turning to discussion of the differential cross sections, it is worth noting that while the fully-differential cross section is five dimensional, the graphically presented cross sections represent projections onto one dimension, and therefore have been integrated over four kinematic variables. If the cross section extends into regions that are not covered by the acceptance of the experiment then the result of integrating the (correctly) measured cross sections will not necessarily resemble the result of integrating the real cross section over the full kinematic range.

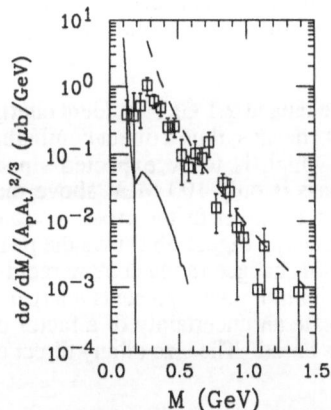

Figure 4. DLS data. Projected mass spectrum for 4.9 GeV p+Be
collisions. Dashed curve fitted through KEK data at 12 GeV[7].
Solid curve - calculation of Dalitz background.

Figure 4 shows the DLS data from p+Be collisions at 4.9 GeV, projected onto the pair-mass variable. (All data to be shown have been normalized by $(A_p A_t)^{2/3}$ to give an effective nucleon-nucleon cross section.)[10] The solid curve represents a calculation of the yield of pairs from Dalitz decays. For masses above 200 MeV, this contribution is always less than 10% of the measured yield. Thus there is indeed a direct, continuum signal at 4.9 GeV. In this data one observes a structure at the ρ mass, and, for masses above 300 MeV, good agreement in shape with the dashed curve which is a fit of the KSS model[8] to the KEK data at 12 GeV.[7] A smaller yield is to be expected at our lower incident energy. The turnover of the cross section for masses below about 275 MeV, however, is a completely

421

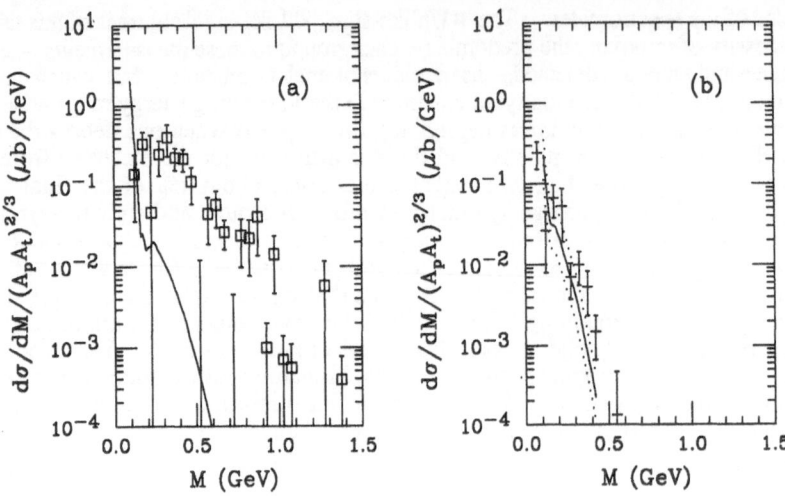

Figure 5. DLS p+Be data. Projected mass spectra at a) 2.1 GeV, and b) 1.05 GeV.
Solid curve - calculation of Dalitz background.

unexpected feature, although it is not in disagreement with earlier measurements, which either
have little data in this mass range, or are limited by a nearby threshold (in the case of mu-pair
measurements). Although the appearance of the structure is enhanced by subtraction of the
calculated Dalitz contribution, we report only the measured cross sections because the Dalitz
calculation has uncertainties that are not related to this experiment. The width of this feature
is somewhat larger than the experimental resolution; the significance of the structure will be
discussed later.

Figure 5a shows the p+Be data at 2.1 GeV incident energy. The measured yield is well
above that calculated for Dalitz decays, thus a direct continuum yield exists at this energy as
well. No ρ signal is seen, which is to be expected since the available energy in the
nucleon-nucleon center of mass is only 100 MeV above the ρ mass. There is also some
evidence in this spectrum of a turnover of the cross section near 275 MeV, although the
statistics are somewhat poorer here. Figure 5b shows the p+Be data at 1.04 GeV. Not only
has the character of the spectrum changed (a much more rapid decrease with increasing mass)
but the calculated Dalitz contribution now represents a larger fraction of the measured yield.
The curves in the figure indicate an uncertainty of a factor of two in the cross sections on
which the Dalitz calculation is based. The size of any direct continuum signal at this energy
is therefore quite uncertain.

At the two higher energies, the $P_t{}^2$ dependence of the cross section is similar to that
observed at energies above 10 GeV. At 1.04 GeV both the statistics and the P_t range are too
limited to permit any conclusion.

I now return to a discussion of the significance of the apparent turnover in the 4.9 and
2.1 GeV cross sections for masses below about 275 MeV. To quantify the statistical
significance of the turnover one must calculate the effects of the experimental acceptance
upon some 'structureless' cross section and compare this to the measurements. To account
for the effects of the dependence of the acceptance on variables other than mass, a usable
model must give a complete kinematic description of the cross section. While the evaluation
of the acceptance is straightforward, the choice of an appropriate model for the cross section
is not. We have found that parameterizations of presently available models are not
sufficiently constrained; they do not allow us to give a meaningful value to the statistical
significance of the turnover.   The question of the agreement between models and all existing

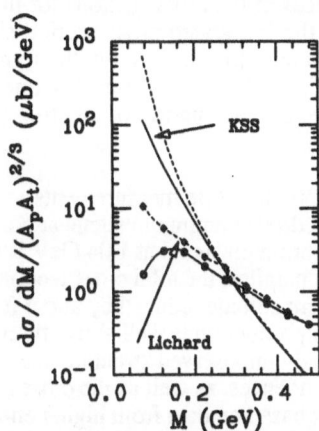

Figure 6. Projected mass spectrum calculated from Ref. 15, and from the KSS model (Ref. 8) with parameters of Ref. 7. Solid curves show effect of including effect of DLS acceptance.

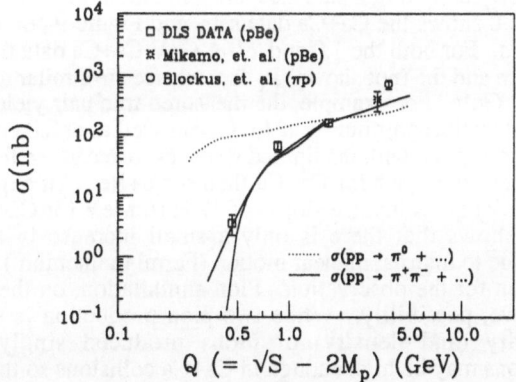

Figure 7. Integrated cross section vs. available energy in nucleon-nucleon center of mass. Curves show total cross section for inclusive production of: single pions (dots); and pion pairs (solid). Pion cross sections have been normalized to the DLS data point at $Q \approx 2$ (ie divided by about $\alpha^2/3$).

data needs to be addressed on a wider scale. We can show (Fig. 6) that for two models (that of Ref 15, and that of Ref 8 with parameters from Ref 7), the experimental acceptance has little effect on the shape of the mass spectrum for M > 150 MeV.

Figure 7 shows the integrated cross sections for the DLS measurements as well as those of Blockus, et al.[10], and Mikamo, et al.[7]. Also shown for comparison are two curves representing the energy dependences of the cross sections for inclusive production of single pions and pion pairs. Note that the integrated cross section shows a sharp drop near the threshold for two-pion production and does not follow the rather flat production of single pions down to lower energies. This behavior and the structure observed at twice the pion suggest that, in this energy range, the production of electron pairs may be mostly due to annihilation of pairs of real pions.

To summarize our findings from the p+Be measurements:
- We find clear evidence for a direct continuum signal at both 4.9 and 2.1 Gev.
- The evidence for a direct continuum signal at 1.04 GeV is less convincing.
- The integrated cross section parallels the total cross section for inclusive production of pion pairs with a magnitude reduced by about ($\alpha^2/4$).
- At energies where two-pion production is well above threshold, the mass spectrum of electron-pairs shows previously unobserved structure near $2m_\pi$. Other than this structure, the mass and P_t dependences, as well as the cross section for ρ production, are consistent with expectations based on data from higher energies.

The existence of the electron-pair signal at these energies makes possible the use of pairs as a probe in the study of A+A collisions at the Bevalac and SIS; the suggestion that this yield may be from annihilation of pions makes this possibility exciting. These data also demonstrate some of the advantages of using a detector with large acceptance for this type of measurement.

As stated earlier, some first steps have been made toward making measurements for A+A collisions. The pair statistics obtained from our first Ca+Ca runs are given in Table 2. The effects of the increased pion multiplicity on the relative magnitude of the like-sign pair background and the consequently larger statistical error in the subtracted data are apparent in these numbers. Figure 8 shows the Ca+Ca data sets, and Figure 9 compares the 1.0 GeV data for p+Be and Ca+Ca. For both the 1.0 and 2.0 A GeV Ca+Ca data the general features of both the mass spectrum and the (not shown) P_t dependence are similar to those obtained in p+Be collisions above 2 GeV. For example, the measured true pair yield from Ca+Ca lies well above the calculated Dalitz contribution at 1.0 Gev as well as at 2.0 GeV, which was not true for p+Be. Furthermore, even with the limited statistics, one can see that for the 1.0 GeV data the average pair-mass is higher for Ca+Ca than for p+Be. An exponential fit to the spectra (for M > 200 MeV) gives inverse slopes of 125±16 MeV for Ca+Ca and 71±18 for p+Be. A calculation shows that there is only a small increase in average available center-of-mass energy due to internal nuclear motion (Fermi momentum), and thus does not offer a ready explanation for the observation. Pion annihilation, on the other hand, is an interesting, if speculative, possibility. While two-pion production is suppressed at this energy, the multiplicity (and density) of pions produced singly in independent nucleon-nucleon collisions may be high enough in Ca+Ca collisions so that the annihilation mechanism can account for the increased yield at higher masses. If this hypothesis were correct higher statistics would show a ρ peak not seen in p+Be or in nuclear collisions with very small A. We find that, for total yield at 1.0 A GeV, the ratio of Ca+Ca to p+Be shows a dependence $(A_pA_t)^{1.0\pm0.1}$. (Note that this is in disagreement with the $(A_pA_t)^{2/3}$ scaling that we have used in reporting effective nucleon-nucleon cross sections from the measured data.)

Table 2. DLS Pair Statistics for Ca+Ca data

T (A GeV)	# Unlike-Sign	# Like-Sign	# True
2.0	94	45	49±12
1.0	731	476	255±35

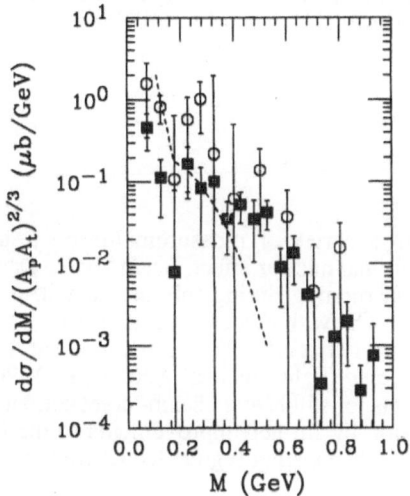

Figure 8.   DLS data. Projected mass spectra for Ca+Ca collisions at: 2.0 A GeV
(circles); and 1.0 A GeV(solid). Dashed lines are calculated Dalitz yields.

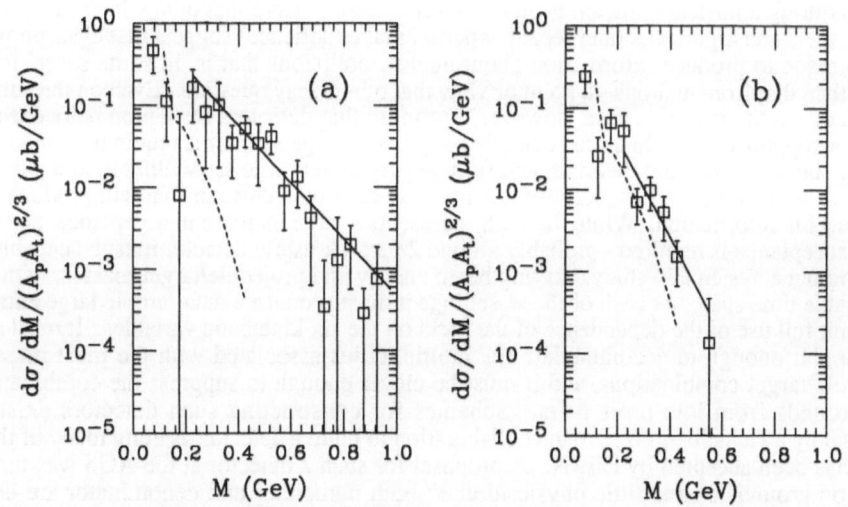

Figure 9.   DLS data. Projected mass spectra at 1.0 A GeV for collisions of Ca+Ca (a),
and p+Be (b). Solid lines are fitted exponentials, and dashed lines are
calculated Dalitz yields.

These first data show the possibility of experiments using e^+e^- pairs as a probe of nuclear collisions at beam energies for which the full mass range is presently available. Much better statistics will be required to permit detailed study of the collision process.

Because of the importance of making comparisons between theory and experiment using the full dependence of the yield on all kinematic variables, and because of the necessity to account for the experimental acceptance in making these comparisons, we will make both published data and the experimental acceptance available to interested parties in computer readable form. At present the data and acceptance are kept in a three dimensional table with variables (M, P_t, Y).

FUTURE

In 1989 we will begin a series of measurements on p+p and p+d collisions to differentiate between hadronic and nuclear effects, and to try to extract the contribution from bremsstrahlung. Heavy ion running during this period will be devoted to acquiring a high-statistics data set for Ca+Ca (with associated multiplicity), and to pushing the system toward higher projectile/target masses. While the present system is well suited to measurements of p+N, p+A and 'light' nuclear systems, in 1990 and beyond it may be possible to make measurements for still heavier beam-target combinations and to improve the achievable statistical precision by further improvements to the instrumentation - eg, the replacement of one or more of the existing segmented Cherenkovs by RICH (Ring-Imaging CHerenkov) detectors.

In its first two years of operation the DLS has demonstrated the existence of useful yields of electron pairs for incident energies down to 2 GeV in the case of p+Be, and down to 1.0 A GeV for Ca+Ca, and it has also demonstrated both the feasibility and the advantages of using a device with large solid angle. For the long term future of studies of relativistic nuclear collisions, it may be the latter fact that is the more important. There are quite general, fundamental arguments that the sensitivity of lepton pairs to the evolution of the system formed during a nuclear collision is qualitatively different from that of any hadronic probe. To me, the same arguments (and recent experimental experience) suggest that electron pairs can be made to produce information about nuclear collisions that is, in some sense, more useful than that from hadrons - a point of view that others may question. Even on the simple ground of qualitative difference, however, I propose that detectors for lepton pairs deserve parity of support with the hadronic detector programs. Large support is required - not only in funding, but in effort from theorists, and from experimenters who are willing to join existing groups or to form new ones - because it is not easy to build a detector that will produce this 'more useful' information. While the DLS represents a large increase in acceptance, an even larger acceptance is required - probably around 2π sr. A useful detector must be capable of carrying out a systematic study, varying beam energy and projectile/target masses, within a reasonable time span. At each of these settings it must produce a data sample large enough to permit full use of the dependence of the yield on the six kinematic variables. It must also be granular enough to accommodate the multiplicities associated with the most massive projectile/target combinations, and it must be clever enough to suppress the combinatoric backgrounds from low mass pairs. Schemes for constructing such detectors exist: A proposal by a Heidelberg-Weizmann collaboration to build a detector meeting many of these goals has been accepted by CERN. A proposal for such a detector at the AGS was turned down on grounds of 'too little physics/dollar'; both numerator and denominator are being worked on. Both SIS and the Bevalac could use such a detector. The primary limitation is manpower, those of working enthusiastically in this field are actively seeking discussions, advice, calculations, techniques, and collaborators.

REFERENCES

1. A. Maki et al., Phys. Lett. 106B:423 (1981), and references therein
2. A. Browman et al., Phys. Rev. Lett. 37:246 (1976)

3. S. D. Drell and T. Yan, Phys. Rev. Lett. 25: 316 (1970)
4. R. Rueckl, Phys. Lett. 64B: 39 (1976); N.S. Craigee and H.N. Thompson, Nucl. Phys. B141:121 (1978)
5. B.J. Bjorken and H. Weisberg, Phys. Rev. D 13:1405 (1976); V. Cerny, et al., Phys. Rev. D 24:652 (1981)
6. T. Goldman, M. Duong-Van, and R. Blanckenbeckler, Phys. Rev. D20:619 (1979)
7. S. Mikamo, et al., Phys. Lett. 106B:428 (1981)
8. K. Kinoshita, H. Satz, and D. Schildknecht, Phys. Rev. D 17:1834 (1978)
9. S. Mikamo, in Tokyo INS Symposium 1979, 362 (also KEK-Preprint-79-27)
10. R. Stroynowski, Phys. Rept 71:1 (1981)
11. D. Blockus, et al., Nucl Phys. B201:205 (1977)
12. M.R. Adams, et al., Phys. Rev. D 27:1977 (1983)
13. The collaboration which has done the work presented here: J. Bystricky, J. Carroll, J. Gordon, T. Hallman, G. Igo, P. Kirk, G. Krebs, E. Lallier, G. Landaud, A. Letessier-Selvon, L. Madansky, H. Matis, D. Miller, C. Naudet, G. Roche, L. Schroeder, P. Seidl, Z.F. Wang, R. Welsh, and A. Yegneswaran
14. G. Roche et al, Phys. Rev. Lett. 61:1069 (1988); C. Naudet, et al, submitted to Phys. Rev. Lett.; A. Letesier-Selvon, et al, submitted to Phys. Rev. C; G. Roche, et al , submitted to Phys. Lett. B
15. P. Lichard, private communication; V. Cerny, et al., Phys. Rev. D24:652 (1981)

STATUS OF THE SIS/ESR – PROJECT AT GSI

P.Kienle

Gesellschaft für Schwerionenforschung mbH
D-6100 Darmstadt, West Germany

This is a report on the status of the construction of the SIS/ESR– accelerator complex and the experimental equipment at GSI as of February 1989.

1 Introduction, Overview

At GSI, heavy ion research will be extended to relativistic energies. Fig. 1 gives an overview of the heavy ion accelerator complex under construction [1]. It consists of an upgraded UNILAC used as an injector into a medium energy (1–2 GeV/u) heavy ion synchrotron SIS 18 [2,3] which is connected with a storage cooler ring ESR [4] of half the circumfence of SIS 18.

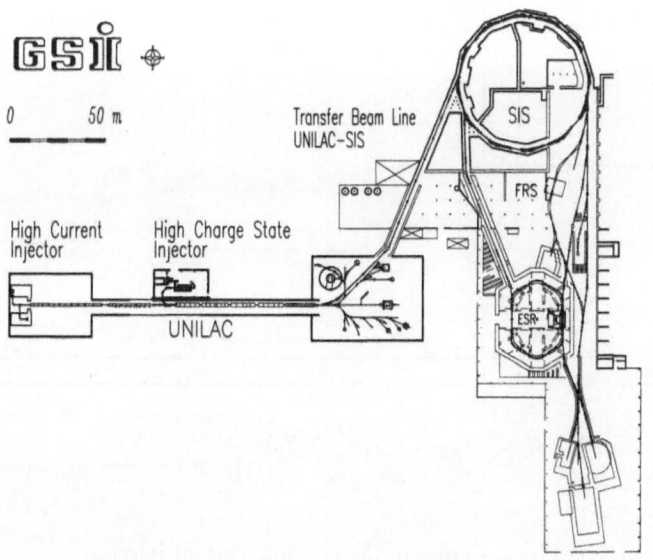

Fig.1. Layout of the upgraded UNILAC, SIS, ESR. The fragment separator (FRS) and the new experimental area are also indicated.

The combination of these two rings will allow to produce completely stripped heavy ion beams up to U^{92+} with the highest possible phase space densities, achieved by various beam cooling techniques. In addition SIS/ESR will provide beams of radioactive nuclei in the energy range from several MeV/u up to 1–2 GeV/u, again cooled to the highest possible phase space densities. The beams of the ESR may be used either circulating with high currents or extracted with a great variety of time structures and intensities. They may also be reinjected into SIS for further acceleration or deceleration. There will be a large experimental area with several experiments set up on beams from both SIS and ESR. Further experimental areas are located directly behind SIS, between SIS and ESR, and around the ESR. In future one can think of injecting the high phase space density beams of completely stripped ions into superconducting collider rings with small apertures, modest size and cost to achieve very high cm–energies ($\geq$ 20 GeV/u).

2 Accelerators

2.1 UNILAC Upgrade

Very recently we changed our injector concept into the UNILAC in such a way that we can run a truly independent low energy program with a free choice of ion species and energy parallel to a low duty–factor high current injection cycle into SIS 18. The SIS injection is based on recently developed high intensity ion sources [5] for low charge states (U^{2+}) which will be accelerated by 27 MHz RFQ structures up to 130 keV/u and after stripping injected straight into the second Wideröe tank. This high current injector will be operated with a duty–factor of 1%, which is sufficient for synchrotron injection. It can provide 100–1000 times more injection current than the present UNILAC.

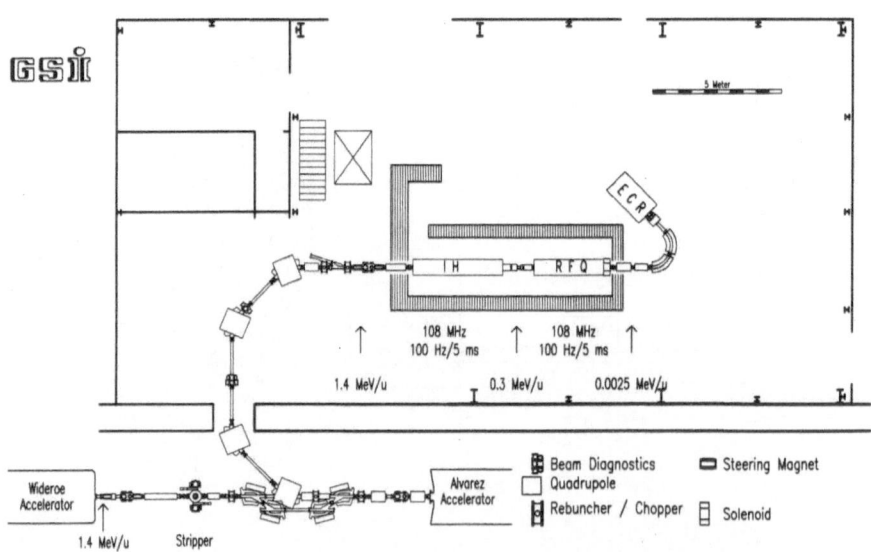

Fig.2. Layout of the new low–current injector.

For the low energy UNILAC–program we construct an independent injector (Fig.2) [6], which consists of a 14.5 GHz ECR–source, a RFQ linac for energies up to 300 keV/u, followed by an interdigital line structure up to the injection energies of the Alvarez section (1.4 MeV/u). These structures will be operated at 108 MHz and a duty–factor of 50%. The U^{28+}-current is specified as $5\mu A$, which is more than an order of magnitude larger than the one presently available. It also has an improved microstructure delivering a pulse each 9 ns, which is favourable for coincidence experiments with fast detectors. In addition it can be debunched, thus achieving dc beams, if desired.

2.2 Heavy–Ion Synchrotron SIS

Since April 1985, a heavy–ion synchrotron (SIS 18) is under construction. It has a bending power of 18 Tm and a circumfence of 216.72 m.

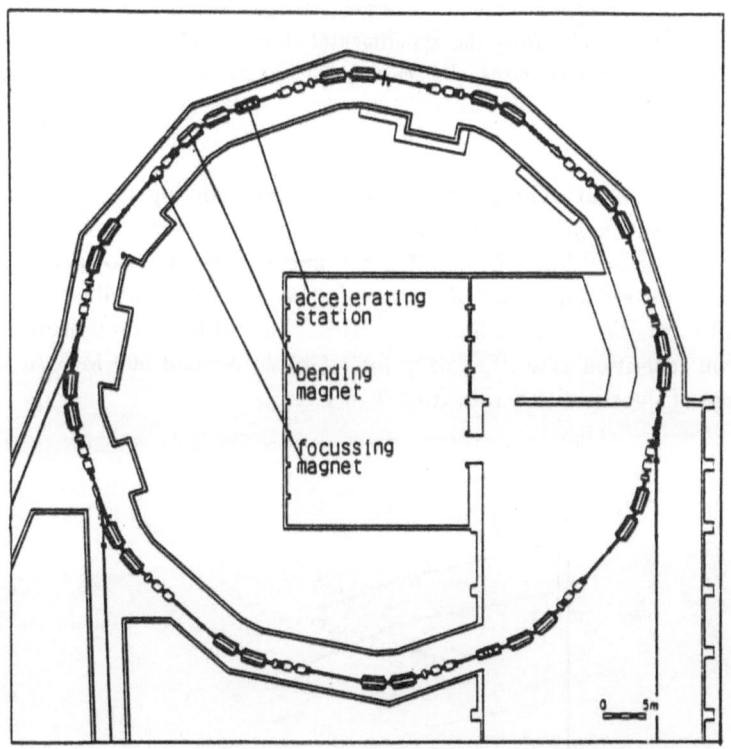

Fig.3. Layout of the synchrotron for heavy ions (SIS 18).

The heavy ion beam accelerated in the UNILAC up to 11.4 MeV/u, and stripped to an adequate high charge state for the desired energy and intensity, is injected into SIS 18 during 10 to 30 turns and then accelerated with a repetition rate between 3 Hz (up to 1.2 T) and 1 Hz (up to 1.8 T) to maximum energies, depending on the charge states of the ions as shown in fig.4.

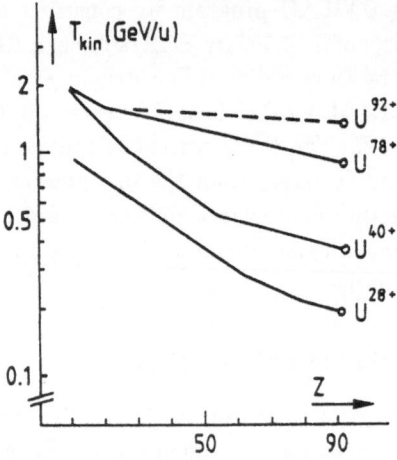

Fig.4. Maximum achievable energies at SIS 18 as a function of nuclear charge. The energies are given for a gas or a foil–stripper at an energy of 1.4 MeV/u, resulting in a relatively low degree of ionization. If a second stripper at 11.4 MeV/u is added or if completely ionized particles from the experimental storage ring ESR are reinjected into the synchrotron, higher energies can be achieved.

For uranium ions at a charge state of q=78, after stripping at 11.4 MeV/u with a foil target, 1 GeV/u is achieved as maximum energy. The maximum beam intensities from SIS 18 are shown in fig.5 for Ne– and U–ions of various charge states, depending on the stripping procedure, as a function of their specific energies. The decrease of intensities towards higher energies is caused by a small decrease of the synchrotron repetition rate. The drop for 1 GeV/u Ne and 500 MeV/u U is due to a change of the repetition rate from 3 to 1 Hz.

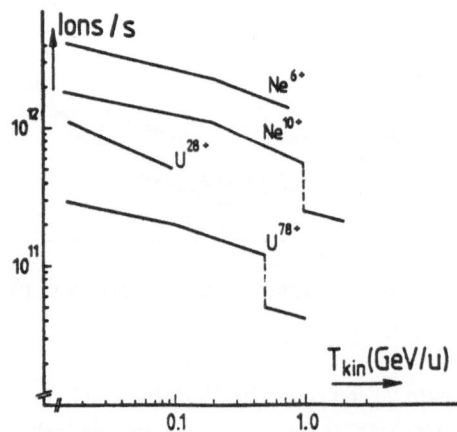

Fig.5. Beam currents for various charge states of Ne– and U–ions, obtained by different stripping procedures as a function of the energy. The intensity drops for Ne^{10+} and U^{78+} are due to a decrease of the repetition rate from 3 Hz to 1 Hz.

Between SIS and ESR the beam may be stripped once more to the highest desired charge state. The ESR with a bending power of $B\rho = 10$ Tm allows to store ions up to U^{92+} with the following maximum energies: Ne^{10+} (834 MeV/u), Ar^{18+} (709 MeV/u), Kr^{36+} (656 MeV/u), Xe^{51+} (609 MeV/u), and U^{92+} (556 MeV/u). The uranium ions can be fully stripped at this energy in a Cu-target of 100 mg/cm^2 thickness with an efficiency of 60% [7]. The stripping yield increases strongly with decreasing nuclear charge, thus one expects a yield of 70% for Pb^{82+}-ions (574 MeV/u), and already 100% for Xe^{54+} (609 MeV/u). Alternatively one can install a reaction target for projectile fragmentation. The favourable kinematic focusing of the products around the beam direction and velocity allows an effective mass separation in the projectile fragment separator (FRS; see sect.(3.1) below) between SIS and ESR, followed by accumulation of radioactive beams with the ESR, which accepts beams with $\Delta p/p = \pm\,0.5\%$ and transverse emittances of 20 πmm mrad.

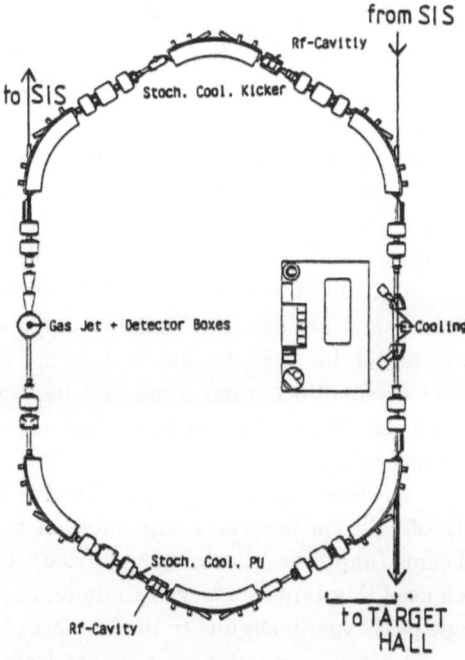

Fig.6. Layout of the experimental storage ring (ESR) for heavy ions.

The whole SIS-ring has been installed in the tunnel and connected to power supplies and cooling circuits. The transfer line between UNILAC and SIS was taken in operation in fall of 1988. On November 23rd, 1988 at 3.10 pm an argon beam was injected into SIS for the first time. The experimental program at SIS is expected to start in December 1989, three years after the start of the construction.

2.3 The Storage and Cooler Ring ESR

The ESR (Fig.6) with a circumfence of 108.36 m has two 9.5 m long straight experimental sections, in one of which an electron cooling device will be installed. The other four straight sections will be used for the installation of rf cavities, and slow and fast extraction elements. The rf cavities are used for acceleration, deceleration, and especially also for bunching the beam together with the electron cooling for reduction of the occupied longitudinal phase space volume. With the fast extraction system of the ESR one can transfer a highly ionized and cooled beam back to SIS 18 for further acceleration or especially also deceleration. The optics of the ring allows three modes of operation: one with a moderate dispersion along the ring specially suited for the accumulation of beams with large momentum spread ($\Delta p/p = 1\%$) and emittance ($E_{hv} = 20$ πmm mrad), one with zero dispersion in the straight sections, which allows multi-charge operation ($U^{89+} - U^{92+}$), and one with large dispersion to accomodate two beams of slightly different momenta, which then may be brought to merge with a well defined angle of about 100 mrad [8]. This can be used to study collisions of *two* highly ionized beams at fixed target equivalent energies of up to 7.2 MeV/u and an energy definition better than 10%.

The most important facilities of the ESR are various cooling devices which can be applied complementary. For secondary beams with low phase space density stochastic pre–cooling may be used. For cooling to very high phase space density, electron cooling of completely stripped heavy ions is foreseen in an interaction zone of 2 m length. A "cool" electron beam of 5–10 A is focussed within an area of 5 cm diameter collinearly along the ion beam at the corresponding average velocity. For cooling of ion beams between 30 MeV/u and 560 MeV/u, electron energies in the range of 16.5 keV and 310 keV are required. With an electron beam current density of up to 1 A/cm^2 and ion beams of initially $\Delta p/p = 0.1$ % and 4 πmm mrad cooling times of 30 ms for U^{92+} at 500 MeV/u are expected. Heavy ion beams with emittances as small 0.1 πmm mrad and momentum spreads of less than 10^{-5} may be produced. Space charge effects limit the number of ions to be cooled in a circulating beam [9].

While the cooled beam circulates in the ring, it may be used in the second straight section for the study of collision processes with internal targets, which may be atomic or electron beams (unpolarized or even polarized), gas jets or fibres. For all experiments which need thin targets a large gain in luminosity may be achieved compared with a single pass experiment due to the increase of the circulating beam current ($\approx 2 \times 10^6$). Also the interaction of collinear laser and electron beams with the circulating ions of high intensity and small momentum spread may be favourably studied.

The ESR is constructed parallel to SIS. The magnets, power supplies, UHV systems, and the electron cooling device are under construction. In summer 1988 the installation of the ESR magnets has begun with the goal to start with commissioning and cooling experiments after SIS has started to be used for experiments. Already the initial experimental program at the ESR will be quite diversified. It will start with electron cooling of heavy ion beams to find out the limits of phase space densities which can be achieved. The ultimate aim will be to reach a plasma parameter $\Gamma = (Z^2 e^2/a)/kT$ (with Z being the charge of the ion, a the Wigner–Seitz radius, and kT the temperature) as large as possible in order to come close to a critical value ($\Gamma_c \approx 170$) at which a transition to a crystalline state should occur. Unfortunately the lattice of the ESR will introduce shears due to the bending of the beam in the dipole magnets so high that a complete order may not be achieved.

Another goal is the cooling of radioactive beams which are delivered from the fragment separator. These beams which have low phase space densities, will be pre–cooled by stochastic cooling and then brought to still lower temperature by electron cooling.

By performing Schottky–scans of the revolution frequency of the radioactive cooled beams, mass determinations with an accuracy of 10^{-5} to 10^{-6} should be possible. For very short lived activities mass measurements by time of flight techniques in an achromatic operation mode of the ESR are proposed.

With cooled radioactive beams various nuclear reaction experiments are proposed. One class being precision reaction spectroscopy experiments at energies up to 200 MeV/u, using inverse reaction kinematics. Proton, deuteron or $^{3,4}He$–targets may be bombarded with cooled radioactive beams. By measurement of the emission angle and the energy of the light reaction products, reaction spectroscopy with a resolution of about 50 keV may be carried out on radioactive nuclei.

With completely ionized nuclei a new decay mode, namely β–decay to bound states, may become observable in the ESR. Such processes may occur in highly ionized plasmas of stars and thus may be used as thermometers. Electron capture is the inverse process, therefore by studying bound state β–decay one may be able to measure interesting weak coupling matrix elements.

3 Experimental Facilities

3.1 Projectile Fragment Separator (FRS)

SIS will provide high–intensity beams of relativistic heavy ions. Secondary beams of radioactive isotopes can be produced by projectile fragmentation and electromagnetic dissociation. The projectile fragments are emitted in forward direction with velocities close to those of the projectiles, so that they can be separated in flight and injected into the storage ring ESR. The separator for projectile fragments (FRS, see fig. 7) is planned to separate isotopes up to uranium according to A and Z at specific energies in the range between 0.1 and 1.0 GeV/u. The principle of the isotopic separation is based on a combination of magnetic analysis (Bρ) and energy loss (ΔE) of the fragments in matter.

The keys for the separation are an achromatic magnetic system, characterized by a high resolving power independent of the velocity spread of the fragments (Δv/v = 1%) and a profiled degrader at the dispersive focal plane, providing the separation in Z and A. The separator, located in a beam transfer line from SIS to ESR (see fig.1) can deliver separated beams of projectile fragments either to the experimental area or to the ESR.

The FRS consists of 4 magnetic dipole stages with focussing quadrupoles. The production target is positioned at the entrance of the FRS. Radioactive beams with intensities up to $5 \times 10^8/s$ can be produced using target thicknesses of $\approx 1g/cm^2$. The primary beam has to be separated from the fragments at the focal plane of the first dipole section. Because all fragments have similar velocities the first sepa-

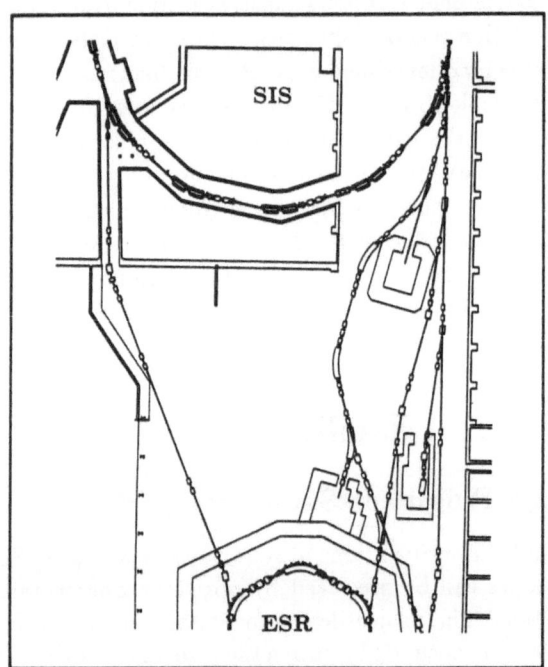

Fig.7. Layout of the fragment separator (FRS).

ration selects only those with a certain A/Z-ratio i.e. all fragments with the same magnetic rigidity are focused on the degrader. The energy loss of the fragments in the degrader provides the additional selection needed for the separation of a nucleid. The shape of the degrader is chosen to preserve the velocity achromatism of the system. The separation quality at the final focus is strongly dependent on the ion–optical resolution of the separator, the thickness and perfection of the degrader and the energy of the fragments. Separation in A and Z of all fragments up to mass 240 seems possible with low background.

The physics program at the FRS is expected to become very diversified. First we will focus on the study of the fragmentation process especially with heavy masses including Coulomb–dissociation using the relativisticly enhanced field of fast ions. Besides the measurement of A and Z distributions the study of the momentum and energy transfer on the fragment should shed light on the reaction dynamics. With separated isotopes detailed nuclear structure studies of heavy neutron rich nuclei should be posssible. Then of course high energy radioactive beams may be produced and used for reaction studies, in particular also in combination with the ESR, in which they may be cooled and decelerated.

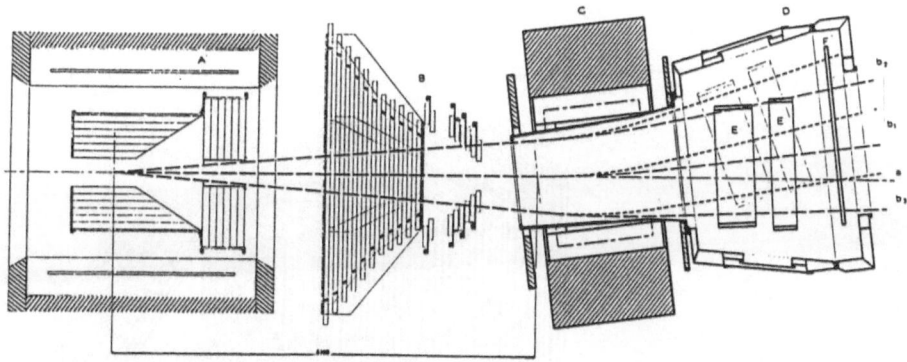

Fig.8. Layout of the complete 4π-detector including ALADIN.

A very different class of reactions which may be favourably investigated with the FRS are fusion reactions at high energies using inverse kinematics, like for example $^{12}C(p;\gamma,\pi^0)^{13}N$. The heavy fusion products emitted in a small forward cone may be identified and completely momentum analyzed with 100% detection efficiency. Thus, very rare processes can be investigated. Another interesting field is connected with the proposed study of Δ–production in quasielastic collisions, for which the FRS may be used as a high resolution spectrometer. Very rare processes like subthreshold production of K^- and antiprotons can advantageously be studied at the FRS.

3.2 4π–Detector

For the study of central collisions an advanced 4π-detector for charged particles including a forward–spectrometer (ALADIN, see sect.(3.3) below) a large
BaF_2–detector array (TAPS, see sect.(3.5) below) for high energy photon spectroscopy and a large area neutron detector (LAND, see sect.(3.6) below)is under
construction. This device is designed to measure the complete momentum flow
($d^3\sigma/d\vec{p}$) for all charged particles originating from a hard collision, which will allow to analyze in substantial detail the collective nuclear matter flow first observed
in exclusive experiments by Gustafson et al. [10].

A schematic lay–out of the complete detector system is shown in fig.8. The target
is placed in a large solenoid (2.4 m diameter and 3.34 m length) with superconducting coils, which produce a uniform magnetic field ($\Delta B/B\leq2.5\%$) up to 0.6 T. The
target is surrounded by a central drift chamber and a barrel of plastic scintillators
for time of flight measurements. There is additional space for Cherenkov–detectors,
which may be necessary for K^+–identification. In forward direction between 7°
and 30° a drift chamber which measures the transversal momentum components
of the particles and a "spaghetti–type" time of flight detector is placed. It consists
of 512 rods of plastic scintillators with cross sections of $20 \times 24 mm^2$ with fast
photomultipliers on both sides. They are arranged in an octogon shaped roof–like
structure as indicated in fig.9.

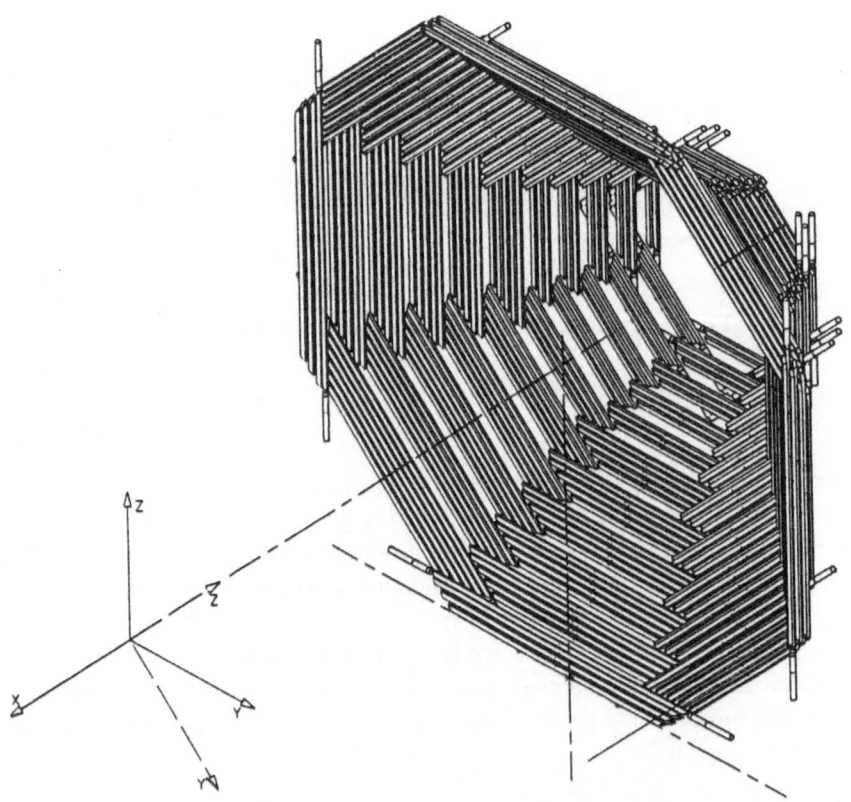

Fig.9. Perspective view of the forward plastic wall.

For a better Z–identification of larger fragments emitted in forward direction, ionisation chambers also are included.

Thus, the 4π–detector as a whole allows a momentum, velocity and energy loss determination of all charged particles with high granularity. All particles, including K^+–mesons can be identified and their momentum be determined on an event–by–event basis. To complete such an analysis a forward–spectrometer is added in the forward direction which covers the angular range between 0° and 5°.

3.3   Forward–Spectrometer (ALADIN)

The Forward-Spectrometer will be built from an already existing magnet which has been obtained from CERN as a temporary loan (MNP21). It has an aperture of $1.5 \times 0.5 \ m^2$ and a bending power of 2.3 Tm (see fig.8). This spectrometer, capable of detecting and identifying nuclear fragments up to the largest masses and momenta to be expected at SIS will complement the 4π–facility in the forward direction ($\pm 2°$ vertically,$\pm 5°$ horizontally).

As a tracking and Z–identification device a multiple sampling position sensitive ionisation chamber will be used. Time of flight is measured with a wall of 2×100 vertical plastic scintillator rods of 110 cm length.The momentum resolution of the forward spectrometer should be better than 1%, the ionisation chamber gives a complete Z resolution up to the heaviest atomic number and from the long time of flight (10 m) a good velocity resolution is expected.

The forward spectrometer will be used to study the mechanisms of multifragmentation. A special application will be the investigation of exclusive multifragmentation using inverse reaction kinematics.

3.4   K–Spectrometer

A double focussing QD magnetic spectrometer (fig.10) is being built to be installed in a separate cave in the experimental hall (see fig.1). Its primary purpose is to study in detail meson production in energetic collisions between nuclei. The compact design is especially matched to the requirements for kaon detection with short flight path ($\approx 5m$), large solid angle (20 – 35 msr), wide momentum accep-

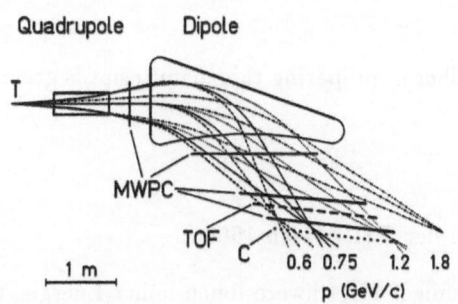

Fig.10. Schematic picture of the K-spectrometer.

tance ($\pm 30\%$), maximum momentum at 1.2 GeV/c (1.8 GeV/c at 10 msr), and modest momentum resolution ($\approx 1\%$ without and $\approx 10^{-3}$ with raytraycing). A focal plane length of about 1.5 m allows the efficient use of the detectors necessary for particle identification and raytraycing, involving wire–chambers, time–of–flight scintillators, aerogel and water Cherenkov detectors and segmented calorimeters for particle decay. While the primary purpose for the construction of the spectrometer is the measurement of kaons, it can as well be considered a general purpose magnetic spectrometer for other hadrons and for leptons. Its large solid angle also allows the study of two–particle (e.g. $\pi - \pi$) correlations.

3.5   Two Arm Photon Spectrometer (TAPS)

The production of γ–rays, π^0- and η^0–mesons will be studied with a Two Arm Photon Spectrometer (TAPS) consisting of four to six arrays of 64 BaF_2-crystals, each being 12 radiation lengths deep, which are arranged in 2 or 3 tower structures. High energy γ–spectroscopy may be a useful probe to investigate the temperature and possibly the energy density of the hot nuclear matter in an unambiguous way. It may also be possible to study directly the production and decay of barionic resonances. At higher bombarding energies the combinatorial background of many γ–rays from π^0–dacay may prevent single photon spectroscopy or will make it very difficult.

3.6   Large Area Neutron Detector (LAND)

A large area neutron detector (LAND) will be installed at the SIS- facility. It consists of a structure of subsequent layers of converter material (Fe, thickness 5 mm) and active plastic material (BC408, 5 mm thick) with a total thickness of 1 m. The total area of 2×2 m^2 is subdivided into 200 separate cells (paddles) of $200 \times 10 \times 10$ cm^3. Subsequent layers of the paddles are arranged perpendicular to each other. Each detector is read out by two 2" photomultipliers at the small front faces of the paddle.

The LAND will be used to study extremely peripheral collisions where multiple excitation of the Giant Dipole Resonance may populate exotic high–lying nuclear states, which will decay mainly by neutron evaporation. In these experiments the neutron detector will be placed at 0° and operated together with ALADIN. For the investigation of central collisions it may also be operated at various angles in conjunction with the 4π–detector.

Acknowledgement

The help of Dr.H.Ströher in preparing this manuscript is gratefully acknowledged.

References

[1] Die Ausbaupläne der GSI, March 1984

[2] SIS–Ein Beschleuniger für schwere Ionen hoher Energie, GSI–Bericht 82-2

[3] K.Blasche, D.Böhne, B.Franzke, H.Prange, 1985 Particle Acc. Conf., Vancouver, IEEE Trans. NS 32 (1985)

[4] B.Franzke et al., Zwischenbericht zur Planung des Experimentier Speicherrings (ESR) der GSI, GSI–SIS–INT/84-5, August 1984, and Information about ESR, GSI–ESR–TN/87-02

[5] R.Keller et al., Proc. Int. Ion Engineering Congress, Kyoto (1983)

[6] N.Angert, internal GSI–report

[7] H.Gould et al., Phys. Rev. Lett. 52 (1984) 180

[8] B.Franzke, Ch.Schmelzer, GSI–Scientific Report 1984, p.341

[9] I.Hofmann, GSI–Scientific Report 1985, p.387

[10] H.A.Gustafson et al., Phys. Rev. Lett. 52 (1984) 1590

HEAVY IONS AT SATURNE

Jacques Arvieux

Laboratoire National Saturne
Cen Saclay
91191 Gif-Sur-Yvette Cedex, France

INTRODUCTION

Saturne is a synchrotron accelerating particles of momentum p and charge z up to $p/z = 4$ GeV/c. Final energies are 3 GeV for protons, 2.3 GeV for deuterons, 5.6 GeV for ^{3}He and 1.15 GeV per nucleon for heavy ions with a charge to mass ratio of 0.5. Heavy ions range from Carbon to Argon. There are 3 sources and 2 injectors available.

SOURCES

The sources are shown schematically in fig. 1. HYPERION is a source housed in a 400 kV platform producing 500 µA of polarized protons and deuterons. AMALTHEE is a 800 kV source producing 30 mA of unpolarized light ions (p, d, ^{3}He, ^{4}He). DIONE is the heavy ion source held at 12 kV x A. The percentage of time requested by users is 60% for polarized particles, 20% for heavy ions and 20-30% for unpolarized light particles.

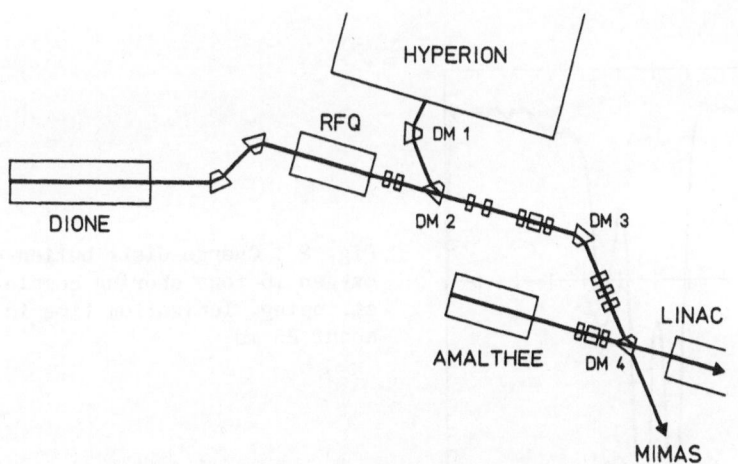

Fig. 1 . The three sources of SATURNE (see text)

CHOICE OF A HEAVY ION SOURCE

A synchrotron is a pulsed machine for which the injection phase is very short (less than a millisecond) compared to the overall cycle of the machine (1 to 3 secondes). The final energy is directly related to the charge state z of the particle accelerated (in fact the final momentum is proportional to z). So the ideal choice is a source which would be pulsed and which would give the largest number of ions in the highest charge state during the shortest time possible.

Our choice has been an EBIS type of source [1] producing highly charged ions, which are injected directly into Saturne without further stripping, contrary to ECR sources which produce large intensities of heavy ions but ion low charge states. Charge states obtained with ^{16}O ions are shown in fig. 2.

Ions with one positive charge are produced in a small ion source and then injected into DIONE where they cross an intense beam of electrons which are confined by a 5.6 T superconducting field. When they are fully stripped (up to Neon 10^+) or when some equilibrium between different charge states is achieved (e. g. Argon 16^+) the ions are extracted from DIONE, accelerated to 187.5 keV per nucleon by a RFQ structure and injected into MIMAS.

EBIS sources are characterized by 3 main quantities :
1.   the electronic density J produced by the electron gun ;
2.   the confinement time necessary to attain a given charge state ;
3.   the number of ions produced and their charge repartition.
Points 2 and 3 depend directly on the electronic density J. Simulations showed that we could hope to get J = 10 000 A.cm^{-2} leading to the obtention of up to 10^{11} charges in less than 10 ms for Argon ions. In fact our present limitation, due essentially to density distribution and non-linear plasma effects is of the order of a few 10^9 charges per pulse in about 100 ms.

THE MIMAS INJECTOR

The first injector of Saturne was a Linac built in 1968. It is still used for unpolarized light ions but it has been phased out for heavy and polarized ions operation, for which it has been replaced by a new injector and storage ring MIMAS, which came into operation in october 1987. MIMAS

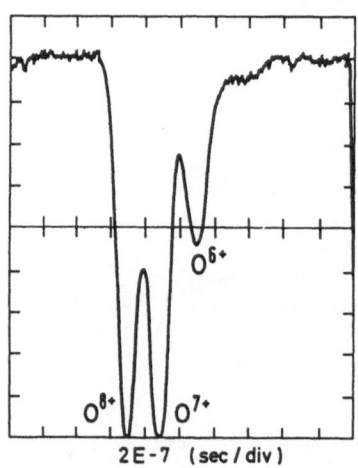

O^{6+}

O^{8+} O^{7+}

2 E-7 (sec / div)

Fig. 2 . Charge distribution of oxygen 16 ions showing complete stripping. Ionization time is about 25 ms

is a synchrotron of radius 5.85 m (1/3 of Saturne radius) which is seating inside the Saturne main ring (fig. 3). It can store many pulses (up to 5 until now) from DIONE or a long pulse (800 μs) from the D. C. polarized source Hyperion, on a low energy palier (at 187.5 keV x A), before a one turn injection into Saturne. This method requires a very good vacuum (5 x 10^{-11} t in MIMAS, 10^{-9} t in Saturne) which is obtained by backing MIMAS to 350 °C and Saturne to 80°C for about one week. The final intensity is given in table 1 for different ions.

HEAVY ION PHYSICS AT SATURNE

Only one short list of subjects will be given since most of these topics will be discussed at this school.
- Central collisions induced by heavy ions from ^{4}He to ^{40}Ar with the DIOGENE detector : equation of state of nuclear matter, interferometry, Π^+/Π^- production [2].
- Elastic and inelastic nucleus-nucleus scattering [3, 4].
- Reaction mechanisms above 100 MeV A : partial fusion, multifragmentation [5].
- Charge-exchange reactions induced by heavy ions : Δ-excitation, G. T. resonances, missing strength [6].
- Cooperative production of pions and kaons below threshold : compression of nuclear matter [7].

FUTURE OF HEAVY ION PHYSICS

In 1988 we have accelerated krypton 30^+ beams in MIMAS and they will be accelerated in Saturne in spring 1989. A first tentative experiment

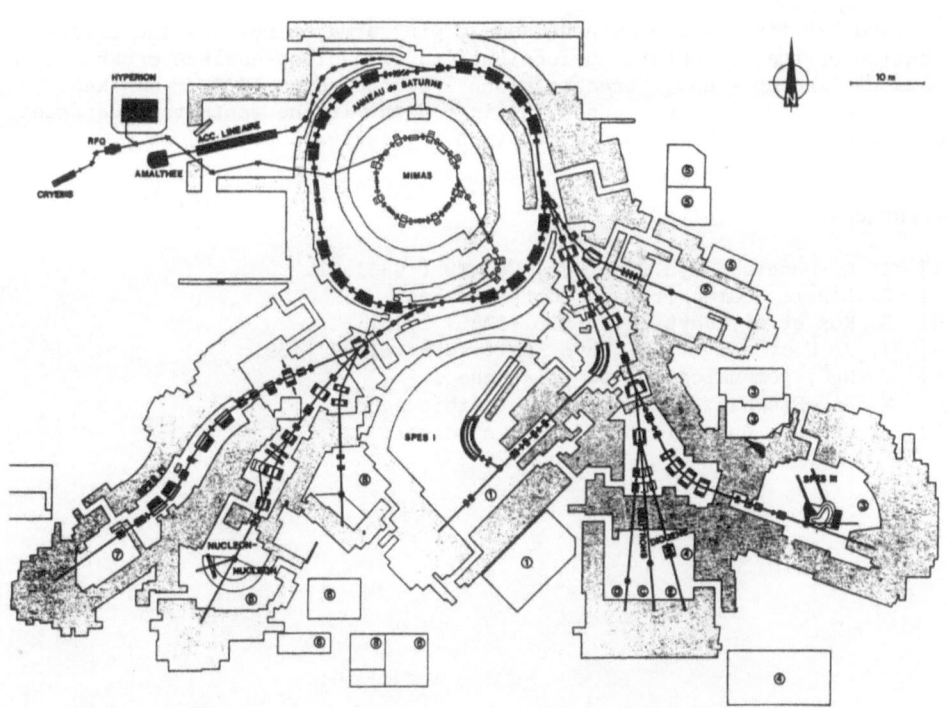

Fig. 3 . Injector storage ring MIMAS and SATURNE main ring

Table 1. Heavy ions available at Saturne

Ion species	Source exit (charges)	Number of injected pulses	Intensity in SATURNE
N^{7+}	7.10^9		
C^{6+}	6.10^9	5	$5.5.10^9$
Ne^{10+}	10^9	4	$1.4.10^9$
Ar^{16+} Ar^{17+} Ar^{18+}	$1.8.10^9$ 6.10^8 2.10^7	3	$1.8.10^9$
Kr^{30+}	10^8	1	3.10^7 (MIMAS only)

using Kr-beams is scheduled for 1989. In fall 1989 we will also have a beam of unpolarized ^{6}Li followed by polarized ^{6}Li in 1990. The first experiment will be the ($^6\overrightarrow{\text{Li}}$, ^{6}He) charge-exchange reaction on nuclei, aiming at a clean separation of the transverse and longitudinal excitation in Δ-production.

When one looks towards 1991 and over, the competition with SIS which will have larger energies (up to 2 GeV per nucleon for light heavy ions), larger intensities and higher masses (up to ^{238}U) will become increasingly difficult. Our intention is to play the complementarity card by making the best use of the unique set of Saturne high resolution spectrometers (SPES 1 to 4).

Another field where Saturne should play a major role is the determination of the elementary nucleon-nucleon or nucleon-nucleus cross sections like np → npγ ("bremstrahlung") scheduled in 1989 or pp →KΛN near threshold, which are essential ingredient if one wants to understand photons or subthreshold kaon production in heavy ion collisions.

REFERENCES

[1] E. D. Donets, Phys. Scripta, Vol T3 (1983) 11.
[2] D. L'Hôte, communication to this school.
[3] S. Kox et al, Phys. Rev. C35 (1985) 1678.
[4] J. Y. Hostachy et al, Phys. Lett. B184 (1987) 139.
[5] C. Ngô, communication to this school.
[6] M. Roy-Stephan, communication to this school.
[7] E. Grosse, communication to this school.

PION POLARISATIONS IN THE RELATIVISTIC DIRAC-BRUECKNER MODEL

Fred de Jong, Bernard ter Haar and Rudi Malfliet

Kernfysisch Versneller Instituut
Zernikelaan 25
9747 AA Groningen, The Netherlands

In the Dirac-Brueckner approach to Nuclear Matter of [1] the effect of the medium on the propagation of a nucleon in that medium is described by means of a mean field containing a Lorentz scalar and a Lorentz vector part. This mean field is calculated in the Brueckner ladder approximation using a One-Boson-Exchange potential. In conventional Brueckner theory this contribution is believed to be the dominant one. Also the propagation of the exchanged mesons is affected by the medium. This involves diagrams of higher order in the hole-line expansion, so by arguments from conventional Brueckner theory one expects a relatively small contribution. In the following we will study the medium effects on the pion propagator and investigate its importance.

The effective pion propagator, denoted by $D(k)$, is found by solving the corresponding Dyson equation:

$$D(k) = (k^2 - m_\pi^2 - \Pi(k) + i\varepsilon)^{-1} \tag{1}$$

$\Pi(k)$ is the proper pion self-energy (polarisation). We will approximate this by calculating its first order contribution and incorporating higher order effects due to the so-called particle-hole interaction [2] phenomenologically. The effective Lagrangian, containing the OBE interaction that is fitted to nucleon-nucleon data, provides the vertices, form-factors and coupling constants. We have:

$$\Pi(0,\vec{k}) = \Pi^{(1)}(0,\vec{k}) / \left[1 - \frac{g'}{k^2} \Pi^{(1)}(0,\vec{k}) \right]$$

$$\Pi^{(1)}(0,\vec{k}) = i\int \frac{d^4p}{(2\pi)^4} \left\{ \mathrm{Tr}\left[G^N(p+k)\Gamma_{\pi N N} G^N(p)\Gamma_{\pi N N} \right] \right. +$$

$$\left. 2*\mathrm{Tr}\left[G^N(p+k)\ \Gamma_{\pi N \Delta}^\nu\ G_{\nu\nu}^\Delta(p)\ \Gamma_{\pi N \Delta}^\nu \right] \right\} \tag{2}$$

The expression (2) for $\Pi^{(1)}$ (note that for the present purpose we can set $\omega = 0$) can be worked out to give for the nucleon-nucleon contribution:

447

$$\mathrm{Re}\ \Pi_{NN}^{(1)}(\vec{k}) = - \frac{g_\pi^2}{m_N^2} \left[\frac{\Lambda_{NN}^2}{\Lambda_{NN}^2 + \vec{k}^2} \right]^2 \int \frac{d^3p}{(2\pi)^3}\ \theta(k_F - |\vec{p}|)$$

$$\times \left[\frac{1}{E_p^*} + \frac{1}{E_{p+k}^*} \right] \left[\vec{p}\cdot\vec{k} + \frac{2(m_N^*)^2 \vec{k}^2}{\vec{k}^2 + 2\vec{p}\cdot\vec{k}} \right] \qquad (3)$$

For the nucleon-Δ contribution we obtain:

$$\mathrm{Re}\ \Pi^{(1)}(\vec{k}) = - \frac{4}{3} \frac{f_{\Delta\pi}^2}{m_\pi^2} \left(\frac{\Lambda_\Delta^2}{\Lambda_\Delta^2 + \vec{k}^2} \right)^4 \int \frac{d^3p}{(2\pi)^3}\ \frac{\mathrm{Tr}[p, p+k]}{E_{p+k}^\Delta\ E_p^*}\ \frac{\theta(k_f - |\vec{p}|)}{E_p^* - \Sigma_0 - \Sigma_{p+k}^\Delta}$$

$$\mathrm{Tr}[p, p+k] = \frac{4}{3} \left[-\vec{k}^2 + \frac{(\vec{k}\cdot(\vec{p}+\vec{k}))^2}{m_\Delta^2} \right] (m^{*2} - \Sigma_0 E_p^* + m_\Delta m^* - \vec{p}\cdot\vec{k}) \qquad (4)$$

A few comments have to be made here: the use of the pseudo-vector πNN coupling leads to a form that has the correct non-relativistic limit, this in contrast to the result with the pseudo-scalar coupling. Compared to the non-relativistic result there are two effects that reduce the polarisation: the lower effective mass reduces the nucleon-nucleon contribution and the dipole form-factor on the πNΔ-vertex reduces the nucleon-Δ contribution. The choice of g' is a little ambiguous. The particle-hole interaction couples, in principle, differently to the NN and NΔ contributions, it could be density dependend and we do not include self-consistency effects in our approach. A review of the literature [3] shows that the calculation of g' is far from settled. Mostly values of 0.4 to 0.7 are quoted, the theoretical values tend to lie in the lower part of the interval while the experimental values lie in the upper part of the interval. Motivated by the indications for additional screening upon inclusion of the induced interaction [4] we choose g' conservatively and set it equal to 0.6. Since the dependence of the calculated quantities on g' turns out to be rather weak in the range $0.5 < g' < 0.7$, we think this simplification will not affect the general trends we observe.

In the evaluation of the nucleon self-energy we have to be careful to avoid double countings. These originate from the fact that one obtains the self-energy (a 2-point function) by closing the 4-point G-matrix with a hole line. We solved this problem by subtracting from the Dirac-Brueckner G-matrix the appropiate diagrams. This defined a modified G-matrix, denoted G', from which the nucleon self-energy can be calculated without double countings. Clearly for 4-point observables like the effective cross-sections we should use the full G-matrix. With this procedure we can calculate, again self-consistently, the nucleon self-energy.

We calculated the Equation of State for $g' = 0.6$. We observe a small shift in the saturation point towards the empirical saturation point of about 1 MeV upon inclusion of pion polarisations. One should note that the mean field and the effective mass are shifted in the order of a few ten MeV. Also we calculated effective cross sections including pion polarisations, in the same way as in ref. 5. These are increased by

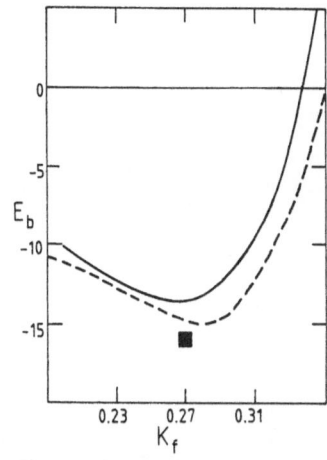

fig 1. Binding energy per nucleon vs Fermi-
momentum. The full line is the result
without pion polarisations. The dashed
line is the result including pion polarisations with g' = 0.6. The empirical saturation point is also
indicated.

a factor 2-3 compared to the result without pion polarisations, both
include the ladder type modifications of the Brueckner prescription. We
think an explanation for this can be found in the fact that the one-pion
exchange contributes significantly to the cross sections, but only a few
ten MeV to the nucleon self-energy. One may think that the higher order
terms tend to cancel. So in this way an enhancement of the pion
propagator, as accomplished by the inclusion of pion polarisations, can
lead to a drastic change in the cross section (a function of T^2) and
only a moderate shift in the self-energy.

references

[1] B. ter Haar and R. Malfliet, Phys.Rep. 149(1987)207
[2] S.O. Baeckmann and W. Weise, chapter 28, in: "Mesons in Nuclei",
    North Holland (1979)
[3] F. de Jong, B. ter Haar and R. Malfliet, to be published in Phys.
    Lett. B (1989)
[4] W. Dickhoff and H. Müther, Nucl. Phys. A473(1987)394
[5] B. ter Haar and R. Malfliet, Phys. Rev. C36(1987)1611

CUT-OFF AND EFFECTIVE MESON FIELD THEORY

M. Jaminon[*], G. Ripka[**] and P. Stassart[*]

[*] Université de Liège, Institut de Physique B5, Sart Tilman
4000 Liège 1, Belgium.   [**] C.E.N. de Saclay, Physique
Théorique, 91191 Gif-sur-Yvette Cédex, France

INTRODUCTION

QCD is known to reduce to an effective meson field theory when the number of colors is large. The baryons are soliton configurations of the meson fields. At low energy, chiral symmetry determines much of the meson Lagrangian. One such effective meson Lagrangian is the Nambu Jona-Lasinio (NJL) model.[1] The constituent quark mass emerges from a spontaneous symmetry breaking. An explicit symmetry breaking term is added to ensure that the pion, which otherwise is a massless Goldstone boson, has a nonzero mass. Since the NJL Lagrangian is not renormalizable, the regularization procedure becomes part of the model. In the present paper, we investigate the behavior of the constituent quark mass in the vacuum with the choice of the regularization scheme and with the value of the corresponding cut-off.

THE NAMBU JONA-LASINIO MODEL

The NJL Lagrangian[1] can be written as

$$L_{NJL} = \bar{q}\,(-\,i\slashed{\partial} + \varphi U)q + \frac{a^2\varphi^2}{2} - \frac{tr}{2\nu}\frac{a^2\varphi m}{2}\,(U + U^+) \tag{1}$$

where

$$U = e^{i\gamma_5\,\vec{\theta}.\vec{\tau}} \quad ; \tag{2}$$

φ represents the scalar field that will be associated with the σ meson while $\vec{\theta}$ is the pseudoscalar field associated with the π meson ; m is the explicit current quark mass breaking term.

Integrating over the quark fields q, one obtains an effective Lagrangian. We require the effective action to be stationnary with respect to the first order variation of the fields. Then, substracting its vacuum value ($\varphi = \varphi_0$, U = 1), we obtain a logarithmically divergent expression that can be used to calculate φ_0 in the vacuum, as well as the mass of the σ meson and the quark-meson coupling constants.[2,3] Here, we only look at the constituent quark mass φ_0. We examine two different regularization schemes :
(i) The "hard" regularization scheme consists in setting a cut-off on the 3-dimensional momentum. (ii) The "soft" regularization scheme follows the procedure of Ref. 4 that amounts here to a Pauli Villars substraction.[2]

THE CONSTITUENT QUARK MASS

The behavior of $\varphi_0(\Lambda)$ for the vacuum is shown in Fig. 1. The curve $\varphi_0(\Lambda)$ has two branches : (i) the lower branch where φ_0 decreases when Λ increases ; (ii) the upper branch where φ_0 varies in the opposite way. On the latter, the "hard" cut-off regularization scheme (dashed curve) yields values of φ_0 such that $\varphi_0 > \Lambda$. This would be difficult to explain within a Dirac sea picture and should be handled with care. In the "soft" regularization scheme (solid curve), $\varphi_0 \to \Lambda$ when $\Lambda \to \infty$.

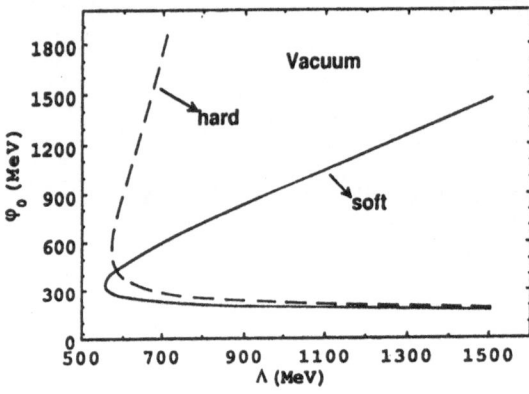

Fig. 1.

Dependence upon the cut-off Λ of the constituent quark mass φ_0 in the "soft" (solid curve) and "hard" (dashed curve) regularization schemes.

On the lower branch, both schemes exhibit a similar behavior. The conventional value $\varphi_0 \approx 300$ MeV is reached for $\Lambda = 580$ MeV ("soft") or $\Lambda = 700$ MeV ("hard") on this lower branch.

CONCLUSIONS

We studied the behavior of the constituent quark mass φ_0 on a wide range of the cut-off value for two different regularization schemes. The cut-off value is the sole free parameter of the model. We observed that the curve $\varphi_0(\Lambda)$ has two branches. Moreover, there exists a minimum value of the cut-off under which no constituent quark mass can be obtained. In the "soft" regularization scheme, this minimum corresponds to a value of φ_0 close to 300 MeV.

To look for nuclear matter saturation, one has to add a Fermi sea of quarks[2] together with a vector field responsible for short-range repulsion. It will be worth to investigate the behavior of such a system with respect to the cut-off definition in order to establish whether the features observed for a fixed cut-off[5] rely on it or whether they are more general properties.

REFERENCES

1. Y. Nambu and G. Jona-Lasinio, Phys. Rev. 122:354 (1961).
2. M. Jaminon, G. Ripka and P. Stassart, submitted to Nucl. Phys.
3. M. Jaminon, G. Ripka and P. Stassart, to be published.
4. C. Itzykson and J.B. Zuber, "Quantum Field Theory", McGraw-Hill, New York (1980).
5. V. Bernard, U.G. Meissner and I. Zahed, Phys. Rev. Lett. 59:966 (1987). V. Bernard and U.G. Meissner, Nucl. Phys. (in press).

A NON-PERTURBATIVE TRANSPORT THEORY

FOR NUCLEAR COLLECTIVE MOTION

Helmut Hofmann

Physik-Department TUM
8046 Garching/ West-Germany

INTRODUCTION

We address the problem of slow collective motion being parametrized by collective variables, which are to be introduced as shape degrees of freedom. We apply the picture of Bohr and Mottelson according to which the coupling to the intrinsic degrees of freedom is approximately given through the deformed shell model potential. This picture has to be completed by a suitable application of Strutinsky's idea for renormalizing the total energy. As the basic theoretical means we describe global collective motion in a locally harmonic approximation.

THE AVERAGE MOTION TREATED SELF-CONSISTENTLY

Self-consistency can be incorporated by describing the local motion in an adapted version of RPA. We use a form similar to the one given in[1] and in chapter 8 of[2] for the case of genuine vibrations around a stable minimum. In this way one ends up with response functions like: $\chi_{qq}(\omega) = (k^2\chi(\omega))/(1 + k\chi(\omega))$. In our case there appears an implicit dependence on temperature and the collective variable, for both the intrinsic response function $\chi(\omega)$ as well as for the effective coupling constant k, with:

$$\frac{1}{k} = -\frac{\partial^2 f}{\partial Q_0^2} - \chi(\omega = 0) \tag{1}$$

and $f(T, Q)$ being the free energy. As we want to describe dissipative motion, there has to be a microscopic damping mechanism. The poles of the response function determine the possible local frequencies. We are dealing with processes which are "slow" on the intrinsic scale. They will not proceed through modes related to giant resonances. (Think of fission, for instance, for which barriers calculated for fast modes would turn out much higher than the "adiabatic" ones). Therefore, we explicitly care for the low frequency modes only. An analysis of the pole structure allows to compute the transport coefficients for inertia M, friction γ and stiffness C (for details see e.g.[3,4,5]), to appear in a differential equation. It turns out that their values differ considerably from the previous suggestions (see e.g.[6]) of the zero frequency limit (ZFL), obtained through an expansion of $\chi(\omega)$ around $\omega = 0$:

$$\left(\frac{1}{k} + \chi(0)\right) + \omega\left(\frac{\partial\chi}{\partial\omega}\right)_{\omega=0} + \omega^2\left(\frac{1}{2}\frac{\partial^2\chi}{\partial^2\omega}\right)_{\omega=0} = -C(0) + \omega i\gamma(0) + \omega^2 M(0) = 0 \tag{2}$$

One reason for such a behaviour can be found in the observed large damping, which makes the poles move away from the real axis and, thus from $\omega = 0$.

THE HAMILTONIAN FOR THE EXTENDED SYSTEM

The picture described above has to be extended to fluctuations. As the first prerequisit we need a Hamiltonian for the system of collective plus intrinsic degrees of freedom. For nuclear physics this is not a trivial problem and in the past usually was swept under the rug whenever transport equations had been derived (see e.g.[6] and section 11.1 of[2]). Recently a Hamiltonian has been suggested and discussed in[7] within the locally harmonic approximation:

$$\hat{\mathcal{H}}_a = \hat{H}_0 + k\hat{P}\hat{\hat{F}} - \frac{\beta}{k}(\hat{Q} - Q_0)\hat{F} + \frac{\hat{P}^2}{2m_0} + \frac{1}{2k^2}(2\beta + k)(\hat{Q} - Q_0)^2 \qquad (3)$$

It consists of an intrinsic part H_0 representing the mean field at some given Q_0, an unperturbed collective part, plus two(!) coupling terms. Notice that all elements of this Hamiltonian are given (the coefficient β finally gets fixed through the frequencies of the local average motion). That it is appropriate for our problem has been demonstrated in[7] both for undamped as well as for average damped motion, provided the coupling terms are treated to infinite order.

THE MODIFIED PERTURBATION SCHEME

The fluctuations ought to be treated consistently with average motion. Therefore we need to take over this non-perturbative approach to the derivation of the transport-equation itself. This was achieved recently in[8] by applying basic features of the Nakajima-Zwanzig projection technique, but avoiding simple-minded perturbation theory. This is another essential point in which the new formulation differs from the old one obtained in[6] and reported in[2]. Within this new method it was possible to prove consistency with the quantal version of the fluctuation-dissipation theorem applied to collective motion, albeit for the fully non-Markovian situation.

THE DIFFUSION COEFFICIENTS

Non-Markovian equations are hard to solve. One would like to reduce them to differential form, also for the fluctuations. For the diffusion coefficients the ZFL unavoidably must lead to the Einstein relation, $D_{pp} = \gamma T$. But this is a high temperature limit which cannot hold true: the collective frequencies are just too large. Their absolute values may well range to a few MeV[5] with T usually being of the same magnitude or smaller. So we obviously must avoid the ZFL, which is a difficult task, much more complicated than anything related to average motion. One of the reasons can be found in the complex nature of terms appearing in the equations for the second moments. Their structure looks like: $\int_{-\infty}^{+\infty} ds\tilde{g}(s)\left(\frac{1}{2}tr\hat{d}(t)[\hat{q}(-s),\hat{q}]_{\pm}\right)$ with $\tilde{g}$ representing either an intrinsic response or correlation function. The collective factors look like response and correlation functions as well. But notice it is the non-equilibrium density which appears. The way out of this dilemma chosen in[8] was to assume the fluctuation-dissipation theorem to be fulfilled for our local oscillators even after our reduction to the lowest poles. In this way one was able to generalize the Einstein relation not only to the quantal case, but also and more important, to situations of strong damping. It was found that the new coefficients turn over into the old ones only for temperatures larger than about 2 MeV, depending on the values of M, γ, C.

REFERENCES

1. A. Bohr and B. R. Mottelson, "Nuclear Structure", Vol. II (Benjamin, London, 1975)
2. P. J. Siemens and A. S. Jensen "Elements of Nuclei: Many-Body Physics with the Strong Interaction", Addison and Wesley, 1987
3. A. S. Jensen, J. Leffers, K. Reese, H. Hofmann and P. J. Siemens, Phys. Lett. 117B (1982) 5
4. A. S. Jensen, J. Leffers, H. Hofmann and P. J. Siemens, Physica Scripta Vol. T5 (1983) 186
5. S. Yamaji, H. Hofmann and R. Samhammer, Nucl. Phys. A475 (1988) 487
6. H. Hofmann and P.J. Siemens, Nucl. Phys. A275 (1977) 464
7. H. Hofmann and R. Sollacher, Ann. Phys. 184 (1988) 62
8. H. Hofmann, R. Samhammer and G. Ockenfuß, Nucl. Phys., to appear

TIMESCALE OF PARTICLE EMISSION USING NUCLEAR

INTERFEROMETRY

D. Ardouin[a], P. Lautridou[b], D. Durand[a], D. Goujdami[b], F. Guilbault[a], C. Lebrun[a], A. Peghaire[c], J. Quebert[b], and F. Saint-Laurent[c]

a) Laboratoire de Physique Nucléaire de Nantes, 2 rue de la Houssinière 44072 Nantes Cedex 03 France
b) Centre d'Etudes Nucléaires de Bordeaux-Gradignan, Domaine du Haut-Vigneau 33170 Gradignan France
c) Laboratoire GANIL, BP 5027, 14021 Caen Cedex France

I - INTRODUCTION

Thirty years ago, Goldhaber[1], et al. discovered that pairs of identical pions produced by a proton-antiproton interaction showed a higher probability for close emission than pairs of opposite charge. The GGLP effect was later interpreted as a manifestation of Bose-Einstein statistics, which allows a determination of space-time characteristics of the emission source by interferometry procedure. This method has been applied in numerous studies of meson correlations with a large variety of energies up to ultra-relativistic nucleus-nucleus collisions. The probability of detecting two particles with four-momenta p_1, p_2 in the same event can be expressed as:

$$P(p_1, p_2) = \int |\psi_{12}(p_1, p_2)|^2 \rho(r_1) \rho(r_2) d^4 r_1 d^4 r_2 \qquad (1)$$

Using a plane wave symmetric amplitude ψ_{12} for the system,

$$P(p_1 p_2) = 1 + |\tilde{\rho}(q)|^2 \qquad (2)$$

where $\tilde{\rho}(q)$ is the Fourier transform of the source distribution which depends only on the relative four-momenta $q = p_1 - p_2$. For a gaussian distribution (size R, lifetime τ), Yano and Koonin[2] get

$$P(p_1 p_2) = 1 + \exp\left(-q^2 R^2/2 - q_0^2 \tau^2/2\right). \qquad (3)$$

Another largely used expression of a correlation function (C.F.) has been developed by Kopylov[3] assuming an incoherent emission from a uniform distribution of point-like oscillators with lifetime τ :

$$\text{C.F.} = 1 + [3 J_1(qR)/qR]^2 \times \left[1 + (q_0 \tau)^2\right]^{-1} \qquad (4)$$

In the case of $\pi - \pi$ correlations, the size R thus determined from the width of the C.F. lies always close to the projectile size except for high multiplicity events for which larger sizes and larger lifetimes may be inferred.

In the case of baryon correlations, the antisymmetrisation requirement yields: $P(p_1 p_2) = 1 - |\tilde{\rho}(q)|^2$. This approach is not useful if final state interactions develop. Describing the scattering state of two protons, Koonin[4] has provided a method to analyze the C.F. by

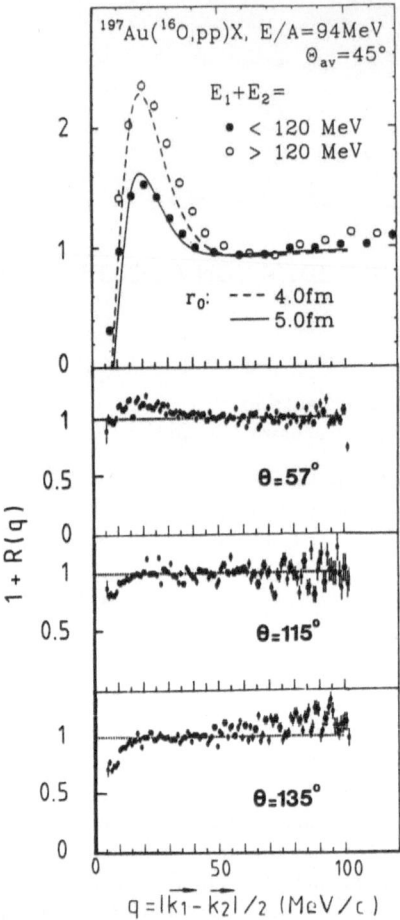

Figure 1 . Comparison of proton-proton correlation functions measured at $\theta = 45°$ (Ref. 8) and $\theta = 57, 115$ and $135°$ (present work).

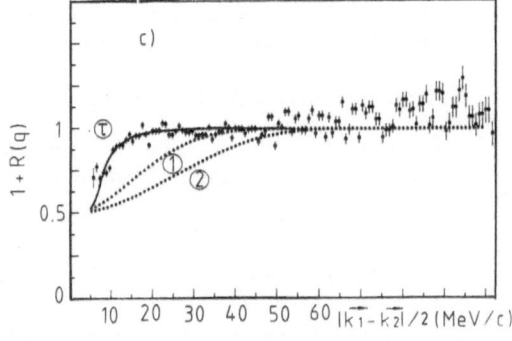

Figure 2 . Projection of the correlation plot along the q axis - The curve labelled τ is a fit with a lifetime $\tau = 1.1 \times 10^{-21} s$ (see text). Curves labelled 1 and 2 use zero lifetimes for a gaussian and spherical source respectively.

studying the height of a peak originating from the nuclear interaction. This method which has been also largely used, neglecting the finite lifetime effects, gives a source size of the order of the target size. However some interesting trends are revealed when triggering on high multiplicity[5-6] or low kinetic energies[7]: in thoses cases, an increasing apparent source-size is observed. This may be due to the neglect of lifetime effects yielding a larger apparent size, associated with last steps of the interaction processes. Several possible scenarios such as expanding sources or long average emission times have to be considered.

In order to infer the respective role of quantum, nuclear and coulomb effects, it seems interesting to measure particle correlations at very backward angles. Indeed, the sequential nature of the emission from a thermalized system is expected to damp the final state interaction processes. In addition, the knowledge of the system size allows to better isolate lifetime effects.

II - CORRELATION DATA AT BACKWARD ANGLES

We have measured proton-proton correlations at average angles varying from 57° to 145° using the $^{16}0 + Au$ reaction at $E/A = 94$ MeV. Figure 1 shows, as expected above, a disappearance of the attractive nuclear interaction peak at $q = 20$ MeV/c. We thus tentatively interpreted our C.F. in the frame of quantum statistical interferometry (formula equivalent to (4) for fermions). By studying the two-dimensional ($q = p_1 - p_2$, $q_0 = \varepsilon_1 - \varepsilon_2$) variations of the C.F., we were able to estimate a lifetime

$$\tau_0 = \left(1.1 \, ^{+0.9}_{-0.1}\right) \times 10^{-21} s.$$

for the average emission of two protons (Fig.2). This determination may be altered and under-estimated by the neglect of final state interactions. However, looking at composite systems like (d,d) and (p,d), for which flat structures of the C.F. are observed[9], allows to estimate a weak contribution of these interactions. In conclusion, these experimental conditions seem to demonstrate a possible access to lifetime estimations of particles emission and thermalisation processes in heavy-ion collisions.

REFERENCES

1) Goldhaber et al., Phys. Rev. Lett. 3 (1959) 181.

2) Yano et al., Phys. Lett. 78 B (1978) 556.

3) Kopylov et al., Sov. J. Nucl. Phys. 19 (1974) 215.

4) S. Koonin, Phys. Lett. 70 B (1977) 43.

5) Gustafsson et al., Phys. Rev. Lett. 53 (1984) 544.

6) Kyanowski et al., Phys. Lett. B1 81 (1986) 43.

7) Pochodzalla et al., Phys. Rev. C 35 (1987) 1695.

8) Chen et al., Phys. Rev. C 36 (1987) 2297.

9) Ardouin et al., Proceedings of the Int. Workshop on Nuclear Dynamics, Bad-Honnef, RFA, Oct. 1988 - Nuclear Physics A.

DETERMINATION OF TIME SCALES IN INTERMEDIATE ENERGY REACTIONS*

Uli Lynen

GSI, Darmstadt

D-6100 Darmstadt, FRG

The collision between two heavy ions at intermediate and high energies is complicated by the fact that in general not all nucleons participate to the same degree in the interaction. This can be taken into account by attributing the decay products to different sources, representing the highly excited interaction region, and the comparatively cold residual nuclei, respectively. In intermediate energy collisions, especially in the case of small impact parameters, an additional difficulty arises by the fact that these sources are not separated in space but will remain in close contact with each other for some time. In order to compare the results deduced from different decay products we must therefore take into account the different emission times of these ejectiles. In the following two investigations will be discussed where time scales or time hierarchies have been determined.

Isotopic Yield Ratios

In the experiments at the CERN-SC beams of different N/Z-ratios, e.g. ^{12}C and ^{18}O were available, so that the isotopic yields of light ejectiles could be measured as a function of the N/Z-ratio of both the target and the projectile. If we assume that during the early phase of a collision the interaction region is composed of a similar number of nucleons from the projectile and the target (1:1 source) whereas at a later time the N/Z-ratio of the source is that of the combined system, then for an

*The experimental data presented in this talk were obtained in collaboration with: K.D. Hildenbrand, W.F.J. Müller, H.J. Rabe, H. Sann, H. Stelzer, W. Trautmann, R. Trockel R. Wada (GSI), N. Brummund, R. Glasow, K.H. Kampert, R. Santo (Univ. Münster), W.M. Eckert, J. Pochodzalla (IKF Frankfurt) and D. Pelte (Univ. Heidelberg).

asymmetric projectile-target combination where both nuclei have different N/Z-ratios, the N/Z-ratio of the interaction region will change with time. By measuring the isotopic yields of different isotopes as a function of the N/Z-ratio of the projectile and the target nuclei, respectively, the emission times of different ejectiles can be determined relative to each other.[1] These results, which are summarized in figs. 1 and 2 confirm that high-energetic H and He nuclei are emitted at an early time from a 1:1 source, whereas intermediate mass fragments (IMF) are formed at a later time, when N/Z-equilibration has been achieved.

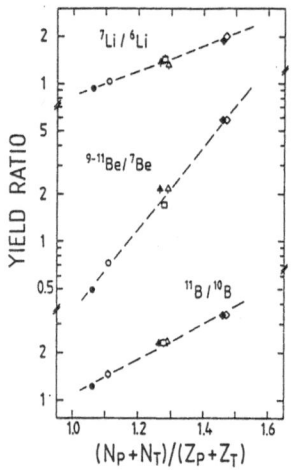

Fig.1

Isotopic yield ratios of different IMF observed with ^{12}C (full symbols) and, ^{18}O (open symbols) on targets of ^{58}Ni (circles), ^{64}Ni (squares), Ag (triangles) and Au (diamonds) at an energy of E/A = 84 MeV.

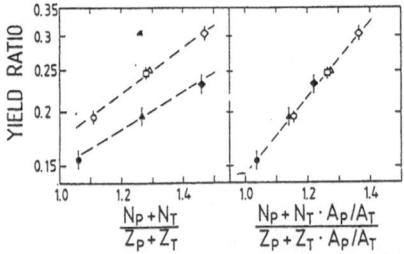

Fig. 2

t/p-ratio plotted over the N/Z-ratio of the combined system (left part) and of a subsystem consisting of equal numbers of nucleons from the projectile and the target. The incident energy was E/A = 84 MeV and the same symbols are used as in fig. 1.

Small angle correlations

The time delay between the emission of two IMF can directly be determined by measuring the correlations between two fragments. If the fragments are emitted simultaneously then small relative velocities are strongly suppressed by the Coulomb interaction between both fragments, whereas for large time delays only a small sup-

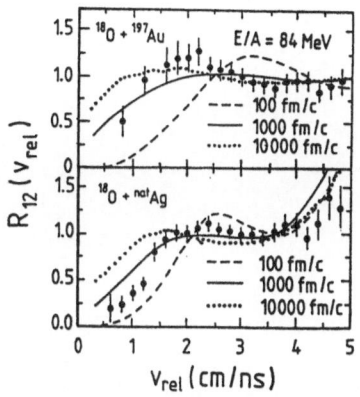

<u>Fig. 3</u>

Correlation function of two IMF.

pression due to recoil effects is expected. This method has been used[2] for the analysis of two fragments with $Z \gtrsim 10$ observed in the reaction $^{18}O + ^{nat}Ag, ^{197}Au$ at $E/A = 84$ MeV. The results are shown in fig. 3.

The observed time delay of ~ 500 fm/c is rather large and does not confirm a sudden breakup as assumed in most multifragmentation models. In order to clarify whether long emission times are a general feature of the emission process of IMF or whether the results are just due to the comparatively small incident energy and the large mass asymmetry of the two colliding nuclei, experiments with heavier nuclei at high energies are important.

<u>References</u>

[1] R. Wada et al., Phys. Rev. Lett. <u>58</u> (1987) 1829.
[2] R. Trockel et al., Phys. Rev. Lett. <u>59</u> (1987) 2844.

FRAGMENT PRODUCTION AND NUCLEAR FREEZEOUT

W.G. Lynch

National Superconducting Cyclotron Laboratory
and Department of Physics and Astronomy
Michigan State University, East Lansing, MI 48824

Investigations of reaction dynamics[1] and of the statistical properties of hot nuclear systems are both significantly influenced by uncertainties in the fragment production mechanism. For studies of hot systems, complex fragment emission is of particular interest since it could provide important clues about the properties of highly excited nuclear matter, such as the occurence on adiabatic instabilities related to the liquid-gas phase transition of nuclear matter. To distinguish between the many models for fragment production, one must assess whether the process proceeds through a binary or multifragment mechanism and determine the excitation energy, density and degree of thermalization which prevail as the system proceeds from breakup to thermal freezeout.

Relative populations of quantum states should reflect the intrinsic excitation energy of the emitting system at freezeout.[2-4] If states of emitted fragments are thermally populated at low density, the ensemble of emitted fragments may be approximated by a thermal distribution characterized by an "emission temperature", T. The relative population, $R(T)=N_i/N_k$, of states i and k becomes:

$$R(T) = (2J_i+1)/(2J_k+1) \times \exp\left[-(E_i-E_k)/T\right] = R(\infty) \times \exp(-\Delta E/T) \ . \quad (1)$$

Here, we will mainly discuss emission temperatures extracted for fragments produced in fast non-compound processes. Emission temperatures of $T\approx4-5$ MeV have been extracted from pairs of widely separated ($\Delta E > T$) particle unbound states in non-compound ^{4}He, ^{5}Li, and ^{8}Be fragments at angles significantly greater than the grazing angle where contributions from projectile fragmentation are negligible. These temperatures are surprisingly insensitive to the incident energy over the range of E/A=35-94 MeV.[2] Temperatures of 3-4 MeV, have also been extracted from the decay of high lying Y-ray transitions in non-compound fragments produced in the ^{32}S+Ag reaction at E/A =22.3.[4]

Within grand canonical fragmentation models, the insensitivity of the emission temperature to the incident energy requires that fragment freeze-out occurs at nearly constant temperature, $T\approx4-5$ MeV, rather than at constant density as is frequently assumed. A number of theoretical investigations of expanding nuclear systems arrived at qualitatively similar conclusions.[5-10] Some of these calculations surprisingly predict that all fragments including a target-like residue will be emitted at a common fragmentation temperature.[5-7]

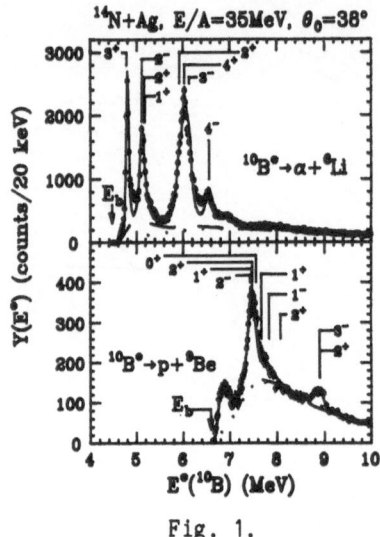

Fig. 1.

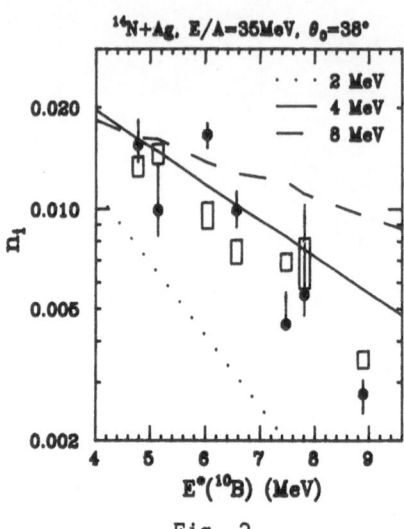

Fig. 2.

Stringent tests of statistical models become possible when one measures the population probabilities of a large number of states in a single fragment and investigates whether they follow predicted statistical distributions. Such a test was performed[11] for the ^{14}N+Ag reaction by measuring the decays ^{10}B* → ^{9}Be+p and ^{10}B* → ^{6}Li+α with a newly developed[12] high resolution hodoscope. Fig. 1 shows the measured coincidence yields, Y(E*), as a function of excitation energy in ^{10}B.

The extracted population probabilities, shown in Fig. 2, exhibit large deviations from the monotonic exp(-ΔE/T) dependence on excitation energy expected from statistical arguments. The curves in the figure show the results of calculations in which the sequential feeding[13] from known states of heavier fragments are taken into account. In these calculations, states were assumed thermally populated with temperatures T=2, 4, and 8 MeV, prior to sequential decay. The shaded bands in the figure show the results of more elaborate calculations which consider also the decay of continuum states of primary fragments, populated at a temperature of 4 MeV, with excitation energies as large as E*/A=5 MeV. The widths of these bands indicate uncertainties due to unknown spins and parities. The measured populations of particle unbound states of ^{10}B could not be reconciled with thermal fragment distributions at the time of emission. Since the mass of ^{10}B is relatively close to that of the projectile, non-statistical production mechanisms cannot be excluded with certainty. However, even within statistical models, the populations of asymptotic states could exhibit significant deviations from thermal distributions if the decay configurations are determined at transition states which differ strongly from the asymptotic states. In binary fission, for example, level crossings occurring during the descent from the saddle point to the scission point could lead to non-thermal populations of final states even if the states at the saddle point were statistically populated.

references

1. W.G. Lynch, Ann. Rev. Nucl. Sci. 37, 493 (1987) and references therein.
2. Z. Chen, et al., Phys. Rev. C36, 2297 (1987); Phys. Lett. B197, 511 (1987).
3. J. Pochodzalla, et al., Phys. Rev. Lett. 55, 177 (1985).
4. H.M. Xu, et al., Phys. Lett. B182, 155 (1986), and to be published.
5. A. Vicentini, et al., Phys. Rev. C31, 1783 (1985).
6. R.J. Lenk and V.R. Pandharipande, Phys. Rev. C34, 177 (1986).

7. T.J. Schlagel and V.R. Pandharipande, Phys. Rev. C36, 162 (1987).
8. K. Sneppen and L. Vinet, Nucl. Phys. A480, 342 (1988).
9. W.A. Friedman, Phys. Rev. Lett. 60, 2125 (1988).
10. D.H. Boal, Phys. Rev. Lett. 62, 737 (1989).
11. T.K. Nayak, et al., Phys. Rev. Lett. 62, 1021 (1989).
12. T. Murakami, et al., Nucl. Instr. and Meth. A275, 112 (1989).
13. D.J. Fields, et al., Phys. Lett. B187, 257 (1987).

Δ EXCITATION IN NUCLEI BY HEAVY ION CHARGE-EXCHANGE REACTIONS

Michèle Roy-Stéphan

I.P.N. Orsay - 91406 Orsay Cedex - France

An extensive program on charge-exchange reactions is developed at Laboratoire National SATURNE. It consists of experiments on $(^3\text{He,t})$, $(\vec{d},^2\text{He})$ and heavy ion charge-exchange . In the incident energy domain around 1 GeV per nucleon, the spin-isospin modes are selectively excited. The spin-isospin response is concentrated in two excitation energy domains : - At low excitation energy, several particle-hole states contribute , the Gamow-Teller resonance and higher multipolarity spin-flip resonances. - Around 300 MeV, a nucleon from the target is excited into a Δ(1232) resonance. The strength in the Δ sector has the same order of magnitude as in the nuclear sector. In the case of Δ excitation, one π exchange is expected to be dominant and to induce collective or at least coherent attractive effects.

Heavy ion charge-exchange experiments around 1 GeV per nucleon gave evidence of Δ excitation in nuclei by means of heavy ions[1]. Systematic measurements have been performed with ^{12}C ,^{16}O and ^{20}Ne beams at 900 MeV per nucleon and with ^{12}C beam at 1100 MeV per nucleon on ^1H, ^2H, ^{12}C, ^{89}Y and ^{208}Pb. Preliminary measurements with ^{40}Ar beam have been performed. These reactions are very peripheral, therefore, the angular distributions are very narrow and forward peaked. Integrated cross sections have been measured, using the spectrometer SPES4 at 0^0 with ± 14 mrad horizontal and vertical aperture.

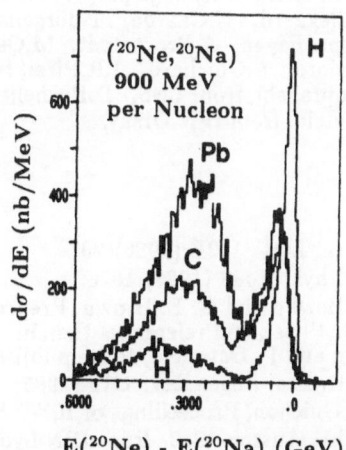

Fig. 1. Integrated cross sections versus energy transfer, for $(^{20}\text{Ne},^{20}\text{Na})$. One sees particle-hole excitations at low energy transfer and Δ - hole excitations around 300 MeV.

Several models have been developed to predict heavy ion charge-exchange cross sections, some of them are further applied to subthreshold π production[2]. From a theoretical point of view, the $(^{12}C,^{12}N)$ reaction is the simplest, since it proceeds via a single transition : the Gamow-Teller transition to the ground state of ^{12}N , the only ^{12}N bound state. In reference 2 this reaction is described by a coherent mechanism where the projectile and the target undergo collective spin-isospin excitations. For Δ - hole and for particle-hole excitations, the interaction is one meson exchange. This model explains very well the experimental cross sections in the nuclear and in the Δ sector for the reaction $(^{12}C,^{12}N)$ on the proton and on ^{12}C . The cross section dependence on the incident energy is a consequence of the variation of the nucleon-nucleon interaction on one side, and of the projectile form factor on the other side. The absorption is very strong, it has several consequences : On one hand, the cross section is sensitive to the various components of the interaction, on the other hand, the reaction is an extreme surface one. But the local density, in the overlap zone, is $\rho \simeq \rho_0/2$, which is larger than the density of the target active region in light ion charge-exchange . Heavy ion charge-exchange are then well suited to observe medium effects.

In the energy loss spectra, we observe a shift from the free Δ peak (reaction on the proton), to the Δ peak corresponding to the reaction on nuclei $(A \geq 12)$: for example, 70 MeV in the case of ^{208}Pb $(^{12}C,^{12}N)$ at 1100 MeV per nucleon . This shift depends on the projectile-ejectile system. In contrast to the (p,n) type reactions, in the (n,p) type reactions, it also depends on the mass of the target, this might be related to the isospin dependence of the Δ - nucleus interaction. The energy shift for Δ in nuclei is observed in all charge-exchange reactions around 1 GeV per nucleon, from (p,n) to $(^{40}Ar,^{40}K)$. Part of the shift is an effect of the slope of the projectile form factor versus energy transfer combined with the Δ broadening in nuclei. But neither this effect, nor distortions[3] can account for the whole shift. In reference 2 the Δ energy is reproduced if one takes account of attenuated medium effects corresponding to $\rho/\rho_0=1/2$ and if one assumes that the energy which is needed to create the Δ mass, is transferred to the whole target nucleus. In the $(^3He,t)$ case, the shift may be explained by assuming quasi-free Δ production but larger renormalization of the Δ mass and width in medium[4]. In another calculation, the (p,n) and $(^3He,t)$ spectra are reproduced by taking into account attractive collective effects (the pion branch) induced by π exchange[5]. The consistency of this approach with real π and γ results has been checked. However polarization measurements in $(\vec{d},^2He)$ do not agree with this picture. π exchange is not as dominant as expected, and shorter range contributions are stronger than expected. A coincidence experiment will clarify this problem by measuring the branching ratios for various decay modes of the Δ in nuclei.

ACKNOWLEDGEMENTS

The SATURNE heavy ion charge-exchange program has involved a Franco-Scandinavian collaboration : C.Ellegaard, C.Gaarde, T.Jörgensen, J.S.Larsen, B.Million, from Niels Bohr Institute Copenhagen, A.Brockstedt, M.Osterlund, from Lund University, D.Contardo, J.Y.Grossiord, A.Guichard, J.R.Pizzi, from IPN Lyon, R.Ekström, P.Radvanyi, J.Tinsley, P.Zupranski, from LNS, D.Bachelier, J.L.Boyard, T.Hennino, J.C.Jourdain, B.Faure-Ramstein, from IPN Orsay.

REFERENCES

1) D. Bachelier et. al., Phys. Lett. 172B (1986) 23.
   M. Roy-Stéphan Nucl. Phys. A488 (1988) 187c.
2) C. Guet, M. Soyeur, J. Bowlin and G. E. Brown, Preprint Saclay PhT/88-160 to be published in Nucl. Phys., and references therein.
3) T.Udagawa, S. W. Hong and F. Osterfeld, to be published.
4) H. Esbensen and T. S. H. Lee, Phys. Rev. C12 (1985) 1966.
5) J. Delorme and P.A.M. Guichon, Proceedings of 10eme Session d'Etudes Biennale de Physique nucléaire, Aussois 1989. Rap. IPN Lyon LYCEN 8902 (1989) C.4.1.

MOMENTUM DEPENDENT POTENTIALS

IN RELATIVISTIC HEAVY ION COLLISIONS[†]

Volker Koch

Institut für Theoretische Physik, Universität Gießen
D–6300 Gießen, West Germany

A major goal in the study of relativistic heavy ion collisions is to measure the nuclear equation of state (EOS). The interpretation of the experimental data such as the transverse flow are based on simulations of the Boltzmann equation, which combine the nuclear mean field and individual nucleon-nucleon collisions as dynamical input. It has been found that most observables are affected by both ingredients[1,2].

The major difficulty in describing relativistic heavy ion collisions is that we have to deal with a highly nonequilibrium state at about twice nuclear matter density. Therefore in medium corrections and nonequilibrium effects have to be taken into account seriously. Several authors[1,2,4] for instance have proposed in medium corrected nucleon-nucleon cross sections which range from $0.7 \, \sigma_{free}$[3] to $2 \, \sigma_{free}$[1,4] at twice nuclear matter density. This implies a rather large uncertainty for the determination of the EOS from flow data.

Yet there is another, maybe even bigger uncertainty connected with the momentum dependence of the mean field[5] as also mentioned by Welke et al.[6]. The mean field potentials used in the calculations commonly consist of a Skyrme type density dependent part and a term which governs the momentum dependence[5]. The parameters of this latter term are adjusted such that for a given EOS the energy dependence of the nucleon-nucleus (n–N) potential is fitted.

A relativistic heavy ion collision schematically can be described by two nuclei overlapping with relative velocity corresponding to the c.m. energy. For this situation the resulting mean field potential very strongly depends on the

---

[†]Work supported by BMFT and GSI Darmstadt.

functional form of the momentum dependent term. Let us illustrate this by considering the following two energy functionals:

$$V_1(\rho) = \frac{A_1}{2}\frac{\rho^2}{\rho_0} + \frac{B_1}{\sigma+1}\frac{\rho^{\sigma+1}}{\rho_0^\sigma} + \frac{C_1}{\rho_0}\iint d^3p\,d^3p'\, \frac{f(r,p)f(r,p')}{1 + \left[\frac{p-\langle p'\rangle}{\Lambda}\right]^2}$$

$$V_2(\rho) = \frac{A_2}{2}\frac{\rho^2}{\rho_0} + \frac{B_2}{\sigma+1}\frac{\rho^{\sigma+1}}{\rho_0^\sigma} + \frac{C_2}{\rho_0}\iint d^3p\,d^3p'\, \frac{f(r,p)f(r,p')}{1 + \left[\frac{p-p'}{\Lambda}\right]^2}$$

The first functional is the one used by Gale et al.[2] and we use their parameters for the "lightweight" equation of state. Here the momentum dependent potential depends on the difference between the momenta of a test particle and the average momentum of the surrounding medium. The second functional differs from the first only in the momentum dependent term, which now depends on the momentum difference between the test particles themselves. The parameters are adjusted such that it leads to the same EOS and energy dependence of the n–N potential as given by the first potential (fig. 1,2) This is different for a 800 MeV/u heavy ion collision as shown in fig. 3. There the potentials differ by more then 80 MeV in the energy range of interest.

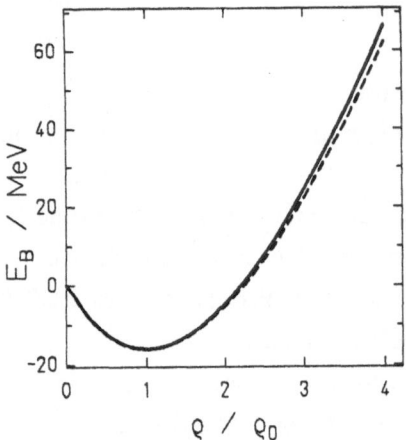

Fig. 1.   Equation of state. The full line corresponds to the functional W_1 the dashed line to W_2.

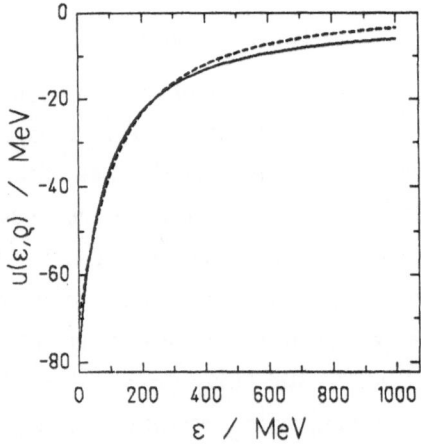

Fig. 2.   Energy dependence of the nucleon–nucleus potential.

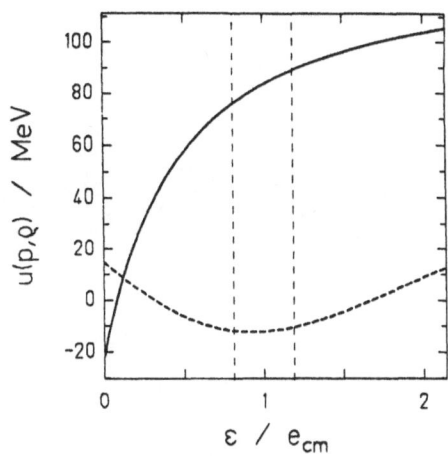

Fig. 3. Single particle potential for a 800 MeV/u collision. The vertical lines indicate the energy range most nucleons will be in.

Of course the second functional seems to be the more natural choice as it resembles an exchange term in local density approximation. Yet a theoretically even more refined parameterization once again might lead to a different result. Therefore, it would be very helpful to have some guidance from more microscopical approaches like the Dirac–Brueckner or so, which could calculate the single particle potential for the given dynamical situation.

Since all observables related to the EOS so far result from a subtle interplay of compressional effects, n–n collisions and momentum dependence of the mean field, uncertainties like the ones outlined above have to be settled, before quantitative conclusions on the equation of state of hot dense nuclear matter from heavy ion collisions are possible.

REFERENCES

1) G.F. Bertsch et al., Nucl.Phys.A490 745 (1988)
2) G.F. Bertsch et al., Phys.Lett.189B 384 (1987)
3) B ter Haar and R. Malfliet, Phys.Repts.149 207 (1987)
4) F. de Jong and R. Malfliet, these proceedings
5) C. Gale et al., Phys.Rev.C35 1666 (1987)
    J. Aichelin et al., Phys.Rev.Lett.59 1926 (1987)
    B. Blättel et al., Phys.Rev.C38 1767 (1988)
6) G.M. Welke et al., Phys.Rev.C38 2101 (1988)

CLASSICAL MODELS OF HEAVY-ION COLLISIONS

T.J. Schlagel

Department of Physics, University of Illinois at
Urbana-Champaign
1110 West Green Street, Urbana, IL   61801   USA

We have performed computer simulations of collisions between gold nuclei at E_{lab} = 200, 400, and 600 MeV/A using classical molecular dynamics (MD).[1] Two classical models giving nuclear matter with incompressibilities K = 283 and 546 MeV are used, to study the sensitivity of reaction observables to the equation of state. Central and off-center collisions are studied to simulate the impact parameter dependence.

Observables based on the transverse momentum generated in collisions have been studied extensively.[2] In particular, the azimuthal distribution with respect to the reaction plane[3,4] is sensitive to the EOS at high energies (Fig. 1). The eigenvalues of the kinetic energy flow tensor[5] are also sensitive to the EOS. Plots of the eigenvalues can be made on an event by event basis, without using multiplicity cuts, and give an unambiguous signature of the EOS (Fig. 2).

The Vlasov-Nordheim (VN) equation has been used to study heavy-ion collisions and examine the effects of the nuclear equation of state. Recently, an accurate method to solve the VN equation has been developed by Lenk and Pandharipande[6], in which energy and momentum are conserved. The basic assumption of the dynamics being dominated by mean-field and pair collision terms, inherent in the VN equation, can be tested in the classical limit by comparisons with results of exact MD. In preliminary studies, we find that stopping in MD is reproduced in VN by using a hard sphere cross section consistent with the pair potentials used in the simulations. At low energies, both MD and VN lead to fragmentation (Fig. 3).

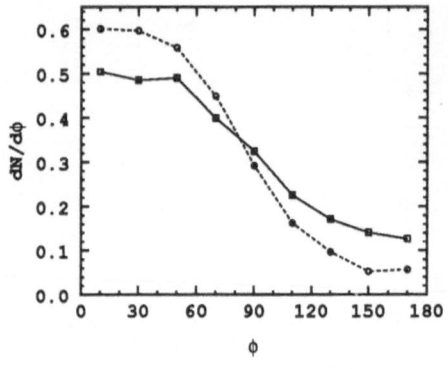

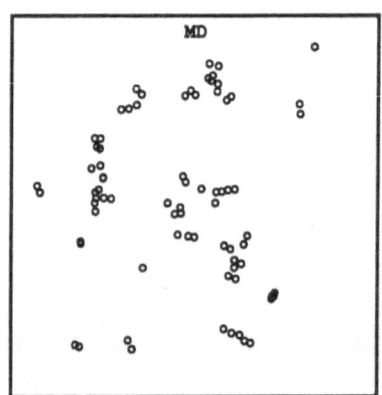

Fig. 1. Azimuthal angular distributions for charged particles in the rapidity region $0.5 < (y/y_p)cm < 0.8$ and in the multiplicity region $M > 0.8 M_{max}$ for the reaction Au (600 MeV/A) on Au for K=283 MeV (solid) and K=546 MeV (dashed).

Fig. 2. The larger eigenvalues of the kinetic energy flow tensor, λ_1 and λ_2, versus the smallest eigenvalue, λ_3. The result of the K=283 MeV model are shown with pluses, while those of the K=546 MeV model are shown with circles.

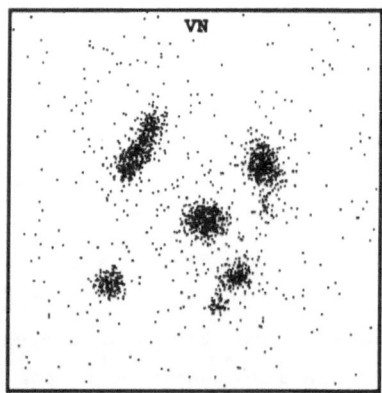

Fig. 3. Comparison of 50+50 collisions at $E_{cm}=0$. Both MD and VN lead to fragmentation.

The simulations were carried out at the National Center for Super-computing Applications of the University of Illinois. This work was supported by the National Science Foundation under Grant PHY84-15064.

REFERENCES

1. T. J. Schlagel and V. R. Pandharipande, in progress.
2. K. G. R. Doss, et al., Phys. Rev. Lett. 57, 302 (1986).
   B. V. Jacak, Nucl. Phys. Lett. A488, 325c (1988). H. H. Gutbrod, et al., Phys. Lett. B216, 267 (1989).
3. G. M. Welke, M. Prakash, T. T. S. Kuo, S. Das Gupta and C. Gale, Phys. Rev. C38, 2101 (1988).
4. P. Danielewicz and G. Odyniec, Phys. Lett. 157B, 146 (1985).
5. M. Gyulassy, K. A. Frankel and H. Stöcker, Phys. Lett. 110B, 185 (1985).
6. R. J. Lenk and V. R. Pandharipande, submitted to Phys. Rev. C.
7. T. J. Schlagel, R. J. Lenk and V. R. Pandharipande, in progress.

RELATIVISTIC MOLECULAR DYNAMICS AND THE RELATIVISTIC

VLASOV EQUATION

Michael Schönhofen, Hans Feldmeier, and Manuel Cubero

Gesellschaft für Schwerionenforschung (GSI)
Postfach 110552
D-6100 Darmstadt, West Germany

In the following we present results from the numerical solution of the classical relativistic equations of motion for nucleons with non-radiating meson fields, derived in ref.[1], for the collision $^{16}O + ^{16}O$ at $E_{beam} = 600 A MeV$ and using 20 and 100 test-particles per nucleon. The coupling strengths and meson masses which enter the calculation are taken from table 5 in ref.[2]. The interaction between the test-particles is given by the covariant two-particle Yukawa potentials which appear in the small acceleration approximation [1]. We use a parabolic cutoff at short distances to simulate a finite size of the test-particles. For deriving the Vlasov equation one supposes smooth densities and hence smooth fields. Therefore, a numerical condition on the density of test-particles is that it is large enough to ensure the mean distance to be smaller than the cutoff distance. A physically motivated value of the cutoff distance would be about twice the nucleon radius ($\simeq 1.2 fm$).

A calculation with 20 test-particles per nucleon using the cutoff $c_V = c_S = 0.8 fm$ shows that the two ^{16}O nuclei pass through each other and finally disintegrate. If one performs the same calculation but with 100 test-particles per nucleon and rescales the cutoff distance such that the ratio to the mean distance between the test-particles in the system remains constant, the results are not changed. To illustrate this, we have plotted in fig.1 for a head-on collision the momentum space distribution of the test-particles for both calculations at the final time $t = 20 fm/c$. The two momentum distributions are clearly separated but their width is increased compared to the initial spread of about $\pm 1.4 fm^{-1}$ (Fermi motion). Especially in the transverse direction we see a broadening.

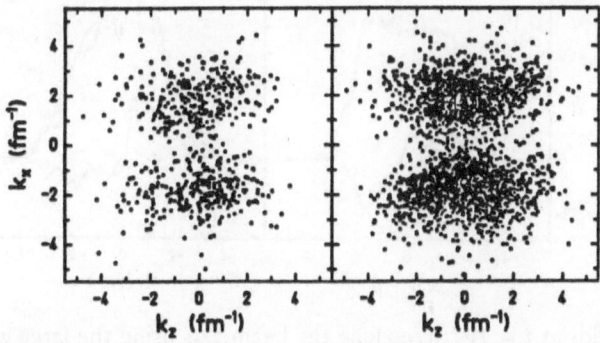

Fig.1. Distribution of final momenta of a central collision of two oxygen nuclei for $c_V = c_S = 0.8 fm$ with N (no. of test-particles per particle) = 20 (left hand side) and for $c_V = c_S = 0.46 fm$ with $N = 100$ (right hand side).

A different result is obtained in the case of 100 test-particles per nucleon when we use a nucleon-nucleon potential with a small cutoff parameter $c_V = 0.2fm$. The smaller the cutoff distance c_V for the vector field is, the more pronounced is the repulsive core of the potential. In this case the momentum distribution of the test-particles (fig.2) shows much less transparency than in fig.1, i.e. the stopping power is considerably enhanced. One observes a filling of the central region near momentum zero.

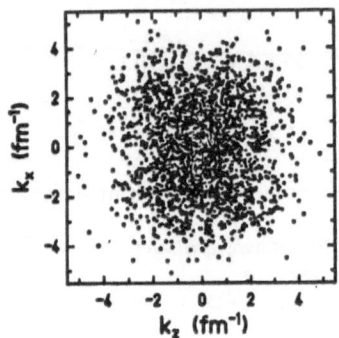

Fig.2. Distribution of final momenta of a central collision of two oxygen nuclei for $c_V = 0.2fm, c_S = 0.8fm$ with $N = 100$.

The different behaviour in the time evolution of the same collision can be understood by looking at the meson fields as a function of position on the beam axis. Fig. 3 shows the vector and scalar meson fields along the beam axis at the time of highest nucleon density ($t = 10fm/c$). The vector and scalar fields are smooth for $c_V = 0.46fm$ (left-hand side of fig.3) whereas for the small cutoff parameters $c_V = 0.2fm$ the fields exhibit large fluctuations around their mean value. This implies that the forces, in which the derivatives of the fields enter, are much stronger than in the pure mean field calculation. Because the derivation of the Vlasov equation assumes smooth fields we interpret the calculation with $c_V = 0.8$ as the solution of the Vlasov equation whereas the latter case corresponds to the solution of the Vlasov equation with a fluctuating Langevin force added. As one increases the number of test-particles per nucleon and keeps the value of the cutoff parameters fixed, the fluctuations and with them the stochastic force disappear.

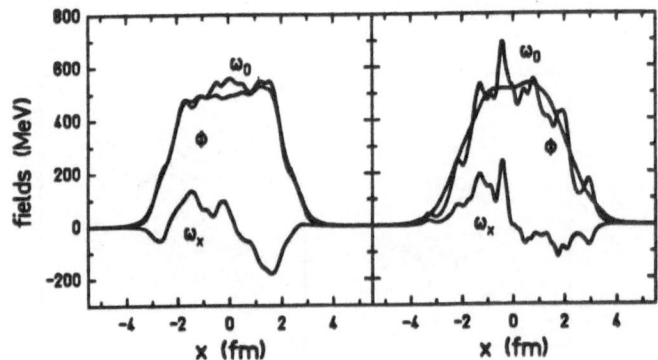

Fig.3. Meson fields at $t = 10fm/c$ along the beam axis using the large cutoff $c_V = c_S = 0.46fm$ (left-hand side) and the small cutoff $c_V = 0.2fm$ but $c_S = 0.8fm$ (right-hand side); both with $N = 100$.

REFERENCES

1. H. Feldmeier et al., Proc. Int. Workshop on Nuclear Dynamics at Medium and High Energies, Bad Honnef 1988, to be published in Nucl. Phys. A and GSI-preprint 88-63

2. B. Serot and J. D. Walecka, Adv. Nucl. Phys. 16 (1986) 1, J. W. Negele and E. Vogt eds. (Plenum Press)

STABILITY CONDITIONS AND VIBRATION MODES OF SELF-CONSISTENT VLASOV SOLUTIONS WITH DIFFUSE SURFACE

S.J. Lee, E.D. Cooper, H.H. Gan, and S. Das Gupta

Physics Department, McGill University
Montreal, P.Q., H3A 2T8, Canada

The Coulomb force as well as the diffusivity of the nuclear surface are expected to play an important role in peripheral heavy ion collisions at low and intermediate energies.[1] Boltzmann-Uehling-Uhlenbeck (BUU) calculations should incorporate these effects. This has prompted us to consider static Vlasov solutions which have diffuse surfaces. Such solutions can be obtained by using a finite range force such as an Yukawa interaction[2] or a gradient dependent interaction.[3] Here we report a short summary of the properties of the static Vlasov solution with a diffuse surface through a Yukawa force. Details will be published elsewhere.[4]

The Wigner transform[5,6] of time dependent Hartree-Fock (TDHF) equation for density matrix gives, in a classical limit ($\hbar \to 0$), the Vlasov equation:

$$\frac{\partial}{\partial t} f(\vec{r}, \vec{p}, t) + \frac{\vec{p}}{m} \cdot \vec{\nabla} f(\vec{r}, \vec{p}, t) - \left[\vec{\nabla} U(\vec{r}, t) \right] \cdot \vec{\nabla}_p f(\vec{r}, \vec{p}, t) = 0, \tag{1}$$

where $\vec{\nabla}$ and $\vec{\nabla}_p$ mean the gradients in $r-$ and $p-$space respectively. Here we have considered the single particle Hamiltonian to be $p^2/2m + U(\vec{r}, t)$ with a self-consistent mean field potential U. For the static Vlasov solution, $\partial f/\partial t = 0$. Wigner transform of the density matrix condition ($\rho^2 = \rho$) of TDHF determines the form of Wigner function to be, again in a classical limit,

$$f(\vec{r}, \vec{p}, t) = \frac{4}{(2\pi\hbar)^3} \theta(p_F(\vec{r}, \hat{p}, t) - p). \tag{2}$$

Here $\hat{p}$ is the unit vector of $\vec{p}$ and p_F becomes $p_F(\vec{r})$ for a static nucleus.

One of the simplest interactions which can give diffuse surface in a classical limit can be represented by the potential energy

$$V = \sum_i \frac{C_i}{\sigma_i + 1} \int d^3r \; \rho^{\sigma_i+1}(\vec{r}) + \frac{1}{2} \int d^3r \; d^3r' \; \rho(\vec{r}) v(\vec{r}, \vec{r'}) \rho(\vec{r'}). \tag{3}$$

Due to Eq. (2), the kinetic energy is $T = C \int d^3r \rho^{5/3}(\vec{r})$ for a static nucleus where $C = (3\hbar^2/10m)(3\pi^2/2)^{2/3}$. Minimizing the energy functional with respect to the density variation, we get an equation for the self-consistent density in a classical limit which is a self-consistent solution of static Vlasov equation. The details are in Ref. 2. The second order variation of energy functional gives a stability condition for self-consistent Vlasov solutions. This stability condition can be cast in the form of eigenvalue equation:

$$\int d^3r' \; S(\vec{r}, \vec{r'}) \delta\rho(\vec{r'}) = \epsilon \; \delta\rho(\vec{r}), \tag{4}$$

$$S(\vec{r}, \vec{r'}) = \left[\frac{10}{9} C \rho^{-1/3}(\vec{r}) + \sum_i C_i \sigma_i \rho^{\sigma_i - 1}(\vec{r}) \right] \delta(\vec{r} - \vec{r'}) + v(\vec{r}, \vec{r'}). \qquad (5)$$

These eigenvalues ϵ must be all real positive.

On the other hand, integrating the Vlasov Equation Eq. (1) over momentum gives the continuity equation and integrating over $\vec{p}$ after multiplying $\vec{p}/m$ to Eq. (1) gives the time variation of the current density. Combining these two equations, we get dispersion relation for small amplitude vibration.[4] Neglecting quadrupole deformation of the local Fermi sea, the dispersion relation is reduced to a wave equation for the density vibration in a nucleus:

$$\frac{\partial^2}{\partial t^2} \delta\rho(\vec{r}, t) = \frac{1}{m} \vec{\nabla} \cdot \left[\rho(\vec{r}) \vec{\nabla} \left(\int d^3 r' \; S(\vec{r}, \vec{r'}) \delta\rho(\vec{r'}, t) \right) \right]. \qquad (6)$$

The same stability matrix $S(\vec{r}, \vec{r'})$ appears in this wave equation and the dispersion relation which leads us to understand the relationship between stability (time independent picture) and vibration modes (time dependent picture). Considering $\delta\rho(\vec{r}, t) = e^{i\omega t} \delta\rho(\vec{r})$, we have

$$\omega^2 \int d^3 r' \; d^3 r \; \delta\rho(\vec{r'}) S(\vec{r'}, \vec{r}) \delta\rho(\vec{r})$$

$$= \frac{1}{m} \int d^3 r \; \rho(\vec{r}) \left| \vec{\nabla} \int d^3 r' \; S(\vec{r}, \vec{r'}) \delta\rho(\vec{r'}) \right|^2. \qquad (7)$$

Since the right hand side of Eq. (7) is positive definite, the stability condition is related to the requirement of the eigenfrequencies ω of vibration mode to be real.

As a conclusion, we have shown that the stability of the self-consistent density of a static Vlasov equation with finite range force is related with requiring the eigenfrequency of time dependent vibration mode to be real. This is same as in TDHF case where the stability of HF density is related with real eigenvalues of RPA matrix. The numerical calculations[4] of Eq. (4) and Eq. (6) show that the lowest eigenmodes of stability matrix for each multipole are the same as the corresponding eigenmodes of vibration and are quite close to the transition density in an inhomogeneous irrotational and incompressible fluid[5].

Acknowledgement

Discussions with N. de Takacsy are gratefully acknowledged. This work was supported in part by NSERC and in part by the Quebec Department of Education.

REFERENCES

1. H.H. Gan, S.J. Lee, and S. Das Gupta, Phys. Rev. C 36:2365 (1987).
2. S.J. Lee, H.H. Gan, E.D. Cooper, and S. Das Gupta, submitted to Phys. Rev. C.
3. R.J. Lenk and V.R. Pandharipande, preprint.
4. S.J. Lee, E.D. Cooper, H.H. Gan, and S. Das Gupta, to be published.
5. P. Ring and P. Schuck, "The Nuclear Many–Body Problem," Springer-Verlag, New York (1980).
6. G.F. Bertsch and S. Das Gupta, Phys. Rep. 160:189 (1988).

SIDEWARD FLOW OF CHARGED PARTICLES AND NEUTRONS

IN HEAVY ION COLLISIONS

D. Keane,[a,b] B.D. Anderson,[a] A.R. Baldwin,[a] D. Beavis,[b] S.Y. Chu,[b] S.Y. Fung,[b]
M. Elaasar,[a] G. Krebs,[c] Y.M. Liu,[d,b] R. Madey,[a] J. Schambach,[a] G. VanDalen,[b]
M. Vient,[b] S. Wang,[d,b] J.W. Watson,[a] G.D. Westfall,[e] H. Wieman,[c] and W.M. Zhang[a]

[a] *Department of Physics, Kent State University, Kent, Ohio 44242*
[b] *Department of Physics, University of California, Riverside, California 92521*
[c] *Lawrence Berkeley Laboratory, Berkeley, California 94720*
[d] *Department of Physics, Harbin Institute of Technology, Harbin, People's Republic of China*
[e] *Cyclotron Laboratory, Michigan State University, East Lansing, Michigan 48824*

To probe the early, high density stage of a heavy-ion collision, it is essential to focus on observables that are minimally distorted during the subsequent processes of expansion, decay of excited states, and chemical freeze-out. It is now widely accepted that correlations characteristic of fluid-like collective behavior ("flow") fulfill these requirements. Such correlations signal the release of compressional energy, and provide a relative measure of the peak nuclear pressure generated in the collision.

Table 1 summarizes all currently available results from the Bevalac streamer chamber for transverse flow, p_x, *i.e.*, transverse momentum/nucleon projected onto the event reaction plane.[1] The tabulated data are based on a total of over 10^5 fully reconstructed events, and are averaged over parti-

TABLE 1. Transverse momentum/nucleon (MeV/c) in the reaction plane, p_x, averaged over forward rapidities ($y_r \gtrsim 0.7$), for streamer chamber samples with a minimum bias trigger and a multiplicity cut which selects ~25% of the inelastic cross section.

Beam energy: (GeV/nucleon)	0.4	0.8	1.2	1.8
Ne+NaF	25		43±5	53±5 [†]
Ar+KCl		50±4 [§]	65±5	95±5 [§]
La+La		72±6 [§]		
U+U		85±10 [†]		
Ne+BaI$_2$				*
Ar+BaI$_2$			120±10	
Ar+Pb	60±7	130±7 [§]		*

§ Central trigger data[1,4] from the GSI/LBL group.
† These U+U collisions were at 0.9 GeV/nucleon.
‡ These Ne+NaF collisions were at 2.1 GeV/nucleon.
* Analysis in progress.

cles with center-of-mass rapidity $y_r = y/y_{beam}$ above 0.7; in this kinematic region, p_x is relatively large and can be determined with minimal experimental biases. As noted in the caption, seven of the entries (the U.C. Riverside data[2,3]) are based on events selected using a minimum bias trigger and a high multiplicity cut, while the other four (the GSI/LBL data[1,4]) were selected using a central collision trigger. Within the quoted uncertainties, these two selections correspond to a comparable degree of centrality. In mass-symmetric systems, the data indicate that $\langle p_x \rangle$ increases monotonically with both mass and beam energy. The forthcoming results for Ar + Pb at 1.8 A GeV (a collaborative effort between Heidelberg and Riverside) imply a more complex beam energy dependence for Ar + heavy; in particular, the large difference between Ar + KCl and Ar + heavy at lower energies is substantially reduced at 1.8 A GeV. This unexpected finding provides an important insight into the dynamics of the collision process and the behavior of the high-density zone.[5]

VUU model[6] predictions have been generated for Ne + NaF and Ar + KCl at 1.2 and 1.8 A GeV, for U + U at 0.9 A GeV, and for Ar + BaI$_2$ at 1.2 A GeV. For a variety of reasons, comparisons with the streamer chamber data have not led to unambiguous conclusions about the stiffness of the nuclear equation of state.[2] Nevertheless, the model predicts the correct p_x scaling behavior among the six systems mentioned above. It has not yet been established whether this or any other existing model can reproduce the observed variation of p_x with beam energy for Ar + heavy; such a test should prove particularly stringent.

In 1988, a Kent State-Michigan State-Berkeley collaboration carried out a major new Bevalac experiment, E848H, designed to measure collective flow among neutrons. Data were obtained for collisions of Au on Au at beam energies of 75, 150, 250, 400 and 650 A MeV, and for Nb on Nb at 400 A MeV. The principal detector components are an array of 18 neutron counters sampling laboratory polar angles θ between 3° and 90°, and a 184-element time-of-flight (TOF) wall located 4.5 m from the target, which detects all charged particles emitted between $\theta = 2°$ and $\sim 30°$ (about half or more of the charged particles that would be observed in an ideal 4π detector). The neutron detectors determine the energy (through time-of-flight) for the emitted neutrons at each angle, as in a particle-inclusive experiment; the TOF wall provides an estimate of the azimuthal angle of the reaction plane for each event, and thereby tags each detected neutron with its angle relative to that plane, $\phi - \phi_R$. Thus, absolute triple-differential neutron cross sections $d^3\sigma/dE \, d\theta \, d(\phi - \phi_R)$ can be determined.[7]

This experimental configuration has several advantages over conventional 4π detectors. Instead of attempting to cover the entire available phase space and detect, identify, and measure the momentum of every emitted particle — a formidable goal that inevitably leads to compromises — the E848II approach requires only that all the available *rapidities* be sampled. This capability is adequate for investigating collective flow as long as a supplementary 4π (or almost 4π) detector with relatively modest performance is available to determine ϕ_R for each event, as explained above. In E848II, the neutron detectors identified neutrons unambiguously at all rapidities, and measured their energy over the entire range from about 50 MeV to the highest values encountered at the Bevalac. Neither the streamer chamber, the Plastic Ball,[8] nor Diogène[9] has an equivalent capability for any baryonic species; moreover, neutrons are not subject to Coulomb interactions, which sometimes complicate the interpretation of experimental results. Thus, the data from E848H will not only complement the earlier experiments, but will also provide some of the most detailed information to date on collective flow.

For most of the duration of the experiment, the trigger was set to select only events with a high multiplicity in the TOF wall; for instance, the recorded data for Au + Au at 150 A MeV correspond to $\sim 7\%$ of the total geometric cross section. The dispersion in determining the event reaction plane for this system is about 48° — a sufficient resolution for investigating the signatures of compressed nuclear matter.[7] One of the objectives of experiment 848H is to go beyond studies of p_x, and to probe the more detailed structure of the absolute triple-differential cross sections;[7,10] however, it is appropriate

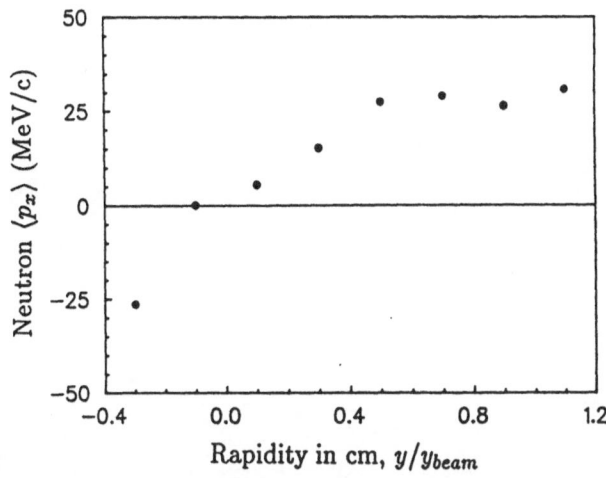

Figure 1. Neutron $\langle p_x \rangle$ as a function of rapidity for Au + Au at 150 A MeV. Statistical errors are smaller than the symbol size; systematic uncertainties are not shown.

to first examine the more familiar p_x parameter. Figure 1 shows preliminary results for $\langle p_x \rangle$ versus y_r for Au + Au at 150 A MeV. Further details of these preliminary results are given elsewhere.[11]

Acknowledgement

This work has been supported in part by the U.S. Department of Energy and the National Science Foundation.

REFERENCES

1. P. Danielewicz and G. Odyniec, Phys. Lett. **157B**, 146 (1985).
2. D. Keane *et al.*, Phys. Rev. C **37**, 1447 (1988), and *Proc. of the 8th High Energy Heavy Ion Study, Berkeley, 1987*, LBL-24580, p. 165 (1988).
3. M. Vient, Ph.D. thesis, University of California, Riverside, 1988; M. Vient *et al.*, to be published.
4. P. Danielewicz *et al.*, Phys. Rev. C **38**, 120 (1988).
5. H. Ströbele *et al.*, to be published.
6. H. Kruse, B.V. Jacak, and H. Stöcker, Phys. Rev. Lett. **54**, 289 (1985).
7. G. Fai, W. Zhang, and M. Gyulassy, Phys. Rev. C **36**, 597 (1987).
8. A. Baden *et al.*, Nucl. Instr. and Meth. **203**, 189 (1982).
9. J.P. Alard *et al.*, Nucl. Instr. and Meth. **A261**, 379 (1987).
10. G.M. Welke *et al.*, Phys. Rev. C **38**, 2101 (1988).
11. R. Madey *et al.*, Bevalac Newsletter (1989).

DILEPTON RADIATION FROM HOT NUCLEAR MATTER

AND NUCLEON-NUCLEON COLLISIONS

Charles Gale

Theoretical Physics Branch
Chalk River Nuclear Laboratories
Chalk River, Ont.
Canada, K0J-1J0

Still very little is known about the properties of excited nuclear matter in temperature and density regions far removed from equilibrium. Such conditions are achieved in the laboratory when heavy nuclei collide at high energies. Nuclear matter, which is initially in its ground state, is first heated and compressed. A phase of cooling and decompression then follows and the system finally evolves to its asymptotic state. The final configurations of this strongly interacting many-body system have been analyzed using a variety of methods. Inclusive spectra of protons and composites have been measured[1] and particle production channels have also been investigated[2]. At present, some promising analysis methods are liked to variables reflecting the momentum flow during the collision. These are the kinetic flow tensor[3] and the transverse momentum distribution[4]. However, the unambiguous extraction of the nuclear equation of state from such measurements is not without technical difficulties[5]. Recently, much attention has been drawn to the observation of direct photon production[6]. The advantage of observing an electromagnetic signal from a strongly interacting many-body system is that it will travel relatively unperturbed from the production point to the detector. Depending on the angle and on the energy of the emitted quanta, this should provide information on nuclear stopping power[7] and on the dynamics of baryon-baryon cascading in hot and dense nuclear matter[8].

The interest of measuring lepton pairs over photons is clear when one concentrates on annihilation reactions. In this discussion below, we restrict our attention to dielectron pairs of invariant mass greater than 100-200 MeV. In the incident energy range of a few GeV/A (relevant to LBL, GSI/SIS and the AGS), the dominant dilepton-producing mechanisms will be neutron-proton bremsstrahlung $(n\,p \rightarrow n\,p\,e^+\,e^-)$ and pion-pion annihilation $(\pi^+\,\pi^- \rightarrow e^+\,e^-)$[9]. Furthermore, it can be shown[9] that the rate for $\pi^+\,\pi^-$ annihilation depends strongly on the pion dispersion relation $\omega_\pi(k)$ in nuclear matter, thereby allowing

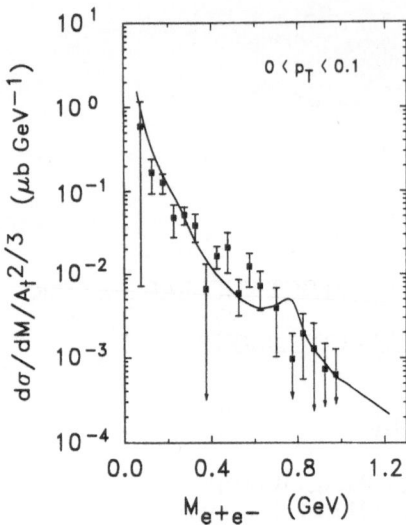

Fig. 1 The invariant mass spectrum for transverse momenta less than 0.1 GeV/c. The bremsstrahlung and delta decay contributions are summed and corrected by the DLS acceptance. The result is the solid line. Data are from the DLS collaboration.

the possibility of measuring it in heavy ion collisions. Therefore, there is hope that we will finally be able to study the behavior of pions in dense, high temperature nuclear matter without fear that the signal will be absorbed upon expansion.

The Dilepton Spectrometer (DLS) collaboration has undertaken a program of measuring direct lepton pair production in nucleon-nucleon, nucleon-nucleus ans nucleus-nucleus collisions at the Lawrence Berkeley Bevalac. The first experimental results for dielectron production in p + Be collisions at 4.9 GeV have recently been obtained[10]. Since the clear signature of a many-body effect involves the ruling out of elementary nucleon-nucleon contributions, it is desirable to theoretically address those first results. Since Be is a relatively small system, one would not expect the the formation of hot and dense equilibrated nuclear matter in a reaction induced by a 4.9 GeV proton. We thus consider the bremsstrahlung contribution, evaluated in the soft photon limit, and the radiative decay of the delta into a nucleon and a virtual photon which internally converts to an electron-positron pair. We consider the process $NN \rightarrow \Delta X \rightarrow e^+e^- X'$. The formalism follows straightforwardly[11] and involves the structure functions of the delta; in the spirit of the Vector Dominance Model, we let the photon transform into a ρ meson before coupling to the delta. The resulting nucleon-nucleon contribution to the dilepton spectrum is displayed on figure 1 for a bin of low transverse momentum. The agreement is quite reasonable. For higher p_T's it is still satisfactory although less spectacular[11]. This is attributed to the poor performance of the soft photon approximation in the high M/high p_T limit.

In closing, we believe that the current set of DLS p + Be measurements can be accounted for in terms of nucleon-nucleon interactions. Work is in progress to go beyond the approximations involved in the present calculations and to obtain a more accurate nucleon-nucleon cross section for dilepton production. Combined with dynamical calculations, this will permit the precise extraction of the many-body manifestations. Clearly, this type of probe has a bright future and will benefit from the coming generation of high intensity machines. We will finally be able to tackle the problem of pion and delta dynamics in hot and dense nuclear matter which has eluded us up to now.

Acknowledgement

The author would like to acknowledge an ongoing and fruitful collaboration with J. Kapusta .

REFERENCES

1. S. Nagamiya et al., Phys. Rev. C24:971 (1981); S. Hayashi et al., Phys. Rev. C38:1229 (1988).
2. R. Stock, Phys. Rep. 135:259 (1986).
3. M. Gyulassy, K.A. Frankel and H. Stöcker, Phys. Lett. 110B:185 (1982).
4. P. Danielewicz and G. Odyniec, Phys. Lett. 157B: 146 (1985); P. Danielewicz et al., Phys. Rev. C38:120 (1988).
5. See, for example, "Proceedings of the 8th high energy heavy ion study", J.W. Harris and G.J. Wozniack eds., LBL-24580, Berkeley (1988).
6. K.B. Beard et al., Phys. Rev. C32:1111 (1985); E. Grosse et al., Europhys. Lett. 2:9 (1986); N. Alamanos et al., Phys. Lett. 173B: 392 (1986).
7. J.I. Kapusta, Phys. Rev. C15:1580 (1977); M.P. Budiansky et al., Phys. Rev. Lett. 49:361 (1982); D. Vasak et al., Nucl. Phys. A428:291c (1984).
8. H. Nifenecker and J.P. Bondorf, Nucl. Phys. A442:478 (1985); C.M. Ko, G. Bertsch and J. Aichelin, Phys. Rev. C31:2324 (1985); W. Cassing et al., Phys. Lett. 181B:217 (1986); W. Bauer et al., Phys. Rev. C34:2127 (1986); K. Nakayama and G. Bertsch, Phys. Rev. C34:2190 (1986).
9. C. Gale and J. Kapusta, Phys. Rev. C35:2107 (1987); Phys. Rev. C38:2659 (1988).
10. G. Roche et al., Phys. Rev. Lett. 61:1069 (1988).
11. Charles Gale and Joseph Kapusta, Preprint CRNL-TP, December 1988.

KAON AND PION PRODUCTION IN THE REACTION SILICON (14.5 AGeV) ON GOLD

H. Sorge

Institut für Theoretische Physik, Johann Wolfgang Goethe-Universität
Postfach 111932, D-6000 Frankfurt am Main, West Germany

At the AGS several experiments with Silicon projectiles of kinetic energies between 10 and 14.5 AGeV on various targets have been performed in the last two years [1,2]. The experimentally measured transverse energies are compatible with a high degree of nuclear stopping in central collisions indicating thermalization in the colliding system [1]. The observed enhancement in the K^+/π^+ and K^-/π^- ratios for Si(14.5AGeV) on a gold target compared to pp collisions at the same C.M. energy gave rise to speculations about the onset of quark gluon plasma formation, because strangeness enhancement is discussed to be a signature for the QGP [3]. However, this problem had been addressed in a simple thermal picture only, leaving aside the complex dynamics of an ultrarelativistic heavy ion collision. To take the dynamics of the collision into account we considered the kaon and pion production in the framework of a microscopic phase space approach which is purely based on hadron physics.

The model which was used to calculate the multiplicity and rapidity distributions together with the transverse momentum spectra of the different particle species is dubbed Relativistic Quantum Molecular Dynamics (RQMD). The main features of the RQMD model are described elsewhere [4]. The RQMD approach combines classical propagation of the hadrons (molecular dynamics) with quantum effects as stochastic scattering and particle decay.

In the RQMD model it is assumed in general that pions and other mesons are not directly produced, but result from the decay of excited hadrons which are produced in inelastic collisions. As only exceptions kaons can be directly produced in low energy collisions of two nucleons according to the measured cross sections. The different resonance production probabilities are fitted to the existing experimental data. This has been done for nucleon nucleon, pion nucleon, pion pion and kaon nucleon scattering [7].

Let us at first look at the medium energy collisions. The resonances which can be produced as an excitation of an ingoing particle are the nucleonic and the Δ resonances, the ρ meson and the K^* meson. The resonances can decay into various channels, for instance the nonstrange baryons (with masses below 2 GeV/c) $\Delta', N' \rightarrow N + \pi, \Delta + \pi, N + \varrho, N + \eta$ and $Y + K$ with the branching ratios and the mean life times determined by inspection of the experimental data [6].

Resonances with masses higher than a certain cutoff value (for non-strange baryons 2GeV/c^2) which may be produced in high energetic collisions decay jetlike via emission of the produced hadrons. For this high mass hadron decay we use a phenomenological string decay model in the RQMD model [9]. The hadrons which contain the original quarks of the decaying jet system can interact afterwards without time delay, because these quarks are "dressed". In a simple constituent quark model their interaction probability remains the same before and after the jet system decays. The other hadrons produced in the jets need some time before they are on-shell and can interact with other particles. This leads to the inside-outside picture for fast moving particles. In the RQMD calculations a mean time of 1fm/c in the repective rest system is assumed before one of these secondary particles can rescatter.

Because we are interested in looking at the produced strangeness in nucleus nucleus collisions, the following reactions and their available measured cross sections have been implemented [8]: $K + N \rightarrow K^* + N$, $\pi N \leftrightarrow Y K$, $N \overline{K} \leftrightarrow \pi Y$, $N N \rightarrow N Y K$, $N N \rightarrow N N K \overline{K}$ and $K^+ n \leftrightarrow K^0 + p$. A major part of the kaons should be produced in the interactions between the nonstrange mesons π, ρ, η and ω. Because experimental information is missing in these cases we always used the same constant value above two kaon mass threshold, for instance for $\pi^+\pi^0 \mapsto K^+ + \overline{K}^0$ the cross section 1 mb. The other cross sections are analogously given according to additive quark model arguments.

In Fig.1 the measured pseudorapidity distribution for charged particles is compared with the results of the RQMD calculations. Note that the RQMD calculations have been done at fixed impact parameter (b=1 fm) whereas for the experimental data the multiplicities were averaged over the "central events". This means that only the seven percent highest multiplicity events of all measured events have been selected as central corresponding to a maximum impact parameter around 2.5 fm when a simple geometric picture is applied. This simplification in the simulation of the experimental trigger has been done to minimize CPU time. The comparison between calculations and experimental data shows that the total multiplicities and the general shape of the η-distribution of the charged particles can be reproduced quite well by the RQMD calculations.

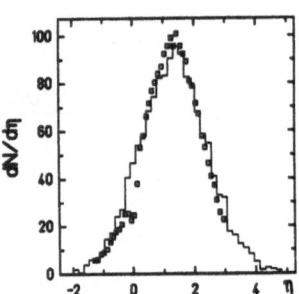

Fig.1   Charged particle pseudo-rapidity distribution in Si(14.5 AGeV) on Au: comparison between the RQMD results (histogram) for fixed impact parameter b=1fm and the experimental data points for central events.

To prove whether the collision dynamics is modelled realistically in the RQMD calculations, one has to look at more exclusive observables like the different particle yields. The E802 group recently measured the charged kaon and pion yields in a limited region of phase space [2]. However, an absolute scale is missing in the experimental yields. The experimental results give values of around 20% for the integrated K^+/π^+-ratio and around 5-6% for the integrated K^-/π^--ratio. To compare our results with the experimental data we simulated the complicated acceptances of the experimental detectors by a simple cutoff in rapidity ($1.0 < y < 1.5$) and transverse momentum ($p_t > 0.3 GeV$) in the calculations. The RQMD calculations of the π^+, K^+ and the π^-, K^- spectra in this rapidity region are shown in Fig. 2.

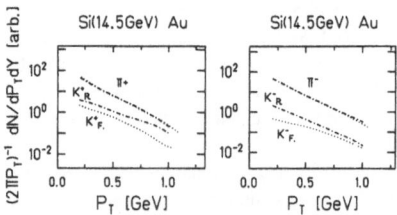

Fig.2 p_t spectra for K^+ (K^-) and π^+ (π^-) in the rapidity bin $1 < y < 1.5$: comparison between RQMD and FRITIOF results. Both π spectra are normalized on the same value.

Integrated they give nearly the same ratios as the experimental ratios above. They are compared with results from an independent hadron fragmentation model (FRITIOF) in a way that the π^+ (π^-) spectra are overlayed. This is done to see the clear difference in the K/π-ratios in both models. Note that the FRITIOF and the RQMD calculations give practically the same slopes in the π^+ and π^- spectra. One can clearly see that FRITIOF fails to explain the observed K/π-ratios.

The basic physical idea to explain the enhancement of the K/π ratios when comparing the RQMD and the FRITIOF results is the following: In an independent fragmentation scheme like FRITIOF which was devellopped by the LUND group [10] it is assumed that the projectile nucleons move on straight lines through the target thereby hitting all target nucleons which are geometrically within a tube whose section is given by the inelastic nucleon nucleon cross section. The excited projectile and target nucleons decay independently afterwards. This means that the produced hadrons do not interact in their hadronic environment any more (infinite mean free path). In the FRITIOF model one therefore gets nearly the same ratio of K^+/π^+ and K^-/π^- as in elementary nucleon nucleon collisions at the same C.M. energy. The main difference between the RQMD model and an independent fragmentation scheme with respect to the K/π ratios is that the interaction of the produced hadrons with the other hadrons is taken into account in the RQMD model (rescattering effect). As the RQMD calculations show the nonequilibrium aspect of the ultrarelativistic heavy ion collision must be taken into account. In the first stages of the collision the rapidity distribution of the interacting hadrons is far from thermal equilibrium which may lead to high relative momenta between colliding hadrons. Therefore the meson-baryon and meson-meson channels give an important contribution to the secondary kaons [11].

Conclusion: The K^+/π^+ and K^-/π^- enhancement in the Si(14.5 GeV) on Au experiment can be explained in a hadronic model when rescattering and the nonequilibrium dynamics are taken into account. The explanation of the kaon production by purely hadronic processes leaves no space to consider the kaon enhancement as a signature of quark gluon plasma formation in this AGS-experiment.

References

[1] P. Braun-Munzinger et al.: Z.Phys. C38(1988) 45

[2] P. Vincent: talk presented at the conference Quark Matter '88, Lenox(USA), to be publ. in Nucl. Phys. (1989)

[3] N.K. Glendenning, J. Rafelski: Phys.Rev. C31(1985) 823; J. Kapusta, A. Mekijan: Phys.Rev. D33(1986) 1304; T. Matsui, B. Svetitsky, L.D. McLerran: Phys.Rev. D34(1986) 783,2047; K. Kajantie, M. Kataja, P.V. Ruuskanen: Phys.Lett. B179(1986) 153; P. Koch: Z.Phys. C38(1988) 269

[4] H. Sorge, H. Stöcker, W. Greiner: Ann. Phys. 1989 (in print); talk presented at the conference Quark Matter '88, to be publ. in Nucl. Phys. (1989)

[5] A. Komar: Phys.Rev. D18(1978) 1881; J. Samuel: Phys.Rev. D26(1982) 3475,3482; H. Sorge: Ph. d. thesis, Univ. Bremen (1987)

[6] Particle Data group: LBL-100 Revised UC-34d (1982)

[7] J. Bystricky et al: DPhPE 87-03,Saclay 1987; G. Bertsch, M. Gong, L. McLerran, V. Ruuskanen, E. Sarkkinen: Phys.Rev. D37(1988) 1202; A.M. Rossi et al.: Nucl.Phys. B84(1975) 269; C.B. Dover, G.E. Walker: Phys.Rep.Lett. 89(1982) 1

[8] H.W. Barz, H. Iwe: Nucl.Phys. A453(1986) 728; J. Randrup, C.M. Ko: Nucl.Phys. A343(1980) 519

[9] B. Andersson, G. Gustafson, G. Ingelman, T. Sjöstrand: Phys.Rep. 141(1983) 31; X. Artru: Phys.Rep. 141(1983) 147

[10] T. Sjostrand: Comp.Phys.Comm. 39(1986) 347; B. Andersson, G. Gustafson, B. Nilsson-Almqvist: Nucl.Phys. B281(1987) 289

[11] R. Mattiello, H. Sorge, H. Stöcker, W. Greiner: subm. to Phys.Rev.Lett. 1989

PARTICIPANTS

Adorno, Amerita, Universita di Catania, Dipartimento di Fisica, 57 Corso Italia, I-95129-Catania, Italy.

Ardouin, Daniel, Université de Nantes, Laboratoire de Physique Nucléaire, 2 rue de la Houssinière, F-44072-Nantes Cedex 03, France.

Arvieux, Jacques, Laboratoire National Saturne, CEN de Saclay, F-91191 Gif-sur-Yvette Cedex, France.

Bedau, Christoph, Institut für Kernphysik, Schlossgartenstrasse 9, D-6100-Darmstadt, Germany.

Berenguer, Maria, Institut für Theoretische Physik, Universität Frankfurt/M, Robert-Mayer-Strasse 8-10, Postfach 11 19 32, D-6000-Frankfurt am Main 11, Germany.

Bondorf, Jacob, Niels Bohr Institutet, Blegdamsvej 17, DK-2100-København Ø, Denmark.

Carroll, Jim, Lawrence Berkeley Laboratory, 50/348, Berkeley, CA 94720, USA.

Cheynis, Brigitte, Institut de Physique Nucléaire de Lyon, 43 Boulevard du 11 Novembre 1918, F-69622-Villeurbanne Cedex, France.

de Jong, Fred, K.V.I., Zernikelaan 25, 9747 AA Groningen, The Netherlands.

Durand, Dominique, Université de Caen, Laboratoire de Physique Corpusculaire, Boulevard du Maréchal Juin, F-14032-Caen Cedex, France.

Faure-Ramstein, Béatrice, Faculté des Sciences d'Orsay, Institut de Physique Nucléaire, BP N°1, F-91406-Orsay Cedex, France.

Fiolhais, Carlos, Departamento de Fisica da Universidade, P-3000-Coimbra, Portugal.

Flocard, Hubert, Faculté des Sciences d'Orsay, Institut de Physique Nucléaire, BP n°1, F-91406-Orsay Cedex, France.

Gale, Charles, Theoretical Physics Branch, Chalk River Nuclear Laboratories, Chalk River, Ontario, Canada K0J 1J0.

Grégoire, Christian, State University of New-York at Stony-Brook, Department of Physics, Nuclear Theory Group, Stony-Brook, NY 11794-3800, USA.

Grinberg, Maurice, Institute for Nuclear Research and Nuclear Energy, Boul. Lenin 72, Sofia 1784, Bulgaria.

Grosse, Eckart, GSI, Planckstrasse 1, D-6100-Darmstadt 11, Germany.

Guerréau, Daniel, GANIL, BP 5027, F-14021-Caen Cedex, France.

Gulminelli, Francesca, Universita degli Studi di Milano, Dipartimento di Fisica, Via Celoria 16, I-20133-Milano, Italy.

Hofmann, Helmut, Physik Department, TU München, D-8046-Garching, Germany.

Jørgensen, Thomas, Niels Bohr Institutet, Blegdamsvej 17, DK-2100-København Ø, Denmark.

Kampert, Karl-Heinz, Institut für Kernphysik, Wilhelm-Klemm Strasse 9, D-4400-Münster, Germany.

Keane, Declan, Kent State University, Smith Laboratory of Physics, Kent, Ohio 44242-0001, USA.

Kienle, Paul, GSI, Postfach 11 05 41, Planckstrasse 1, D-6100-Darmstadt 11, Germany.

Koch, Volker, Justus-Liebig-Universität Giessen, Institut für Theoretische Physik, Heinrich-Buff-Ring 16, D-6300-Giessen, Germany.

Lee, Suk-Joon, McGill University, Department of Physics, Ernest Rutherford Physics Building, 3600 University Street, Montréal, PQ, H3A2T8, Canada.

Lynch, William, NSCL/Cyclotron Laboratory, Michigan State University, East Lansing, MI 48824-1321, USA.

Lynen, Uli, GSI, Postfach 11 05 41, Planckstrasse 1, D-6100-Darmstadt 11, Germany.

L'Hôte, Denis, CEN de Saclay, IRF-DPhN-ME, F-91191 Gif-sur-Yvette Cedex, France.

Mahaux, Claude, Institut de Physique B5, Université de Liège, Sart-Tilman, B-4000-Liège 1, Belgium.

Malfliet, Rudi, KVI, Zernikelaan 25, 9747 AA Groningen, The Netherlands.

Marcos, Saturnino, Departamento de Fisica Moderna, Facultad de Ciencias, Universidad de Cantabria, Avda. de los Castros, s/n, E-39005-Santander, Spain.

Martin, Cécile, Faculté des Sciences d'Orsay, Institut de Physique Nucléaire, BP n°1, F-91406-Orsay Cedex, France.

Miskowiec, Dariusz, Jagellonian University, Reymonta 4, PL-30059-Krakow, Poland.

Ngô, Christian, Laboratoire National Saturne, CEN de Saclay, F-91191-Gif-sur-Yvette Cedex, France.

Ngô, Hélène, Faculté des Sciences d'Orsay, Institut de Physique Nucléaire, BP n°1, F-91406-Orsay Cedex, France.

Panagiotou, Apostolos, Univ. Athens, Dept. Nuclear and Particle Physics, Panepistimiopolis, 157-71 Athens, Greece.

Pandharipande, Vijay, University of Illinois at Urbana-Champaign, Dept of Physics, 1110 W. Green Street, Urbana, IL 61801, USA.

Pollacco, Emanuel, CEN de Saclay, IRF-DPhN-BE, F-91191-Gif-sur-Yvette Cedex, France.

Ramos, Angels, Universitat de Barcelona, Departament d'Estructura I Constituens de la Matèria, Diagonal, 647, E-08028-Barcelona, Spain.

Roy-Stephan, Michèle, Faculté des Sciences d'Orsay, Institut de Physique Nucléaire, BP n°1, F-91406-Orsay Cedex, France.

Sapienza, Piera, INFN-LNS, Viale Andre Doria, Angolo Via S. Sofia, I-95123-Catania, Italy.

Sauvestre, Jean-Etienne, CEN de Saclay, IRF-DPhN-BE, F-91191-Gif-sur-Yvette Cedex, France.

Schlagel, Tom, University of Illinois at Urbana-Champaign, Dept of Physics, 1110 W. Green Street, Urbana, IL 61801, USA.

Schönhofen, Michael, GSI, Postfach 11 05 52, Planckstrasse 1, D-6100-Darmstadt 11, Germany.

Senger, Peter, GSI, Postfach 11 05 52, Planckstrasse 1, D-6100-Darmstadt 11, Germany.

Serot, Brian, University of Indiana, Dept of Physics, Bloomington, IN 47405, USA.

Siemens, Philip, Oregon State University, Dept of Physics, Corvallis, Oregon 97331, USA.

Sneppen, Kim, Niels Bohr Institutet, Blegdamsvej 17, DK-2100-København Ø, Denmark.

Sorge, Heinz, Institut für Theoretische Physik, Universität Frankfurt/M, Robert-Mayer-Strasse 8-10, Postfach 11 19 32, D-6000-Frankfurt am Main 11, Germany.

Soyeur, Madeleine, CEN de Saclay, IRF-DPhG-SPhT, F-91191-Gif-sur-Yvette Cedex, France.

Spina, Maria-Helena, Universität Regensburg, Institut für Theoretische Physik, D-8400-Regensburg, Germany.

Stassart, Pierre, Institut de Physique B5, Université de Liège, Sart-Tilman, B-4000-Liège 1, Belgium.

Stöcker, Horst, Institut für Theoretische Physik, Universität Frankfurt/M, Robert-Mayer-Strasse 8-10, Postfach 11 19 32, D-6000-Frankfurt am Main 11, Germany.

Suraud, Eric, GANIL, BP 5027, F-14021-Caen Cedex, France.

Tamain, Bernard, Université de Caen, Laboratoire de Physique Corpusculaire, Boulevard du Maréchal Juin, F-14032-Caen Cedex, France.

Thies, Michael, Institut für Theoretische Physik III, Universität Erlangen-Nürnberg, D-8520-Erlangen, Germany.

Valette, Olivier, CEN de Saclay, IRF-DPhN-ME, F-91191-Gif-sur-Yvette Cedex, France.

Waldhauser, Béla, Institut für Theoretische Physik, Universität Frankfurt/M, Robert-Mayer-Strasse 8-10, Postfach 11 19 32, D-6000-Frankfurt am Main 11, Germany.

Welke, Gerd, State University of New-York at Stony-Brook, Dept of Physics, Nuclear Theory Group, Stony-Brook, NY 11794-3800, USA.

White, Gary, Northwestern State College, Dept of Physics, Natchitoches, Louisiana 71457, USA.

INDEX

Boltzmann equation, 137-139
  generalized, 153-155
  Uehling-Uhlenbeck, see
    Vlassov-Uehling-Uhlenbeck
Brueckner-Hartree-Fock
    approximation, 16-19
Delta resonance, 167-184, 447-449
  in heavy ion charge exchange
    reactions, 469-470
  in relativistic heavy ion
    collisions, 337-338
Dirac-Brueckner approach
  for nuclear matter, 83-101
  pion polarization in, 447-449
Dyson equations, 80, 84
  generalized, 151-153
Effective mass of the nucleon,
    6, 114-115, 123-124
  in relativistic mean
    field theory, 43, 57
Exp S method, 9
Fermi gas, 2, 4
  hot non interacting, 104-106, 269
Green functions, 9, 75-80, 82, 148, 173
  advanced, 77, 149
  anticausal, 77, 149
  causal, 76, 77, 149
  Keldysh, 76, 77, 152
  mixed, 78
  retarded, 77, 149
  Wigner representation, 150
Hard photon production in
  heavy ion collisions, 206, 406-410
Hartree-Fock approximation,
    10-11, 80-81, 216
Hierarchy
  BBGKY, 135-136
  of Green functions, 9, 79, 136
Hole-line expansion, 9, 22-25

Hot nuclear matter, 113-119
  equation of state of, 118-119,
    268-272
  mean field calculation of,
    57-59
  flowing, 50-57
  growth of fluctuations in,
    125-127
  variational theory of, 107-108
Hot nuclear gas, 120-122
Hot nuclei
  definition, 187
  experimental characterization,
    190-207
  limiting temperature of,
    215-219
  properties, 207-215
  thermalization time, 188-189
Hugenholtz-Van Hove theorem, 5, 9
Intra nuclear cascade calculations
  for heavy ion collisions, 332-333,
    345
  for pion-nucleus collisions, 174
Kadanoff and Baym equations,
    150-152
Kinetic equations, 97
  for gases, 139-144
Landau-Vlassov equation,
  see Vlassov-Uehling-Uhlenbeck
    equation
Loop expansion, 60-65
Many-body problem
  non relativistic, 7-33
  relativistic, 40-70, 74-89, 165
Mean-field approximation, 41
Multifragmentation of nuclei
  experimental data, 231-242,
    299-300
  onset of, 222-223
  theoretical models, 242-261,
    274-283, 356-361

Nuclear forces, 19-20, 74, 108-113
Nuclear matter
    definition, 2
    empirical properties, 2-6, 91-93
    equation of state in relativistic
        mean field theory, 43-46
    low temperature, 122-125
    non relativistic theory, 7-33
    phase diagram, 119
    relativistic theory, 40-68, 83-97
    single particle energies in,
        93-97
    see also Hot nuclear matter
Pion absorption in nuclei,
    167-185
Polarization operator in
    Dyson equations, 80-82
Quantum hadrodynamics,
    38-70,
Quantum molecular dynamics,
    347-349
Reaction matrix, 18
Relativistic heavy ion collisions,
    287-318, 343-386
    composite particle production,
        296-299, 365-370
    dilepton production, 417-426,
        487-489
    energy spectra, 295-296
    experimental techniques,
        288-290, 326, 406, 420, 429-440,
        443-446
    flow in, 300-316, 371-381,
        396-402, 483-485

kaon production, 414-415,
    491-493
multiplicity distributions,
    291-292
particle correlations, 329-330
pion production, 330-338,
    381-386, 491-493
    subthreshold, 410-414
rapidity distributions, 292-294
SATURNE, 443-446
SIS/ESR, 429-440
Thermodynamics, 103-104
    covariant, 46-50
Thompson equation, 87-89
Time dependent Hartree-Fock,
    144-145, 345
Time scales in heavy ion collisions,
    188-189, 209, 457-459,
    461-463
T-matrix, 86-89, 155-157, 173
Transport theories
    for classical systems, 134-147
    for non relativistic
        quantum systems, 147-161
    for nuclear collective motion,
        453-454
    for relativistic systems of
        fluctuating fields, 165
Variational principle, 10, 21, 107
Vlassov equation, 136-137
Vlassov-Uehling-Uhlenbeck
    equation, 96-97, 156-157, 258-260,
        345-347, 391-402
Walecka model, 40-43
Wigner transform, 150